Radioactive and Stable Isotope Geology

Radioactive and Stable Isotope Geology

H.-G. Attendorn

Institute of Geology and Palaeontology, and Museum of the Westfälische Wilhelms University Münster, Germany

and

R. N. C. Bowen

formerly of the Institute of Geology and Palaeontology, and Museum of the Westfälische Wilhelms University Münster, Germany

CHAPMAN & HALL

London · Weinheim · New York · Tokyo · Melbourne · Madras

Published by Chapman & Hall, 2–6 Boundary Row, London SE1 8HN, UK

Chapman & Hall, 2–6 Boundary Row, London SE1 8HN, UK

Chapman & Hall GmbH, Pappelallee 3, 69469 Weinheim, Germany

Chapman & Hall USA, 115 Fifth Avenue, New York, NY 10003, USA

Chapman & Hall Japan, ITP-Japan, Kyowa Building, 3F, 2-2-1 Hirakawacho, Chiyoda-ku, Tokyo 102, Japan

Chapman & Hall Australia, 102 Dodds Street, South Melbourne, Victoria 3205, Australia

Chapman & Hall India, R. Seshadri, 32 Second Main Road, CIT East, Madras 600 035, India

First edition 1997

Typeset in 10/12 Palatino by Thomson Press (India) Ltd, Chennai

Printed in Great Britain by Hartnolls Ltd, Bodmin, Cornwall

ISBN 0 412 75280 8

A catalogue record for this book is available from the British Library

Library of Congress Catalog Card Number: 96-85529

∞ Printed on permanent acid-free text paper, manufactured in accordance with ANSI/NISO Z39.48-1992 and ANSI/NISO Z39.48-1984 (Permanence of Paper).

Contents

Preface

Isotopic (nuclear) geology constitutes an exact, quantitative branch of the Earth sciences which has expanded rapidly to cover a wide spectrum of applications since the establishment of its basic principles by the early 1950s, in part following seminal researches by H. C. Urey, H. A. Lowenstam, S. Epstein, T. Mayeda and numerous others. This immense progress, accelerating in recent years, led to the application of isotopes in attempting to resolve a variety of geochemical and geological problems in the Earth sciences. Lunar exploration too provided rocks for analysis and their examination stimulated refinements in mass spectrometry later used for terrestrial materials as well. New geochronometric methods were devised and include those based on the radioactive decay of ^{147}Sm to ^{143}Nd, ^{176}Lu to ^{176}Hf, ^{187}Re to ^{187}Os and ^{40}K to ^{40}Ca, as well as others depending upon the production and distribution of cosmogenic radionuclides such as ^{26}Al, ^{10}Be and ^{36}Cl.

The impact of all these developments has been tremendous and shed light on such diverse topics as the origin of igneous rocks, where the isotopic compositions of neodymium, strontium, lead and hafnium suggest that magmas from the Earth's mantle are often crustally contaminated. The isotopic compositions of carbon, oxygen and sulphur have proved important in elucidating aspects of petrogenesis. The study of environmental isotopes such as tritium and radiocarbon has been invaluable, not only as regards revealing the climates of the past using palaeotemperature analyses, but also in understanding the ice volume effect, the stratigraphy of ice and snow and the dating of sediments and volcanics, together with the isotopic composition of sea water, both now and through Earth history. Isotopic analyses of the ^{18}O composition of foraminifera from Caribbean sea-bottom cores formed the basis for C.

Emiliani's 16 isotopic stages originally believed to reflect warmer (odd-numbered) and cooler (even-numbered) time intervals, later shown by W. Dansgaard, N. J. Shackleton, N. D. Opdyke and others to reflect seawater isotopic changes resulting from the growth and decay of continental ice sheets. The relationship between the isotope record and global ice volume was utilized as a stratigraphic tool by members of CLIMAP (Climate Long Range Investigation and Mapping) examining deep-ocean cores and associated isotopic data. Indeed, quantitative techniques have been used in isotopic chronostratigraphy, firstly by D. F. Williams and D. Trainor as an innovative approach to locating hydrocarbon source rocks and their oil and gas reservoirs in deep-water tracts. An appropriate schema was later presented by T. E. McKenna and others who treated every isotopic record as a time series and used common analyses to develop criteria for identifying common events in records. The method has been shown to be useful, although it would be improved if more was known about the palaeoevolutionary behaviour of foraminifera. Another factor requiring further scrutinay relates to vital effects resulting from disequilibrium fractionation revealed by oxygen and carbon isotopic analyses of living benthic foraminifera such as *Heterostegina depressa*, a phenomenon thought to be species specific and one found in all Recent species of planktonic foraminifera, which are in isotopic disequilibrium with respect to ^{13}C values in equilibrium calcite. Such vital effects could have a considerable bearing on palaeotemperature measurements, but the present state of play is indicated by W. S. Broecker's remark that this possible physiological control on the carbon and oxygen isotopic compositions of the skeletons of some marine organisms is not understood (although differences in fractionation factors between different species of foraminifera attest to its existence). At present the authors are actively investigating the matter and some preliminary results have already been published.

The preceding, necessarily brief, discussion illustrates both the academic interest of researches into isotopes in the Earth together with the actual and potential practical applications of data so obtained. This book is intended to update its predecessor and considers significant researches from 1988 to its date of completion, relevant information being included, albeit sometimes in abbreviated form. Whereas in earlier times, it was feasible to discuss unclassified reports as well, the vast increase in research (unfortunately most of it in excessive detail contributing little or nothing to the advancement of understanding) has made this impossible, hence the authors have attempted to provide basic information useful to both specialists and non-specialists in the field of isotope geology, occasionally adding their own, clearly identified, comments. Those needing more information should consult the literature. The bibliography in this book relates mainly to the post-1988 period, except for some classic publications. Constraints of space have kept figures to a minimum, but those which were deemed indispensible appear in suitable parts of the text. The structure of the book is essentially in two parts, the first dealing with radiometric dating and the second mostly with stable isotope geology.

Finally, on a matter of style, the authors would have preferred names of elements to be given in full thoughout the text (except in equations) rather than only occasionally which was the publishers' choice.

R. N. C. BOWEN
January 1997

Eon	Era	Sub-era	Period	Epoch		Age	Ma
Phanerozoic	Cenozoic	Quaternary		Holocene			0.01
				Pleistocene			2
		Tertiary	Neogene	Pliocene		Piacenzian	
						Zanclian	3.1
				Miocene	3	Messinian	
						Tortonian	
					2	Serrayallian	11.3
						Late Langhian	
					1	Early Langhian	14.4
						Burdigalian	
						Aquitanian	
			Palaeogene	Oligocene		Chattian	24.6
						Rupelian	38.0
				Eocene	3	Priabonian	
					2	Bartonian	
						Lutetian	
					1	Ypresian	50.5
				Palaeocene		Thanetian	
						Danian	60.2
	Mesozoic		Cretaceous K	K_2	Senonian	Maastrichtian	
						Campanian	73
						Santonian	
						Coniacian	87.5
						Turonian	
						Cenomanian	91
				K_1		Albian	
						Aptian	113
						Barremian	
					Neocomian	Hauterivian	125
						Valanginian	
						Berriasian	138
			Jurassic J	J_3	Malm	Tithonian	
						Kimmeridgian	150
						Oxfordian	
				J_2	Dogger	Callovian	163
						Bathonian	
						Bajocian	179
						Aalenian	
				J_1	Lias	Toarcian	188
						Pliensbachian	
						Sloemurian	200
						Hettangian	213
			Triassic Tr	Tr_3		Rhaetian	
						Norian	
						Camian	225
				Tr_2		Ladinian	
						Anisian	238
				Tr_1	Olenekian	Spathian	
						Smithian	
					Induan	Dieherian	
						Griesbachian	248
	Palaeozoic		Permian P	P_2		Tatarian	
						Kazanian	
						Ufimian	258
				P_1		Kunqurian	
						Artinskian	268
						Sakmarian	
						Asselian	286

*Quaternary (Pleistogene)

Chronometrically calibrated

Chronostratigraphic scale.

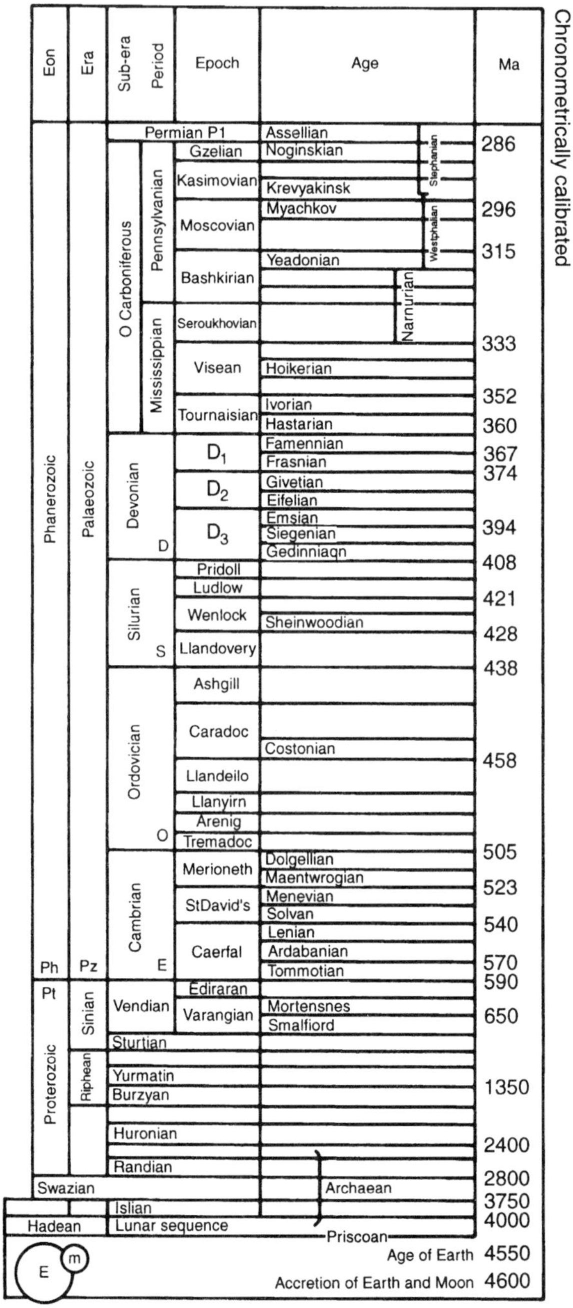

Chronostratigraphic scale contd.

IA	IIA	IIIB	IVB	VB	VIB	VIIB	VIII			IB	IIB	IIIA	IVA	VA	VIA	VIIA	0
1 **H** 1.0079																	2 **He** 4.00260
3 **Li** 6.94	4 **Be** 9.01218											5 **B** 10.81	6 **C** 12.011	7 **N** 14.0067	8 **O** 15.994	9 **F** 18.998403	10 **Ne** 20.17
11 **Na** 22.98977	12 **Mg** 24.305											13 **Al** 29.98154	14 **Si** 28.0855	15 **P** 30.97376	16 **S** 32.06	17 **Cl** 35.453	18 **Ar** 39.948
19 **K** 39.0983	20 **Ca** 40.08	21 **Sc** 44.9559	22 **Ti** 47.90	23 **V** 50.9415	24 **Cr** 51.996	25 **Mn** 54.9380	26 **Fe** 55.847	27 **Co** 58.9332	28 **Ni** 58.71	29 **Cu** 63.546	30 **Zn** 65.38	31 **Ga** 69.735	32 **Ge** 72.59	33 **As** 74.992	34 **Se** 78.96	35 **Br** 79.904	36 **Kr** 83.80
37 **Rb** 85.467	38 **Sr** 87.62	39 **Y** 88.9059	40 **Zr** 91.22	41 **Nb** 92.9064	42 **Mo** 95.94	43 **Tc** 98.9062	44 **Ru** 101.07	45 **Rh** 102.9055	46 **Pd** 106.4	47 **Ag** 107.868	48 **Cd** 112.41	49 **In** 114.82	50 **Sn** 118.69	51 **Sb** 121.75	52 **Te** 127.60	53 **I** 126.904	54 **Xe** 131.30
55 **Cs** 132.9054	56 **Ba** 137.33	57* **La** 138.9055	72 **Hf** 178.49	73 **Ta** 180.947	74 **W** 183.85	75 **Re** 186.207	76 **Os** 190.2	77 **Ir** 192.22	78 **Pt** 195.09	79 **Au** 196.9665	80 **Hg** 200.59	81 **Tl** 204.37	82 **Pb** 207.2	83 **Bi** 208.9804	84 **Po** (209)	85 **At** (210)	86 **Rn** (222)
87 **Fr** (223)	88 **Ra** 226.0254	89** **Ac** (227)	104 **Rf** (260)	105 **Ha** (260)	106 (263)												

KEY TO CHART

Atomic number → 50
Symbol → **Sn**
Atomic weight → 118.69

Transition elements (IIIB–IIB); Group VIII (Fe, Co, Ni columns)

*Lanthanides	58 **Ce** 140.12	59 **Pr** 140.9077	60 **Nd** 144.24	61 **Pm** (145)	62 **Sm** 150.4	63 **Eu** 151.96	64 **Gd** 157.25	65 **Tb** 158.9254	66 **Dy** 162.50	67 **Ho** 164.9304	68 **Er** 167.26	69 **Tm** 168.9342	70 **Yb** 173.04	71 **Lu** 174.967 ± 0.003
**Actinides	90 **Th** 232.0381	91 **Pa** 231.0359	92 **U** 238.029	93 **Np** 237.0482	94 **Pu** (242)	95 **Am** (243)	96 **Cm** (247)	97 **Bk** (247)	98 **Cf** (251)	99 **Es** (254)	100 **Fm** (257)	101 **Md** (258)	102 **No** (259)	103 **Lr** (260)

Numbers in parentheses are mass numbers of most stable isotope of that element

Perodic Table of the Elements.

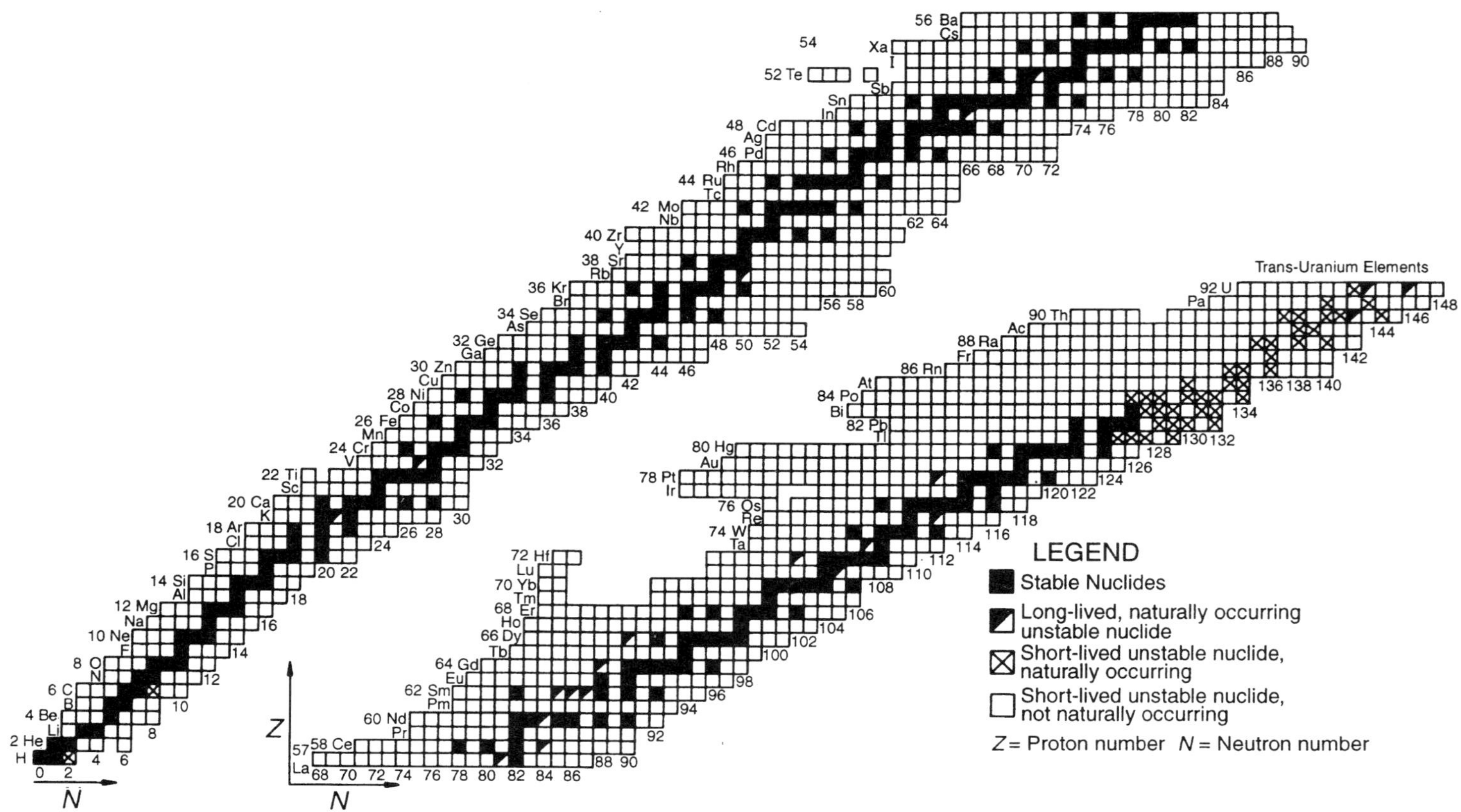

Chart of the Nuclides.

Part One
General Introduction

1 Introduction

> Nature is but a name for an effect,
> Whose cause is god.
>
> *William Cowper, 1731–1800 (The task)*

The discovery of isotopes dates back to the beginning of the century, when Boltwood (1907) found radioactive elements with the same chemical properties but different masses, and three kinds of radioactivity of which the α and β radiation had to be seen as charged particles, for it could be influenced by electrical fields. The occurrence of different masses at the same position of the periodic system, i.e. the same number of protons but different numbers of neutrons (which were discovered first in 1930) could be explained by the different results of radioactive decay. The emission of an α particle (now known as a heliumcore) shifted the resulting element two places to the left in the periodic system and, if β-decay then occurs twice (in this case the β^-) the element was twice shifted back, reaching its original position again with a different number of neutrons (which were discovered after 1930). Soddy (1914) called these elements isotopes. Richards and Lembert (1914) found different atomic masses for Pb and Thomson (1914) separated the ion beams of neon in combined electric and magnetic fields, first proving the existence of isotopes of non radioactive atoms and establishing the principles of modern mass spectrometry.

Urey predicted, on theoretical grounds, that there should be a difference in the vapour pressures of the isotopes of H. The interest in this particular element arose from a suspicion that it might have naturally occurring isotopes. An experiment was effected in order to detect ^2_1H and ^3_1H by spectroscopic methods in the residual volume of gas made by evaporating about 6 L of liquid H. Deuterium was in fact shown to be present, but tritium was not recorded. Urey called the heavy H deuterium (from *deuteros* = second) because it has a mass almost twice that of ordinary hydrogen. At that time it was still not known why, since neutrons were not discovered until a year later.

The study of isotope ratios within the Earth sciences started after World War II when H. C. Urey considered the possibility that the stable isotopes of O may be fractionated by natural processes and that the amount of such fractionation is a function of temperature. This idea led to the development of the oxygen isotope method and the concomitant invention of a tool for paleothermometry. Much subsequent work has been carried out since these first steps, showing the restrictions and on the other hand prospects of the method, which has led to various new applications of stable isotope investigations in the Earth sciences.

Now, the problems no longer arise from the 'accuracy' of the different types of mass spectrometers in use, but rather from the interpretation of the resulting data and sometimes from the sample preparation. For instance, it has turned out that palaeotemperature studies are not as reliable as was formerly thought. The various determining parameters (isotope ratios of seawater, salinity, temperature, mixing, evaporation, biological fractionation or life habits) are mostly unknown in the case of these palaeostudies. A given function $f(v, w, x...)$ therefore entails constancy of variables, assuming an empirical value, e.g. by the use of standard mean ocean water (SMOW) or a normal salinity of 35‰. Additionally, differences occur in the different modifications of $CaCO_3$ (calcite/aragonite), where the latter has been found either enriched or depleted in $\delta^{18}O$ and $\delta^{13}C$.

The fractionation of istopes is well known through experimental studies at various temperatures in different phase systems as for example carbonate–water, phosphate–water, water–vapour and many more. The fractionation factor α is the ratio between the isotopic proportions of the different phases of a system which is in equilibrium. For example, the following equation is relevant to the water–vapour equilibrium at 0°C:

$$\alpha_{w-v}{}^{18}O = \frac{R\,{}^{18}O_w}{R\,{}^{18}O_v} = 1.0115,$$

so that, if water with an isotope ratio of 2.0449×10^{-3} attains an equilibrium with its vapour at 0°C, the isotope ratio in the vapour will be calculated as shown by the following equation:

$$R^{18}O_v = \frac{2.0449 \times 10^{-3}}{1.0115} = 2.2017 \times 10^{-3}.$$

The fractionation factor, investigated by Bigeleisen (1965), is often expressed as $\alpha = 1 + \epsilon/1000$ where ϵ is the fractionation. This is possible because fractionation factors are close to 1, except for H, where α is significantly greater (Table 1.1) due to the large relative mass differences of the isotopes. Generally, the fractionation factors increase with decreasing atomic mass units. The basic assumption which follows from the fractionation processes, described in detail by, for instance, Hoefs (1987); and Fritz and Fontes (1986), is that only isotope ratios attained by equilibrium reactions can give reliable information.

Non-equilibrium fractionation and the parameters already mentioned can influence both the isotope ratios and the presumed conditions, such as the isotopic composition of ocean water. Therefore the problem for today's scientists is mainly how to understand and estimate the influence of the various combined processes controlling the fractionation of isotopes.

The isotope ratios are usually expressed in δ-notation as follows:

$$\delta = \frac{R_{\text{sample}} - R_{\text{standard}}}{R_{\text{standard}}} \times 1000\ (‰),$$

where R is the measured isotope ratio of the sample and a standard sample. A well-understood ecosystem and a considerable knowledge of interactions occurring at physical and chemical interfaces as, for instance, at air–water, water–sediment and solid–liquid boundaries, are therefore as mandatory as the distribution of matter within the different parts of the biosphere and lithosphere. The aims of the following chapters are to summarize the state-of-the-art and to show the difficulties scientists have in solving what may be a vital puzzle and to supply further information within the various fields of application of isotope geoscience.

Table 1.1 Fractionation factors for $\delta^{18}O$ and D at various temperatures

T(°C)	^{18}O	D
−5	1.0121	1.116
0	1.0115	1.108
5	1.0110	1.101
10	1.0105	1.095
15	1.0100	1.088
20	1.0096	1.082
25	1.0092	1.077
30	1.0088	1.072
35	1.0084	1.067
40	1.0091	1.062

1.1 MASS SPECTROMETRY

The principles of mass spectrometry have been described to a greater or lesser extent by many authors who have written on isotope geology and, in a book intended to interest scientists specializing in isotope research, it might be thought jejune to repeat the subject and so 'carry coals to Newcastle'. However, this is untrue, as the book may well appeal to other interested people. In addition, writing for specialists in various fields of isotope geology normally entails including much specialized knowledge. Hence, this book should be regarded as a summary of newer data, its contents deriving from precisely those specialists to whom it is in part addressed. It can thus be a source for a particular specialist who may read those chapters covering topics in which he is not specialized.

1.1.1 Historical Background

The first attempts to analyse canal rays were done by Thomson (1912) after the discovery of these positively charged rays by Goldstein. The mass spectrograph was simple in construction, as can be seen from Fig.1.1, with an ion source (IS), two slit diaphragms focusing the ion beam (SL_1 and SL_2), a combined magnetic and electrical field and a photographic plate (P). The lines of force in both fields run parallel and the beam is vertical to the field lines. The limited resolution of this apparatus, however, meant that it could not even distinguish the isotopes of H and O. Thomson (1914) showed that neon has two isotopes ^{20}Ne and ^{22}Ne, but he did not succeed in separating them by fractional diffusion. An increase in the possible resolution had to be achieved by better focusing of the ion beams and by adjusting of the electrical and magnetical fields.

Other mass spectrographs were designed in the following years by Dempster (1918) and Aston (1919), who was a student of Thomson.

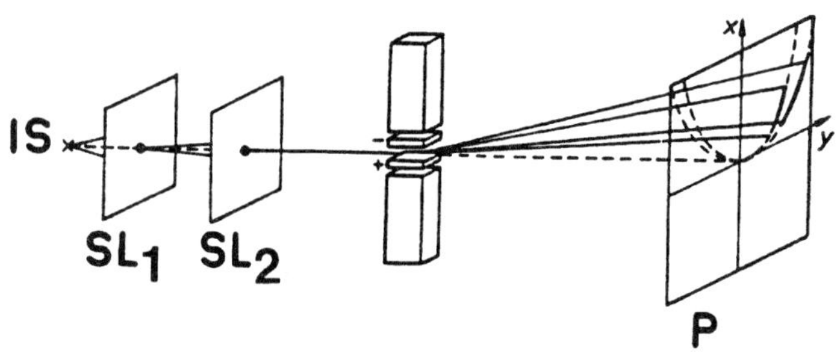

Fig. 1.1 Schematic diagram of the first spectrograph (after Thomson, 1912).

These had several advantages over the Thomson apparatus, but Aston's mass spectrograph was still of a similar design (Fig. 1.2), although it had higher precision and a better resolution of $m/Q = 130$. With this configuration, the ions of the same charge are focused on the same spot on the photographic plate (P), independently from the acceleration in the ion source (IS). The focused ion beam (diaphragms SL_1, SL_2) first passes through an electrical field and after paralleling the resulting beam in SL_3 passes through a magnetic field (B). Both fields are vertical to each other, thus inducing the beam to remain on one plane. With this mass spectrograph, Aston could prove the existence of isotopes and calculate several isotope masses with reasonable precision.

The first real mass spectrometer for measurements with high precision was constructed by Dempster. It only worked with a magnetic field and a new kind of ion source providing a similar energy for the exiting positive ions, which were subsequently focused by electrodes and accelerated by a potential difference (Fig. 1.3). Emitted from the ion source (IS) and focused and accelerated by the electrodes (E), providing a

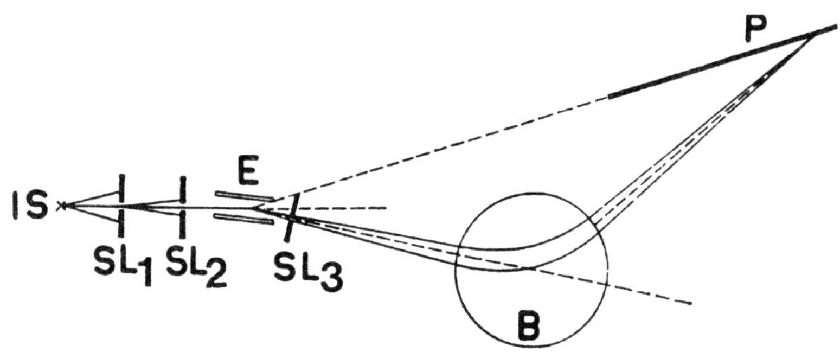

Fig. 1.2 Schematic diagram of the first spectrograph (after Aston, 1919).

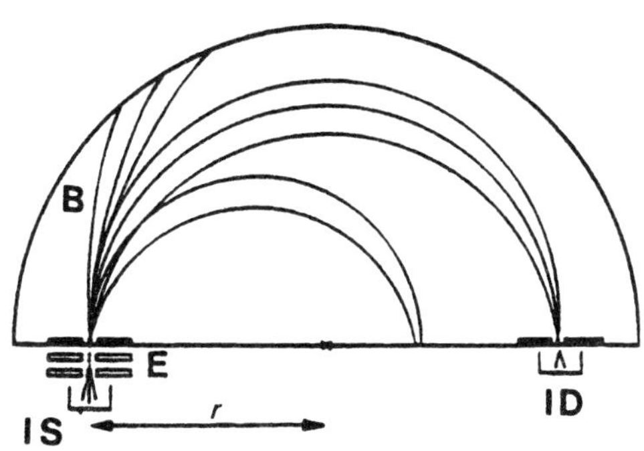

Fig. 1.3 Schematic diagram of the first spectrometer (after Dempster, 1918).

potential difference, the ion beam runs on a circular path through the magnetic field (B) and the ion detector (ID), a great improvement compared to photographic plates, detecting the ion stream with a radius r. Further improvements in this type of mass spectrometer were made in the 1930s by Bainbridge and Herzog. Nier (1940) developed a mass spectrometer, which also obtained the mass spectrum of a beam of ions, employing appropriately disposed magnetic and electric fields. As the deflection of a particular ion in a magnetic or electric field depends upon the ratio of its mass to its electric charge (m/e), a continuous variation of the field strength enables different types of ions to be detected without moving the detector.

1.1.2 A Moving Charged Particle in a Magnetic Field

The following clarification summarizes the almost endless repetitions found in the literature. Generally the particles which are of interest in connection with mass spectrometry are single positively charged ions. Negative or doubly charged positive ions follow the same equations but are not relevant. The ions are produced with cathode rays in the ion source of the mass spectrometer. Cathode rays are high-speed electrons emitted from a cathode with a high direct current in a vacuum < 1 pascal. Naturally the vacuum in an ion source of a mass spectrometer is better, avoiding pollution of the sample gas and allowing free path for the particles. The cathode rays remove an electron by collision and a molecule of the sample gas therefore acquires a single positive charge. It is also possible that an ion recombines with an electron. Doubly charged ions will normally not be produced, because the necessary ionization energy is lacking.

The charge of the resultant ions is positive due to the lack of an electron with the elementary charge of $e = 1.60217733 \times 10^{-19}$ C (Coulomb), which is also the charge of a proton and therefore of the resulting ion. If the ion passes through a magnetic field with the velocity v, it experiences a force F which is expressed by

$$F = ev \times B.$$

The force is maximal if the direction of the field is perpendicular to the speed vector ($V \perp B$) as it is provided by the arrangement of the flight tube and the magnet.

Since a molecule of the sample gas is an ion it will be accelerated by a potential difference following the equation:

$$E = eU = \tfrac{1}{2}mv^2.$$

It will have energy E, the product of the elemental charge and the potential difference or acceleration voltage U, and also the product of the mass divided by two and multiplied by the velocity squared. The acceleration voltage depends on the type of mass spectrometer, operates at several kV and normally can be varied. The variation of the voltage in a constant magnetic field will have the same effect as the variation of the magnetic field at a constant voltage. Older types of mass spectrometers, which had permanent magnetic fields, needed a variable acceleration voltage in order to scan the isotope range.

When the ions, which have different masses depending on their isotopic composition, leave the source through the exit slit, they will have the same kinetic energy but different velocities as

$$v = \left(\frac{2eU}{\mathrm{m}}\right)^{\frac{1}{2}} \quad \text{or} \quad eU = \frac{mv^2}{2}.$$

Entering the magnetic field, the ions are deflected into curved paths, and the process is described by

$$Bev = m\frac{v^2}{r}.$$

with the magnetic induction $B = \mu_0 H$ (the product of the permeability in the vacuum μ_0 and the magnetic field strength H), and r the radius of the circular path. The elimination of the velocity term by the combination of the above equations leads to

$$\frac{2eU}{m} = \frac{B^2e^2r^2}{m^2},$$

and

$$\frac{m}{e} = \frac{B^2r^2}{2U}.$$

If the magnetic induction B is measured in $\mathrm{Vs\,m^{-2}}$ ($1\,\mathrm{Vs\,m^{-2}} = 1$ tesla $= 10^4$ gauss), the radius r in centimetres, the mass m in amu, the potential difference U in volts and the elementary charge e in units of electronic charge, the equations can be solved to determine either the radius r or the magnetic induction B. The principles of ion motion in magnetic or electric fields are the bases for the design of mass spectrometers. If a radius of a flight tube for different masses is calculated, B and U remain variable, so that within the fixed geometrical arrangement of the flight tube and the ion current collectors, different masses can be scanned. Although originally one collector was used to measure different isotopic masses in a row, this is not a sufficiently accurate technique. Nowadays multiple collectors are employed for simultaneous measurements of all masses. For instance, the radii of the masses 44, 45 and 46 of CO_2 will be slightly different to the masses of sulphur dioxide, and therefore need a different collector arrangement. Generally, the radius increases with the mass of an ion, and if B and U are kept constant, it is proportional to the square root of the mass as

$$r \sim m^{0.5} .$$

The difference of the radii of ions from the same compound but with different masses due to the isotopic composition increases with the relative mass difference and is consequently the largest for D/H.

1.2 ISOTOPE EFFECTS AND FRACTIONATION PROCESSES

The chemical properties of isotopes from a nuclide are determined by the configuration of electrons in the outer electronic shells or better the electron orbits, whereas the physical properties mainly depend on the mass of the nucleus. The configuration of electrons for the different isotopes of an element is the same but the mass concentrated in the nuclei differs. However, one cannot really separate chemical and physical properties but must speak of physicochemical processes with the result that the chemical behaviour might also be slightly influenced. This does not result in different reaction products if a reaction occurs, but the reaction times may vary as well as the isotope composition, due to isotope exchange in a molecule and, for example, such chemical or physical characteristics as vapour pressure or reaction times. Some physical constants for hydrogen and water are given in Table 1.2. The differences in these physical constants are due only to the isotopic composition of the molecules and slightly different values are noted for H–D, H–D – ^{18}O etc. Apparently the large relative mass difference for hydrogen and deuterium causes the biggest mass-dependent fractionation effects.

The origin of the shift of the physical and thermodynamical characteristics of the molecules is a problem of quantum mechanics, which describes in this case the behaviour of matter on the atomic scale and the molecular forces. As may be seen from Table 1.2, the bond strength for

Table 1.2 Physical constants for H, D and light and heavy water

	H_2	D_2	$H_2{}^{16}O$	$D_2{}^{16}O$
Density ρ 25°C (g × 1)$^{-1}$	0.0899		0.99707	1.1045
Boiling point, 105 Pa (K)	20.4	23.6	373.0	374.42
Melting point, 105 Pa (K)	13.0	18.6	273.0	276.82
Dielectric constant	1.228[a]	1.277[a]	78.54[b]	78.25[b]
Ionization potential (eV)	15.427	15.46	12.6	12.6
Bond length 298 K (Å)	0.7461	0.7416	0.958 angle	104.45
Bond strength (KJ mol^{-1})	436.002	443.546	Diatomic	–

[a] At 20 K. [b] At 25°C.

the H_2 and D_2 molecules differs as a result of the different isotope masses. The light molecule has a weaker bond than the heavy one and the interproton distance is longer for the light H_2 molecule. Generally, the different properties arising from the isotopic composition have consequences which are discussed below.

1.2.1 Energy States of a Molecule

On the lowest scale, isotopes are exchanged within diatomic molecules e.g. the H_2 molecule, which may be used as an example. The bond of the molecule can be described through wave mechanics and the total potential energy $E(p)$ derives from the electrostatic forces between the nuclei, the electrons and the nuclei and the electrons. It can be expressed as

$$E(p) = V_0 + \sum_{i=1}^{N} V_i(r_i) + \sum_{i<j}^{N} V_{ij}(r_{ij}), \qquad (1.1)$$

where V_0 is the electrostatic interaction between the protons and V_i is the potential energy of the ith electron as a functions of its core distance r. Several energy states are shown in Fig. 1.4, where the interaction between the atoms is presented with a diagram giving the energy states as a function of the core distance. The minimum shows the equilibrium distance of the cores balanced by attraction and repulsion. The attraction increases with the core distance until the dissociation energy is reached and the atoms separate (Fig. 1.4, stage 1). It may be seen from the curve that the potential energy increases as the cores are forced to approach.

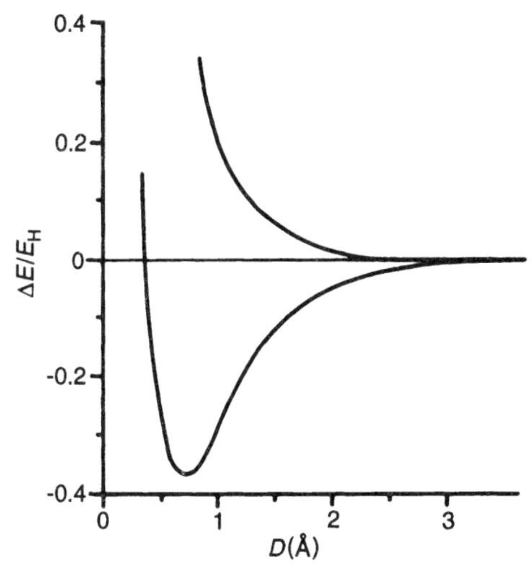

Fig. 1.4 Potential energy E (kJ mol^{-1}) of two H atoms forming a molecule.

Under normal conditions there will always be a distance between the cores. Starting from the minimum (or so called zero point), although there is no zero-energy, even at the absolute zero temperature, the right part of the curve is of interest. There are several discrete energy states between the lowest potential energy and the dissociation energy.

To calculate the total potential energy the above mentioned interactions have to be taken into account. This will be shown here partially for hydrogen in order to explain why hydrogen molecules exchange isotopes. The term V_0 of equation (1.1) can be considered as constant if the thermal movement of the protons is neglected. V_0 can then be expressed as

$$V_0 = \frac{e^2}{4\pi \epsilon_0 r_{xy}},$$

with e being the elementary charge ($1.60217733 \times 10^{-19}$ C) and ϵ_0 the electric field constant ($8.854187817 \times 10^{-12}$ A s V^{-1} m^{-1}) for the cores x and y with a core distance xy. These cores are atoms with the electrons a and b and thus the total energy for the electron–core interaction will be:

$$\sum_{i=1}^{2} V_i(r_i) = -\frac{e^2}{4\pi\epsilon_0} \times \left(\frac{1}{r_{xa}} \frac{1}{r_{xb}} \frac{1}{r_{ya}} \frac{1}{r_{yb}}\right),$$

with xa, xb, ya and yb as the distances of the electrons to the cores. The distance of the electrons a and b additionally provides the energy

$$V_{ab}(r_{ab}) = \frac{e^2}{4\pi\epsilon_0 r_{ab}}. \tag{1.2}$$

Figure 1.4, which shows an asymmetric curve, should be compared to Fig. 1.5, which represents the potential energy of an harmonic oscillator.

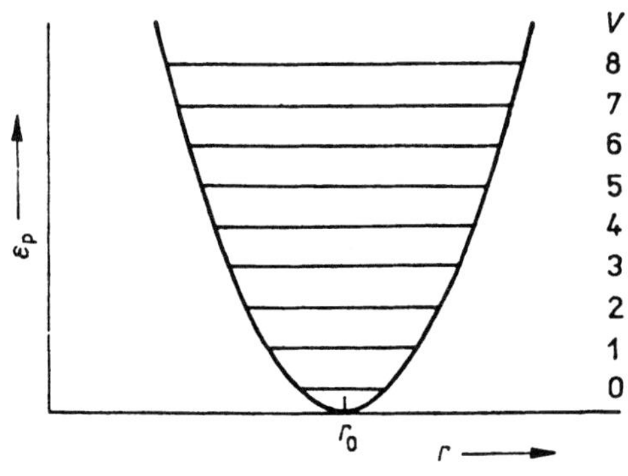

Fig. 1.5 Energy of the harmonic oscillation of a diatomic molecule around the zero-point.

The symmetrical vibration can be described by a parabolic curve with equidistant energetic levels. They must be equidistant, for the vibrational energy $E(v)$ of an harmonic oscillator follows:

$$E(v) = (n + \tfrac{1}{2})\,\hbar w_0 \,. \tag{1.3}$$

The parabolic graph is only realized in nature for small vibrations and supposing an elastic potential, the backwards directed force can be expressed as

$$F = -D\,(r - r_0)\,.$$

Thus the vibrations are those of an harmonic oscillator (analogous to Hooke's law) and the elongation is

$$x = x_0 \cos w_0 t\,,$$

therefore the phase integral will be

$$\oint p\mathrm{d}q = \int_0^{2\pi/w_0} (-m w_0 x_0 \sin w_0 t)\,(-x_0 w_0 x_0 \sin w_0 t\ \mathrm{d}t) = \pi m w_0 x_0^2\,, \tag{1.4}$$

with the impulse $p = -m w_0 x_0 \sin w_0 t$ and q as the location x. So far, the mass m is a factor influencing the calculations and thus the zero-point energy when molecules with different isotopes vibrate about their centre of gravity. From the above equation follows

$$E(g)_0 = \tfrac{1}{2}\,\hbar w_0 = \tfrac{1}{2}\,h\nu\,, \tag{1.5}$$

with $E(g)_0$ as the total vibrational energy at the zero point, v as the circular frequency ($\nu = 2\pi w_0$, $\hbar = h/2\pi$) and h being the Planck constant (6.626076×10^{-34} Js). The derivation of the equation (1.5) will not be given here, but refers to the physical behaviour of a linear harmonic oscillator. However, the equation shows that a molecule at the zero-point will have the energy $E_0 = 1/2hv$, which also derives from the Heisenberg uncertainty principle as the oscillator cannot have a defined location and a defined impulse, thus never being in a state of rest.

Regarding the chemical bonds, the behaviour at zero-point energy is of great importance. The quantum theoretical explanation of homeopolar bonds is very complicated, but in principle two approaching hydrogen atoms offer the electrons a common area of potential which is bigger than that of a single proton. At a certain distance the lowering of the energetic level of the electrons becomes bigger than the repelling force of the cores. The resulting energy difference ΔE is the energy of the bond. This simplified explanation omits several factors, such as the electron spin which has to be opposite for binding stage.

Normally the oscillation is not harmonic and different energy states of a hydrogen molecule exist. The binding energy can be calculated using the Schrödinger equation. The Austrian Erwin Schrödinger formulated the mathematical theory of wave mechanics and won the Nobel Prize in 1933 together with Paul Dirac, who combined both Einstein's theory of relativity and Schrödinger's wave mechanics. Although the hydrogen molecule is the simplest case, the calculations are relatively complicated. Therefore, it is not feasible here to discuss the relevant calculations, particular as matters become more complicated in systems having more than two electrons or solid phases where the lattice vibrations must be considered. Resulting from the approximation that the wave function is expressed through the product of two single electron functions ($\psi = \psi_1 \psi_2$), the total energy E of the molecule orbital $\psi \pm$ (two electrons with opposite spin) can be expressed as

$$E_{\pm} = 2E_{\pm^*} + \frac{e^2}{4\pi \epsilon_0 r_{ab}} = 2\epsilon + \frac{e^2}{4\pi \epsilon_0 r_{ab}} + \frac{2}{1 \pm S}\left[-A + B \pm (-C + D) \right],$$

where $E_{\pm}$ is the total energy of the hydrogen molecule, $E_{\pm^*}$ the energy of an isolated hydrogen atom, is given by

$$E_{\pm^*} = \epsilon + \frac{1}{1 \pm S}\left[-A + B \pm (-C + D) \right],$$

which leads to the term 2ϵ as the first on the right side of the equation. $e^2/4\pi \epsilon_0 r_{ab}$ is the repulsion energy of the hydrogen atoms (see equation (1.2)). The terms S, A, B, C and D are integrals describing different interactions. S is the value for the overlapping of the single ψ functions; A describes the electrostatic interaction between the electron of atom x and core y (vice versa); B describes the interaction between the electrons while C and D represent interactions between the single particles.

As the brief explanations above only refer to a few aspects of a large field, the description cannot satisfactorily deal with quantum mechanics and the nature of chemical bonds. However, the aim of this short excursion into the field of physics is twofold; namely to show that the lack of explanations in books about isotope geology is due to the impossibility of briefly describing the subject and to note that it is sufficient to know that the mass differences and resulting zero-point energy cause isotope effects on the molecular level.

Lighter isotopes have weaker bonds than heavier ones (Table 1.2) and thus react more easily during chemical reactions. In addition to exchange processes on the molecular level, kinetic and biological fractionation are the main processes which can cause isotope enrichment or depletion. They will be discussed briefly in the following paragraphs.

1.2.2 Isotope Separation in the Laboratory

Artificial enrichment of isotopes dates back to 1914, when Thomson first tried to separate the isotopes of neon, but successful enrichment of isotopes was first achieved in the 1930s. Although isotopes exchange in the course of chemical processes due to differences in physicochemical properties, the separation of isotopes utilizing only chemical methods will not be successful because of their similarity in chemical behaviour. Thus the different properties arising from mass differences have to be utilized. It is obvious that a method for separating isotopes with 100% purity is of interest in various scientific fields. For instance, separation methods lead to the production of standards which are widely used or enable the application of tracers. Apart from direct methods of isotope separation (to which belong, among others, the mass spectrograph and photochemical excitation), the enrichment of isotopes is of great importance. The effectiveness of the isotope enrichment is given by the enrichment factor

$$R_{\mathrm{I}} = N_{\mathrm{I}}/(1 - N_{\mathrm{I}})/N_0/(1 - N_0)\,,$$

for the first fractionation step (with R = enrichment factor and N_0, N_{I}... for further reaction steps). Naturally, the reactions need time, and running these processes at temperatures as low as possible increases the efficiency because the fractional velocity differences are small at higher temperatures. A second fractionation process would increase the enrichment as follows:

$$R_{\mathrm{II}} = N_{\mathrm{II}}/(1 - N_{\mathrm{II}})/N_0/(1 - N_0)\,.$$

Consequently, an enrichment factor α can be defined as

$$\alpha = \frac{R_{\mathrm{I}}}{R_{\mathrm{II}}}\,.$$

If the enrichment process is repeated N times the enrichment becomes R^N. Thus an appropriate enrichment can be achieved for small values of R if $R^N \approx 100$, which means that, if R lies between 1.1 and 1.01, between 48 and 463 steps are necessary to achieve a nearly 100% enrichment. Several enrichment methods, of which a few will be described briefly, are possible and have been applied successfully since the 1930s. Gaseous diffusion is based on the fact that, from a mixture of gases the one with the lighter molar mass or the lighter isotopes diffuses more rapidly than the heavier one, because the velocity of diffusion is inversely proportional to the square root of the molecular weight or

$$R \sim \left(\frac{h}{1}\right)^{1/2}.$$

In 1932 Hertz separated ^{22}Ne and ^{20}Ne utilizing fractional diffusion and recirculation of the gas through a series of porous tubes, and the method was also applied to obtain large-scale enrichments of ^{235}U from uranium hexafluoride gas.

More effective and faster, however, was the method of thermal diffusion, invented in 1938. In a vertical cylinder with a diameter of several millimetres and a length of several metres, an electrically heated wire was positioned along the axis and the walls of the cylinder were cooled, thus providing a constant temperature difference. The method combines thermal diffusion with thermal convection and, whereas the light molecules move upwards along the hot wire, the heavy molecules sink down the cooled walls. The enrichment factor can be expressed as

$$\alpha = e^{lF(m_2 - m_1)/(m_2 + m_1)},$$

where m are the different masses, l the length of the cylinder and F a function which includes, for example, properties of the gas (viscosity) and the cylinder (wall temperature). This method is also applicable to liquid phases.

Isotope fractionation during exchange reactions is due to the fact that the heavy isotopes usually react slower than the light ones. It can be used for isotope separation. This is especially effective for light elements with a large relative mass difference. H.C. Urey was the first to use this method for the partial separation of several isotopes of carbon, sulphur and nitrogen. For example, the equilibrium reaction for ammonia ($NH_{3\ gaseous}$) and the dissociated ammonium ion ($NH^+_{4\ aqueous}$) is given as

$$^{15}NH_3 + {}^{14}NH_4+ \Leftrightarrow {}^{14}NH_3 + {}^{15}NH_4^+,$$

with an equilibrium constant of $k = 1.023$, and thus the equilibrium shifts to the right side of the equation. This leads to an enrichment of the heavy ^{15}N isotope in the liquid which facilitates the separation of ^{15}N by fractional distillation. Similarly to diffusion, the rate of evaporation is greater for the light isotopes, the heavy isotope concentrating in the liquid phase. This is due to the differences in the mass-dependent molar evaporational enthalpy which is, for instance, 46.5 kJ mol^{-1} for heavy water and 40.7 kJ mol^{-1} for water. Thus the resulting vapour pressure is different and, consequently, if water evaporates, the D_2O becomes enriched in the liquid and the H_2O goes into the vapour. To a lesser extent, the same effect produces differences for $H_2^{18}O$ and $H_2^{16}O$, of which the latter evaporates more easily.

Less important methods for artificial enrichment of isotopes are, for instance, electrolysis and centrifugation. Electrolysis is mainly useful for H and has become an industrial method for the production of heavy water. As the light isotope ^{1}H becomes disengaged at the cathode, D will be concentrated in the water. However, this method is not satisfactory for other elements. Centrifuging, on the other hand, is mainly effective at low temperatures and the gases used should have mass differences from 1 to 4 amu (atomic mass units). For a single run in a centrifuge, the separation can be described as

$$\alpha = e^{(m_h - m_l)\,(v^2/2kt)} ,$$

with m_h and m_l being the mass differences of the heavy and light isotopes, v the peripheral velocity and k and T being the Boltzmann constant ($k = 1.380658 \times 10^{-23}$ JK^{-1}) and the absolute temperature. During centrifuging, the light isotope concentrates at the rotor axis, whereas the heavy isotope will move to the periphery. As the enrichment does not depend on the mass ratio but on the absolute mass differences, a good enrichment of isotopes can be achieved at very low temperatures.

1.3 SPECTROSCOPIC METHODS

As a number of isotopes have been discovered by spectroscopic methods, some aspects of the different spectra may be discussed briefly. The shift in the spectra of a molecule or atom is caused by both mass effects and nuclear volume. Thus the isotopic composition influences the spectra of a molecule. Generally, in a molecule the shift of an electron from one energy level to another causes variations in the rotational and vibrational energy. Thus, in contrast to an atom, the absorption or emission of energy does not cause a spectral line, but a number of spectral bands. If the total energy T^{tot} of a diatomic molecule – to use the example of the H_2 molecule – is to be expressed, different contributions must be taken into account.

All these energy contributions are quantified and consist of

- E_{electron}, due to the electronic structure;
- $E_{\text{vibration}}$, due to vibrations along the intercore axis;
- E_{rotation}, due to rotational states rectangular to the intercore axis.

Thus it follows that E_{total} can be expressed as

$$E_{\text{tot}} = E_e + E_v + E_r .$$

This is not true, however, for changes in the energy level caused by absorption or emission. Changes have to be expressed by a different equation, for the quantified energy ($\Delta E = h\nu$) (either absorbed or emitted) must be calculated and thus the expression becomes

$$h\nu = E_{\text{tot}} - E' = (E'_e - E^1_e) + (E_v - E^1_v) + (E_r - E^1_r) .$$

It might be possible to imagine a molecule vibrating and rotating using a dumbbell as a comparable object, but the different forces do not really become obvious. However, if both weights were to be connected by a rubber band, it would be clear that vibrations and rotations change with an input or loss of energy, as well as with two different weights at the ends of the rubber band. This is a very simplified picture since the rubber would change its nature as well and the electrons also have a little mass, but it might be a visual aproach. This is to say that the vibrational quantum has a magnitude of 0.1 to 0.01 eV and thus produces spectra at wavelengths of 10 to 100 μm of the infrared. For most elements the isotope exchange due to the molecule energy is dependent only on this vibrational energy. An exception is the H_2 molecule, as its rotational quantum is very large (1.51×10^{-2} eV) compared to others. The rotational quantum is normally of a magnitude of 10^{-2} to 10^{-6} eV and the quantum is inversely proportional to the moment of inertia of the molecule. Thus, as the moment of intertia of the H_2 molecule is the smallest of all, its rotational quantum is the largest and must also be taken into account for molecular isotope exchange.

The pure rotational component of a diatomic molecule is given as follows:

$$E_{\mathrm{r}} = \frac{h}{8\pi^2 I} J\,(J+1)\,,$$

with h as the Planck constant, I as the moment of inertia and J as the rotational angular-momentum quantum number. From the quantum mechanics for waves it follows that J has discrete values of $0, 1, 2, \ldots, n$. Thus the difference between two energy levels is

$$\Delta E_{\mathrm{r}} = E_{\mathrm{r},J+n} - E_{\mathrm{r},J} = \frac{h^2}{8\pi^2 I} 2\,(J + N) = h\nu_{\mathrm{r}}\,.$$

These explanations will not be extended further as they are a matter of physics and it is only of interest to know that the spectroscopic data (obtained by optical methods or microwave spectroscopy) enabled nearly exact determinations of isotope masses to be made. As the intensity of the rotational bands is also a function of temperature, it need only be mentioned that the bands are clearer at low temperatures because the influence of vibrational energy is low. In general, the temperature is a determining parameter in all isotope reactions and, at very high temperatures, fractionation factors approach zero. This means that the mass differences become ineffective in a highly energetic environment. Nevertheless, as the vibrational energy has been mentioned, a short review of rotational–vibrational spectra will close this chapter.

As the vibrational quanta are usually greater than the rotational ones, they dominate the band spectra. The vibrational frequencies are mass dependent too and are inversely proportional to the square root of the reduced mass m_0 and thus

$$h\upsilon \equiv (m_0)^{\frac{1}{2}} = \left(\frac{1}{m_1} + \frac{1}{m_2}\right)^{\frac{1}{2}}.$$

It is again demonstrated that the mass difference of isotopes is the factor determining the isotope exchange.

All the isotope fractionations described in Chapter 3 are basically dependent on the mass differences within the chemical compounds built of different isotopes of an element. Apart from H, only the varying vibrational energy causes the exchange of isotopes within these compounds, because the binding energy differs and thus the evaporational enthalpy, the dissociation energy and the vapour pressure are also different. Nevertheless, the vibrational energy is temperature- and, to a far lesser extent, pressure dependent. If one imagines a pressure as high as in neutron stars, this might cause a core distance approaching zero and thus an asymptotic increase of the potential energy of a molecule tending to be infinite. The absolute zero temperature of *c.* −273 C should lead to the zero-point or no movement but this is not true as has been seen. The increase of temperature first leads to evaporation and, if driven too far, to the dissociation of a molecule. Under such conditions, however, isotope effects can be neglected. Although physics is able to explain lots of processes occurring in the environment, it should be stressed that the calculations for a diatomic molecule are the simplest ones. Not only do they become more complicated when more atoms are involved, but they become less precise. The psi-functions, for instance, describe processes we do not really understand. It is not known why an electron sometimes acts as a particle and sometimes as a wave. In addition, the nature of magnetic or gravitational fields is not understood. Only their effects can be described.

Although isotope fractionation is not yet completely understood, especially in biological systems, the empirical and experimental data provide tools for the conditions in which we live.

1.4 ISOTOPE FRACTIONATION

As already discussed, isotopic fractionation results from minute differences in the physicochemical properties of the isotopically different molecules. Isotope fractionation has been discussed in detail by, for example, Bigeleisen and Mayer (1947), Urey (1947) and Hulston (1978).

Apart from the already mentioned exchange reactions, kinetic fractionation is of major importance in natural processes. It occurs throughout physical processes (see also section 1.2.2) and during chemical reactions. Additionally, biological fractionation has to be taken into account, although the reason for the fractionation is the same from the physical point of view. However, active transport by enzymatic pumps or diffusion through biological membranes in organisms may cause isotopic fractionation different from the expected equilibrium.

Generally, the equilibrium fractionation is of interest to the geoscientist. For example, the palaeotemperature determination from mollusc shells is based on the fact that the marine organisms precipitate the $\delta^{18}O$ of their shells in isotopic equilibrium with the surrounding sea water. The equilibrium isotopic fractionation occurs between components similar to the normal chemical equilibrium and is mainly dependent on the temperature, the vibrational frequencies and, under normal conditions only to a negligible extent, on the pressure. Biological functions may sometimes also serve as a 'catalyst'. A few aspects of isotope fractionation processes are reviewed in the following section.

1.4.1 Exchange Reactions and Kinetic Fractionation

Kinetic isotope effects take place when the rates of chemical reactions or physical processes of isotopic species differ from one another. The isotopic fractionation due to kinetic or equilibrium reactions is a function of the vibrational frequencies of the molecules or atoms. In equilibrium the isotopes redistribute and exchange in such a manner as to obtain minimum free energy in the system. Problems in the realistic interpretation of the results often arise, if the observed isotope ratios are the result of different reaction steps and pathways or different parameters controlling the isotope ratios.

For instance, O isotope ratios from mollusc shells could be due to temperature but also to salinity changes. As different chemical compounds naturally have different fractionation factors, the theoretical and experimental considerations to determine the equilibrium fractionation were mandatory for successful isotope geoscience. As the calculations on the basis of statistical quantum mechanics are very difficult, experiments under known conditions have been a valuable source of knowledge. Thus the interpretation of isotope ratios is mainly based on empirical results. Although all these methods have some limitations and experiments rarely consider the whole complexity of natural processes, isotope data are an amazing tool in helping to understand geological and biological proceedings.

The amount of theoretical and experimental data is enormous and can only be summarized briefly as follows. The system best understood is the carbonate–water system, on which the method for the determination of palaeotemperatures is based. The fractionation factor α can be expressed as

$$\alpha = \frac{3(C^{18}O_3^=) + 2(C^{16}O^{18}O_2^=) + (C^{16}O_2^{18}O^=)}{3(C^{16}O_3^=) + 2(C^{16}O_2^{18}O^=) + (C^{16}O^{18}O_2^=)} \Big/ \frac{(H_2{}^{18}O)}{(H_2{}^{16}O)} .$$

This reduces to

$$H_2{}^{18}O_{(liquid)} + 1/3\, Ca\, C^{16}\, O_{3(solid)} \Leftrightarrow 1/3 Ca\ C^{18}\, O_{3(solid)} + H_2{}^{16}O_{(liquid)} .$$

The early work of, for example, McCrea (1950) and Clayton (1961), describes the problems in determining the fractionation factors and, if samples are prepared with phosphoric acid, the reaction yields only two thirds of the total O from the carbonate. For the solid carbonate phase, lattice vibrations have also to be taken into account and McCrea (1950) gave a short mathematical description of the basic problems in calculating the different contributions of vibrational, rotational and translational frequencies. Nonetheless, the experimentally found fractionation factors are the result of the quantum mechanical properties and the thermodynamic behaviour; thus the differences from expected equilibrium found in samples are subject to explanations as well as the cycles of isotopes. If the reactions for two compounds are generally written as

$$nA_l + mB_h \Leftrightarrow nA_h + mB_l ,$$

with n or m moles of A + B as compounds containing l = light or h = heavy isotopes, the equilibrium constant is expressed as

$$K = (A_h/A_l)^n/(B_h/B_l)^m ,$$

where the equilibrium constant K is a function of temperature and the parentheses contain the molar ratios of the compounds in question. It is certainly possible to calculate the value of K by utilizing the rules of statistical quantum mechanics (Urey, 1947, and others), although experimental data are preferable. With regard to the aforementioned, the calculations may slightly differ from experimental results, even if spectroscopic data for the compounds were available.

As the pressure dependence of the equilibrium is small and can usually be neglected, the temperature is the main factor controlling the isotope fractionation. However, the isotope ratio measured is dependent on other parameters as well. If, for instance, the influence of salinity on C and O isotope ratios in mollusc shells is considered, increasing salinity causes increasing $\delta^{13}C$ and $\delta^{18}O$ values (Epstein and Mayeda, 1953; Clayton and Degens, 1959). Decreasing salinity has the opposite effect and leads to lower isotope ratios. It should, however, be stressed that the term isotope fractionation is used on two different levels: (1) the exchange reactions as given above, where isotopes are redistributed within the molecules during chemical reactions and (2) processes such as evaporation, diffusion, condensation or dilution, where the molecules containing the heavy or light isotopes are either enriched or depleted. The latter processes are more dependent on other parameters. It is easily imaginable that evaporation may depend on the altitude and related atmospheric pressure, on the surface area of a water body, or on the exchange of air over the water. Thus the problems in interpreting isotope ratios are less dependent on the understanding of exchange processes, and are due more to our lack of knowledge, e.g. to what extent the isotope ratios of seawater are due to temperature, salinity, mixing or ice volume effects.

The fractionation factor defined as

$$\alpha_{(A-B)} = R_A/R_B\,,$$

of the two compounds A + B is related to the equilibrium constant as follows:

$$\alpha_{(A-B)} = K^{1/n}\,,$$

if n atoms are exchanged between the randomly distributed isotopes of the components A + B. Apart from H, with its large relative mass difference, the fractionation factor is close to unity and can be expressed as

$$\alpha = 1 + \epsilon/1000\,,$$

where the fractionation ϵ is related to α as

$$\epsilon_{(A-B)} = (R_A/R_B - 1) \times 10^3 \approx (\alpha - 1) \times 10^3,$$

where the result is expressed in per mil and the δ-value is then

$$\delta‰ = (R_{\text{sample}}/R_{\text{standard}} - 1) \times 10^3,$$

if a sample is measured against a standard.

The above equations are common knowledge, but exchange reactions and isotope enrichment or depletion have to be divided, as there are different controlling processes. Dealing with two problems, the equilibrium can either be obtained through exchange of isotopes within the molecules or, on another level, molecules with different isotopic composition can be equilibrated, enriched or depleted. Consequently, isotope investigations also measure different effects.

A mollusc shell, considering the fractionation in equilibrium with sea water as correct, reflects the exchange between the dissolved ions and precipitated solid molecules. The isotopic equilibria in carbonaceous systems have, for example, been studied by Usdowski *et al.* (1991) and more recently by Zhong and Mucci (1993) and Usdowski and Hoefs (1993). The latter authors re-examined the work of McCrea (1950) and showed that the $\delta^{18}O$ ratio of rapidly precipitated $CaCO_3$ also depends upon the pH value of the solution. The relation found permits the calculation of fractionation factors for the individual carbonate species.

The sea water itself has, on the other hand, an isotopic composition achieved by, for example, enrichment or dilution processes. Thus our knowledge about the various equilibrium fractionations enables us to obtain information about these waters. On the other hand, they do not tell us how these waters got their isotopic compositions (e.g. by rainfall, mixing, evaporation, etc.) and that is the crucial problem in interpreting the measured values.

Although the equilibrium fractionation is generally assumed, disequilibrium has to be considered too and often isotope ratios are not in equilibrium as expected. Therefore research must deal with the question of how to explain the observed isotope ratios. The crux of the matter is that the original isotope ratios, i.e. the distribution of molecules with different isotopes, is dependent on many more factors than the temperature and, according to the system under observation, the incorporation of other data (i.e. palaeontological, mineralogical or botanical) is essential for the interpretation of many isotope results.

This also counts if enrichment or depletion results are measured, as for instance on ice cores. The isotopically light water bound in ice is a result of a precipitation process due to the different vapour pressures of $H_2{}^{18}O$ and $H_2{}^{16}O$. Both fractionated evaporation and condensation can be mathematically described by the Rayleigh function and have been summarized by Hoefs (1987) and Bowen (1988). However, the application of this function is a mere approximation if describing natural processes.

Experimental data may not be far from naturally observed ones on a global scale but certainly depend on the different morphology and the temperature zones involved. Dansgaard (1964) has studied the precipitation of H and O caused by equilibrium processes such as condensation and evaporation and this is different to the exchange of isotopes within molecules in being an 'exchange' of molecules between the phases, i.e. water and vapour. However, each exchange is due to kinetic processes, as kinetics are from the physical point of view due to the energy of movement, which, on the other hand, is mass dependent. Thus the vibrational energy and the vapour pressure are mass dependent.

Unfortunately, the terminology used in isotope geoscience seldom divides the different levels on which isotope separation occurs. For example, if the exchange of isotopes within molecules is under normal conditions not dependent on the pressure, the evaporation of molecules with different vapour pressures shows this dependence. Although this might be unimportant for atmospheric processes and the evaporation or condensation of water, within magmatic systems an imagined gas phase might be under high pressure and also reach its pressure of saturation.

However, to divide the two levels between exchange due to the vibrational frequencies and enrichment/depletion due to, for example, differences in vapour pressure is a question of terminology and thus is not discussed further.

Hydrogen with its large relative mass difference and its widespread natural occurrence, for example, exhibits a great range in stable isotope ratios. δD variations of more than 400‰ have been reported by Dansgaard (1964) and an even wider range is found in extraterrestrial matter. Bottinga (1969) calculated fractionation factors for the exchange of deuterium between hydrogen, water vapour and methane over the temperature interval from 0 to 700°C. Some fractionation factors for different systems are listed in Table 1.3, where the values from Dansgaard (1964) give fractionation factors for the enrichment of the heavy hydrogen during condensation. The use of different notations for the fractionation factor is due to the fact that α is related to α as

$$\delta_A - \delta_B \approx 10^3 \ln^{\alpha}_{A-B} ,$$

and the natural logarithm ln of the fractionation factor α also shows a temperature dependence at high temperatures as

$$\ln \alpha \approx 1/T^2 .$$

This expression with T as the absolute temperature in kelvin is, however, valid only at temperatures > 1000 K or if the vibrational frequencies are low at low temperatures.

Table 1.3 Fractionation factors for H and D for different systems

System	*T* (°C)	*Notation*	*Reference* [a]
H_2O vapour–hydrogen	0	ln $a = 1.395$	1
	25	ln $a = 1.253$	1
	100	ln $a = 1.943$	1
	500	ln $a = 1.325$	1
CH_4–H_2	0	ln $a = 1.387$	1
	25	ln $a = 1.232$	1
	100	ln $a = 0.893$	1
	500	ln $a = 0.255$	1
H_2O vapour–CH_4	0	10^3 ln $a = 7.3$	1
	25	10^3 ln $a = 21.8$	1
	100	10^3 ln $a = 49.9$	1
	500	10^3 ln $a = 69.6$	1
Glauconite–H_2O	St	$a = 0.93$	2
Smectite–H_2O	St	$a = 0.94$	2
Kaolinite–H_2O	St	$a = 0.97$	2
Gibbsite–H_2O	St	$a = 0.985$	3

[a] 1 = Bottinga (1969), 2 = Savin and Epstein (1970), 3 = Lawrence and Taylor (1971).

H and O fractionation and precipitation have been extensively studied by Craig and Gordon (1965). They found the following abundances of molecules for mean seawater in ppm:

$$H_2{}^{16}O \quad 1\,000\,000$$
$$H_2{}^{18}O \quad 2000$$
$$HDO \quad 320$$

and the processes leading to enrichment or depletion of a molecular species are mainly evaporation and condensation, due to the differences in vapour pressure. HDO has a lower vapour pressure than $H_2{}^{18}O$ and is consequently enriched in the remaining liquid phase if evaporation occurs. The opposite effect, the slightly easier condensation of HDO in clouds, leads on the other hand to a depletion of HDO in the remaining vapour. As the atmosphere of the Earth has typical patterns of wind drift (Fig. 1.6b) due to the rotation and the Coriolis force, the stepwise or fractional precipitation leads to lower HDO ($H_2{}^{18}O$) isotope ratios with increasing latitude (Fig. 1.6a).

These δ-values are estimated from numerous analyses and the resulting isotope ratios are certainly an average of rain or snow precipitated under different condensation / evaporation conditions, depending on latitude, altitude and surface (water / land). Minor, almost negligible, separation of water molecules is due to small differences in the freezing points. The behaviour of O is naturally very close to that of H if the above effects are considered, although the amount of enrichment or depletion is smaller due to the relative isotopic mass differences.

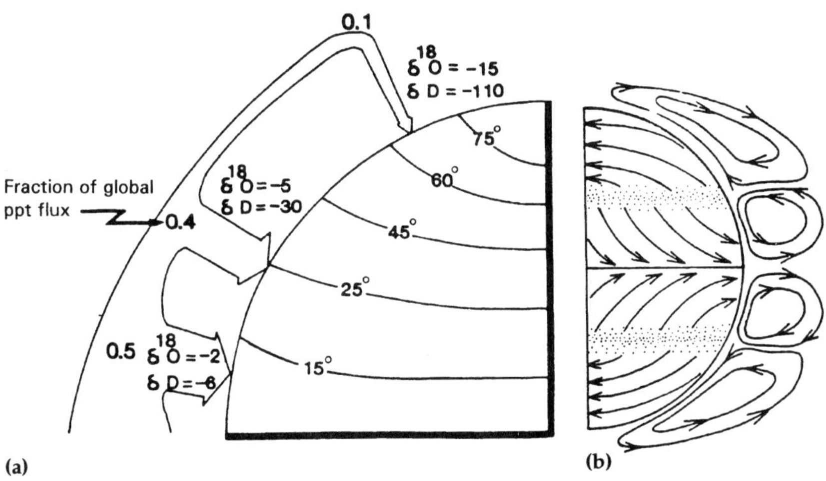

Fig. 1.6 (a) Latitude-dependent variations of H/D and $^{18}O/^{16}O$ ratios (after Craig and Gordon, 1965) and (b) circulation of air.

As a nearly omnipresent element, other fractionation effects and isotope exchange are also of importance for H. They occur, for instance, in plants and mineralogical systems and thus will be discussed in other chapters. The enrichment or depletion of O isotopes is close to that of H if kinetic effects such as evaporation or condensation are considered.

1.4.2 Biological Fractionation

The biological fractionation of isotopes by marine organisms is still poorly understood. The isotope ratios of C, for example, are only enriched by 0.035‰ in the system HCO_3^-–$CaCO_3$ if the temperature increases 1° C. Thus a change of 1‰ would necessitate a temperature change of *c.* 28° C. It is therefore assumed that metabolic processes have a great influence on the δ^{13} C ratios of marine organisms, as was recognized early by Urey *et al.* (1951). For example, some foraminifera, algae and corals show isotope values in isotopic disequilibrium (Erez, 1978; Woodruff, Savin and Douglas, 1980; Wefer, 1985). These deviations from the expected equilibrium are usually assumed to be constant and thus the isotope ratios can be corrected. On the other hand, very little is known about the metabolism of most species which might serve for scientific research and the enrichment or depletion of O or C isotopes might be due to various effects. However, two main processes must be considered: the vital effects in the narrow sense which occur within or close to the organism and, in the wider sense, the effects caused, i.e. by symbionts influencing the isotopic composition of the sea water, both leading to isotope fractionation in disequilibrium. The former are probably due to exchange reactions between CO_2 respirated and HCO_3^- dissolved. Such effects are shown, for instance, by Crinoidea, Asteroidea and Echinoidea (Weber and Raup, 1966; Weber, 1968). Marine carbonate shell-secreting arthropods also belong to this group, although a temperature equation has been formulated for Balanoidea by Killingley and Newman (1982) as

$$T(°C) = 22.2 - 4.65\,(\delta_C - \delta_W) + 0.14\,(\delta_C - \delta_W)^2.$$

This equation takes an average δ^{18}O enrichment of c. 1.3‰ into account, whereas C isotope ratios have been found to be depleted. Other arthropods such as Insecta, Crustacea and Merostomata show, for instance, no systematics for O and H isotopes of their chitin skeletons (Schimmelmann and DeNiro, 1986) and therefore their value for palaeoecological research remains unclear.

Additonal problems arise from the fact that the annual growth periods are mainly unknown and are possibly influenced by environmental conditions. As for Bivalvia, the vital phase and therefore growth stops if the species-dependent temperature for biological activity becomes too low. The pearl oyster *Pinctada martensii,* for example, hibernates at temperatures lower than 13°C (Kobayashi and Watabe, 1959). In addition, shell secretion and shell dissolution are reversible processes. The latter might occur in unknown quantities if the shells are closed and the quickly increasing pH value is buffered by the dissolution of shell $CaCO_3$. This process (see scheme in Fig. 1.7) might also lead to a non-equilibrium fractionation. Over- or under-saturation of ions within the pallial fluid might well be compensated by active enzymatic transport.

In foraminifera, enrichment or depletion of O and C isotopes has been found as well as equlibrium fractionation. Relationships between isotope fractionation and shell size, depth habitats during age, rate of shell growth and temperature and fertility are among explanations for observed disequilibria. This broad field has been summarized, for example, by Berger (1981) and Wefer (1985).

McConnaughey (1989a, b) studied isotopic disequilibrium and divided patterns due to kinetic (during CO_2 hydration or hydroxylation) and metabolic (enrichment or depletion reflecting the isotopic composition of the DIC (dissolved inorganic carbon)) effects, the latter caused by respiration and photosynthesis.

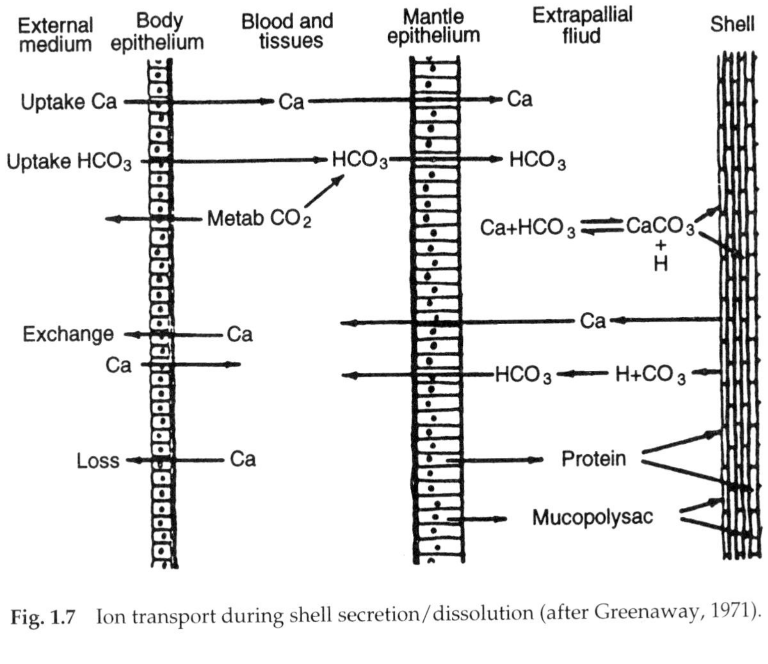

Fig. 1.7 Ion transport during shell secretion/dissolution (after Greenaway, 1971).

1.5 SECONDARY CHANGES OF ISOTOPE RATIOS

Equilibrium fractionation is a prerequisite if mass spectrometry is applied to samples of either chemical or biological origin. Apart from fractionation in disequilibrium, as sometimes caused by metabolic processes, rapid fractionation may also lead to isotope ratios different from the expected equilibrium, simply because there is not enough time to exchange isotopes and thereby to reach the lowest energy level of the chemical system in question.

However, neglecting these exceptions, secondary changes of originally equilibrated isotope ratios occur, for example, during diagenesis of sediments or through interaction of hydrothermal fluids with igneous rocks. Naturally, the grade of alteration depends on the permeability of the sediment or rock body, which, at first, gives access to waters of different origin and different isotopic composition. It is obvious that during rock history, rocks can be exposed to such altering conditions several times and thus it is not easy to judge whether an isotope signal is attributable to environmental changes or to one or more secondary alterations.

Diagenetic effects, although ignored for a long time, have now been extensively studied, for example by Veizer and Hoefs (1976), Brand and Veizer (1980, 1981) and Killingley (1983). As Patterson and Walter (1993) demonstrated, a rapid syndepositional diagenesis leads to dissolution and reprecipitation and consequently the original isotope ratios will be influenced by the circulating pore water. Usually, diagenesis leads to an enrichment of the light isotope for both O and C, but the processes involved are complex and may change rapidly with the distance from the sediment–water interface. The dissolution of calcium carbonate and the decay of calcium carbonate and the decay of organic matter cause a mixing of isotopically heavy CO_2 due to the dissolution of organic and sedimentary carbonate and light ^{12}C, released during the decomposition of the organic residuals. On the other hand, in anoxic environments an enrichment of ^{13}C of pore water was observed, due to the isotopically light methane produced by bacteria during methanogenesis (Anderson and Arthur, 1983). The fact that anoxic conditions may occur within the range of burrowing depths of organisms regionally in shallows when clay is sedimented, while contrariwise nearby sandy sediments have normal conditions, may explain the so-called strange isotope ratios which are measured sometimes in fossils.

As changes of the isotopic environment may occur within the burrowing depth of organisms, an influence of the milieu on living specimens cannot be excluded, supposing the organism also exchanges isotopes through the mantle and siphon surface (Attendorn, 1993).

It is obvious that the influence of secondary processes increases with the age of the samples and the grade of diagenesis or possibly even metamorphism. Therefore the problems caused by diagenetic effects must be understood in order to interpret isotope measurements of past fossils or sediments. Numerous investigations tackle this problem (e.g. Hamilton *et al.* 1987; Kaufman, Hayes and Klein 1990; Kaufman *et al.*, 1991; Smith and Dorobek, 1993; and many others).

An aspect not so much considered is how isotope exchange varies the δ-values. If sediment grains are influenced, solution or solid liquid exchange may affect the outer rims. Of major importance then must be the permeability of the sediment and, therefore, the grain–pore water boundaries should show an alteration different from that of the grain–grain boundaries, only because there is no access to fluids with different isotopic composition. Another question would be whether an isotopic change of the outer zones of, for example, a fossil shell, would cause a migration of isotopes from, say, heavy to light in a solid state diffusion.

A new technique for isotope measurements, described later, should enable such investigation in controlled experiments. And why should one not expect that a part of the shell of an ancient fossil is not diagenetically altered in its isotope ratios?

2

Nucleosynthesis: *Fons et origo* of the chemical elements in the universe and on Earth

Current cosmological opinion has it that around 15 Ga ago, a violent explosion occurred from a region smaller than a dime at a moment of infinite temperature and, within a fraction of a nanosecond, quarks coalesced to form protons, neutrons and other hadrons about 10^{-5} s after the event which has been termed the Big Bang. Protons then fused into the nuclei of He atoms and nucleosynthesis in this primaeval matter/radiation soup is thought to have ended after 10^2 s. After the formation of matter, expansion took place and was accompained by rapid cooling to about 10^8 times that of the Sun's core, by which time the forces of nature acquired their present properties. After another thousandfold expansion, all measurable matter occupied a region about as large as the solar system. Remaining free quarks were restricted to protons and neutrons and, following a further thousandfold expansion, atomic nuclei were created, including nearly all the He and D now existing. The entire process to this point probably took about a minute and conditions were still too hot for electrons to be captured by these nuclei. It may have taken another 300 000 years for neutral atoms to appear, the universe then being some thousand times smaller than now. Neutral atoms cohered into gas clouds from which stars evolved and, when the universe reached one-fifth of its present size, fledgling galaxies emerged. Most of the heavy elements from which terrestrial planets arose were generated by nuclear reactions in stars when the universe was half its present size. The solar system originated about 5 Ga ago when it was two-thirds of its present size. Through time, star formation has consumed much of the gas in galaxies, so that stellar populations are diminishing. In the primordial plasma, gravitational forces augmented ripples which drew the densest regions together into galaxies and voids and these have left traces apparent today. In April 1992 NASA's COBE (Cosmic Background Explorer) satellite detected them in cosmic background radiation which, in some parts of the sky, is slightly brighter than elsewhere. Although the extra radiance is a mere 0.001%, Stephen Hawking described it as the discovery of the 20th century, if not of all time. This is because such 'hotter' spots suggest the seeds of galaxies and their cosmic microwave background reflects the largest and oldest structures in the universe. Its beginning involved a continuous alteration in the elemental structure of matter, carbon, nitrogen and other elements essential to life on Earth being synthesized in the interiors of stars now long dead. As Peebles *et al.* (1994) indicated, understanding this process of genesis and development is a major achievement entailing the concept of a universe expanding in an almost homogeneous manner from the dense initial state. Of course there are uncertainties in the Big Bang theory, for instance precisely how galaxies form, and whether expansion will go on indefinitely or slow down, reverse and culminate in a big crunch, but the model remains the best so far devised. It is interesting that its name was first used pejoratively by Fred Hoyle when he, Thomas Gold and Hermann Bondi propounded an alternative hypothesis, that of continuous creation. A variant of this was presented by Hoyle (1994) when he suggested that a 'creation field' of negative energy pervades the universe and brings new matter into existence. This is envisaged as occurring sporadically, induced by intense gravitational fields possibly located in the dense centres of galaxies. Positive outbursts of matter would take place, this contradicting the views of those who favour centrally placed black holes into which matter disappears. If correct, the Big Bang would become redundant, the universe existing always with matter continuously, but intermittently, forming within it. However, standard Big Bang cosmology remains by far the majority view, though it must be borne in mind that the idea of space and the distribution of matter relate, the expansion of galaxies showing the unfolding of space accompanied by a waning in density. Development of the Big Bang model became possible after astronomical studies at the start of the 20th century revealed paradoxes àpropos the ages of stars and planets. Kirshner (1994) noted that parallel computations about that of the Sun assumed that it is energized either by chemical combustion or by gravitational shrinkage, and hence could only shine for some millions of years. Later researches on natural radioactivity on Earth furnished evidence on the ages of the chemical elements, the slow decay of U into stable Pb implying that the Earth must be several billion years old. The disparity between these incompatible signals was resolved through advances in nuclear physics which showed that nuclear reactions produce extremely large energy yields following the famous formula $E = mc^2$. Such yields exceed those of chemical reactions by a factor of at least a million, so that a star burning hydrogen can have a life span of as much as 10 Ga.

The chemical elements are substances composed entirely of atoms with the same atomic number, that is to say having the same number of protons (Z). This proton number corresponds to the electron number on which chemical behaviour depends, chemical combinations taking place between atoms by sharing or transferring outer electrons in appropriate shells, electrons in the same shell having the same principal quantum number, n. Such shells are designated by the letters K to P (equivalent to values of n from 1 to 6) in order of increasing distance from the nucleus. The number of electrons in each shell is restricted by Pauli's exclusion principle. Each electron can be characterized by values of four quantum numbers and the principle states that no two electrons in an atom can have the same set of four quantum numbers. It is of fundamental importance in the theoretical development of the Periodic Table and, incidentally, applies also to all fermions (elementary particles conforming to Fermi–Dirac statistics, the numbers of which are conserved throughout all nuclear reactions and which are divided into baryons and leptons, groups distinguished by the fact that members of one cannot transform into members of the other–all fermions have spin½). Because of this, the Pauli exclusion principle influences the combining of quarks to form elementary particles. Within each electron shell, $2n^2$ electrons or less can be contained and energy sub-levels (subshells) exist according to their orbital angular momentum (represented by their azimuthal quantum number, l). Separate subshells are distinguished by the letters s,p,d and f, these corresponding to values of l from 0 to 3. Sharing or transferring electrons is described by the electronic theory of valency which assumes that certain arrangements of outer electrons in atoms, for instance octets where the outer shell has eight electrons, are stable and tend to form by one or the other method. The principal types of linkage are electrovalent bonds and covalent bonds. The former result from transfer, the atom losing an electron becoming a positive ion and that gaining one becoming a negative ion. The latter ensue when a pair of electrons is shared, one being supplied by each atom.

An atom is specified by its atomic number and its neutron number, their sum being the nucleon or mass number and the nucleus of a type of atom identified by its atomic number and mass number is termed a nuclide (which also defines the type of atom to which such a nucleus belongs). The Periodic Table mirrors the grouping of electrons in atomic shells, electrons in the same shell having the same value of n. The atomic weight of an element is its relative atomic mass, r.a.m., namely the ratio of the average mass per atom of a specified atomic composition of that element to one-twelfth of the mass of an atom of $^{12}_{6}C$. The natural isotopic composition is assumed unless otherwise stated. Since elements possess isotopes, atomic weights can never be whole numbers. For instance, the atomic weight of H is 1.00794, that of O is 15.9994 and that of C is 12.011, reflecting the mixtures of the isotopes involved in each case. Isotopic masses have been established by mass spectrometric measurements and the total number of nuclides is about 1700 of which only 260 are stable. These all make up the Periodic Table and nearly all have at least two or more naturally occurring isotopes, though a few have only one and some elements are not found on Earth in nature.

Radioactivity is a common phenomenon encountered in the great majority of nuclides. The weak nuclear force is carried by W and Z gauge particles which last for about 1×10^{-18} s and pass their lives in atomic nuclei where they trigger radioactive decay, for example in the element U. Whether or not a fifth force, a postulated short-range correction to Newtonian gravitation, exists or not is significant simply because, if it does, it would require instability in the proton. However, since the associated decay time would have to be of the order of 1×10^{18} years, i. e. orders of magnitude longer than the age of the universe, the proton could be considered stable for all practical purposes even if the fifth force were to be confirmed. Searching for it has been a rewarding exercise in many ways, according to Maddox (1992), leading to innovations in the design of torsion balances and their emplacement on cliffs and dry dock edges as well as increased sensitivity in measuring gravitational attraction between masses separated by laboratory distances.

Most known radionuclides do not occur naturally because of their extremely rapid decay rates, but they can be produced artificially in nuclear reactions in the laboratory. Radioactive decay is accompanied by the emission of either α or β-particles and/or gamma rays. The commonest disintegration mode is β-decay in which a neutron converts to a proton with the emission of an electron and an antineutrino or in which a proton changes to a neutron with the emission of a positron and a neutrino. As Podosek (1994) indicated, more than 45 years earlier it had been shown that $\beta\beta$ decay is possible and a handful of other nuclides have since been discovered to undergo it. A β-decay entails unit change of atomic number, but no change of mass number. All neutrons outside a nucleus are unstable to β-decay, a form of weak interaction. The commonest type of beta decay is that involving neutron to proton conversion. Alpha particles (helium nuclei) are emitted only by certain radionuclides of the heavier elements and, when α-decay happens, the atomic number of the daughter nucleus is two less than that of its parent, the mass number dropping by four units. Gamma-ray emissions accompany α- or β-emissions when the daughter nucleus is formed in an excited (metastable) state. Excitation occurs when energy is added to a nucleus, atom or molecule, transferring it from its ground state to a higher energy level. The excitation energy is the difference in energy between the ground state and the excited state.

Capture is another process taking place when an atomic or nuclear system obtains an extra particle; as an illustration when electrons are acquired by ions or neutrons by nuclei. Radiative capture is a special case ending in the emission of gamma rays only. K-capture entails absorbing an electron from the inner, K-shell of an atom by its nucleus to form a different nuclide as in the following reaction: ${}^{7}_{4}Be \rightarrow {}^{7}_{3}Li$. During such K-capture, the atom becomes excited and usually decays thereafter by emitting an X-ray photon. Gamma rays (gamma radiation, γ-rays) comprise electromagnetic radiation of the nature of, but with shorter wavelength than, X-rays (10^{-10} to 10^{-13} m). They are emitted by excited atomic nuclei during decay to a lower excitation stage. Gamma-ray photons have an energy range of 10^3 to 10^6 eV. And the photon is a quantum of electromagnetic radiation with zero rest mass and energy equal to the product of the frequency of the radiation and the Planck constant h ($=6.626076 \times 10^{-34}$ Js), the latter being a universal constant relating the frequency of a radiation with its quantum of energy. Photons are generated when a particle having an electrical charge changes its momentum, both in collisions between nuclei or electrons and in the decay of certain nuclei and particles.

Radioactive decay embraces the transformation of a radioactive substance (the parent) into its decay products (the daughters). The decay (or disintegration) constant refers to the probability of the decay of an atomic nucleus per unit time which characterizes a radioactive nuclide. It determines the exponential decrease with time t of the activity A given by $A = A_0 e^{-\lambda t}$, where A_0 is the activity when $t = 0$ and λ is the decay constant. The reciprocal of the decay constant is the mean life. Decay schemes used in radiometric dating are those of the ${}^{238}U$, ${}^{235}U$ and ${}^{232}Th$ families and all end in stable Pb isotopes through numerous daughters of which some have half-lives of geological magnitude. This is the reason why some geological deposits can be dated by measuring the degree to which daughter nuclides have reached secular, radioactive equilibrium with their parents. The main requirement is that the parent deposited in the presence of a known deficiency or excess of one or more of the daughters. The optimal situation is that of a parent emplaced in the complete absence of daughters so that the ratio of daughter to parent increases with time to a limit of unit activity ratio, i.e. attains secular equilibrium.

Radioactivity units include the becquerel, the gray and the sievert, the derived SI units of activity, absorbed dose of ionizing radiation and dose equivalent respectively. The becquerel, Bq, refers to the activity of a radionuclide decaying at an average rate of one spontaneous nuclear transition per second and it replaced the curie. The gray, Gy, refers to the energy in joules absorbed by 1 kg of irradiated material and it replaced the rad. The sievert, Sv, is the dose equivalent when the absorbed dose of ionizing radiation multiplied by a stipulated dimensionless factor yields 1 joule per kilogram. The relevant dimensionless factors are specified by the International Commission on Radiological Protection. The sievert replaced the rem.

Nucleosynthesis, the origin of the chemical elements in the cosmos and on Earth, is probably best deciphered from the chemical compositions of stars, of which some are older and some younger than the Sun, which is about 5 Ga old and can be expected to continue radiating energy for many more billions of years. Nuclear burning of stars produces energy and that from the Sun permits life to flourish on Earth, that from other stars elsewhere in the Milky Way galaxy and in other galaxies no doubt allowing it on many other planets. The terrestrial biosphere is dependent on solar energy originating from the nuclear fusion of H and He in the interior of the Sun. In the course of this, H nuclei (protons) fuse into He nuclei (two protons and two neutrons), a process accompanied by the emission of two positrons, two neutrinos and energy. In older stars the central cores shrink and become both hotter and denser while their outer layers expand to form red giants. New nuclear reactions then become possible, for instance two He nuclei may collide and forge an ephemeral form of Be with four protons and four neutrons. When a helium nucleus collides with this, C is created and, if another He is added, O arises. The unstable Be so made and B also stem from the spallation of heavier elements in the interstellar medium (ISM) as well. Lithium too may originate in this way in addition to deriving from the Big Bang. Lithium, Be and B have much smaller nuclei than that of C and they are approximately a million times less abundant. When the red giant has lasted for a few hundred million years, it becomes unstable and ejects a planetary nebula as well as sometimes emitting graphite. Eventually it coagulates into a white dwarf sustained by the quantum repulsion of its free electrons, with its interior C and O practically crystalline.

Even more diverse nuclear reactions occur in stars far more massive than the Sun. While solar-type stars accumulate C and O which remain in white dwarfs, the interiors of massive stars fuse C nuclei to produce neon and magnesium and the fusion of O produces S and Si. The Si combusts into iron and intermediate processes form a number of elements up to Fe. Iron is the most tightly bound nucleus, has an abundance peak at its atomic number 26 and above it binding energies diminish to higher atomic numbers with some subsidiary peaks, e.g. that of Ni. The expenditure of energy is required to form heavier nuclei, which means that iron-cored stars cannot engender energy by fusion. There are also peaks associated with nuclei containing 52, 82 or 126 neutrons. Nuclear structure theory predicts that such nuclei should be especially stable and these numbers are termed 'magic' numbers. The overall decline in chemical abundances up the Periodic Table may be illustrated by the values for H and U (elements 1 and 92). For each U atom, there are 100×10^9 H atoms. Ongoing energy radiation at a stupendous rate eventually ends in a supernova explosion with the core caving in within 1s to change either into a black hole or into a neutron star. Type II supernova explosions are significant in supplementing the stock of chemical elements in the universe. Such bursting stars throw out their outermost layers which are not burnt, then emitting He resulting from H buring together with C, O, S and Si representing the products of further burning. Behind the expanding shock wave, new elements are synthesized, intense heat promoting nuclear reactions impossible in steadily burning stars. Some are radioactive, others are stable elements heavier than iron. Neutrons bombard Fe nuclei to form Au which then transforms into Pb. In its turn, Pb is bombarded to form elements up to U. Such post-Fe elements of the Periodic Table are very rare in the universe.

The concept of such nuclear reactions producing heavy elements was tested in February 1987 when a supernova, SN 1987A, exploded in the Large Magellanic Cloud. A star 20 times more massive than the Sun, Sanduleak—69° 202, observed in 1986, disappeared. Monitoring of the events by the International Ultraviolet Explorer in 1987 and later by the Hubble Space Telescope gave information which intimated that this star was at one time a red giant. Subsequent images taken in 1994 with the corrected Hubble showed rings around the supernova. The inner one is made of material lost when the star was a red giant and it is revealed by the flash of ultraviolet light from the supernova. Burning products occur in a central point barely resolved by Hubble, but thought to be expanding at the rate of $3\,000 \text{ km s}^{-1}$. No neutron star has been observed to date. These observations certainly suggest that the theoretical model for the origin of the chemical elements involving supernova explosions is largely accurate.

Unlike such supernovae, white dwarfs have their products imprisoned and a study of 189 of them near the Sun was reported by Edmunds (1993) after 10 years of research. F- and G-types were examined and they do not differ very much from the Sun, Edmunds suggesting that the latter has what he called a 'typical' elemental composition. His sampling was done from earlier photometric surveys so as to provide stars of which the ages could be assessed, these encompassing almost the whole history of the galactic disc and a range of a factor of 20 in heavy elements abundances. The work was made possible by using more sensitive detectors on modern spectrographs together with better models of stellar atmospheres giving an accuracy of about 12% to 25%. In the past, iron abundances have been used to infer general heavy element abundances of 'metallicity' in stars. Edmunds pointed out that this is because they provide numerous spectral lines in F-, G- and K-type stars which are the easiest to analyse. He added that, subject to qualification, to a first approximation, old disc stars in the solar neighbourhood have low metallicity, while young ones have high metallicity. Ratios of other elements to iron as a function of overall metallicity could shed light on past nucleosynthetic processes.

As regards the interstellar medium in the galaxy, gas concentrations occur in the spiral arms where optical examination is frustrated by dust. However, such dust, although it absorbs much transient light, protects H atoms from ultraviolet light and so permits H_2 molecules to form. Other molecules form as well and these include H_2O, CO, NH_3 and a number of others. In May 1994 the amino acid glycine ($CH_2(NH_2){\cdot}COOH$) was provisionally detected in a star cloud at the centre of the galaxy called Sagittarius-B2. Complex organic molecules in space may be due to ion–molecule reactions which can proceed fairly rapidly at extremely low temperatures. Baggott (1994) described relevant researches at the University of Rennes in France which cleared the way to study chemical reactions involving atoms and free radicals at temperatures of from 5 to 10K, these being in molecules which included O_2, NH_3 and hydrocarbons such as acetylene, ethylene and ethane. Theoretically, chemical reactions should slow down as temperatures fall, but the opposite effect was recorded and this is surprising because increasingly fewer reactants should have sufficient energy to stimulate them at such temperatures. Indeed, I. Smith at the University of Birmingham went so far as to assert that chemists will have to revise their theories in order to account for the observed effect. Astronomical implications were clear because temperatures not far above absolute zero ($0K = -273.15°C$) are characteristic of the interstellar medium. Ion–molecule reactions were believed to be catalyzed by the surfaces of minute dust grains and reactions involving only neutral molecules were thought to be too slow to be significant. These conceptions have been disturbed by the new findings and the influence which they have on matching a model with the observed abundances. This entailed faster reaction speeds of some critical neutral–neutral reactions and its predictions failed to match, although tallying with experimental observations if certain other types of reaction are assumed to be slow. The matter is not yet resolved, but some hold that there may be a link between French and British results and the identification of large molecules such as glycine and fullerenes in space. As mentioned above, glycine was tentatively identified by Lewis Synder and others of the University of Illinois at Urbana-Champaign in the United States.

That the interstellar medium comprises dust and gas has been known for many years, but the question whether there is a diffused intergalactic medium or not remained open until the advent of the Hubble which recently produced the first possible evidence of it. An observation of an absorption feature impressed on the far ultraviolet spectrum of the high redshift quasar Q0302-003 implies that the intergalactic medium is a highly ionized plasma of hydrogen and helium. As Lanzetta (1994) indicated, the search for an appropriate quasar commenced in 1990 using the Faint Object Camera on the Hubble and far ultraviolet spectra of 25 high redshift quasars were recorded. From these, Q0302-003 was selected as optimal following a spectrum obtained on December 4 1992, this proving inconclusive because of the then still-existing difficulty of spherical aberration. The necessary additional data were derived from a new observation on January 24 1994 after the COSTAR corrective optics were installed on the space telescope by the crew of the Space Shuttle Endeavor. The information so secured was excellent, the ultraviolet spectrum of Q0302-003 giving a strong signal traceable form 2 000 Å to 1 300 Å (sic). However, it showed an abrupt discontinuity around 1 300 Å and no signal under this, this discontinuity occurring at the wavelength predicted for the helium 304 Å transition at the redshift $z = 3.286$ of Q0302-003. From this it was inferred that, since there was close accord between the observation and the prediction together with the absence of signal below the discontinuity, the absorption feature probably results from smoothly distributed singly ionized He and may well constitute the signature of the diffuse intergalactic medium. There are some caveats, however, so the matter is not yet as completely resolved as could be wished. One is that the absorption feature is caused by Lyman continuum absorption from some foreground source of neutral hydrogen at a redshift near $z = 0.42$. But, as Lanzetta pointed out, the chance of this occurring is very small. Another is that the feature results from the numerous discrete 'Lyman-α forest' clouds comprising the 'forest' of Lyman α-absorption observed at high redshifts. The objection to this is that the known population of Lyman-α forest clouds could not produce the absorption feature noted unless the ratios of singly ionized helium to neutral hydrogen in them were unbelievably large. A so far undetected population of very tenuous Lyman-α forest clouds could produce the feature, but such objects might well fill most of the intergalactic space and even constitute the intergalactic medium. Lanzetta concluded that the observation is most probably exactly what it seems to be, i.e. the diffuse intergalactic medium. As for its significance, he mentioned that it will open up to further study the material from which galaxies are made and offers the first direct evidence that the intergalactic medium is a highly ionized plasma of H and He as well as the challenge to locate the sources of the ionization. Most importantly, it reinforces the Big Bang model of nucleosynthesis which predicts that helium and other light elements were produced in the very young universe and should be apparent at the very largest redshifts. He added that the next step might be to go to lower redshifts and shorter ultraviolet wavelengths where the difficulty of foreground neutral H is not so critical, but this will necessitate the next generation of far-ultraviolet satellite space telescopes. Neutral H in the Milky Way galaxy obscures the extragalactic universe at wavelengths under 912 Å, hence the quest for the helium 304 Å is always going to be restricted to redshifts which satisfy $(1+z) \times 304$ Å > 912 Å, or $z > 2$. So unless and until another probe of ionized intergalactic medium is forthcoming, knowledge of the intergalactic medium in the furthest reaches of the universe will remain greater than that pertaining to this medium near to the Milky Way galaxy.

Stellar nucleosynthesis furthers accretion of the α-elements (Mg, Si, Ca and Ti) in stars and supernovae, essentially through the incremental addition of helium nuclei and, like oxygen, they have a higher abundance relative to iron in low-metallicity, old stars than in younger, high-metallicity ones. This implies that, when synthesis of elements commenced with the liberation of freshly made α-elements and O from type II supernova explosions of short-lived massive stars, a pause occurred until type Ia supernovae, deriving from longer-lived and lower mass binary systems contributed to the Fe-group elements. Slight variations in the behaviour of the α-elements might prove to be important in evaluating nucleosynthetic theory. Edmunds referred to an apparent decrease in α-element excess relative to Fe with increasing galactic radius, which may mean that the inner parts of the galaxy formed more rapidly than its periphery. It certainly accords with early observations of high Mg to Fe ratios in stars in the galactic bulge and may well verify the idea that Mg possesses a high relative abundance centrally in galaxies which underwent rapid formation. The metallicity of the galactic disc probably increased slowly during its development and took a course between a square root and a linear increase with time, according to Edmunds. The decline in the metallicity of the gas of the disc correlates with increasing radius and he inferred a similar radial decrease in the progenitor material of those stars of which birthplaces have been estimated from current motions. Perhaps the scatter in metallicities from stars formed at similar times and at about the same distance from the centre of the disc, itself resembling the overall increase in metallicity with time, is a real spread. However, the heavy element ratios show less scatter and Edmunds suggested that all the various nucleosynthetic locations contributing different elements to the initial stellar abundances amalgamate their products well. This would eliminate any possibility that inhomogeneities accumulate as a result of stars forming in regions contaminated by the presence of a few supernovae. Also, while the metallicity dispersion could be a consequence of an inhomogeneous accretion of metal-depleted gas coming from outside the disc, a notion often included in models of galactic chemical evolution, it is likely that those processes mixing nucleosynthetic products into the disc would incorporate any such gas as well. As regards other element ratios, curiously anomalous data were cited, for instance barium and other elements formed by neutron capture on a gradual timescale in giant stars do not show an anticipated fast increase in abundance with rising metallicity, perhaps because the latter phenomenon reduces the bombarding neutron flux in crude compensation for the increasing availability of heavy target nuclei. In addition sodium, which on nucleosynthetic grounds should vary markedly with metallicity, apparently remains in an almost constant ratio to iron.

As for the abundances of elements heavier than Fe in stars, about 50% of them are probably produced by rapid neutron capture reactions, the so-called r-process. In this, capture takes place on timescales far shorter than typical β-decay lifetimes. And sequential neutron captures promote the production of extremely neutron-rich isotopes which later β-decay to stable atomic nuclei. Thus the r-process makes a significant contribution to the elemental composition of the galaxy, even though it is not well understood because the associated abundances are very small and hence difficult to observe astronomically. Nevertheless, β-decay rates have been calculated for some of the important radionuclei engendered by the r-process and they constrain neutron density, temperature and duration of the relevant r-process environment. It has been demonstrated that the latter might involve a high neutron density β-flow equilibrium environment. The β-flow equilibrium is the rough equality of the product of abundance and β-decay rate for different nuclei and the neutron source is made in the star, the r-process giving the type II supernova rate. In consequence, low-mass type II supernovae would constitute the most probable location for the r-process. Mathews and Cowan (1990) suggested that these could be divided into two categories. One includes those in which the r-process elements are produced as primary products directly in a star. The other is that in which these elements are secondary in nature, requiring heavy 'seed' nuclei acting to capture neutrons and thus make r-process elements. In the first category are high-temperature, neutron-rich environments such as shock or jet material ejected from neutron-rich supernova cores or ejecta from a binary neutron star tidal disruption. In the second category are less neutron-rich and lower-temperature environments such as nova outbursts, shock-induced explosive helium or carbon burning, the core He flash in low-mass stars and neutrino inelastic scattering in the outer envelopes of core-bounce supernovae.

With reference to lighter elements, F is one of the small number of which the stellar source is still sub judice. Mathews (1992) gave two reasons. One is that the sole stable isotope, ^{19}F, is fragile, and hence easily destroyed by proton capture reactions at stellar temperatures. The second is that it is very difficult to find the element in stars due to a shortage of useful F lines in the visible part of the spectrum. Additionally, while most elements of similar mass to F result from the late phases of thermonuclear burning in massive stars, F is demolished so rapidly that it usually fails to participate in usual stellar evolution. A recent advance entailing deriving F abundances on the surfaces of a number of red giant stars from the infrared rotation–vibration lines of the HF molecule has solved the quandary arising from the lack of F lines in the visible part of the spectrum alluded to above. Such giant red stars are classified according to spectral class (redness) and, in order of increasing redness, are designated by the letters K, Ba, M, Ms, S, Sc, N and J. The different spectral types relate to age and evolution along the asymptotic giant branch. Such stars become progressively redder and more luminous as thermonuclear burning occurs in shells outside an inert C-O core simultaneously with extension of the outer envelopes. In them, He burning, in which three He nuclei fuse into a C nucleus, takes place in a shell so thin that, as energy is freed, the temperature rises faster than the star can expand and cool. The ensuing thermonuclear runaway reaction is stopped only by the development of a deep convection zone which scoops material from the thermonuclear burning areas to the surfaces. If this hypothesis is valid, sporadic He-burning shell flashes as such stars evolve will enrich their surfaces in the products of thermonuclear burning. With respect to them, measurements showed that F abundances exceed those of the Sun by a factor of 30.

The production and survival of F at He-shell flash temperatures has been explained by a reaction sequence entailing the capture of an α-particle (He nucleus) by ^{14}N to produce radioactive ^{18}F which then decays to ^{18}O. Proton capture by this leads indirectly to ^{19}F thus:

$$^{14}\mathrm{N}(\alpha,\gamma)\,^{18}\mathrm{F}\,(\beta^+)\,^{18}\mathrm{O}(p,\alpha)\,^{15}\mathrm{N}\,(\alpha,\gamma)\,^{19}\mathrm{F}\,.$$

The protons derive from a couple of reactions involving ^{13}C and ^{14}N. Mathews asserted that F is not destroyed by proton capture as a result of two factors, namely that few protons are present and that the element is removed from high-temperature regions through circulation. And the fact that fluorine is not destroyed by the ^{19}F $(\alpha,\mathrm{p})\,^{22}$Ne reaction was regarded as implying that a constraint exists on the temperature and neutron source reaction for the slow neutron capture (s-process) nucleosynthesis of heavy elements. At temperatures sufficiently high to produce neutrons through ^{22}Ne(α, n)^{25}Mg in He-shell flashes, F would be destroyed. Therefore ^{22}Ne cannot be the neutron source for the s-process.

As for protons, in 1991 Mathews alluded to six new nuclear isotopes near the limits of stability for proton-rich nuclei in the region extending from Ga to Sr and suggested that these may redefine this region which is termed the 'proton drip line'. He also suggested that they might intimate a different evolution for models describing the generation of astronomical X-ray bursts from high-temperature thermonuclear H burning in accreting neutron stars. Indeed their actual existence defines the balance between the attractive nuclear and the repulsive electromagnetic forces in atomic nuclei. The nuclear force is slightly repulsive between two protons or two neutrons, but attractive between a proton and a neutron. Consequently, there is a limit on how many more protons than neutrons or vice versa can be added to a nucleus. The matter is exacerbated by the electromagnetic Coulomb repulsion among protons which attempts to thrust the nucleus apart. The limits on how many protons or neutrons can be added to nuclei in fact comprises the proton and neutron drip lines. The newly made isotopes under discussion possess almost equal numbers of both types of particle, and hence have the strongest nuclear binding. However, so many protons are present that the Coulomb repulsion practically cancels the nuclear force. Consequently, the nuclei are delicately bound and Mathews declared that it is difficult to predict theoretically which nuclei will be bound and which not.

Such isotopes can be made by inducing a collision between a heavier, proton-rich nucleus and a lighter one, a process accompanied by the stripping away of nucleons from the heavier nucleus and consequent production of new isotopes which recoil from the momentum. In the particular case discussed by Mathews (1991), the technique was to produce proton-rich isotopes by bombarding a ^{58}Ni target with high-energy ^{78}Kr ions, the recoil products then being focused after collection through a sequence of magnetic lenses and their masses measured. Resultant mass spectra demonstrated the existence of the six, previously unknown, isotopes which are ^{61}Ga, ^{62}Ge, ^{63}Ge, ^{65}As, ^{69}Br and ^{75}Sr. Two of these are especially interesting, viz. ^{65}As and ^{69}Br, because prior mass extrapolations were unable to show whether they have or do not have bound states for the last protons. The new researches have extended the known proton drip line and also placed rigorous constraints on models for the properties of symmetric ($z = z$) nuclei at the limits of stability. Mathews indicated that the data may have a bearing on the thermonuclear burning of H at temperatures in excess of 10^9 K. While high temperatures are thought to exist naturally in the cores of massive stars at the end of their lives, these are depleted in H. According to him, a high-temperature, H-rich environment necessitates a more exotic setting, for example the thermonuclear bursts connected with novae, supernova shocks, accreting white dwarfs and neutron stars, and also even the Big Bang itself in models which involve inhomogeneous distribution of primordial nuclear matter. The highest temperatures are depicted in models attempting to explain astronomical type 1 X-ray bursts in terms of material gradually accreted onto the surface of a neutron star from an adjacent companion star. As soon as the pressure and density of the accreted material exceeds the threshold for nuclear reactions, thermonuclear energy rapidly heats the material to high temperatures in the intense gravitational field of the neutron star. The associated thermonuclear burning is characterized by rapid proton captures (the rp process) from carbon, nitrogen and oxygen nuclei up to the proton drip line. Wallace and Woosley (1984) noted that this process would cease after formation of the nucleus ^{64}Ge if ^{65}As could not be produced by addition of another proton. The half-life for the β^+-decay of ^{64}Ge is relatively long at 64 s in comparison with the timescale for the thermonuclear burst. It can be inferred that this would allow enough time for the rp process to cool down below the threshold for hydrogen burning as it waited for the decay of ^{64}Ge. As ^{65}As and ^{69}Br are known to be stable against the loss of a single nucleon, this may permit the rp process to continue beyond ^{64}Ge through ^{69}Br to heavier nuclei. However, as ^{73}Rb was not recorded in the measurements given it may be beyond the proton drip line. Mathews interpreted the new waiting point for the rp process of probably the 17 s decay half-life of ^{72}Kr as indicative of energy generation in the rp process at a faster rate, the available H being consumed more completely, than in earlier models for X-ray bursts. He stated also that this more complete burning could be a problem in explaining some cases of multiple X-ray bursts which have been observed to occur 5-10 min apart, these quiescent time intervals being far too short to allow accretion to accumulate a critical mass. Woosley and Weaver (1984) suggested that the repeated X-ray flashes could be caused by re-ignition of H left over when the rp-process ceased. This could happen through a convective instability. Of course, if the H burning during the first burst is more complete, the ignition of successive bursts would become more unlikely.

In sum, Mathews claimed that the effect, if any, of the new isotopes on the models remains uncertain. Particle-stable ^{65}As and ^{69}Br may not be stable enough to contribute to the rp process. Even if ^{65}As is bound, the binding energy of its final proton would have to exceed some hundreds of KeV for it to be at all significant in X-ray bursts. Otherwise the excitation of ^{65}As by background thermal photons would photodissociate the nucleus prior to its decay to ^{65}Ge and allow the rp process to proceed. The same would be the case for ^{69}Br but, as few events have been observed involving either this or ^{65}As, the inference is that both nuclei are bound weakly. Mathews believed that quantifying the effects of the new isotopes on models for such X-ray bursts needs accurate measurement of their binding energies, either from the β^+-decay endpoint energy (which is difficult) or by making the nuclei in two-body reactions utilizing radioactive ion beams, e.g. $^{58}\mathrm{Ni} + {}^{10}\mathrm{C} \rightarrow {}^{3}\mathrm{H} + {}^{65}\mathrm{As}$. To achieve such reactions, appropriate radioactive ion-beam accelerators, when constructed, must be used, (Livio *et al.*, 1989). Clearly, atomic nuclei are created and destroyed in many astrophysical environments, the abundances of the various nuclei offering insights on these environments through the history of the universe. Relevant data would extend back to the Big Bang. However, taking into account the vast timescale involved and the complex, multitudinous events which occurred during its course, matters would be simplified were the approach to be modified. Unfortunately, some aspects cannot be changed, for instance that He and D both originated at the begining (although enhancement of He and partial deterioration of D occurred subsequently). Nonetheless, it can be stated that the generally accepted Big Bang model entails a homogenous primordial universe with a density for ordinary baryonic matter less than 10% of that required for a 'flat' universe, which is one containing only sufficient matter to prevent external expansion if correct abundances for He and D are to be produced. In addition, a maximum of four species of light neutrino would be permissible (agreeing with accelerator experiments which constrained the number to less than five).

Hogan (1989) mentioned testing an idea of his and Applegate that small-scale inhomogeneities developed in an early first-order phase transition in the course of the Big Bang (possibly connected with quantum-chromodynamic transition effects). These may have led to some regions becoming neutron rich at the onset of primitive nucleosynthesis–a concept which would replace the standard Big Bang with a nearly flat universe comprising only baryons (nuclear matter) and no 'exotic' matter at all. But this and similar views would produce less He and more D, thus allowing a much higher baryon density and tending to increase Li abundance as well as possibly a primordial enrichment of very heavy elements otherwise believed to be synthesized only in stars.

Ideally the primordial Li abundance should be known. The element occurs in both old and young stars as well as in the interstellar medium. Its abundance in gas and young stars is thought to be about 10^{-9} by mass, old stars having an abundance an order of magnitude lower. The question as to which, if either, of these abundances is primordial remains and it is suggested that either the primordial abundance was low (requiring sudden production of Li uniformly throughout the young galaxy in order to reach its existing high level) or the initial abundance was high (necessitating substantial and uniform depletion through the synthesis of heavier elements in old stars to yield the current low level). Some claim that the latter process is actually occurring in old population I stars, but the depletion effect is inconsistent with a small scatter in the element in some models of stellar evolution. Others have presented models through which early Li enrichment could have taken place.

Olive and Schramm (1992) discussed astrophysical ^{7}Li as a product of Big Bang nucleosynthesis and galactic cosmic-ray spallation. They noted that the main nucleosynthetic product was ^4He produced with a relative abundance by mass of $Y \approx 0.24$ which left a primordial H mass fraction of $Y \sim 76\%$. The ^4He was accompanied by much smaller amounts of D and ^3He with ($D, ^3He/H = 10^{-5}$) (by number). ^7Li is produced with even lower abundances $(^7Li/H) \approx 10^{-10}$ in the standard model. These authors claimed that the fact that the predicted abundances of these light elements are explicable over such a range of almost ten orders of magnitude is a success for the simplest nucleosynthetic model. Futhermore calculations also predict the primordial abundances of other light element isotopes, namely ^6Li, ^9Be, ^{10}B and ^{11}B, which are produced in minuscule quantities as follows: $(^6Li/H) \approx 10^{-14}$, $(^9Be/H) \approx 10^{-18}$, $(^{10}B/H) \approx 10^{-20}$ and $(^{11}B/H) \approx 10^{-18}$, for a baryon to photon ratio of $\eta \approx 3 \times 10^{-10}$ (a value necessary to maintain consistency with the light element abundances). Ever since the observation of the ^7Li plateau, some have argued that the plateau value of $(^7Li/H) \approx 10^{-10}$ corresponds to primordial ^7Li so that this isotope has been a crucial factor in testing consistency for Big Bang nucleosynthesis. Stars comprising this plateau (numbering almost 40 as of 1992) belong to population II halo dwarfs and have low metallicity, $[Fe/H] \leqslant -1.3$, this corresponding to the log abundance by number relative to the solar value. Some drop as low as $[Fe/H] \leqslant -3.5$ and have higher surface temperatures $T > 5\,500$ K. Cooler halo stars possess much ^7Li which accords with stellar models depicting convective depletion at lower, $T < 5\,500$ K, temperatures.

The measured abundances of Be and B in several hot population stars are orders of magnitude above the predicted abundances of these elements from standard Big Bang nucleosynthesis. Olive and Schramm indicated that they do not show a plateau of constant abundance over a wide range of low metallicities and high temperatures, as is the case for ^7Li. It may be inferred that the ^7Li abundance is mostly primordial, but that the Be and B abundances are due to galactic cosmic ray (GCR) spallation reactions superimposed on a far smaller Big Bang component. However, such GCR spallation should produce ^7Li also. Therefore these researchers used Be and B data to subtract from the measured ^7Li abundance an estimate of the quantity generated by GCR spallation–this as a consistency check on the combination of Big Bang nucleosynthesis and GCR spallation. The approach was used for each star in the sample and an estimate of the metallicity-dependent increase of ^7Li due to spallation was added to the baseline. The somewhat reduced primordial ^7Li is, Olive and Schramm suggested, still consistent with Big Bang nucleosynthesis and a single GCR spallation model is able to fit the Be, B and corrected ^7Li abundances for all the stars involved in the study.

Some post-Big Bang events can be investigated through data on the shapes and distribution of extermely distant galaxies identified through high redshift values. Pettini, Boksberg and Hunstead (1989) reported measurements of spectral lines from Zn in galaxies illuminated by bright quasars in their backgrounds, an optimal method because Zn is one of only a few elements which do not become bound into interstellar grains and so become invisible. As seen earlier, the very first stages of nucleosynthesis in the primordial fireball of the Big Bang produced mostly H and He, heavy elements arising later in stellar interiors after the coalescence of stars and galaxies and sometimes being ejected to enrich the interstellar medium when supernova explosions occurred or by star winds resembling the solar wind. Carswell (1989) proposed that some early enrichment of intermediate elements such as C, N and O may have taken place quite quickly through processing in short-lived, high-mass stars. After a billion years or so, Fe-group elements may have emerged in the interstellar medium when longer-lived, lower-mass stars donated appreciable quantities of relevant material. Surmising the real abundances of these from gas-phase ones is complex because most common elements in the material, e.g. C, O, Si and Fe, take part in forming interstellar dust grains, which means that the true abundances exceed those observed from gas-phase abundance measurements. However, as mentioned above, Zn is negligibly depleted into such dust and the problem was that the Zn lines are weak even within the Milky Way galaxy, a weakness accentuated as redshifts increase. Nevertheless, measurements were reported from singly ionized Zn which permitted its abundance relative to H at redshifts $z = 2.8$ and $z = 2.3$ to be determined. It appears that, depending on the rate of slowdown of the expansion of the universe, the chemical contents of material which existed when it was anything from 15% to 30% of its present age can be assessed. Results show that the heavy element abundances obtained from Zn^+ data are about an order of magnitude lower than the value near the Sun. Carswell believed that this might imply that nucleosynthesis in such high redshift systems led to the production of only about 10% of the heavy elements existing now, hence these systems are in an early phase of their chemical evolution. Pettini and others showed that it is possible to compare Zn and Cr abundances, Cr depleting also onto interstellar dust grains. In such early phase systems, measured Cr abundances would be closer to the real values than those obtained in the Milky Way galaxy. In fact, such high redshift systems are probably relatively dust-free and, if so, it should be feasible to derive rather reliable estimates for true abundances of the chemical elements from them, abundances based on the stronger spectral lines of the commoner elements depleted in the galaxy. In this way, galactic chemical evolution will become clearer.

As regards Cr, Mn–Cr isotope systematics were considered in relation to the development of the early solar system by Birck and Allègre (1988). Titanium and Ca isotope anomalies were already known to be widespread in the Allende meteorite's inclusions and these authors alluded to the uncovering of Cr anomalies as well, these latter revealing the presence of ^{53}Mn early in the history of the solar system. They presented high-precision data showing *in situ* decay of this short-lived, extinct nuclide to ^{53}Cr then, noting that the process has a half-life of 3.7 Ma. Since ^{53}Mn is extinct, its former presence in the materials of the early solar system is divulged by variations in the $^{53}Cr/^{52}Cr$ ratio. The investigations included the C2 Murchison, the LL3 Chainpur (in which H and C isotope anomalies were known), the diogenite Tatahouine, the pallasite Eagle Station (containing an O isotope anomaly) and the (E4) enstatite-bearing chondrite Indarch (containing high Mn/Cr sulphides). The authors concluded that the decay of ^{53}Mn to ^{53}Cr in the early solar system offers an extra chronometer with which to constrain the formation times of the small planetary bodies from which the meteorites originated. The complexity of the processes when they formed was demonstrated by comparison of Cr and Ti isotope anomalies showing them to be imperfectly correlated.

Further interesting work on the Murchison meteorite, which fell in Victoria, Australia, in 1969, was discussed by Wright (1993) and he mentioned that this carbonaceous chondrite belongs to a group typically possessing 2% to 3% C by weight. This is mostly in the form of organic materials and, when this carbon complex is isolated, it is found to contain minute grains of diamond, SiC and graphite of presolar origin which may be survivors of the primaeval dust cloud that coalesced around 4.5 Ga ago from the solar nebula. Analytical techniques had to be devised and, according to Wright, the task took almost two decades. The difficulties relate to the tiny amounts of graphite present–about 2 ppm, also to the necessity to analyse individual grains nominally *c*.1×10^{-6}m in size, each comprising *c*.10^{11} atoms of C. In order to analyse a grain completely necessitates the determination of the C stable isotope ratio $^{12}C/^{13}C$ as well as the isotopic ratios of the trace components. These latter include N, O and Mg (the last in order to evaluate the abundance of formerly radioactive ^{26}Al) and of course analyses must be effected on many grains. Wright noted that only rounded graphite grains appear to possess large anomalies in the C stable isotope ratios and Amari *et al.* (1993) obtained a range of from 2 to 7300 (the typical solar system value being 89). As the data fell into a number of groups, the question arises as to how many graphite sources they represent. Graphite may be presumed to have condensed from a stellar source where the carbon/oxygen ratio exceeds 1, but the carbon isotopic compositions of the grains do not agree with those in SiC grains, which is interesting because these latter are thought to contribute most of the carbonaceous dust to the interstellar medium. Isotopic anomalies and abnormalities in the elemental abundances of noble gases in SiC grains of great purity taken from the Murchison meteorite have been recorded, these grains being characterized by deviations in C, N and Si as well in the inert gases. The relevant isotopic ratios are typical of slow neutron capture, the s-process, implying that they may have formed in the extended atmospheres of carbon stars. Among others, Gallino *et al.* (1990) discussed the astrophysical interpretation of the anomalies and considered low mass stars ($1 \leqslant N/M_{\odot} \leqslant 3$) for testing during the thermally pulsing asymptotic giant branch phases. Such stars enrich in C and s-process elements at their surfaces by admixing of material from the partially He-processed region and have been shown to reproduce the distribution of s-process isotopes heavier than atomic mass number $A \sim 80$ in the solar system, i.e. the so-called main component. In their work, Gallino and his associates calculated a series of nucleosynthesis models for these stars of different metallicity, which they defined as the initial mass fraction of all nuclei heavier than He. The reaction scheme employed was based on 450 nuclei from He to Po. They concluded that theoretical models of carbon stars can explain many measured abundances, especially the range of Kr isotopes and the excess ^{22}Ne found in the grains.

Wright recognized that the sources of the SiC grains have to be examined also and mentioned that the C stable isotope ratios in diamonds show only a slight variation from one meteorite to another with the size distribution mirroring a low degree of processing, this implying a single source or at the most two sources. The grains are extremely diminutive (normally 1×10^{-9}m across) so that analyses are carried out on millions of crystals rather than on individuals, and therefore the measured isotope ratios could just be the average of numerous sources of pre-solar dust. Where SiC grains sufficiently large to analyse individually were evaluated, the resulting isotopic data has been diversely interpreted as representing anything from one up to 75 sources. Wright indicated that it is necessary to isolate several grains arising from one source and added that newly discovered aggregates of several hundred SiC grains might suffice for this purpose. For a number of reasons, a solar nebula origin may be ruled out, the aggregation probably having taken place near the source of the grains. However, the overall isotopic compositions of the groups are nearly the same as the average of earlier individual analyses, which suggests that if any aggregation during chemical work removing the grains did not occur, the range of previously recorded isotopic values originated in a single source.

On the graphite grains, those with high C stable isotope ratios call for nucleosynthesis through He burning and two fresh possibilities were put forward by Amari *et al.* One is Wolf Rayet stars at a stage in their evolution when the products of such He burnings ascend to their surfaces. The second is He burning in the outer shells of massive stars before they explode as type II supernovae. Wright pointed out that the supporting isotopic evidence conprising ^{14}N/^{15}N, ^{16}O/^{18}O and ^{26}Al does not permit discrimination between the two. He also stated that the large sizes of some graphite grains probably cannot be accounted for by supernovae and that graphite most likely had several sources. More isotope measurements are required, but Kr isotope data has already shown that some of the graphite grains can be traced to a pulsing star in the asymptotic giant branch while Ne from another subset of grains relates to a nova explosion. One component of carbon in the Allende carbonaceous chondrite has a $^{12}C/^{13}C$ value of 120 which Wright regarded as probably representing the average of a single population of meteoritic graphite, and a further population is distinguishable by a component designated Cα, this having a $^{12}C/^{13}C$ ratio of 66. Both numbers are usable together with average $^{12}C/^{13}C$ values for diamond and SiC in making chemical models of solar system formation. Ongoing researches will assist in increasing understanding of the nucleosynthetic histories of solar system materials. Nucleosynthesis models may be based on the cosmic memories of meteorites, which are valuable in that they allow their validity to be tested in low-mass stars, and the disclosure of isotopic variations in the materials of which they are made has imposed changes on earlier models of the origin of the solar system. The hot Big Bang led to an unexpectedly low entropy universe to which the second law of thermodynamics applies, this implying a gradual running down as energy from nuclear and chemical reactions is transferred to heat and becomes unavailable for conversion. The net result could be what Hermann Helmholtz has called 'the heat death of the Universe', matter collapsing into black holes and maybe at the last a single big crunch. This would represent the end product of the thermodynamic irreversible energy flows in nature occurring in a universe of ever-increasing entropy, which tends towards a maximum corresponding to complete disorder of its particles. The background radiation is a relic of the Big Bang and provides an estimate of the entropy of the early universe, which is 10^{88} units. This has not changed much since, although Jacob Beckenstein and Stephen Hawking have produced a formula noting that the entropy of a black hole is proportional to the square of its mass. This means that as the number and mass of black holes increase, they will make an ever more dominant donation to the total entropy of the universe. A final entropy of about 10^{123} units is envisaged when all matter has collapsed into black holes and the big crunch is reached. The implication is that this state could be realized in *c.* 10 to the 10^{123} times more configurations than the Big Bang.

Adams (1994) provided the above argument and added that the chance of selecting an initial state like the Big Bang from all possible states of the constituents of the universe is only 1 in 10 to the power 10^{123}, appending the comment of Penrose (1989) that the Creator, if there was one, must have a good idea as to how he wanted the universe to begin.

Nucleosynthesis can be examined from the standpoint of the anthropic principle which asserts tautologically that knowledge comes from recently evolved human beings. Maddox (1992) indicated that this means observers are able to perceive, of all possible universes, only that one which is sufficiently old to have allowed natural selection to operate over a time interval of approximately 3 billion years. And this within a reasonably constant ambient temperature. Although a cyclic dialectic is involved, the anthropic approach nevertheless implies that, by considering those conditions necessary for life and how life may have begun, light may be shed on how the universe is constructed and hence on nucleosynthesis. Citing research at the University of Pennsylvania, Moddox noted that all life-forms comprise C, O, H and elements such as P (in DNA), Fe (in haemoglobin) and Co (in vitamin B_{12}), i.e. all life-forms which humanity can recognize as such. This caveat is exemplified in a remark of Thomas Gold's cited by Maddox at a lecture where the speaker alleged that all life is terrestrial. He remarked that 'they might look like rocks'. Pre-solar nucleosynthesis must have functioned in a manner eventually permitting such life-forms to develop on Earth and probably elsewhere in the universe as well. Starting from primordial material (mostly H associated with conceivably a tenth as much He), other elements mainly orginated through nucleosynthetic processes in supernovae and other types of star to produce a variety of which a number are essential for living beings on Earth. Solar system material mainly derived from stellar sources. Maddox states that about 1.9% of the galactic material was produced 20 Ga ago, or possibly 10 Ga ago, such dates depending on the correct value of Hubble's constant, H_0, the ratio of the velocity of recession of the galaxies to the distance of the galaxies being observed measured in kilometres per second per megaparsec (3.26×10^6 light years). There has been considerable controversy regarding this value which can lie anywhere between 55 and 100 km s^{-1} Mpc^{-1}. The reason for this is that, while a galaxy's observed redshift makes it possible to measure the velocity at which it is receding, its distance is more difficult to determine. And this affects inferences as to the age of the universe for which $1/H_0$ constitutes Hubble time and assumes that the rate of its expansion remained constant since the Big Bang. However, this rate may have slowed through gravitational forces and, if so, $1/H_0$ will give an upper limit for the age which will diminish as H increases. Cosmic distances can be assessed by observation of Cepheid variable stars which pulsate with periods from 5 days (for dim ones) to 100 days (for the brightest ones), but such stars are not radiant enough to be visible in distant galaxies. Supernovae are better and constitute standard candles, all thought earlier to produce the same light emissions (as much as a billion suns for some days). If this were so, the inverse square law would be applicable apropos distance because calibrating luminosities in some galaxies where they have been seen provides this parameter. A problem arose with the famous supernova 1937C photographed by W. Baade and a colleague, which turned out to have been fainter than was claimed so that the H_0 value had to be revised from 50 to 65, which implies a somewhat younger universe. A tendency occurred towards high Hubble values which accords with this result, values as high as 90 to 100 being presented. These would give the universe an age of 8 Ga or less, conflicting with evidence that some stars are definitely older than this. Recently gravitational lensing has been applied to estimating the distances of remote galaxies on the basis of the relevant amounts of bending and results from this work suggested a low Hubble value, i.e. an old universe. In October 1994 the distance to the Virgo cluster galaxy M100 was examined from Hubble Space Telescope observations of Cepheids by Freedman *et al.* (1994). They noted that the value is still undecided, although correctly determining it is extremely important because the present figure would show both the expansion timescale and the size scale of the universe. Additionally, it would impose constraints on the density of baryons produced in the Big Bang, the quantity of dark matter and the formation of structure in the early universe. In their work, Cepheids were used because the Virgo cluster of galaxies is near enough for detailed examination, but sufficiently remote to be of interest cosmologically. Regarding the extragalactic distance scale, the Hubble Space Telescope Key Project aimed to yield a value for the Hubble constant accurate to 10%. The results obtained gave a distance of 17.1 ± 1.8 Mpc to the Virgo cluster galaxy M100, which gives a value for H_0 of $80 \pm 17\,\mathrm{km\,s^{-1}Mpc}$. Freedman and her colleagues admit that a weak point in their analysis is that they have measured the distance to just one galaxy based on identification of three Cepheids in it. However, they intend to extend their researches to measure Cepheid distances to a total of *c.* 20 calibrating spirals both in the field and in small groups (e.g. Leo I and Coma I) and including another two galaxies in each of the Virgo and Fornax clusters. This over the two remaining years of the Key Project. They believe that the results embodied in their paper suggest that a Hubble constant accurate to 10% and measured through secondary techniques out to scales of hundreds of megaparsecs has now become a realizable goal.

A dilemma remains and, as Jacoby (1994) described, puts the universe in crisis. Taking the above value for the Hubble constant, this can be compared with other data from the galaxy NGC4571, also in the Virgo cluster, obtained using Cepheids and presented by Pierce *et al.* (1994). Their data gave a value of $87 \pm 7\,\mathrm{km\,s^{-1}\,Mpc^{-1}}$. This result, coupled with the later one of Freedman *et al.* (1994) show that the universe is younger than its oldest stars, a paradox which will compel a reconsideration of how stellar ages are measured. Although the distance to two galaxies in

Virgo is now known, if they are situated in front of the cluster, most Virgo galaxies could be further away and H_0 could still be small. Jacoby believed that the available evidence favoured the galaxies in question lying near the centre of Virgo and being representative of the Virgo cluster. Upcoming data alluded to by Freedman and associates will come from the same region as well as from Fornax, which is a southern galaxy cluster, and from some ten more galaxies intended to act as check points for other techniques. These latter have been utilized previously to assess the Virgo distance, but are not thought to be as reliable as Cepheids. Indeed, Cepheids were originally used to show that spiral nebulae are extragalactic. They comprise relatively young and massive stars with luminosities ~1 000 to 100 000 times greater than the Sun ($-2 < M_v < -7$mag). They are well calibrated, can be identified easily in external galaxies because of their variability and brightness and have measured dispersions in the V- and I-band period luminosity (P–L) relationships amounting only to -0.25 and -0.20 mag, respectively. In addition the underlying physical basis for the Cepheid P–L relation is understood.

Jacoby drew attention to another method distinct from the Cepheid one which has given from type Ia supernovae a value for H_0 of 54 ± 8 and a Hubble Space Telescope team led by Allan Sandage is following more distant check points in order to verify the accuracy of supernovae distances. Yet others assert that type Ia supernovae need more parameters in their calibrations which when used give $H_0 = 67 \pm 7\,\text{km}\,\text{s}^{-1}\,\text{Mpc}^{-1}$, a very interesting result.

Clearly, much remains to be done and the true value of the Hubble constant must be regarded as still sub judice. Possible systematic errors perhaps connected with variable chemical compositions of the Cepheids in different galaxies will have to be discovered and, if possible, eliminated. Nevertheless, Jacoby took the view that most evidence available at present favours a large H_0 and a young universe so that, if clues have been correctly interpreted, controversy about the value will cease and attention will turn to resolving the age crisis. When this is achieved, something new will emerge, either about the ages of the stars or on the nature of the universe.

Of course, other astrophysical matters too remain to be investigated, for instance whether the Milky Way galaxy has a central bar or not and why the black hole believed to lie centrally in it is so quiet. As Gribbin (1994) indicated, such a massive object ought to emit high-energy X-rays, but the initial observations of the galactic centre at these frequencies have detected much less radiation than was anticipated. The most favoured location for the object is a very compact and bright source of radio waves and low-energy X-rays called Sagittarius A. Data on it were collected by the French X-ray telescope Sigma flown on the Russian spacecraft GRANAT. This is senstive to X-rays at energies from 35 to 1300 KeV, but it recorded 4×10^{-7} times less radiation in this range from Sagittarius A than could be expected from a black hole having a mass 10^6 times that of the Sun and swallowing matter at the maximum sustainable rate. Of course, the lack of high-energy X-rays does not automatically imply that the massive black hole is absent. Gribbin cited a hint of what is actually happening from Sigma observations of six separate high-energy X-ray sources near to the centre of the galaxy. They are probably binary systems in which a black hole only a few times more massive than the Sun is orbiting a normal star and stripping matter from it through gravity. It has been indicated that such X-ray sources usually flare up brightly in the high-energy X-ray range for about a month at a time every half century or so. This cycle is connected to the developing of instabilities in the accretion disc. Extrapolating to a black hole a million times the mass of the Sun would imply that it is in the high-energy X-ray emitting stage for only about 1% of the time. The conclusion must be that both normal galaxies such as the Milky Way one and those with active nuclei such as M87 may have black holes which are active for only brief time intervals during their lifetimes, the first category being dormant and the second active.

As well as such matters, much remains to be investigated in the realm of atomic and nuclear physics. For instance, order and chaos in strong fields entail laboratory experiments involving strengths up to 10 T which are small compared with atomic values of 2.4×10^5 T and, on white dwarfs and neutron stars, fields between 10^2–10^5 and 10^5–10^9 T, respectively. As Richards (1988) indicated, the constraint set by relatively weak laboratory fields has been overcome by new techniques allowing work to be done on Rydberg atoms which have a single electron in a highly excited state and are characterized by the principal quantum n which varies from 5 up to 300 in ongoing experiments. However, he opined that increasing n (which defines the energy level or shell in which the electron occurs) has a deeper effect. For large values of n, atoms respond like classical atoms of which the motion is described by the nonlinear equations of Hamilton rather than by the linear wave approach of Schrödinger and the wave function comprises numerous component waves which interfere constructively along classical orbits and destructively away from them. Thus the electron is bound to its classical orbit with its wave-like behaviour suppressed. It tends to act deterministically. Similar results have been observed in light caustics, concentrations of light along rays equivalent to classical orbits. Of course, since the solutions of nonlinear and linear equations are qualitatively different, new effects should occur as n increases and the large n limit is extremely intricate.

Most orbits are chaotic, wandering at random over the entire available energy shell in the case of a classical atom in a strong field. Richards stated that in this 'chaotic sea' there are infinitely numerous periodic orbits of which some are stable and some unstable–although these constitute only a fraction of all possible trajectories. Inferring the periods of relevant classical orbits could be done by Fourier analysis of experimental spectra. But more convincing is comparing absorption spectra of excited H in strong magnetic fields, where a practically perfect correlation between the observed structure and the relevant computed classical orbits (some with bizarre shapes) was observed.

Regarding nuclear considerations, determining the equation of state for nuclear matter at high density entails experiments almost impossible to effect, for which reason it has been suggested using neutron stars. Brown (1988) discussed the astrophysical implications. He stated that the result of supernova explosions is a function of the response of nuclear matter to gravitationally induced compression. Conversely, supernovae offer a test of proposed equations of state. As large star envelopes contain elements from He up to Fe (lighter ones having burned into these through the stellar lifetime), these have to be ejected and the elements from C to Fe plus heavier ones formed through more complex processes. It has been calculated that supernovae occur only if the equation of state of the rather neutron-rich matter is soft up to densities around $4Q_n$, where Q_n is the density of ordinary nuclei. This soft equation of state puts up little resistance against the initial collapse of the star's core, this collapse being arrested only at very high nuclear densities. In the course of the succeeding rebound, the compressed core triggers the explosion which then ejects the envelope. Were this not to take place, it would sink back into the core which, in turn, would subside into a black hole. It has been suggested that the pulsar PSR1957 + 20 is spinning near to its limiting frequency and that the equation of state of its neutron matter is stiff, implying that the star is larger than if this matter were soft. Inferentially, the equatorial matter must be rotating at almost the maximum permissible velocity prior to its being ejected. From this it may be concluded that more rapidly rotating stars would break up. Brown mentioned that this agrees with earlier work on a 35-day periodicity in X-ray intensity from the object Her X-1 interpreted by some as a freely precessing neutron star. An observed wobble frequency is explicable if it possesses a thick crust (requiring the equation of state of neutronic matter to be stiff). However, it cannot be presumed that the equation of state of nuclear matter is soft and that of neutron matter stiff because, if the latter is made too stiff, the following reaction occurs: $n \rightarrow p + e^- + \nu$, with the neutrino escaping. Thus a thermodynamic equilibrium between protons and neutrons is set up which shifts the composition and pressure of the system nearer to that of nuclear matter. The extra electrons exert little pressure. Hence the neutron equation of state cannot be made much stiffer than that for nuclear matter. Possibly the neutron equation of state is soft for densities $\varrho_0 - 4\varrho_0$ and stiff for higher densities. Stellar collapse (preceding a supernova) falls within the first range and, with a soft equation of state in this, most of the neutron star will settle into higher densities. In theory, stiff equations of state allow stable neutron stars with masses $> 2M_\odot$, but none has been detected. Clearly, much remains to be accomplished. A better picture of the chemical history of material in the universe has to be secured, perhaps by probing higher redshifts as well as lower abundance systems in it. By these and other methods, no doubt humanity will eventually be able to comprehend its myriad riveting facets *sub specie aeternitatis.*

3

Mass spectrometry

3.1 MODERN MASS SPECTROMETERS

The technical development in mass spectrometry, due to advanced electronics, sophisticated computer science of both hardware and software and efficient vacuum technology, has improved the precision of the measurements, the time of the performance, the handling, the size of mass spectrometers and most importantly has reduced the amount of sample which is required for analysis. The last factor has enabled scientists to measure growth rings of molluscs or tree-rings with high precision, even if these are subdivided again to attain information regarding annual shifts. Isotope ratios of the smallest parts of samples can be determined, for instance different tissues of plants or, quite innovatively, single compounds by compound-specific isotope analysis. Different types of mass spectrometers have been constructed to be suitable for several applications, and the combination of mass spectrometers with automatic sampling. Preparation lines and/ or other analytical devices such as gas chromatographs has accelerated isotope measurements and widened the fields of application. It should be noted that the development depends both on demands and ideas arising from scientists and the technological innovations and improvements achieved by the manufacturers of mass spectrometers. Similarly to cars, producers such as VG instruments and Finnigan MAT offer their different models with basic equipment which can be extended according to the requirements of the user.

3.1.1 Gas Isotope Ratio Mass Spectrometer (GIRMS)

The basic apparatus without peripheral devices can be divided into four essential parts:

- inlet system;

- ion source;
- mass analyser;
- ion detectors.

The inlet sytem is designed as a symmetrical dual set-up (Fig. 3.1). It enables alternating measurements of sample and standard gases to be made, provides a viscous gas flow and thus prevents a mass separation within the inlet system. In this type of flow, the free path length of the molecules is small (*c.* 10^{-4} m) so that no mass separation due to the different flow velocities can take place. When the gas inlet enters the ion source, an equal amount of gas in the other line (either sample or standard) is pumped off through a waste line. In this manner the flow conditions for sample and standard remain identical throughout the period of measurement. In fact, continuity of gas flow is a basic prerequisite for the attainment of precise and reproducible data. It also affords an equal pressure on both standard and sample sides of the inlet system. This is achieved by variable volumes and, through compressing or expanding the gas at constant temperatue, the pressure is automatically adjusted according to $p \sim v^{-1}$ (Boyle–Mariotte). These volumes can be varied in a range of between *c.* 1 and 20 ml, permitting small sample amount measurements without time-consuming extra manual gas vol-

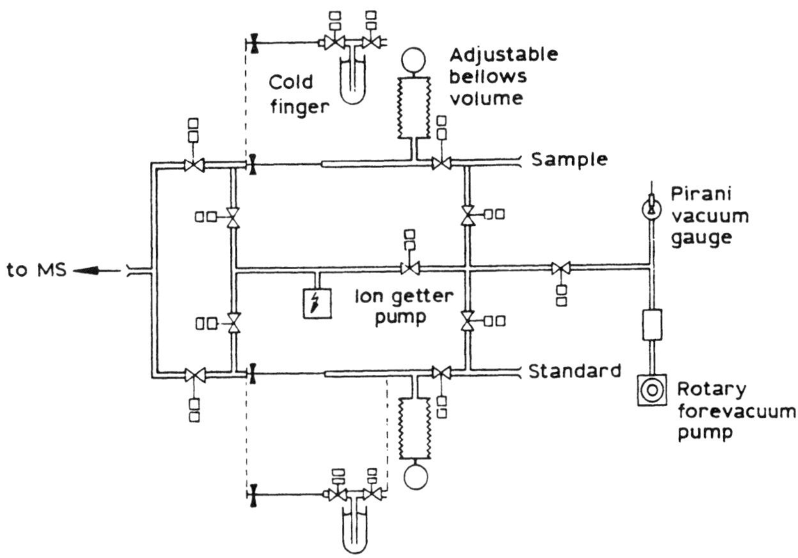

Fig. 3.1 Double inlet system for the GIRMS.

ume cascading. The vacuum for the inlet system and the waste line system is provided with a turbomolecular pump, replacing the ion getter pumps in newer models and commonly a two-stage forevacuum rotary pump. The pump-out time is small, but may vary within a few minutes due to the amount and nature of the gas. Generally, the functions of the mass spectrometer during a measurement are controlled by a computer and therefore the performance, for which the parameters can be chosen as necessary, is time-constant, this also being important for the reproducibility of the measurements. The inlet system as shown in Fig. 3.1 is naturally simplified and, depending on the type of mass spectrometer, optional configurations may be integrated or modifications made. A useful piece of equipment is the microvolume inlet system, which facilitates the precise measurement of very small sample sizes. Whereas samples >70 to 100 bar μl can be measured with the standard inlet system, depending on the type of mass spectrometer, sample sizes might be reduced to >3 to 10 μl if the cold finger inlet is used. Slight differences can occur with respect to the sample gas measured. These values correspond to a few micrograms of, for instance, $CaCO_3$. Assuming the behaviour of an ideal gas, 1 mol CO_2 with a molar weight of 44 g has a volume of *c.* 22.4 l at a pressure of 1 bar. Although this is not exactly true, because the difference from an ideal gas increases with increasing pressure and decreasing temperature, one can nevertheless calculate the amount of $CaCO_3$ necessary with sufficient accuracy.

If very small quantities are measured, either manually or by using a computer-controlled autofreeze system, the sample is transferred into a small storage volume. Therefore it must be condensable on an appropriate substrate of low temperature which is usually liquid N_2 with a boiling point of -195.8°C. Carbon dioxide with a freezing point of -78.5°C can be frozen at such a temperature and sublimates after warming, then being available for the measurement. Multiports for different numbers of samples are available as well as tube crackers. They allow fully automated, unattended measurements, and can also use different experimental procedures.

The sample gas leaves the inlet system by passing through a changeover valve. The switch over from sample to standard is normally computer controlled and the valve itself has a low dead volume and is usually mounted on the ion source, so that the gas pathways are short.

The ion source is the part of the mass spectrometer in which positively charged, monoenergetic ions are formed, accelerated and focused into a narrow beam. The modern ion sources provide a high stability and linerarity of the electron impact together with a very low ($<10^{-9}$ Å) H_3^+ contribution. The gas flow is molecular in an ultrahigh vacuum $<10^{-7}$ mbar and ions may be made thermally or by electron bombardment. In the latter case, the beam of electrons emitted by a heated filament of W or Re is directed between two parallel plates and collimated using a weak magnetic field. Positive ions which result from collisions occurring between the plates are withdrawn by an electric field and further accelerated. Modern mass spectrometers utilize an electron impact (EI) type of ion source. This EI type is optimized for CO_2, SO_2, SF_6 and N_2 measurements, and also for H/D measurements. It may be of a nichrome ceramics design. The gas-tight ionization chamber and collimating magnets which guide the electrons are so arranged that a very high ion yield is obtained, together with unique gas-flow characterisitics. Ultrahigh sensitivity may be achieved by an einzel lens focused at 10 kV accelerating voltage. Einzel lenses provide the excellent linearity and dynamic range. In such instruments only one source is employed for all gases without compromise. An additional benefit is that such an ultrahigh inherent source stability obviates the need for automatic computer-controlled refocusing of the source potentials between sample measurements. Stable ion production is the result of trap regulation of the magnetically focused ionizing electron beam. An ion acceleration of up to 10 kV is attainable and source potentials are controlled from one voltage divider. Manual fine adjustments accessible from outside may be possible, depending on the model in use. Readings of the electron current, which is also variable from outside, and the value of the ionizing voltage are external. Perhaps the most critical specificaton of an ion source is the matter of generation of H_3^+ ions arising from the ion–molecule reaction $H^+ + H_2$... These interfere with the HD (deuterium hydride) peak in H/D measurements. In mass spectrometers made by some manufacturers, this necessitated separate ion sources for H/D and CNOS (carbon–nitrogen–oxygen–sulphur) purposes. When the accelerated ions leave the ion source they enter the mass analyser.

The mass analyser is essentially an ion-separating system which splits up ion beams emerging from the ion source according to their m/e ratios; first-order direction focusing mass analysers are used in stable isotope research. The standard source slit width is 0.5 mm. As an ion beam passes through a magnetic field, the ions are deflected into circular paths of which the radii are proportional to the square root of m/e. In this way, the ions are separated into beams, each with a particular value of m/e.

The ion optical system is very sensitive: attainment of a relative sensitivity as high as 0.5 Å ion current per mbar gas pressure in the analyser depends upon the ion source, which must be gas tight without generating memory effects and possess variable gas-flow facilities, if possible, as well as two einzel lenses for optimum focusing of the ion beam. Multiple 'deep cup' Faraday collectors with different numbers of cups may be employed for single mass spacings up to mass 150. The contribution of the mass-44 intensity to the intensity of mass 45 is under 2×10^{-6}, which obviates any necessity to correct even in the most precise ratio determination with different sample and standard pressure. Figure 3.2 shows an example of an ion optical system.

This approach is developed from that of Nier (1940), who introduced the sector magnetic analyser with deflection occurring in a wedge-shaped magnetic field. Such a sector instrument was valuable because its source and detector were comparatively free from the mass-discriminating influence of the analyser field. H_2/HD ion paths which are detected in an optional H/D collector pass out of the ion optical system in the centre of the magnetic sector: this permits a rapid switchover from measurements of H_2/HD ions or CNOS ions without any kind of mechanical change.

The instruments are usually equipped with an electromagnet and all ions are measured with the 10 kV acceleration potential. The mass range is extended to 280 amu and a very stable magnetic field is attained by a Hall probe-controlled electronic power supply. The EM configuration has the advantage that each mass within the range can be attained and measured under computer control. Thus impurities in a sample can be checked routinely. The ultrahigh vacuum is usually provided by turbomolecular pumps of high capacity and a two-stage forevacuum rotary pump. The vaccum system is equipped with vacuum safety devices and is practically maintenance-free. The turbomolecular pumping system results in high pumping stability and obviates any possibility of contamination resulting from pump oils. A high pumping speed is effective at the ion path. The turbomolecular pump may be air-cooled, thus rendering it quite independent of any external cooling.There is an in-built ionization gauge which is monitored by the computer. Relevant pres-

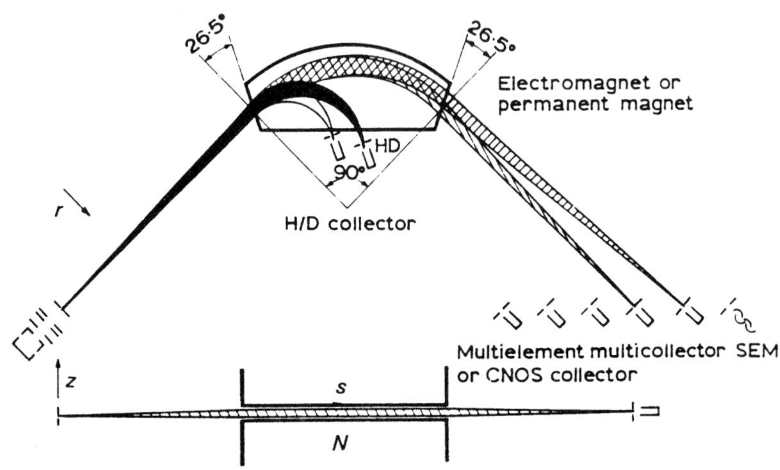

Fig. 3.2 Schematic diagram of an ion optical system. r = radius; z = focusing direction; N and S = north and south poles of magnet.

sures may be printed with analytical data. A considerable advantage is that sample contamination is apparent at a glance.

The measurement of the ion currents is done by the collector system of which a scheme is given in Fig. 3.3.

After passage throught the magnetic field, the separated ions are collected in Faraday cups. The charged ions enter the cup and the charge flows onto the outer surface of the cup, which is connected with an amplifier, because the ion currents are very small. Modern collector systems are very sophisticated and the switching between collectors, necessitated by the various isotope patterns, is under computer control.

3.1.2 External Options For GIRMS System Configurations

The aim of any preparatory work is to produce extremely clean gases such as CO_2 and N_2 from isotopically labelled compounds without isotopic fractionation or dilution by an isotopically different blank, and to achieve a high precision with very small amounts of samples. In order to obtain such clean gases from mixtures, there are four approaches. One is by the already mentioned autofreeze microvolume inlet, and the others relate to the trapping of gases from mixtures, for instance CO_2 and N_2 or both from air or a CO_2, N_2, H_2O and He mixture. Organic compounds containing C or N can be analysed using a gas chromatograph or an elemental analyser combined with the mass spectrometer. Carbonate samples can be prepared automatically. The H/D and CO_2 or water

Sr				84	85	86	87	88	
Nd	142	143	144	145	146	147	148	150	
Pb					204	205	206	207	208
Ca	40			42			43		44
Th					230	232			
B-					42	43			

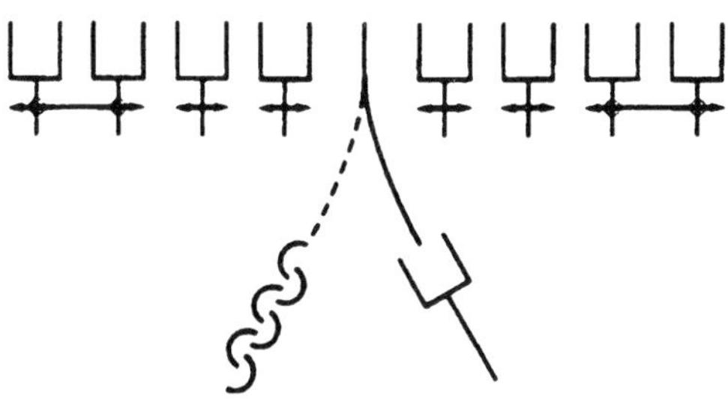

Fig. 3.3 Schematic of a modern multicollector system which might be supplied with up to eight Faraday cups.

samples can be equilibrated automatically in an on-line process. Off-line procedures are suitable for N and H.

The combination of gas chromatographs and elemental analysers with a GIRMS extends the fields of application of mass spectrometric investigations, whereas the automatic preparation lines save time on the one hand and increase the accuracy of the results on the other, for the parameters of the procedure are kept constant.

3.1.3 Preparation of samples for the GIRMS

As the measurement of isotope ratios is performed on gas, solid or liquid, compounds have to be converted into the sample gas by means of different chemical reactions. This can be quite complicated depending on the basic material and the reaction steps involved. On the other hand, on- and off-line sample preparation is available for some common isotope investigations. The preparation methods described give an overview of the principles, and modifications in the lines mostly concern the technical equipment of the arrangement chosen by the laboratory. It has to be stressed that the reactions, especially for O, must be complete to avoid any fractionation and achieve quantitatively correct results. The same care has to be taken with respect to impurities which might occur if temperatures for cooling traps are incorrect and leakage rates of the vacuum line are accidentally too high or reaction times are too long. The pretreatment of the samples by cleaning or separating mineral phases with chemical reagents or by physical means should not affect their isotopic compositions.

3.1.3(a) Hydrogen

The D/H ratios are measured on H_2 gas derived either from water or from hydrous minerals. Water is reduced by reaction with zinc or uranium metal at a temperature of *c.* 800°C. The catalytic process can be described as follows:

$$H_2O + M \xrightarrow{800^\circ C} MO + H_2 ,$$

where M is the metal reacting with water to an oxide plus H_2 gas, which is then trapped in a charcoal-filled vessel and cooled with liquid N_2. A scheme of a sample preparation line as used by Friedmann (1953) and Godfrey (1962) is shown in Fig. 3.4. Water released through the thermal decomposition of samples in the heater (H) passes through the electric furnace (F) and reacts with the metal to produce H_2. The H_2 is then transferred into the evacuated sample tube (S). The amount of gas produced could be calculated knowing the volume and the pressure shown by the manometer (P). A second run can be made through the recycle line (E) if water has remained in the cooling traps and a standard

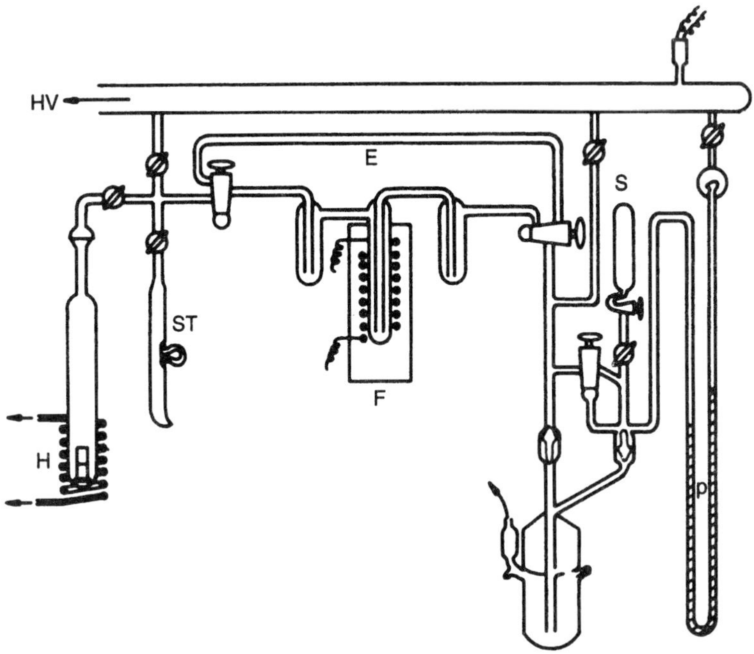

Fig. 3.4 Preparation line of the extraction of H_2 from water and hydrous minerals.

sample inlet (ST) enables runs in order to check the reproducibility of the preparation. Basically similarly constructed devices are available for off-line preparation of H_2 from water, allowing the reduction of water samples of 1 to 5 μl.

Minerals are first degassed for several hours at temperatures of *c.* 150°C to remove adsorbed water. The degassing temperature is far below the dehydration temperature for the minerals and therefore does not affect the bonded H which is to be measured. The thermal decomposition of the degassed samples is carried out at temperatures ranging from several hundred to maximum 1500°C, the process liberating H_2 and H_2O. The H_2 can then be oxidized with CuO at a temperature of 500°C as

$$H_2 + CuO \xrightarrow{500^\circ C} Cu + H_2O,$$

and the water is then reduced as already mentioned.

3.1.3(b) Carbon

The $^{13}C/^{12}C$ carbon isotope ratio is measured on CO_2 gas which can be liberated either from the various different carbonates such as calcite, aragonite and dolomite or from organic matter such as plant tissues. If the isotope ratios of carbonate minerals from shells of organisms, sediments or rocks are required, CO_2 is commonly liberated by reaction with phosphoric acid or, less commonly, by thermal decomposition. The latter is described as follows:

$$MCO_3 \xrightarrow{+T} MO + CO_2 ,$$

where M might be any cation within the different carbonates. The reaction, however, should not produce CO gas. More common, and also applied for the various available on-line carbonate preparation lines, is the reaction with phosphoric acid (H_3PO_4). Carbon dioxide gas is released from carbonate as shown by the reaction

$$3MCO_3 + 2H_3PO_4 \xrightarrow{25-90^\circ C} M_3(PO_4)_2 + 3H_2O + 3CO_2 .$$

The temperature used during the reaction depends on the kinds of carbonate, because dolomite, magnesite and siderite would, for example, react slowly or not at all at 25°C. Automatic devices usually run with minimum temperatures of 70°C and the fractionation factors for the different carbonate systems are also well known.

Pure carbon in the modifications graphite and diamond can be oxidized at high temperatures of *c.* 1000°C using either CuO or O_2 as a reagent.

$$C + 2CuO \xrightarrow{T} 2Cu + CO_2 \quad \text{or} \quad C \xrightarrow[T]{O_2} CO_2 .$$

Preparation lines for the combustion of carbonates and organic matter are shown without further explanation in Figs. 3.5 and 3.6. The presence of O_2 either from a bottle or the atmosphere prevents the simultaneous measurement of C and O isotopes from organic matter and requires a different sample preparation as described, for example, by Epstein, Thomson and Yapp (1977).

3.1.3(c) Oxygen

Oxygen from carbonates and other compounds is usually measured in CO_2 and liberated in the same way as C through reaction with H_3PO_4 as described by McCrea (1950). As only two-thirds of the O_2 reacts to CO_2 the fractionation factor must be known. The concentration of the phosphoric acid used for the reaction sometimes differs in the literature. Concentrations varying from 95% (Mook, 1968) to 105% (Wachter and Hayes, 1985) have been used. The exchange of the O isotopes was studied intensively by the latter authors who used different concentrations and reaction temperatures. According to their results, a concentration of nearly 105% is preferable to minimize the exchange between CO_2 samples and the acid solution. A scheme of a preparation line for the investigation of both (section 3.1.3(b)) C and O isotopes from carbonate material is shown in Fig. 3.7. In this line the CO_2 liberated from the reaction of a carbonate sample and the H_3PO_4 is purified and transferred

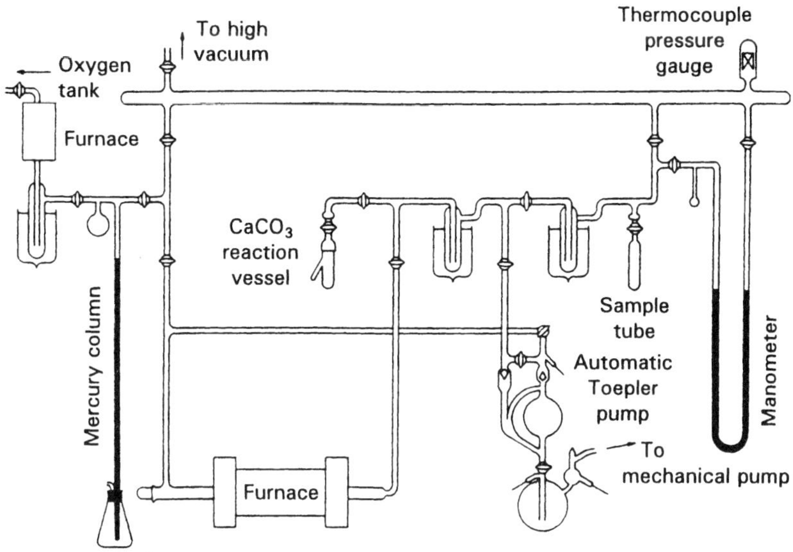

Fig. 3.5 Carbon combustion line (after Craig, 1953).

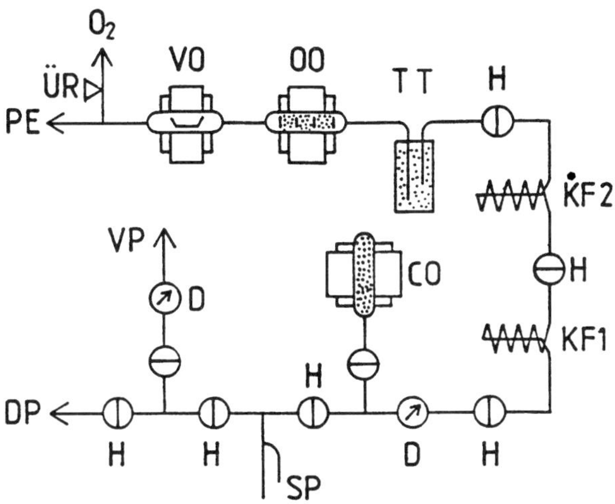

Fig. 3.6 Sample preparation for organic C in an O_2 stream. PE = sample inlet, O_2 = oxygen source, UR = pressure control, VO = sample combustion oven, OO = oxidation oven, Cu = copper reduction oven, TT = sample dryer, H = stopcocks, KF = cooling traps (KF1 = −70°C, KF2 = −160°C), D = pressure gauge, VP and DP = vacuum pumps, SP = spike. (After Hemmann, 1989, based on a method developed by Schleser and Pohling, 1980).

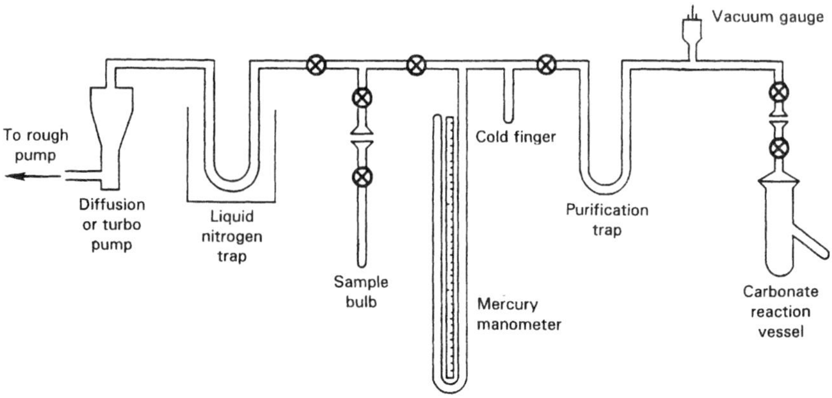

Fig. 3.7 Separation line for O and C measurements from carbonates (after Boutton, 1991a).

into the sample vessels. The reaction vessel filled with carbonate and H_3PO_4 in excess is evacuated through the line and, after reaction in a temperature controlled bath at e.g. 40°C, is fitted again to the separation line. The purification trap is then cooled with liquid N_2 to freeze CO_2 and water. After replacing the cooling trap by a cooled (*c.* −80°C) ethanol (either cooled down with liquid N_2 or mixed with dry ice) the CO_2 sublimes and can be transferred into the sample bulb, whereas the water is retained in the cooling trap. The amount of CO_2 liberated can be measured with the manometer. Other lines, for example, have a reservoir for H_3PO_4 and the sample fitted to the evacuated line is reacted by adding the H_3PO_4 through a valve. In this case the acid and the sample during reaction is temperature controlled. The latter is suitable if carbonate samples with a short reaction time are used, whereas other samples can be reacted overnight.

Oxygen in silicates of both organic precipitated skeletons and minerals is reacted with fluorinating agents e.g. fluorine gas (F_2) or bromine pentafluoride (BrF_5). Several fluoride compounds can be used and the reaction liberating O_2 can be described as follows:

$$SiO_2 + F_2 \xrightarrow[500^\circ C]{10\text{ to }12\,h} SiF_4 + O_2. \tag{3.1}$$

The O_2 produced is usually reacted into CO_2 similar to the above mentioned equation for graphite and diamonds. Figure 3.8 shows the reaction line used by Clayton and Mayeda (1963), which is similar to earlier ones as applied by, among others, Silvermann (1951), Clayton (1955) and Taylor and Epstein (1962).

As may be seen from equation (3.1), the reaction needs time and a temperature of *c.* 500°C. The reaction tubes are attached to the preparation line containing the solid sample. Adsorbed air and water is removed by heating the samples while the vacuum line is open, and a pretreatment with BrF_5 at 100°C and closed tubes should be applied to remove residual oxygen from the metal walls of the tubes, but only when the sample material will not react at this temperature (Tudge,

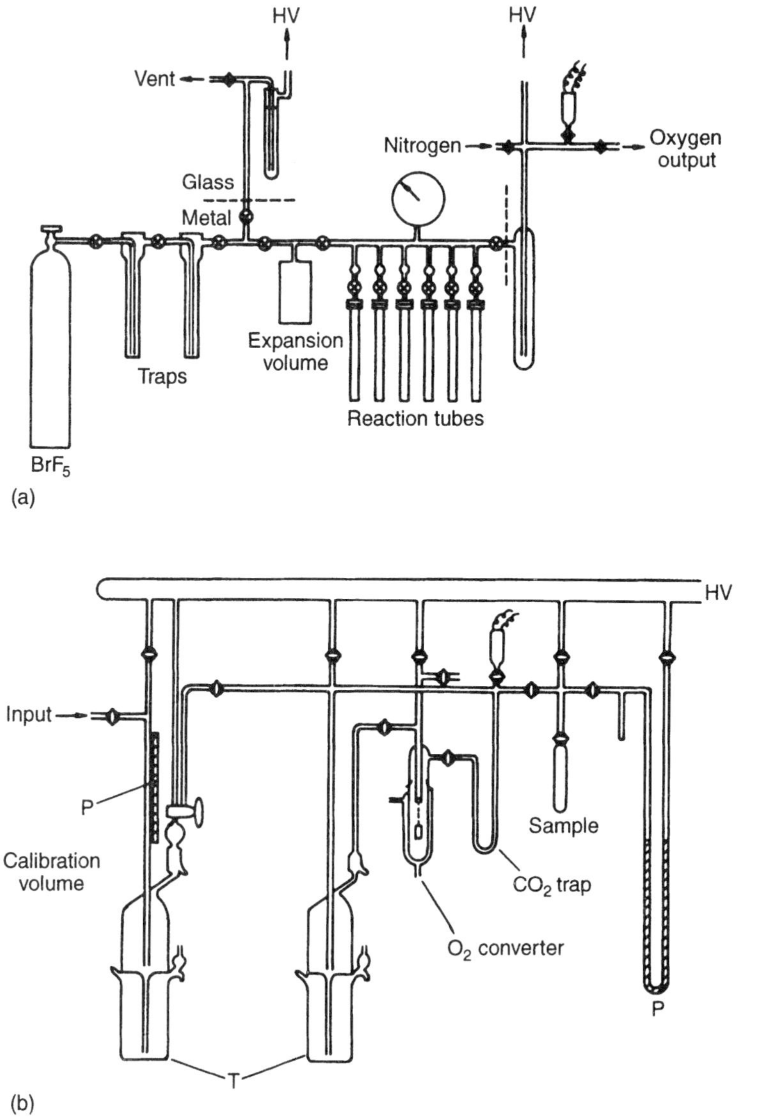

Fig. 3.8 Extraction line for O from silicates. (a) Reaction with the Br components; (b) collection and conversion of O_2. P = pressure gauge; HV = to high vacuum; T = Toepler pump.

1960; Clayton, 1961; Hamza and Epstein, 1980). Minerals that should be handled without pretreatment include hydrated silica, carbonates, phosphates and feldspar. After the line is evacuated, the sample is reacted for several hours to liberate the O_2. The O_2 is then led through a cooling trap, where possible reagents, except the O_2, are frozen out at the temperature of liquid N_2. The O_2, pumped by the Toepler pumps, is then converted to CO_2 with carbon and the resulting CO_2 is transferred into a sample tube after the amount of gas has been measured. It might be added that the method is not used normally for carbonate, but Sharma and Clayton (1965) measured the $\delta^{18}O$ ratios of total O of different carbonates using the BrF_5 method. Although this has been described earlier and the reactions liberate total O in the form of CO_2 and O_2 of which the latter can easily be converted, the method is not commonly used.

A new microanalytical preparation method, especially interesting in the field of mineralogy, has recently been developed for the determination of O isotope ratios from silicates and oxides. This laser-based sample preparation was described by Sharp (1990) and allows the analysis of samples of $< 100\,\mu g$ or spot determinations with a diameter of $< 300\,\mu m$. Thus single mineral grains can be analysed and heterogeneities within the mineral grain can be detected. Former methods of spatial fluorination necessitated samples of several milligrams and therefore a single fine grain could not be analysed.

Additionally, the laser can be positioned exactly on sample spots of interest. The method is based on a combination of laser heating and fluorination techniques shown in Fig. 3.9. The fluorination agent, e.g.

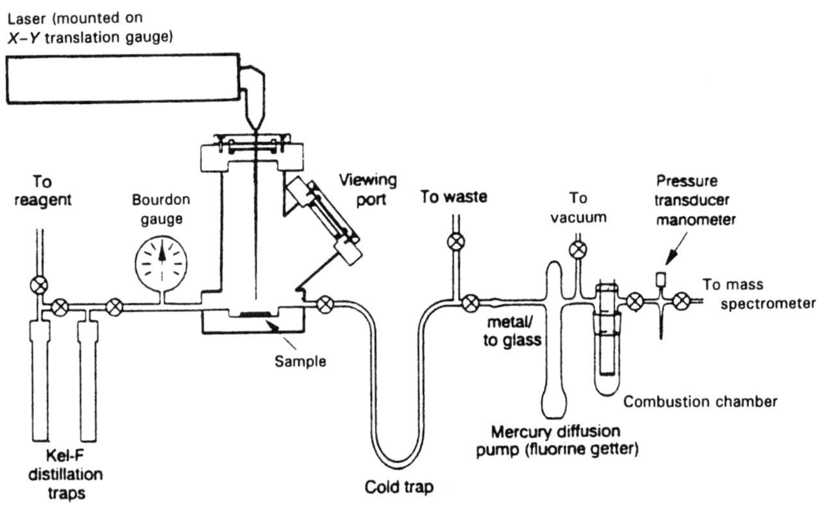

Fig. 3.9 Schematic arrangement of a laser fluorination system (after Sharp, 1990).

BrF_5, is distilled in Kel-F traps and an aliquot up to 0.1 bar is expanded into the sample chamber, where the sample is laser heated. After the reaction, fluorides and condensibles are frozen in the cooling trap and residuals are removed by the diffusion pump. Oxygen is transferred into the combustion chamber and converted to CO_2 which can be passed to the cold finger of the micro-inlet system of the mass spectrometer. Within the sample chamber, the heating laser can be adjusted by a He–Ne locating laser, which is pre-positioned over the sample and can be moved in steps of 1 μm. A more detailed description, including the construction of the reaction chamber etc., is given by Sharp (1990).

Oxygen, from either phosphates as calcium phosphate of organisms (conodonts or bones and teeth) from both marine and terrestrial environments or phosphate compounds in rocks, is liberated in the same way as from silicates using fluorinating agents. However, the original phosphate must first be converted to bismuth phosphate ($BiPO_4$) as described by Tudge (1960) and Kolodny, Luz and Navon (1983). Fluorinating the $BiPO_4$ with bromine trifluoride as

$$BiPO_{4(s)} + 8/3BrF_{3(l)} \rightarrow BiF_{3(s)} + PF_{5(g)} + 4/3Br_{2(g)} + 2O_{2(g)}\,,$$

then requires a temperature of 150°C and the resulting O_2 is converted into CO_2.

Oxygen from sulphates is converted to barium sulphate ($BaSO_4$) and then reduced with graphite at a temperature of 1000°C as follows:

$$BaSO_4 + 2C \xrightarrow{1000^\circ C} BaS + CO_2 + 2CO\,,$$

producing BaS, CO_2 and CO of which the latter is converted to CO_2 either by a glow discharge method or over Ni (Pt) electrodes as follows:

$$2CO \xrightarrow{E} CO_2 + C\,.$$

The resulting amount of CO_2 is then measured.

Oxygen from water samples is measured from CO_2 equilibrated with them. For this purpose, automatically on-line preparation is possible. In the case of HDO equilibrators (Fig. 3.10) H/D measurements can be performed as well. Knowledge of the fractionation factor in the system $CO_2 \leftrightarrow H_2O$ is mandatory for this reaction and has been determined as $\alpha = 1.04115 \pm 5 \times 10^{-5}$ at a temperature of 25°C and a neutral pH of 7. The fractionation factor decreases with increasing temperature and the equilibrium for the exchange of O is reached after several hours.

3.1.3(d) Nitrogen

As the measurements of N isotopes are carried out on N_2 gas, the various chemical compounds containing N must be converted to N_2. A functional diagram for the off-line preparation is given in Fig. 3.11. The preparation device uses ^{15}N from Kjeldahl wet digested samples. The

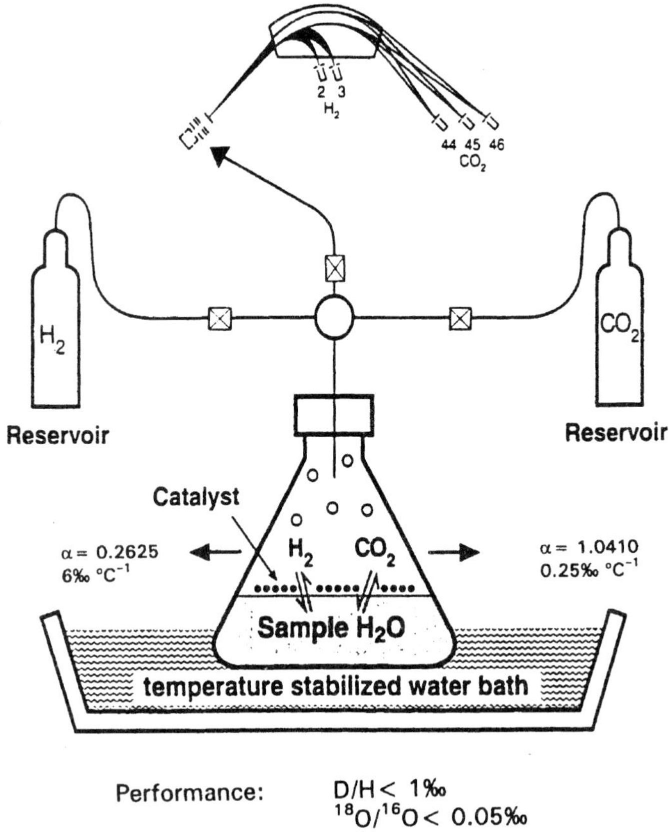

Fig. 3.10 Principle of an HDO equilibrator (after Avak and Brand, 1995).

quantitative reduction of ammonium sulphate $[(NH_4)_2SO_4]$ or ammonium chloride (NH_4Cl) is achieved under vacuum by the reaction with different reagents as sodium hypobromide (NaOBr) or lithium hypobromide (LiOBr).

The Kjeldahl method enables the quantitative determination of the nitrogen content of organic compounds such as proteins. Generally, there is a great variety of N-containing compounds and procedures for the conversion have been described in detail by Bremner and Keeney (1966) and Ross and Martin (1972).

3.1.3(e) Sulphur

The gas on which measurements are performed is usually SO_2. It is produced from sulphides by heating the ground sample in vacuum with an oxidizing agent as Cu_2O at *c.* 1000°C for several minutes. To avoid

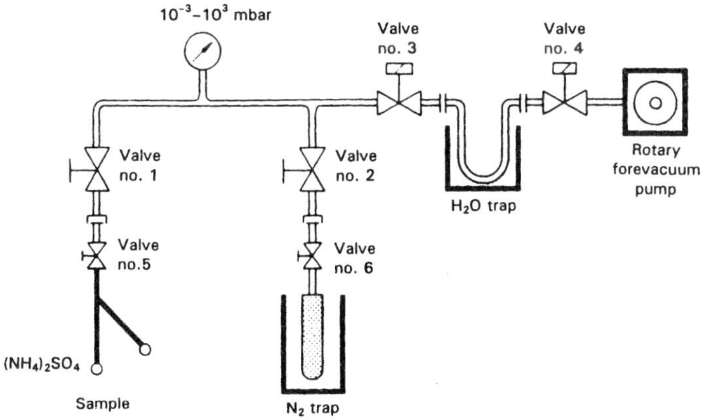

Fig. 3.11 Off-line preparation of N compounds.

corrections due to the $^{18}O/^{16}O$ variations influencing the $^{34}S/^{32}S$ ratios of the SO_2 it might be advisable to convert the sulphur compounds to SF_6 as described by Puchelt, Sabels and Hoering (1971). The advantages of this method are that the sulphur hexafluoride (SF_6) can be prepared quantitatively through reaction with BrF_3 from elemental sulphur as well as from many sulphide minerals, and that the resulting gas is chemically inert and is affected neither by water vapour in the vacuum system nor by adsorption on metallic walls. The following reaction:

$$2FeS_2 + 10BrF_3 \rightarrow 4SF_6 + 2FeF_3 + 5Br_2$$

is possible for many S compounds and the separation of SF_6 is quite easy, due to a large difference in vapour pressure. The extraction line as used by Puchelt, Sabels and Hoering in 1971 is shown in Fig. 3.12. The fluorination products in vessel (A) are attached to traps (C) filled with NaOH pellets. The vessels are then cooled; A and C with dry ice or another cooling agent providing *c.*–70°C where SF_6 slowly sublimes, D and E (glass traps) with liquid N_2 where the gas is frozen again. After cooling the system, the vessel (A) is opened and the gas is frozen out into the glass trap (D). Fluorine and Br compounds are removed through the NaOH within the vessels (C). Further sublimation of the sample to trap (E) and sample bulb (F) purifies the sample gas.

It should be mentioned that care has to be taken if any of the reactions described above are performed. Possible danger does not only derive from the reagents, among which are acids and toxic compounds (i.e. fluorination agents) or from flammable cooling liquids. In the case of glassware incorporated in vacuum reaction lines, implosions could

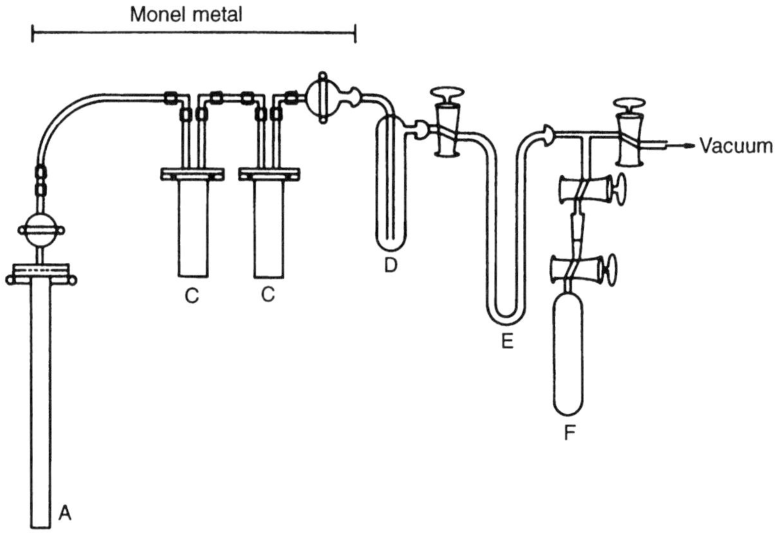

Fig. 3.12 Apparatus for the separation of SF_6.

occur if glass parts have defects. For more detailed information cited articles, even if older, should be used.

3.1.4 Other Mass Spectrometers

Among the various types of mass spectrometers appropriate for analytical research within the geosciences are types mentioned below, although their application sometimes lies mainly in the field of non-stable isotope analysis.

3.1.4(a) Thermal Ionization Mass Spectrometer (TIMS/N-TIMS)

A broad application of isotope ratio measurements is possible through thermal ionization mass spectrometry. Figure 3.13 gives an overview of elements which can be used. Mass spectrometers are used, for instance, in geology, nuclear science and environmental science. Modern instruments such as the VG Sector 54 (Fisons Instruments) or the MAT 262 (Finnigan MAT) are fully automated machines which can measure 13 to 20 samples without operator intervention. Positive ions are produced by evaporation and then ionization of the sample, using two different filaments. As the ionization potential of most elements is $< 8\,eV$, the thermal ionization is widely applicable and also used in some quadrupole mass spectrometers, enabling the effective determination of trace elements together with isotope dilution methods.

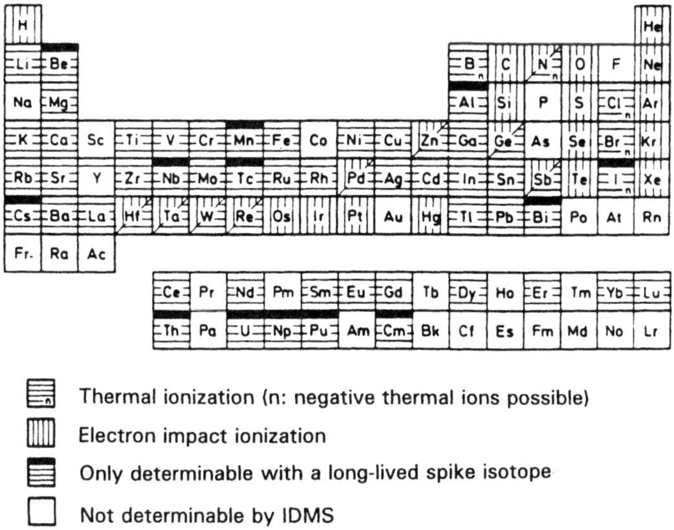

Fig. 3.13 Elements which can be measured by TIMS and N-TIMS.

More conventionally, mass spectrometers can also run in a negative ion mode, enabling more precise analyses of elements which do not easily form positive ions, for instance Os and B. The necessary polarity change is easily done by the controlling computer. Generally, starting from a total number of atoms N as

$$N = n_a + n_i ,$$

the number of non-ionized atoms n_a plus the ions n_i comprises the total sample. Additionally, the volume of the ionization chamber is of importance, since not all the volumes of all possible sites are really available for the electrons. There are repelling forces among the electrons and the space already occupied does not count in the available volume. The relation of the numbers of electrons, ions and atoms to the volume can be given by the Saha ionization equation as follows:

$$\frac{n_e n_i}{n_a} = \frac{1}{V_a} e^{-w/kT} ,$$

with n_e being the number of electrons, V_a being the volume of atoms, e being the elementary charge and W and kT as the ionization energy and Boltzmann constant multiplied by the absolute temperature. The combination of electrons with ions and the production of ions through collision with an electron certainly depend on and increase with the number of electrons per unit volume as well as with the number of ions per unit volume.

The ionization energy depends on the kinds of sample and the ionization leads to the steady production of ions, which is possible since the charged ions are accelerated towards the analyser, thus preventing them recombining with other electrons. The resulting ion current of several hundred mV is then measured. As the light isotopes are preferentially evaporated a correction of this effect might be necessary or a 'total evaporation' technique may be applied. This is generally preferable because of possible inhomogeneities within the sample. The sample-handling systems comprise several system elements such as disposable single, double or triple filament inserts, filament preparation devices and filament preheating and baking devices. The necessary amount of sample material for a single analysis is very small (picograms/nanograms).

As both instruments, the MAT 262 Mass Spectrometer and the VG Sector 54 Mass spectrometer, are highly developed and really sophisticated machines, also offering laser ports for laser resonance ionisation techniques, limitations of space preclude summarizing the technical data, but they are available from the manufacturers.

3.1.4(b) Accelerator-Based Mass Spectrometry (*AMS*)

In recent years, great strides have been made in radioisotope dating as a consequence of the direct counting of appropriate isotopes rather than by measuring the radioactive decay. Bonani *et al.* (1980) indicated that single-atom detection is attainable using an electrostatic accelerator as a component of an ultra-sensitive mass spectrometer. An already existing EN-Tandem accelerator facility at the Swiss Federal Institute of Technology in Zurich was modified so as to measure the isotopic compositions of all elements of interest with $z < 53$. The experimental layout is shown in Fig. 3.14. A new sputter source especially suitable for isotope dating was developed; in this a caesium beam impinges upon the surface of the sputter sample at an angle of 60° to give negative ions. The ion source is combined with a computer-controlled sample exchange mechanism having a magazine with 25 samples, which are exchangeable through a vacuum lock. The negative ions are accelerated to an energy of 60 KeV and mass-analysed by a double-focusing 90° magnet. A rapid change from measuring one isotope to measuring another is effected by varying the energy of the beam through this magnet. The ions are accelerated to the high-voltage terminal (6 MV) where their outer electrons are stripped in a gas stripper. The multiply charged positive ions are then further accelerated to ground potential. Ions with a charge state of 3+ and the right energy are selected in the electrostatic deflector and subsequently analysed by a double-focusing 90° magnet with a momentum resolution $p/\Delta p$ of 500. Rare isotopes are identified by a $\Delta E - E$ gas counter.

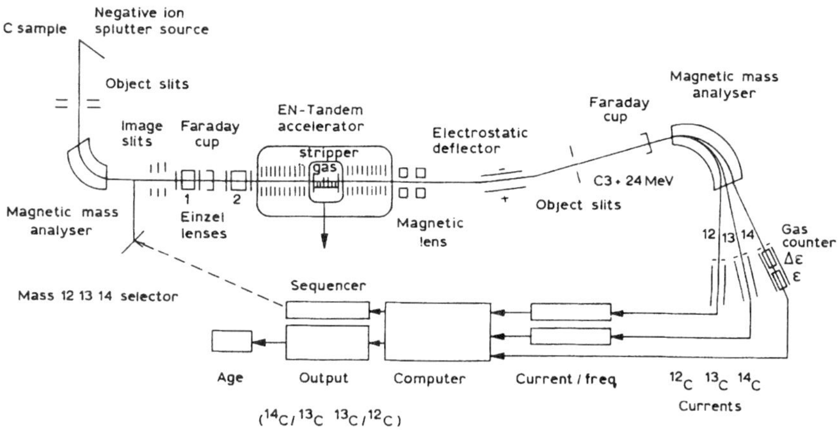

Fig. 3.14 Schematic diagram of the ETH EN-Tandem accelerator used as an ultrasensitive mass spectrometer.

Initial measurements on geological samples were performed at the high-energy end of the accelerator and the C beam for them was produced by a standard sputter source with cone-shaped sample holders. The beam was injected through a 20° magnet with low mass resolution. The ratio of the different C isotopes was measured by injecting and analysing sequentially ^{12}C, ^{13}C and ^{14}C. The switch to a different isotope was effected by varying the field of the 20° magnet and leaving all other steering and focusing components unchanged. The sequence was controlled by computer in such a manner that time intervals of 15 s for ^{12}C and ^{13}C and 150 s for ^{14}C were selected. In order to avoid loading effects of the accelerator, it was necessary to restrict the beam intensity to 100 nA for ^{12}C of charge state 3+. Well-separated peaks for the isotopes named and also for ^{14}N were obtained. The background peaks resulted partly from the low mass resolution of the 20° injection magnet (^{14}N does not form negative ions and presumably $^{14}NH^-$ ions of mass 15 were extracted from the source). All particles reaching the detector had the same momentum.

Additional charge exchange and scattering processes must be taken account of in order to explain the observed background peaks. A set of four samples having known $^{14}C/^{12}C$ ratios determined by β-counting techniques to better than 1% was used for calibration. Unknown samples and standards were measured alternately at 1 h intervals. Calibration samples showed the correct relative dependence within the limits of statistical error (about 3–7%). Fluctuations of up to 4% on the $^{13}C/^{12}C$ ratio resulted from variations in the transmission through the accelerator and, at the same time, limited the accuracy. However, improvements were expected by installing a new source and injection system and this would better the data derived regarding radioisotope dating using an EN-Tandem accelerator.

In a progress report on radiocarbon dating by accelerator mass spectrometry made by Hall (1983), he mentioned that low-energy facilities at the University of Oxford, UK, and in Arizona, USA, were near the stage when actual dates would start emerging. He described the two approaches being used by laboratories: the first was to construct new such low-energy dedicated machines and the second was to adapt existing high-energy accelerators. At that time only one laboratory had both published dates and had an ongoing measurement programme: that of the Zürich/Bern group, namely the Laboratorium für Kernphysik of the Swiss Federal Institute for Technology at Zurich and the Physikalisches Institut of the University of Berne.

The University of Arizona instrument generates negatively charged ions by bombarding a solid sample with a beam of Cs ions which have been accelerated to energies of 30 keV. Ions possessing the required mass are separated magnetically in the first of two mass spectrometers and subsequently accelerated by means of an electric field of several million volts. At that stage, all particles having the same mass as the desired isotope are stripped of some electrons in a differentially pumped gas cell or by collision with a metal foil. The result is that molecules disintegrate into their component monatomic ions. The consequent mixture of positive ions is further accelerated and sorted out in the second mass spectrometer which permits only those ions with the desired mass-to-charge ratio to enter a solid-state nuclear detector capable of distinguishing among isobaric ions on the basis of their energy, rate of energy loss and mass number. The equipment in question is able to identify and count individual ions as they reach the nuclear detector.

Instrumentation of a similar type has also been recently constructed at the University of Rochester and, in Canada, at the University of Toronto. Important researches have been carried out into the abundances of short-lived radioactive atoms which were produced by cosmic rays in the atmosphere and perhaps the best known of these cosmogenic radionuclides is ^{14}C. This is difficult to detect in a conventional mass spectrometer because of its very low abundance and also the masking effect of ^{14}N. The tandem accelerator mass spectrometer avoids the latter by using negatively charged ions in its first stage because ^{14}N is unstable. Such isobaric molecules as $^{12}CH_2$ ($A = 14$) would be difficult to separate from ^{14}C and are destroyed by stripping electrons from them. Consequently, radiocarbon is enhanced relative to the background by a factor of up to 10^{17} during its transit through the instrument. In this manner, tandem accelerator mass spectrometers facilitate the dating of C samples weighing under 1 mg. By contrast, the conventional radiochemical technique necessitates the employment of much greater quantities depending upon the ^{14}C content of the sample. Using AMS, Purser *et al.* (1982) were able to determine that a baby woolly mammoth from Magadan in Siberia died 27000 ± 1000 years BP.

3.1.4(c) Laser Mass Spectrometry

In 1982 it was reported that a laser mass spectrometry facility was being established at NBS (the National Bureau of Standards, Washington, DC, USA) with the aim of developing highly sensitive and isotopically selective atom-counting technology and investigative vaporization, isotopic discrimination and ionization behaviour of inorganic compounds in thermal ion sources. The stimulus for this arises from the projected sensitivities and selectivities for the formation and/or measurement of ions. Enhanced ionization efficiency and sensitivity can result from resonance ionization probability being N_{σ}, where N = photon flux and σ is the absorption cross-section). Consequently, when the product of N and σ exceeds 1, an ion will form with near unit probability. Isotopic selectivity is being exploited for ion formation and ion measurement, and atomic resonance ionization is element specific. Nonetheless, a capability for resolving isotopes masswise with mass spectrometers is necessary in order to complement limited ionization selectivity. The sensitivities and selectivities available from laser-induced ion fluorescence (LIIF) detection systems depend upon the ability to saturate optically the selected atomic ion transition so as to give a measurable number of fluorescence protons from individual ions. Mass-selected high-velocity ions should provide an almost ideal reservoir for using monochromatic laser wavelengths for Doppler-free spectroscopy at resolutions approximating the natural atomic line widths. Such LIIF ion detectors are expected to progress to a capability for 'single ion' detectability and high-abundance sensitivity measurements, as Keller, Bomse and Cremers (1981) indicated.

Laser and mass spectrometry systems were chosen so as to give maximum flexibility to access all except the most difficult part of atomic ionization potentials in the Periodic Table. A 700 mJ pulsed Nd-YAG laser pumping to a tunable dye laser coupled to a frequency doubler generates highly coherent wavelengths down to the near ultraviolet (260 nm). The output of the pulsed laser system may be directed to the ionization region of a time-of-flight (TOF) or a magnetic sector mass spectrometer. Electrons can be removed selectively from specific atoms and molecules utilizing the principles of resonance ionization and related resonance-enhanced multiphoton ionization. A second laser system comprises a continuous wave (CW) Ar ion laser employed to pump a scanning single-mode tunable dye laser. The output of this CW system can be routed either to the ionization region of a mass spectrometer or to the detector region of the magnetic sector instrument. Utilizing CW radiation, ions can be produced using electrostatic stripping of selectively laser-excited Rydberg electrons. The single-mode CW output can be used in addition to produce LIIF of mass-selected ions and for CW laser-induced fluorescence of neutral species in thermal vapours. Vaporization and ionization mechanisms are also being studied in thermal vapour plumes using the TOF mass spectrometer equipped with a 200 MHz transient digitizer to provide a mass spectrum of 4096 5-ns channels. A complete mass spectrum is given for each laser pulse. Successive spectra can be computer-averaged, recorded and displayed sequentially to produce a capability for real-time mass spectral diagnostics.

The formation of atomic ions using resonance ionization necessitates a source of neutral atoms. Single– and multiple–filament thermal ion sources used in a conventional stable isotope mass spectrometer can act as atom reservoirs through at least two mechanisms, as Moore, Heald and Filliben (1978) pointed out. Thus:

$$M^{o}_{(s)} \xrightarrow{\Delta} M^{o}(g);\ M^{o}(s) \xrightarrow{\Delta} n\,(M^{o}_{(g)}) + M^{o}_{(g)} \qquad \text{Single filament} \quad (3.2)$$

$$MX_{(s)} \xrightarrow{\Delta} MX_{(g)} \qquad \text{Sample filament} \quad (3.3)$$

$$MX_{(g)} \xrightarrow{\Delta} M_{(g)} + X^{o}_{(g)} \qquad \text{Ionizing filament} \quad (3.4)$$

Instances of the mechanism (3.2) are the electro-deposited high metals such as Cu and Re and of (3.3) rare earth halides and related high molecular weight inorganic compounds of the alkaline earth elements such as $M_xRe_yO_z(g)$ which dissociate and form atoms, and monovalent cations at sufficiently high ionizing filament temperatures, for example

$$MX_{n(s)} \xrightarrow{\Delta} M^{o}_{(g)} + nX^{o}_{(g)}$$

$$M_xRe_yO_z \xrightarrow{\Delta} M^{o}_{(g)} + Re_2O_{7(g)};\ M^{o}_{(g)} \xrightarrow{\Delta} M^{+}_{(g)} + e^{-}$$

Some elements ionize efficiently enough under normal (Saha-Langmuir) thermal conditions that the relatively low-duty cycle of laser operation (typically 10–20 ns pulses at 10-10^4 Hz) could provide major improvements in ionization efficiencies. Elements of marginal ionizing efficiency would need a closer agreement between vapour production 'on time' and the laser pulse. There are a number of methods available to produce pulsed thermal vapour plumes such as laser ablation and ion bombardment.

With number densities of 10^7–10^{10} cm^{-3}, multiphoton ions have been produced and measured in the Center for Analytical Chemistry at the NBS for residential gas impurities and acetone using the Nd–YAG pulsed laser/TOF mass spectrometer system, and the mass spectra have been characterized and compared with typical electron impact spectra. Researches aim at the multiphoton ionization of NO and the resonance formation of ions such as copper and molybdenum that are difficult to form using conventional methods, and also at element-selective ion formation from mixed transition metals. As well as offering better understanding of the mass discrimination effects produced in the thermal vaporization and ionization of inorganic compounds, the ultimate ionization efficiencies and isotopic selectivities which are potentially available from laser mass spectrometry may well have a large-scale and long-term impact on the ability to measure high isotope abundance ratios of such long-lived radiogenic nuclides as ^{10}Be and ^{129}I.

Although laser mass spectrometry has a broad application in fields like medicine or biological science (qualitative analysis), its use in geosciences is not that widespread. The quantitative analysis of isotopes remains problematic if samples are not completely ionized. The already mentioned combined preparation method of laser heating and fluorination (section 3.1.3) faces the same problem. The understanding of observed heterogeneities within samples requires both the knowledge of the isotopic zonation and of the values of the complete sample.

3.1.4(d) Ion Microprobes

In conventional mass spectrometry it is usually necessary to separate the chemical element to be analysed from other elements in the given sample; this militates against examination of individual mineral grains. Also, it is correspondingly difficult to study the variations in the isotope composition of an element in such a grain. There is in addition the problem of contamination during chemical processing existing and intensifying as the number of atoms in the sample diminishes. In order to avoid these difficulties, the ion microprobe mass spectrometer was developed and permits the examination of small mineral grains in both terrestrial and extraterrestrial samples.

In the ion microprobe a beam of negatively charged oxygen atoms is focused onto the Au- or C-coated surface of the sample. The size of the illuminated spot varies from $1\,\mu m$ to $3\,\mu m$. Ions sputtered from the surface of the sample are accelerated by an electrical field and separated magnetically. The instrument is capable of providing data on both the chemical and the isotope compositions of selected elements in a sample.

The first models of ion probe mass spectrometers encountered problems such as isobaric interference from ionized molecular species and multiply charged ions, inadequate resolution, instablility of ion beams and insufficient precision of measurement, as well as scattering of the primary beam causing the emission of ions from adjacent mineral grains or inclusions outside the illuminated spot.

However, most of these have been resolved and the ion microprobe has become a very powerful tool for studying very small samples. For instance, Huneke, Armstrong and Wasserburg (1983) found large enrichments of stable ^{26}Mg in some Ca–Al-rich inclusions of the carbonaceous chrondritic meterorite Allende. This anomaly results from the decay of parental ^{26}Al in the inclusions, which must have formed in the solar nebula before the short-lived ^{26}Al ($T_{1/2} = 0.716 \times 10^6$a) decayed.

3.1.5 Isotope Dilution

Mass spectrometers, for example the quadrupole mass spectrometer, can be employed in order to determine the concentration of an element in a sample by the technique of isotope dilution. This offers a cost-effective method and the approach is to ascertain the isotopic composition of an element in a mixture of a known quantity of a 'spike' with an unknown quantity of the normal element (Fig. 3.15). The spike may be a liquid or a gas possessing a known concentration of the particular element of which the isotopic composition has been altered by the enrichment of one of its naturally occurring isotopes. The sample to be analysed contains an unknown concentration of the element of which the isotope composition is known. Consequently, when a known quantity of a sample solution is mixed with a known quantity of spike, the isotope composition of the mixture may be utilized in order to calculate the quantity of the element in the sample solution.

Isotope dilution analysis is applicable to all elements having two or more naturally occurring isotopes. The actual isotope enrichment is effected by means of a large electromagnetic separator (calutron) able to separate quantitatively the isotopes of an element. In the case where non-volatile elements have to be analysed, the spike may be prepared as a solution of which the concentration and isotope composition are

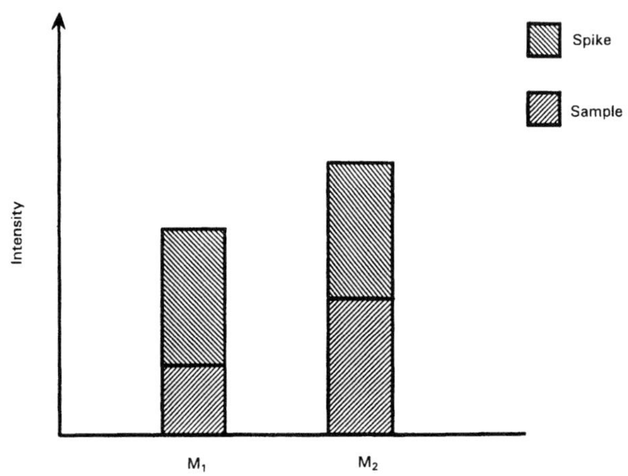

Fig. 3.15 Principle of the isotope dilution method.

verified separately. A known weight or volume of this spike is added to a known weight or volume of the sample solution and the mixture stirred to attain total isotopic homogeneity. The mixture of the normal element and the spike is then analysed on a mass spectrometer to find out its isotopic composition.

The result is expressed in terms of the ratio of the abundances of two isotopes and employed in order to calculate the concentration of the element in question in the sample solution. Isotope dilution has a number of advantages over other analytical methods. It is free of interference from other elements because isobaric isotopes of other elements can be removed chemically prior to analysis in the mass spectrometer. It is extremely sensitive as a result of the fact that the quantity of sample can be increased as the concentration of the desired element decreases. Its high accuracy depends solely upon the calibration of the spike solution. Quantitative recovery of the mixture of spike and normal element is not necessary, provided that isotopic homogeneity has been attained. It must be pointed out, however, that spike solution calibration is subject to errors possibly arising from the potential non-stoichiometric composition of the compound containing the spike. Also, isotopic fractionation may occur during the analysis of the spike in the mass spectrometer and there may be mistakes in weighing and diluting the spike. In addition, the measured value of the isotope ratio of the mixture may be affected by fractionation in the instrument and, if the element of interest has only two naturally occurring isotopes, no correction is possible. It is essential that the normal element and the spike be completely mixed; in some cases this may be difficult, for instance where the element to be determined is found in refractory minerals or is incorporated in an insoluble precipitate during the dissolution of the sample. Another disadvantage is that the concentration of the spike solution may change as a function of time through absorption of the element on the walls of the vessel in which the spike is stored or by the evaporation of water. Then too, the isotopic composition of the spike or of the mixture of the normal element and the spike may alter during processing or the sample because of isotope exchange reactions with the isotopes of that element in the walls of the container. Finally, it must be admitted that the technique is very time consuming, thus discouraging any replication of analyses. Consequently, there may well be inadequate recording of analytical errors.

However, isotope dilution, described in more detail by Faure (1977) and Bowen (1988), is vital to determine ages of rocks and minerals based on radioactive decay. Another significant contribution is made by employing the technique in preference to alternative analytical methods. When elements are determined which possesss more than two naturally occurring isotopes, it is feasible to measure two or more isotope ratios, from which the concentration and the isotopic composition of the normal element can be obtained. This treatment is particularly valuable in examination of the isotopic compositions of Sr and Pb, both of which have four naturally occurring isotopes.

Part Two
Unstable (radioactive) isotope dating

4

Uranium–thorium–lead dating

4.1 GEOCHEMISTRY

Uranium is an element with relative atomic mass of 238.0289 and atomic number 92. A hard white metal, its relative density is 19.05, its melting point is 1 135°C and its boiling point is 4 000°C. It is naturally radioactive and comprises 99.28% ^{238}U (half-life 4.5×10^9 a), 0.71% ^{235}U (half-life 7.1×10^8 a) and 0.006% ^{234}U (half-life 2.5×10^5 a). ^{235}U can sustain a nuclear chain reaction and hence is important in nuclear reactors and nuclear weapons. The main ore is pitchblende (uraninite) which has a variable composition and may be considered as a uranate of uranyl, $2UO_3.UO_2$, together with Th, Zr, Pb, etc., and various quantities of He, Ar, etc. It occurs as a primary constituent of igneous rocks, granites and pegmatites or in high-temperature veins associated with Sn, Cu and Pb minerals. Uranium ores were originally mined as a source of radium and small quantities of the element are used to produce pale yellow or green coloured glass (which fluoresces under UV light). Some uranium oxide is employed for colouring ceramics. An important U ore is carnotite, one of the uranium vanadates ($K_2(UO_2)_2(VO_4)_2.3H_2O$) which is found as a yellow or green-yellow crust or in sandstones and soft aggregates. Uranium is the 48th commonest element in the Earth's crust (2.3ppm) and is much more abundant than some perhaps better-known elements such as Ag, Hg, Cd and I. In fact, its world production (excluding the then USSR) was 37 300 tonnes in 1988 (equivalent to 44 000 tonnes of U_3O_8).

Uranium and the three following elements neptunium, plutonium and americium show oxidation states of (+III), (+IV), (+V) and (+VI). They resemble each other, except that the most stable state drops from U(VI) (VI) to Np(V) to Pu(IV) to Am(III). In the (+III) and (+IV) states, the compounds are similar to lanthanides, but the elements, together with

Th, etc., comprise actinides which include elements with atomic numbers from 89 (actinium) to 103 (lawrencium). The ions formed in the oxidation states (+III), (+IV), (+V) and (+VI) are M^{3+}, M^{4+}, MO_2^+ and MO_2^{2+}, respectively. Oxidation–reduction reactions are rapid between M^{3+} and M^{4+}, or MO_2^+ and MO_2^{2+}as these only entail the transfer of an electron. The oxidation of M^{4+}, to MO_2^{2+} is slow because it involves transfer of O.

Uranium reacts with H_2 even at room temperature to form UH_3, a very reactive hydride often more valuable than the metal in making other compounds. The U–O system is complex because there are several stable oxidation states and the compounds are frequently non-stoichiometric. UO_2 occurs in pitchblende, U_3O_8 is greenish black and UO_3 is orange-yellow. All three are basic and dissolve in acids. UO_3 dissolves in HNO_3 and forms the yellow uranyl ion $[O{=}U{=}O]^{2+}$. Crystallizing this solution gives uranyl nitrate $UO_2(NO_3)_2(H_2O)_n$. The number of molecules of water of crystallization may be two, three or six depending upon whether crystallization occurs from fuming, concentrated or dilute HNO_3. Halides include UF_6 and UCL_6. All are hydrolysed by water.

The ground state electronic configuration of U is $[Rn]5f^36d^17s^2$.

The actinides may be termed inner transition elements or f-block elements (of which thorium is one of interest). Thorium (from Thor, the Scandinavian war god) is a dark grey radioactive metal with relative atomic mass of 232.038 and atomic number 90. Its relative density is 11.72, its melting point is 1700°C and its boiling point is 4500°C. ^{232}Th has a natural abundance of ~ 100% and a half-life of 1.405×10^{10} a. Th is not at all rare and makes up 8.1ppm of the Earth's crust. In fact, it is the 39th most abundant element, coming mainly from monazite sand in which it is present up to 10% as the phosphate, mixed with phosphates of the lanthanides. Th also occurs as uranothorite, a mixed silicate of Th and U, in the Sudbury ores, Canada. The only stable oxidation state is Th(+IV) and the Th^{4+} ion is known both in the solid and in solution. In dilute solutions, hydrated ions $Th(H_2O)_n^{4+}$ exist and, with NaOH, yield a precipitate of $Th(OH)_4$. The oxide, ThO_2, is formed by heating the nitrate or by heating the metal in air. The oxide is white and has the highest melting point of any oxide (3220°C). It is not reactive except for dissolving in a mixture of HF and HNO_3. The other actinide oxides are also refractory.

Anhydrous halides, ThX_4, are formed by direct reaction and also by strongly heating the oxide with the appropriate halogen acid or with CCl_4 at 600°C. The halides are high melting and white. When strongly heated, THI_4 decomposes to the element. The halides hydrolyse in moist air to form oxyhalides, $ThOX_2$. The white colour of Th(IV) compounds is connected with the absence of d or f electrons. The high charge of Th^{4+} favours the formation of complexes. These frequently have high co-ordination numbers and uncommon structures. Th(III) compounds are rare and occur solely as solids. ThI_3 and ThOF have been made, but react with water forming Th(IV) and liberating H_2.

Approximately 500 tonnes of Th compounds are produced each year and the major industrial uses are as follows:

1. ThO_2 containing 1% Ce emits a bright white light when heated in a gas flame and this led to its widespread use in manufacturing incandescent gas mantles, which was very important when gas lighting was the major source of artificial light. Nowadays, this has been mostly replaced by electric light bulbs and fluorescent tubes, although it is still used in caravans and other mobile vehicles. The manufacture of gas mantles is still using half of the Th produced.
2. The Th occurring naturally is practically all ^{232}Th, a non-fissionable isotope. However, if it is irradiated in the outer part of a nuclear reactor, ^{233}U results from the reaction

$$^{232}_{90}Th \xrightarrow{n\gamma} {}^{233}_{90}Th \underset{22\,min}{\overset{\beta}{\longrightarrow}} {}^{233}_{91}Pa \underset{27\,d}{\overset{\beta}{\longrightarrow}} {}^{233}_{92}U\,,$$

   this isotope not occurring in nature (its half-life is 1.6×10^5 a). ^{233}U is fissionable. As more reactor fuel is produced than is consumed by the reactor, it is significant as the basis of the 'breeder reactor'.

The ground-state electronic configuration of Th is $[Rn]6d^27s^2$. Like U, Th belongs to and is the first element of the actinides in which 5f orbitals become progressively filled with electrons. Both elements have similar electron configurations and hence similar chemical properties. In nature they have a tetravalent oxidation state and closely similar ionic radii of 0.105 nm for U^{4+} and 0.11 nm for Th^{4+}, which enable them to substitute for each other. Under oxidizing conditions, U forms the uranyl ion UO_2^{2+} with U having a valence of 6. This forms water-soluble compounds, which makes U a mobile element under such conditions. By contrast, Th is tetravalent and forms compounds usually insoluble in water. Uranium and Th abundances in chondritic meteorites suggest that these are low in the terrestrial crust and mantle. During partial melting and fractional crystallization of magma, both elements concentrate in the liquid phase, thereafter becoming incorporated into more acidic components. Thus granites are enriched in U and Th compared with basic igneous rocks. A progressive geochemical differentiation in the upper mantle may have caused these elements to accumulate in continental crustal rocks in which values for U were recorded as *c.* 3 ppm and for Th as *c.* 8.5 to 17 ppm, compared with ultramafic values for U of 0.001 ppm and for Th of 0.004 ppm. It is interesting that low-Ca granites appeared to be enriched in Th relative to U. Possibly this reflects entry of U into an aqueous phase as uranyl ions at the final stage of crystallization of the magma. The Th/U ratio of sedimentary rocks is similar to that in igneous rocks, although sedimentary carbonates are enriched in U and have a Th/U ratio as low as 0.77, compared with that in granites which varies between 2.8 and 5.7. Common rock-forming silicate minerals have rather low U and Th concentrations, these elements being found mainly in accessory minerals, either as principal constituents or as replacements of other elements. The relevant minerals include uraninite and thorianite, the latter an oxide of Th and U ($ThO_2.U_3O_8$), zircon and thorite (silicates), xenotime and monazite (phosphates) and sphene (a titanite and silicate of Ca, $CaTiSiO_5$).

U has a total of 16 isotopes while Th has 25. ^{234}U,^{235}U, ^{238}U and ^{232}Th are primordial radionuclides, i.e. were created in the build-up of terrestrial matter and are still present today. The rest are radioactive with various forms of radioactive decay (mostly α and negatron). Radioactivity was discovered following experiments on the luminescence of uranyl double sulphate crystals caused by exposure to UV light. This was observed in the walls of cathode ray tubes, which led Becquerel to determine whether U compounds emit X-rays. They do and so does the element. Later Marie Curie demonstrated that this phenomenon also occurs with Th and that U and Th minerals are more active than pure salts of these elements. So natural U ores came under scrutiny as possibly containing more powerful radiation emitters than U. Two new active elements were revealed, namely polonium and radium. Pure Ra was isolated and its ionizing radiation examined by Rutherford. In fact, he recognized that radioactive substances emit radiation comprising α, β and γ components. His work with Soddy on the radioactivity of Th compounds resulted in the development of the theory of radioactive growth and decay.

Attempts have been made to formulate a grand unified theory (GUT) which would combine the strong, weak and electromagnetic interactions into a gauge field theory. Some success has been achieved in the electroweak theory (Glashow–Weinberg–Salam theory), quantum flavourdynamics (QFD), which combines the electromagnetic interaction and the weak interaction and may be considered as the first step towards a GUT. It has been confirmed by particle physics work which culminated in the discovery of the Z particle, the electrically neutral carrier of weak interactions arising from the exchange of photons, the massive W particle and the neutral Z particle between quarks and leptons. W and Z gauge particles last for a mere 1×10^{-18}s and exist only in atomic nuclei where they are believed to trigger radioactive decay.

Heavy radioactive elements may be grouped into four decay series. the common radioactive elements Th, U and Ac are found in nature and belong to three different series named after them. They are the parent members of their respective series and have the longest half lives. They decay by a series of α and β (negatron) emissions to produce successively more stable radioactive elements until at last a stable isotope is reached. All three series end with Pb ($^{206}_{82}Pb$, $^{207}_{82}Pb$ and $^{208}_{82}Pb$). After the discovery of the artificial post-uranium elements, the neptunium series was added and this ends with Bi ($^{209}_{83}Bi$). These four series are designated as follows:

Thorium ($4n$) series;
Neptunium ($4n + 1$) series;
Uranium ($4n + 2$) series;
Actinium ($4n + 3$) series.

The numbers in brackets indicate that the parent and all members of a particular series have mass numbers exactly divisible by four or divisible by four with a remainder of one, two or three. There is no natural cross-linkage between the four series although this can be achieved artificially.

Hahn and Strassmann first split the U atom in Berlin in 1938 and this pointed the way to a chain reaction and a fission bomb. As mentioned earlier, the ^{235}U isotope is fissile and its nucleus breaks into two smaller nuclei if irradiated by thermal (slow) neutrons. About 3×10^{-11} J of energy is thereby released from a single ^{235}U nucleus following the relationship $E = mc^2$, that is to say that 1kg of ^{235}U could produce the energy equivalent of about 3 million tonnes of coal. In other words, nuclear fission liberates millions of times more energy than any chemical reaction could.

If one of the neutrons involved splits another ^{235}U nucleus, a self-perpetuating nuclear chain reaction results and, because three neutrons are evolved per fission, they could split three further ^{235}U nuclei and initiate a branched chain reaction. This might run out of control and liberate energy at an ever-increasing rate until an explosion occurs. However, it is difficult to start and maintain a chain reaction. Neutrons produced by fission are fast neutrons with energies of 2×10^8 kJ mol^{-1}. They have a tendency to escape and are ineffective in promoting further fission. Slow (thermal) neutrons are much more effective and have an energy of *c.* 2 kJ mol^{-1}. They are termed thermal because their energy is equivalent to thermal energy attainable at room temperature. There are two ways of increasing the chance of a chain reaction. One is to slow down fast neutrons by collision with atoms of H, D or C (these constitute 'moderators'). A second approach depends on the fact that a sufficiently large mass of $^{235}_{92}U$ is necessary to make sure that enough neutrons impact another fissile U nucleus rather than escape. Hence there is a critical mass and its size depends on the shape and purity of the material. A sphere has the minimum surface area which minimizes the possibility of neutrons escaping. On the other hand, rods and sheets have much larger areas so that neutrons can escape much more easily. The chance that neutrons will cause fission rises if the proportion of the fissile isotope is increased and hence it is normal practice to enrich the fuel.

Control rods are used to ensure that the reaction does not get out of hand. They absorb neutrons when lowered into the reactor and thus slow down the reaction. To accelerate it, they are raised out of the reactor. Control rods may be made of boron steel, Hf or Cd. An explosion results when the neutron propagation factor exceeds one, in other words when more than one neutron from each fission is effective in causing another fission. To reach this condition, bomb-grade fuel is highly enriched and may contain as much as 80% ^{235}U. However , to reach such an enrichment is difficult since the energy released may melt the radioactive material, which then spreads out and allows more neutrons to escape so that the chain reaction stops. Such spreading out must be prevented by some form of containment if an explosion is to become possible. Temperatures reached compare with those of the Sun and no casing can withstand them.

The radioactivity of a sample is traditionally measured in curies (Ci), 1 curie being that amount of radioactive disintegration in 1g of $^{226}_{88}Ra$. More precisely, it is defined as the quantity of radioisotope yielding 3.7×10^{10} disintegrations per second. Hence, 1 curie of different radioisotopes comprises a widely differing number of atoms. The SI unit of radioactivity is the becquerel (Bq) and this is defined as the amount of a radioisotope giving one disintegration per second. Therefore, $1\,\text{Bq} = 10^{-12}$ Ci.

The radioactivity of U and Th was used to date minerals containing these elements and several aproaches were employed. They included a chemical Pb–U–Th method, a Pb-alpha method, a U–He method and two still in use, namely the U–Th–Pb method and the common Pb method. The abandonment of the first three resulted from the realization that some of the assumptions on which they are based are invalid. For instance, the chemical Pb–U–Th method depended on all Pb in a U mineral being of radiogenic origin with its amount increasing as a function of time. The Pb-alpha method was an extension of this in which the concentration of lead was determined by an optical spectroscope with the U and Th concentrations being measured by the rate of α-emission from the sample. The age was calculated from the following equation:

$$t = \mathrm{c}\,\frac{\mathrm{Pb}}{a},$$

where t is the age in Ma, Pb is the lead concentration in ppm and a is the observed alpha activity due to the radioactive decay of U, Th and their respective daughters. c is a constant of which the value depends on the Th/U ratio of the sample. This approach was formerly widely used in the US Geological Survey to date igneous rocks through accessory minerals such as zircon, xenotime and monazite.

The U–He method assumed that some U- and Th-bearing minerals quantitatively retain He produced in them by the α decay of U, Th and their daughters. Undisturbed crystals of magnetite, zircon and sphene most closely fulfilled this requirement, but even then the degree of retention depends on the nature of the lattice positions occupied by U and Th, the degree of radiation damage in the crystals and the geological histories of the samples. In addition, significant quantities of U and Th in rocks occur on crystal surfaces from which He can more readily escape. Hence U–He dates are underestimates of the ages of samples unless independently confirmed.

As discussed above, the four radionuclides of interest are ^{238}U, ^{235}U, ^{234}U and ^{232}Th. Their decay constants are $1.55125 \times 10^{-10}\ a^{-1}$, $9.8485\ \times 10^{-10}\ a^{-1}$, $2.806 \times 10^{-6}\ a^{-1}$ and $4.948 \times 10^{-11}\ a^{-1}$, respectively. All have α decay modes and five radioactive isotopes of Th occur in nature as short-lived intermediate daughters of ^{238}U, ^{235}U and ^{232}Th. They are, from ^{238}U: ^{234}Th ($T_{1/2}$ = 24.1 d) and ^{230}Th ($T_{1/2}$ = 7.54= 10^4a), the latter also called ionium, Io; from ^{235}U: ^{231}Th ($T_{1/2}$ = 25.5h) and ^{227}Th ($T_{1/2}$ = 18.72d) and from ^{232}Th: ^{228}Th ($T_{1/2}$ = 1.913 a). All are extremely short-lived in comparison with the parent elements.

The radioactive decay of ^{238}U produces the U series which terminates in stable ^{206}Pb and may be expressed as follows:

$$^{238}_{92}U \rightarrow ^{206}_{82}Pb + 8^4_2He + 6\beta^- + Q,$$

where the value of Q is 47.4. MeVatom^{-1} (see also Fig. 4.1) Each atom of ^{238}U undergoing radioactive decay ultimately gives rise to one atom of ^{206}Pb through the emission of eight α-particles and six negatrons. This assumes that the radioactive decay occurred in a closed system as regards U, Pb and all the intermediate daughters, Q representing the total of all the decay energies of the series in MeV. Some of the intermediates undergo branched decay entailing emission of either an α-particle or a negatron. Of course, the end result is the same, that is to say the stable ^{206}Pb (RaG).

The radioactive decay of ^{232}Th produces the Th series which terminates in stable ^{208}Pb and may be expressed as follows:

$$^{232}Th \rightarrow ^{208}Pb + 6^4He + 4\beta^- + Q,$$

where the value of Q is 39.8 MeV atom^{-1} (Fig. 4.2). Each atom of ^{232}Th undergoing radioactive decay ultimately gives rise to one atom of ^{208}Pb through the emission of six alpha particles and four negatrons. This assumes that the radioactive decay occurred in a closed system as regards Th, Pb and all the intermediate daughters, Q representing the total of all the decay energies of the series in MeV. Some of the intermediates

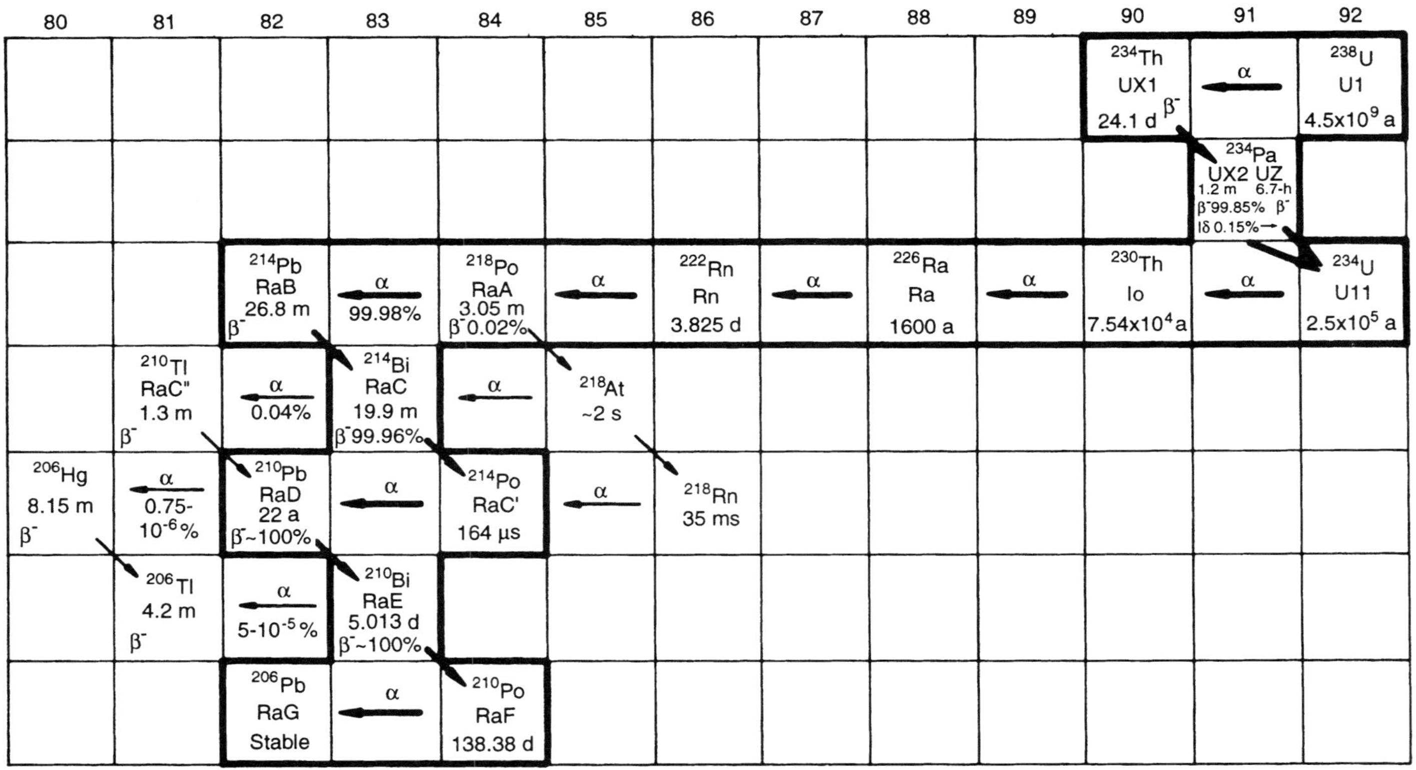

Fig. 4.1 The radioactive decay sequence of the natural U–Ra family of which the exact and systematic name is the (4n+2) family. Its parent is ^{238}U and it includes ^{234}Pb as an intermediate daughter before terminating in the stable ^{206}Pb (radium-G is the radioelement of this nuclide).

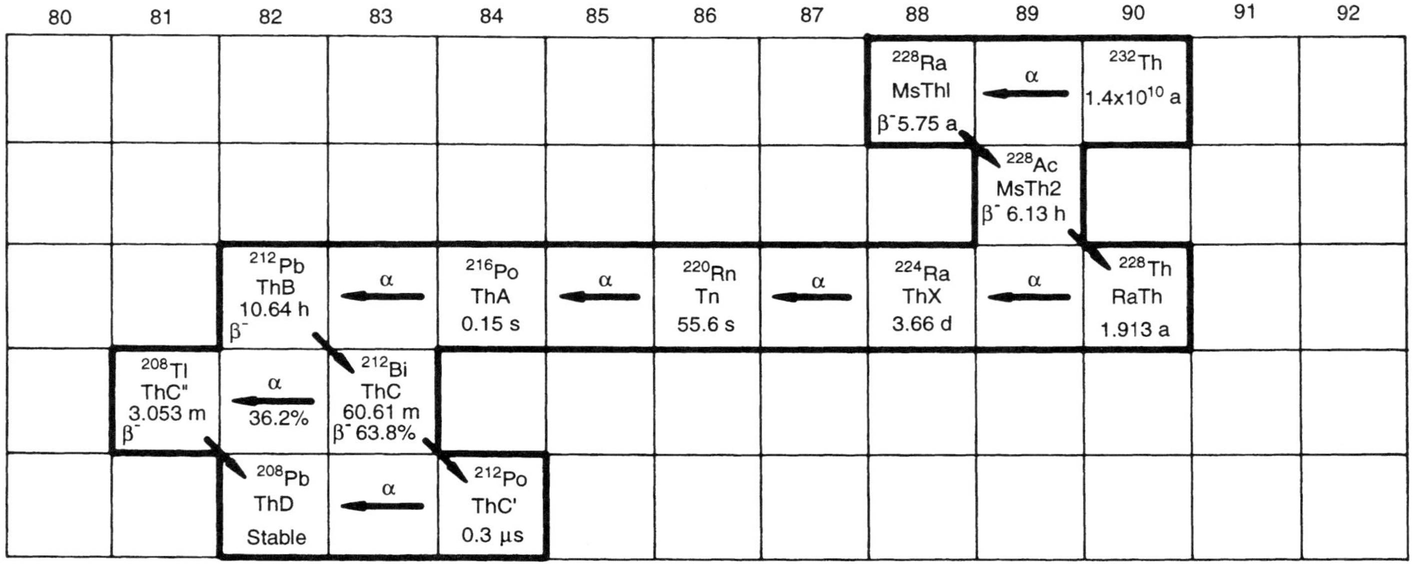

Fig. 4.2 The radioactive decay sequence of the natural Th family of which the exact and systematic name is the A = 4n family. It ends in the stable nuclide ^{208}Pb (radioelement thorium-D).

undergo branched decay, but the final result remains the stable ^{208}Pb (ThD).

Altogether, 42 isotopes of 12 elements are formed as intermediate daughters in the U, Th and Ac radioactive decay series and each series ends in a specific isotope of Pb (^{206}Pb for the U series, ^{208}Pb for the Th series and ^{207}Pb (= AcD) for the Ac series; Fig. 4.3). Because the half-lives of the parent elemental isotopes ^{238}U, ^{235}U and ^{232}Th are longer by many orders of magnitude than those of their respective daughters, the radioactive decay series fulfil the requirements for the establishment of secular equilibrium and the rates of decay of the intermediates equal those of the relevant parents. If the mineral is a closed system showing such secular equilibrium, the rate of production of the stable daughter at the end of a particular decay series is equal to the rate of decay of the parent elemental isotopes. Consequently, it is justifiable to regard the radioactive decay of the U and Th isotopes in minerals having secular equilibrium as if it occurred directly to the respective stable isotopes of Pb. This simplifies the equations necessary to describe the growth of radiogenic lead in U- and Th-bearing minerals.

Lead (Pb from the Latin *plumbum*) is an element with relative atomic mass of 207.2 and atomic number 82. A soft bluish-white metal, its relative density is 11.34, its melting point is 327.5°C and its boiling point is 1 760°C. It occurs mainly as galena, PbS, and it is extracted by roasting the ore in a reverberatory furnace. Compounds are poisonous. The metal is used in the lead accumulator, in alloys and in plumbing while its compounds are employed in the manufacture of paints and in petrol additives (tetraethyl lead, $(C_2H_5)_4Pb$, is utilized to reduce knocking in petrol engines). Ordinary Pb has four naturally occurring isotopes, ^{204}Pb, ^{206}Pb, ^{207}Pb, and ^{208}Pb which have abundances of 1.4%, 24.1%, 22.1% and 52.4%, respectively. There are 28 other isotopes ranging from $^{183}Pb(T_{1/2} = 0.6s)$ to ^{214}Pb ($T_{1/2} = 26.8$m in) and all these are radioactive with very short half-lives mostly of seconds, minutes or hours except for ^{205}Pb with a half-life of 1.5×10^7a and ^{210}Pb with a half-life of 22.3a. ^{204}Pb has a metastable state half-life of 67.2m, the γ-energy of the isomeric transition being 899keV, and the ground state has a half-life of 1.4×10^{17}a. This is so long that ^{204}Pb may be regarded as non-radiogenic although slightly radioactive. Eventually it decays to ^{200}Hg by alpha emission. The following equations may be used to express the isotopic compositions of lead in U- and Th-bearing minerals:

$$\frac{^{206}Pb}{^{204}Pb} = \left(\frac{^{206}Pb}{^{204}Pb}\right)_i + \frac{^{238}U}{^{204}Pb}\,(e^{\lambda_1 t} - 1) \tag{4.1}$$

$$\frac{^{207}Pb}{^{204}Pb} = \left(\frac{^{207}Pb}{^{204}Pb}\right)_i + \frac{^{235}U}{^{204}Pb}\,(e^{\lambda_2 t} - 1) \tag{4.2}$$

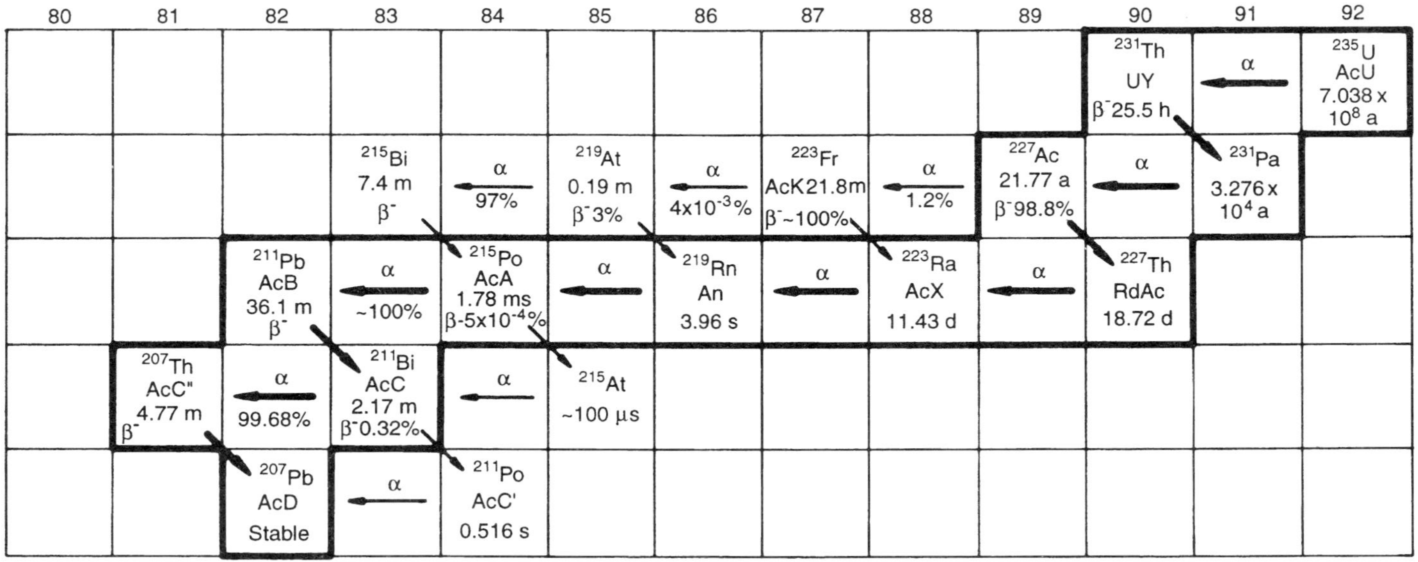

Fig. 4.3 The radioactive decay sequence of the natural Ac family of which the exact and systematic name is the A = 4n + 3 family. It ends in the stable nuclide ^{207}Pb (radioelement actinium-D).

$$\frac{^{208}\text{Pb}}{^{204}\text{Pb}} = \left(\frac{^{208}\text{Pb}}{^{204}\text{Pb}}\right)_{\text{i}} + \frac{^{232}\text{Th}}{^{204}\text{Pb}}\ \text{e}^{\lambda_3 t} - 1) \tag{4.3}$$

where ^{206}Pb/^{204}Pb, ^{207}Pb/^{204}Pb and ^{208}Pb/^{204}Pb are the isotope ratios of Pb in the analysed minerals, the same ratios with the subscript i are those which were present initially when the Pb was incorporated into the minerals when they formed, ^{238}U/^{204}Pb, ^{235}U/^{204}Pb and ^{232}Th/^{204}Pb are the isotope ratios in the minerals when analysed, λ_1 are, λ_2 and λ_3 are the decay constants of ^{238}U, ^{235}U and ^{232}Th, respectively and t is the time passed since closure of the relevant minerals to U, Th, Pb and all intermediate daughters.

The isotope dilution technique is applied to date U- and Th-bearing minerals and determine the concentrations of U, Th and Pb. The isotopic composition of Pb may be assessed using a suitable mass spectrometer. The above equations may be solved for t if reasonable assumed values are taken for the initial Pb isotope ratios when

$$t_{206} = \frac{1}{\lambda} \ln \left[\frac{\frac{^{206}\text{Pb}}{^{204}\text{Pb}} - \left(\frac{^{206}\text{Pb}}{^{204}\text{Pb}}\right)_{\text{i}}}{\frac{^{238}\text{U}}{^{204}\text{Pb}}} \right] + 1.$$

This expression applies to ^{238}U and similar ones can be written for ^{235}U and ^{232}Th.

The resultant data embody three independent dates referring to three radioactive decay series and if they are in accord a viable age for the minerals analysed results. This will be the case if the systems involved were closed to U, Th Pb and all intermediates since cyrstallization and also if the isotopic composition of uranium was not modified in any way, for instance through isotopic fractionation. It is presumed that the values taken for the initial Pb isotope ratios are reasonably accurate, that the decay constants of ^{238}U, ^{235}U and ^{232}Th are accurate and that no analytical or systematic errors are involved.

In practice, discordant dates for U- and Th-bearing minerals are often obtained and indicate that the systems in which they occur were not closed. The main reason for this is loss or gain of U, Th, Pb or intermediates after crystallization took place. If lead was lost, the effect may be minimized by calculating a date based on the ^{207}Pb/^{206}Pb ratio which is not sensitive to this loss. Clearly, minimum sensitivity is displayed where such a Pb loss took place recently and when the lost Pb had the same isotopic composition as that which remained, i.e. isotope fractionation did not occur. From equations (4.1)–(4.3) appropriate equations may be derived to give the following expression:

$$\frac{\dfrac{^{207}\mathrm{Pb}}{^{204}\mathrm{Pb}} - \left(\dfrac{^{207}\mathrm{Pb}}{^{204}\mathrm{Pb}}\right)_{\mathrm{i}}}{\dfrac{^{206}\mathrm{Pb}}{^{204}\mathrm{Pb}} - \left(\dfrac{^{206}\mathrm{Pb}}{^{204}\mathrm{Pb}}\right)_{\mathrm{i}}} = \frac{^{235}\mathrm{U}}{^{238}\mathrm{U}}\left[\frac{e^{\lambda_2 t} - 1}{e^{\lambda_1 t} - 1}\right].$$

The $^{235}U/^{238}U$ ratio is 1/137.88 for U of normal composition found in terrestrial minerals, in lunar rocks and in meterorites. Thus it is feasible to obtain a so-called '207–206' date without any knowledge of the U concentration in the mineral involved. Extrapolating, the following expression is obtained:

$$\frac{\dfrac{^{207}\mathrm{Pb}}{^{204}\mathrm{Pb}} - \left(\dfrac{^{207}\mathrm{Pb}}{^{204}\mathrm{Pb}}\right)_{\mathrm{i}}}{\dfrac{^{206}\mathrm{Pb}}{^{204}\mathrm{Pb}} - \left(\dfrac{^{206}\mathrm{Pb}}{^{204}\mathrm{Pb}}\right)_{\mathrm{i}}} = \left(\frac{^{207}\mathrm{Pb}}{^{206}\mathrm{Pb}}\right)^{+},$$

which gives the ratio of radiogenic Pb-207 to radiogenic Pb-206. The calculation entails subtracting the assumed initial $^{207}Pb/^{204}Pb$ and $^{206}Pb/^{204}Pb$ ratios from the actual measured values. In this manner it is possible to derive 207–206 dates from the isotopic composition of the lead in the mineral being analysed without any need to determine its concentration. The following equation is transcendental and insoluble for t algebraically:

$$\left(\frac{^{207}\mathrm{Pb}}{^{204}\mathrm{Pb}}\right)^{+} = \frac{^{235}\mathrm{U}}{^{238}\mathrm{U}}\left[\frac{e^{(\lambda_2 t)} - 1}{e^{(\lambda_1 t)} - 1}\right]$$

although a computer approach can be adopted whereby a set of successive approximations of t is obtained until a value emerges to solve the equation for a particular $(^{207}Pb/^{206}Pb)^{+}$ ratio to the required level of precision. Alternatively, tables for the calculation of Pb isotope ages exist, e.g. that of Stacey and Stern (1973). In them, t may be derived for any value of the ratio. Where $t = 0$, l'Hôpital's rule is applicable, thus

$$\lim_{t \to 0} \frac{e^{(\lambda_2 t)} - 1}{e^{(\lambda_1 t)} - 1} = \lim_{t \to 0} \frac{2^{e(\lambda_2 t)}}{1^{e(\lambda_1 t)}} = \frac{\lambda_2}{\lambda_1}.$$

At present $t = 0$, and the ratio between radiogenic ^{207}Pb (produced by the radioactive decay of ^{235}U) and radiogenic ^{206}Pb (produced by the radioactive decay of ^{238}U) can be derived as 0.04607. The ratio $(^{207}Pb/^{206}Pb)^{+}$ now equals the ratio of the decay rates of the parent elements. In fact, the current value of this ratio is a minimum which accords with the present low abundance of ^{235}U in U (0.72%).

While U and Th occur in many minerals, only a few of these are suitable for dating using the U, Th–Pb method. The reason is that most are not sufficiently retentive of these elements. Possibly the most retentive is zircon, so it is appropriate to add relevant information on its concentration and purification. The necessary processes are both laborious and time consuming, post-crushing and sieving being followed by recovery using a wet shaking table and the pertinent heavy liquids such as bromoform succeeded in turn by magnetic separation. Krogh (1982a,b) showed that this last procedure may be improved in order to select the least paramagnetic grains, an approach tending to shift data points along discordia towards the point of interaction of concordia on $^{206}Pb^{+}/^{238}U$ versus $^{207}Pb^{+}/^{235}U$ diagrams. The same researcher devised an abrasion technique effected by subjecting zircon grains to a stream of air which tends to remove metamict grains since they are softer than clear, non-cracked ones which are those most suitable for dating. Old cores of zircons can be removed by hand picking after young overgrowths have been removed by abrasion. The above approaches concentrate zircons most nearly concordant which yield optimal U–Pb dates. Also, the elimination of outer layers of zircons by abrasion makes the age of the relevant rock easier to determine because it is no longer necessary to derive this from sets of discordant dates. Other improvements involve analysing fragments of zircon crystals as small as 0.5mg in weight, mounting zircon crystals in the source of a mass spectrometer and using the sensitive high mass resolution ion microprobe called the SHRIMP. This was used in 1984 to obtain a U–Pb geochronology of zircons from lunar breccia.

Zircons in sedimentary rocks originated in sources underlain by igneous-metamorphic complexes in some regions and they well record individually the range of orogenic and magmatic histories of these sources. Gaudette, Vitrac-Michard and Allègre (1981) described North American Precambrian history recorded in a single sample and the high-resolution U–Pb systematics of the Potsdam sandstone detrital zircons. Their data are summarized in Table 4.1

These researchers emphasized that the 1.300 Ga age might imply that the zircons in the Potsdam sandstone found in the Adirondack Moun-

Table 4.1 U–Pb systematics of Precambrian Potsdam sandstone detrital zircons (from Gaudette, Vitrac–Michard and Allègre, 1981)

Zircon variety	*Concordia intercept dates (Ga)*	
	Lower	*Upper*
Bulk sample	0.530	1.300
Brown–elongate	0.325	1.170 Ga ± 100 Ma
Brown–rounded	0.380	1.320 Ga ± 80 Ma
Clear–rounded	0.985 ± 40 Ma	2.160 Ga ± 500 Ma
Clear–elongate	0.985 ± 15 Ma	2.700 Ga ± 500 Ma

tains of New York derived from source rocks of that age, but it is actually an artefact produced from the mixing of zircons of different ages in various proportions. The brown–elongate zircons were the least eroded and their age agrees well with an age of 1.130 Ga from the Ticonderoga gneiss in the Adirondacks. The brown-rounded grains are older and may have come from sources in the Grenville province. The clear zircons are much older and represent sources in the Superior structural province in Canda. In fact, the lower intercept date (0.985 Ga) coincides with the period of high-temperature metamorphism of the Grenville rocks. Hence, they may have been incorporated into pre-Grenville sediments and later undergone episodic Pb loss during the Grenville orogeny. Of course, the lower intercept dates of the brown zircons may have been caused by Pb loss in the course of slow cooling after this orogeny. Still, the 530 Ma date given by a lower concordia intercept of the bulk sample must be regarded as again an artefact of mixing and so does not provide evidence of Pb loss by chemical weathering of the zircons in question during deposition in the Late Cambrian. The above work showed that mixtures of two or more varieties of zircons of different origins coming from one rock sample give misleading results. To avoid this, only single grains should be dated or perhaps small samples of selected grains, both revealing individual or small number geological histories.

Other U- and Th-bearing minerals exist, but none is as suitable for U, Th–Pb dating as zircon which has average concentrations of these elements of 1330 and 560 ppm, respectively. The highest concentrations have been found in pegmatites. Both elements occur through the isomorphous replacements of Zr^{4+} (ionic radius 0.087 nm) by U^{4+} (ionic radius 0.105 nm) and Th^{4+} (ionic radius 0.11 nm), also through thorite inclusions. Clearly such substitution is restricted by the differences in the ionic radii and Pb^{2+} is totally excluded because its ionic radius is 0.132 nm and it has a lower charge. Consequently, zircon does not contain much Pb when it forms, but does have high U/Pb and Th/Pb ratios which makes it a valuable geochronometer. Extraction of U and Pb from zircons is possible using hydrothermal decomposition in a sealed pressure vessel.

Model dates for zircons, apatites and sphenes are sometimes discordant because some of the relevant dating assumptions referred to earlier did not hold. For instance, loss of radiogenic Pb in zircons can be caused by continuous diffusion or through the influence of metamorphic episodes. Unfortunately, it is seldom possible to link a discordant date with a particular geological event. But occasionally the 207 – 206 date may approach the actual age of the zircon And, although discordancy in dating is common, there are numerous examples of concordant dates.

4.2 METHODOLOGY

4.2.1 Theory: Concordia and Discordia, Isochrons, Common, Anomalous and Multistage Leads

Radioactive decay of the naturally occurring uranium isotopes to form stable radiogenic Pb provides two independent geochronometers which are concordant if dating assumptions are valid for the relevant samples. The radioactive decay of ^{238}U to ^{206}Pb as a function of time may be expressed as follows:

$$\frac{^{206}Pb^{+}}{^{238}U} = e^{(\lambda_1 t)} - 1$$

where

$$\frac{^{206}Pb^{+}}{^{238}U} = \frac{\frac{^{206}Pb}{^{204}Pb} - \left(\frac{^{206}Pb}{^{204}Pb}\right)_i}{\frac{^{238}U}{^{204}Pb}}$$

in which λ is the decay constant of ^{238}U and t is the time elapsed since the closure of the mineral involved to uranium. A similar statement can be made for other radioactive decay series. If the procedure is reversed, compatible sets of $^{206}Pb^{+}/^{238}U$ and $^{207}Pb^{+}/^{235}U$ ratios for specified values of t result and constitute the coordinates of points representing U–Pb systems with concordant dates in coordinates of $^{206}Pb^{+}/^{238}U$ as ordinate and $^{207}Pb^{+}/^{235}U$ as abscissa. The last equation and its equivalent for ^{235}U and $^{207}Pb^{+}$ comprise the parametric equations of a particular curve representing the loci of all concordant U–Pb systems. This curve is termed the 'concordia' and is such that $t(^{206}Pb^{+}/^{238}U) = t(^{207}Pb^{+}/^{235}U)$ and all samples with concordant dates must plot on such a curve. Samples with an age of t_1 which lost Pb at some time t_2 plot on a chord connecting t_1 and t_2, their position on this being dependent on the amount of Pb lost. Loss may be episodic or continuous.

Episodic loss may be due to chemical weathering or a metamorphic event. If all radiogenic Pb accumulated up until then is lost, the coordinates of the point representing the system respond by returning it to the origin. Such a U-bearing mineral contains no radiogenic Pb when it crystallizes, the total Pb loss returning the concordia point to the position which it occupied originally. This resets the U–Pb geochronometer and all trace of the earlier history of the system is lost. However, if only some of the radiogenic Pb is lost, the point moves some way along the chord. Systems falling on the chord have discordant dates, hence the chord is termed a 'discordia'. Because minerals usually lose only a small proportion of their radiogenic Pb, resetting is less frequent than positioning on the discordia. It is interesting that varied fractions of Pb may be lost from zircons in the same sample of igneous rock despite the fact that they must have undergone the same conditions. There may be some connection with the actual sizes of the relevant crystals and their U concentrations as well as with radiation damage in them. Smaller grains and those with high U concentrations seem to suffer larger Pb losses than bigger grains so that several zircon fractions may be obtained from one rock sample, these plotting as a set of points along the discordia line. Consequently it is possible to ascertain the position of the discordia by fitting a straight line to data points representing zircons which lost various quantities of their radiogenic Pb. Extrapolation of discordia give two points of intersection with concordia (Fig. 4.4). The first is at the time elapsed since the crystallization of the minerals and the second is at the time elapsed since closure of the system after an episode either of Pb loss or U gain.

Concordia diagrams permit interpretation of the geological histories of U–Pb systems and give information regarding past disturbances of these systems. Thus X on the discordia line in Fig. 4.4 is a measure of the degree to which the daughter-to-parent ratios of the system were affected either by Pb loss or U gain (or loss) where $X = 1/L$. Unless the isotopic composition of added Pb is known, the effect of Pb gain cannot be considered in the model. In addition, Pb loss must have taken place with no discrimination against the Pb isotopes on the basis of their masses. Taking this into account, X can be represented as the ratio of

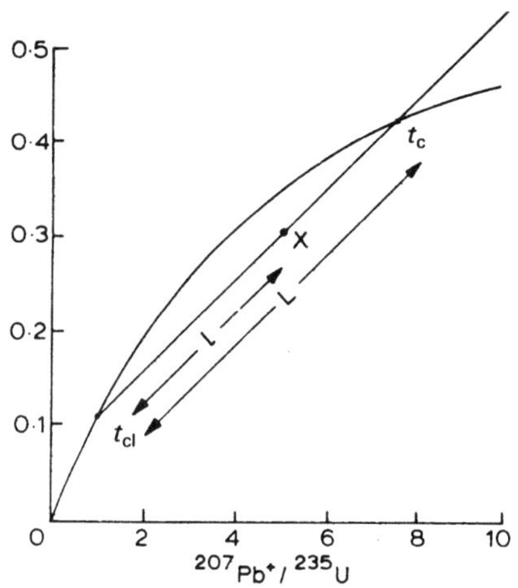

Fig. 4.4 A concordia diagram showing the effects of Pb loss and U gain or loss in U–Pb systems; t_c = time elapsed since the original crystallization, t_{cl} = time elapsed since closure after Pb loss or U gain.

daughter/parent immediately after the change to the ratio of daughter/parent before the change, i.e.

$$X = \frac{Pb_2U_1}{U_2Pb_1},$$

where Pb_1, Pb_2 are the quantities of radiogenic Pb in the system before and after loss and U_1, U_2 are the quanitities of U before and after alteration, respectively.

In a U–Pb system, if half the radiogenic Pb were lost without change in the amount of U, X would be 0.5 and $e = L/2$ and, if the U quantity doubled without any change in the amount of Pb, X would again be 0.5. Clearly, loss of Pb in the system has the same effect as equal gain of U. Thus any U–Pb system lying on discordia betwen t_c and t_{cl} in Fig. 4.4 could have undergone either loss of Pb or gain of U or some combination of both.

Concordia models have been developed in order to interpret the discordance of U–Pb dates of U-bearing minerals. The dilatancy model of Goldich and Mudrey (1972) was one. Minerals undergo radiation damage through the alpha decay of U, Th and their daughters, the extent of this increasing with increasing age and also with the rising U and Th contents of the minerals. The phenomenon has been related to a diffusion parameter D/a^2, where D is the diffusion coefficient and a the effective radius of the sphere. Such damage could produce microcapillaries allowing water to enter crystals, this being retained until uplift and erosion release the pressure on the mineral. Not only does water escape when this occurs, but zircons dilate and lose dissolved radiogenic Pb as well. This might relate to the uplifting and erosion of Precambrain shields' crystalline basement complexes and entail what may be regarded as a relatively recent lead loss in geological terms. This would accord with the view that 207–206 dates often approximate the true ages of U-bearing minerals. Tilton (1960) noted that U minerals from a number of continents lost radiogenic lead 500 Ma to 600 Ma ago, but no global metamorphic episode had been recognized at that time. However, one now is, namely the Pan-African Event which produced extensive occurrences of 500 Ma K–Ar ages in Africa and parallel phenomena in other parts of Gondwana, cf. the Brazilian Event of *c.* 500 Ma ago.

Another model is based on chemical weathering and this affects practically all samples of rocks and minerals from surface outcrops. It was discussed by Stern, Goldich and Newell (1966). Zircons from residual clay formed by the chemical weathering of a gneiss near Redwood Falls, Minnesota, USA, and their dates showed considerable discordancy. These weathered samples may have lost up to 85% of their radiogenic Pb (assuming that the U content did not change). The dates obtained lay on straight line chords commencing from discordia and running to the origin. A third model is a continuous diffusion one, radiogenic Pb having diffused from crystals at a rate governed by the diffusion coefficient D, an effective radius a and the concentration gradient. The crystals are taken as spheres with each having the effective radius and there is presumed to be almost no diffusion of U and intermediates (or, if there is, that the amount is negligible). It is also assumed that the radiogenic Pb diffusion follows Fick's law. The following expression covers the change in Pb concentration with time for any radial volume element in the sphere:

$$\frac{\partial C}{\partial t} = D\frac{\partial^2 C}{\partial r^2} + \frac{2\partial C}{r\partial r} + N_0 e^{-\lambda t},$$

where C is the atom concentration of Pb daughter, N_0 is the initial concentration of U parent atoms, the decay constant of the parent U and t the elapsed time since crystallization. The boundary conditions are:

$$C = O, \quad t = O, \quad \text{all } t, r = a,$$

and a solution was provided by Wasserburg (1954), tables since published giving daughter-to-parent ratios over different values of t and D/a^2 for samples conforming to the diffusion model. Solutions of the diffusion equation generate curves on the concordia diagram which are the loci points representing U–Pb systems of specific ages which underwent continuous Pb loss governed by D/a^2. The curve is practically a straight line in the case of systems with $D/a^2 < 50 \times 10^{-12} a^{-1}$. For greater values, the curve deviates from linearity and approaches the origin.

U–Pb concordia are paralleled by others which involve ^{232}Th in association with either ^{238}U or ^{235}U decay. The concordia curve of ^{208}Pb/^{232}Th and ^{206}Pb/^{238}U ratios is straight because of the similar half-lives of the parent elements. Hence discordia intersections with it are erroneous. However, concordia based on the decay of ^{232}Th to ^{208}Pb and ^{235}U to ^{207}Pb are similar in shape to conventional U–Pb concordia (^{208}Pb/^{232}Th plotted along the ordinate and ^{207}Pb/^{235}U along the abscissa). If no fractionation of either parents or daughters occurs, the response in U–Th–Pb concordias to loss of parents or daughters is analogous to that of U–Pb concordias. However, where the alteration of a U–Th–Pb system involves a preferential loss or gain of parents or disproportionate losses of ^{208}Pb compared with ^{207}Pb or ^{206}Pb, there will be a different response. As U and Th have different geochemical properties, changes in the ratios between them are probable where episodic alteration or continuous loss or gain of parents takes place. Also, ^{208}Pb, ^{207}Pb and ^{206}Pb may be lost at slightly different rates. Such fractionation would induce discordant U–Th–Pb systems to lie on discordia not passing through the origin. But such lower intersections of resultant discordias with U–Th–Pb concordias are insignificant in terms of geological time and merely record change which occurred in the Th/U ratios or Pb isotope ratios through factors alluded to above. As regards U–Pb, Th–Pb and Pb–Pb isochrons, the decay of ^{238}U to Pb in closed systems may be expressed as follows:

$$\frac{^{206}\text{Pb}}{^{204}\text{Pb}} = \left(\frac{^{206}\text{Pb}}{^{204}\text{Pb}}\right)_{\text{i}} + \frac{^{238}\text{U}}{^{204}\text{Pb}}\,\text{e}^{\lambda t - 1},$$

where the ratio with the subscript i is the initial one which entered the system when it formed *t* years ago. Therefore an isochron can be plotted in coordinates of ^{206}Pb/^{204}Pb and ^{238}U/^{208}Pb. Similar equations are applicable to ^{235}U/^{207}Pb and ^{232}Th/^{208}Pb. Three separate isochrons are obtainable. Their slopes show the age of the suite of sample examined if these are derived from closed systems possessing identical initial Pb isotope ratios. U–Pb and Th–Pb isochrons have been used to date samples of granite in the USA, the former USSR and Australia. U–Pb isochrons are not optimal because chemical weathering causes loss of much U from exposed rocks. However, since Th and Pb are retained preferentially, their isochrons are feasible. Where U loss has not occurred, closed U–Pb systems lie on the concordia curve and, if Pb is lost or U is gained, they move towards the origin along a straight-line discordia. If U is lost, movement takes place away from the origin. So rocks which lost varying amounts of U through chemical weathering should lie above concordia and form a straight line passing through the origin if the loss is fairly recent. The age of such rocks is shown by the point of intersection on the concordia curve.

Turning to common Pb dating, common Pb has an isotopic composition which varies with time, a fact which may be used to calculate approximate ages for ore leads. These were called galena ages or conventional ages. The variations result from the radioactive decay of U and Th, entailing a continuous enrichment of the original terrestrial Pb in ^{206}Pb, ^{207}Pb and ^{208}Pb with respect to ^{204}Pb during the geological evolution of the Earth. It may be seen that the ^{206}Pb/^{204}Pb, ^{207}Pb/^{204}Pb and ^{208}Pb/^{204}Pb ratios gradually increase with the decreasing age of the Pb minerals. The radiogenic Pb isotopes are mixed with and added to ore leads through both endogenic and exogenic processes. And at the time when the Pb ores formed, the isotopic composition of their Pb was fixed. It may be inferred that the isotopic composition provides an approximate age for the Pb deposit.

Common Pb occurs in many minerals, as a trace element in minerals with low U and Th contents and in rock-forming minerals such as feldspars (in K feldspar, Pb^{2+} replaces K^{+}). Aston determined the composition of ordinary Pb in 1927 at which time it was already known that its atomic weight was less than that of common Pb associated with U.

The difference was attributed to the presence of radiogenic ^{206}Pb. The content of ^{206}Pb, ^{207}Pb and ^{208}Pb in common Pb may be related to that of ^{204}Pb and the atom number ratio can be expressed as follows:

$$\alpha = \frac{^{206}\mathrm{PB}}{^{204}\mathrm{Pb}} \quad \beta = \frac{^{207}\mathrm{Pb}}{^{204}\mathrm{Pb}} \quad \gamma = \frac{^{208}\mathrm{Pb}}{^{204}\mathrm{Pb}} \,.$$

Every Pb contains a measurable amount of ^{204}Pb and may be represented in a three-dimensional diagram of the above variables α, β and γ. The ^{204}Pb content may be regarded as a measure of the primitiveness of the Pb, not that this means that it can be used as a broad index of age (this is not permissible because geographical trends in the isotopic composition of common Pb may occur). But the older the Pb, the greater the ^{204}Pb content may be, provided that no contamination took place. Houtermans and Geiss derived equations of dating normal galena leads and these give the amount of ^{206}Pb and ^{207}Pb present at the time of formation, p, of a Pb minerals as a function of p. They are:

$$\alpha = \alpha_w + \mu(e^{\lambda_1 w} - e^{\lambda_1 p}) \,, \tag{4.4}$$

$$\beta = \beta_w + \frac{\mu}{137.8}(e^{\lambda_2 w} - e^{\lambda_2 p}) \tag{4.5}$$

and, for ^{208}Pb:

$$\gamma = \gamma_w + k\,\mu\,(e^{\lambda_3 w} - e^{\lambda_3 p}) \,, \tag{4.6}$$

where α is the present-day ratio of ^{206}Pb to ^{204}Pb; β is the present-day ^{207}Pb/^{204}Pb ratio and k is the present-day ^{232}Th/^{238}U ratio in the parent rock of the Pb ore; μ is the present-day ^{238}U/^{204}Pb ratio; w is the time elapsed since the formation of the lithosphere; λ_1, λ_2 and λ_3 are the radioactive decay constants of ^{238}U, ^{235}U and ^{232}Th, respectively; α_w, β_w and γ_w are the Pb isotope ratios in original, i.e. primaeval, Pb and the constant 1/137.8 was taken as the present day ^{235}U/^{238}U ratio (amended value now 137.88), $\in$.

In early researches, Nier had discovered large variations in the isotopic composition of common Pb of different origins. So, by 1939 it was realized that the view that common Pb has a constant atomic weight is untenable, and also that the increase in ^{206}Pb/^{204}Pb ratios is often accompanied by comparable increases in ^{208}Pb/^{204}Pb ratios. In the 1950s the common Pb method was based on two approaches. In the first, the regional crustal variations in the abundances of U, Th and Pb were discarded and a constant Th/U/Pb ratio influenced solely by radioactive decay was assumed for the ore-making parental material. Lead isotope ratios were considered as universal age indexes. However, they are subject to errors due to regional variations. In the second approach, such regional variations were taken into account and Pb isotope ages and U/Pb and Th/Pb ratios were calculated for the source rocks or magmas of the Pb minerals. It was assumed that at some time the isotope composition of Pb was constant, while the U/Th ratio fluctuated in the crust and produced rock melts with differing U/Pb ratios. Hence, the radiogenic Pb isotopes were added later, at different rates, to the primaeval Pb and the $^{207}Pb/^{206}Pb$ ratio was taken to calculate the time of removal of the Pb from contact with U. This method corresponded with the dating of radiogenic Pb in radioactive minerals from the $^{207+}Pb/^{206+}Pb$ ratio.

(α, β) diagrams for constant μ show systems of straight lines of Pb evolution for chemically unaltered U/Pb ratios. They demonstrate how p and μ depend on α and β. Dividing equation (4.5) by (4.4) gives the following:

$$\frac{\beta - \beta_w}{\alpha - \alpha_w} = \frac{1}{137.88} \quad \frac{e^{\lambda_2 w} - e^{\lambda_2 p}}{e^{\lambda_1 w} - e^{\lambda_1 p}},$$

and this is the equation of isochrons, i.e. of a number of straight lines with constant p, which intersect at a point (α_w, β_w) corresponding to the isotopic composition of original, i.e. primaeval, Pb at the time w. Such isochrons can be presented graphically, as in Fig. 4.5.

Efforts have been made to construct quantitative models for the isotopic evolution of Pb in the Earth from which both the age of the planet and that of common Pb minerals could be determined. That of Holmes (1946) and Houtermans (1946) was based on several assumptions. The

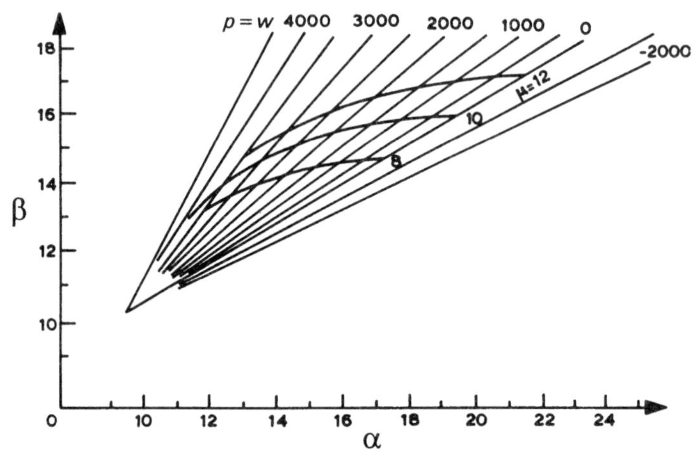

Fig. 4.5 The Holmes–Houtermans model for common Pb (Berne data). The straight lines are isochrons, the curved ones are Pb development lines. The relevant constants used were as follows: $\alpha_w = 9.41$, $\beta_w = 10.27$, $\gamma_w = 29.2$, $w = 4.49 \times 1\,\text{Ga}$, $\epsilon = 1/139$, $\lambda_1 = 1.54 \times 10^{-4}$ Ma^{-1}, $\lambda_2 = 9.80 \times 10^{-4}\,\text{Ma}^{-1}$.

Earth was supposed to have been originally both fluid and homogeneous with U, Th and Pb distributed uniformly with the isotopic composition of this primaeval Pb the same everywhere. Later, the Earth became rigid and small regional differences in the U/Pb ratio arose. In any given area, this ratio changed only through the effect of radioactive decay from U to Pb. Finally, at the time of formation of a common Pb mineral, the Pb was separated from U and Th and its isotopic composition remained constant ever since.

The Holmes–Houtermans model permitted the understanding of differences in α, β and γ for various types of common Pb and also allowed the calculation of the time elapsed since the formation of the Earth's crust together with the time of formation of ores embodying common Pb (at least in some cases). Most accessible Pb comes from igneous and sedimentary rocks in amounts from a few tenths to *c.* 50 ppm. Only a small proportion of crustal Pb is contained in minerals or ores containing more than 0.1% of it (this is also true of U and Th). Hence the incorporation of Pb into a Pb mineral is an extremely rare event and the probability of it recurring in the history of a given sample of Pb is negligible.

In the source rock from which a given sample of common Pb formed by mineralization or ore formation processes, Pb was associated with given quantities of U and Th. It is possible to describe this by means of a chemical milieu index μ, i.e. the atom number of $^{238}U/^{204}Pb$ alluded to earlier, referring to the milieu from which the common Pb sample derived. As the ^{238}U is subject to radioactive decay, its content may be extrapolated to the present as if the mother rock were still in existence. The chemical milieu index μ may be defined as the above ratio. A similar argument is applicable to defining another chemical milieu index of the mother rock of a given sample of common Pb by the ratio of the number of atoms of ^{232}Th to that of ^{238}U extrapolated to the present day in the mother rock from which the sample of common Pb was separated. Designating this index by x:

$$x = (^{232}\mathrm{Th})_{\mathrm{today}}/(^{238}\mathrm{U})_{\mathrm{today}}\,.$$

The Holmes–Houtermans model assumed that all common Pb had an isotopic composition explicable by the addition of radiogenic Pb, i.e. products of the radioactive decay of ^{238}U, ^{235}U and ^{232}Th associated with primaeval Pb when the Earth's crust formed. This primaeval Pb had the relative isotope abundances α_w, β_w and γ_w where w is the age of the lithosphere, as discussed above. The chemical milieu indexes for a given sample of common Pb were taken to have been constant from w to p, the latter being the time of formation of the sample which can be regarded as a model age. The first value of w was *c.* 3×10^9 a.

Impetus to research came when the isotopic composition of minute amounts of Pb from iron meteorites, particularly the iron and troilite phases of Cañon Diablo, were determined. Values of α and β were much lower than those from samples of terrestrial Pb or Pb from stony meteorites. The troilite phase contained Pb, but was almost free of U and Th and the isotopic composition of the Pb was regarded as practically constant since crystallization. Such lead is the least radiogenic available and may approach the isotopic composition of primaeval Pb. If appropriate values of α_w and β_w are assumed and taking into account the isotopic compositions of several relatively recent leads and those from young leads in oceanic manganese nodules, w can be calculated as (4.5Ga $\pm$ 0.05Ga). The Holmes–Houtermans model is shown in Fig. 4.5. In this, α_w, β_w represents primaeval Pb and from this point emerge several curved lines related to given values of μ, these are Pb development lines. The straight lines are isochrons for various values of w, the inclination of each line characterizing equal age, which implies that all common leads of equal age (independent of the chemical milieu index) fall on a straight line. If different values for p are selected, the isochrons diverge up to $p = 0$, this latter being common Pb formed at present. Hence, p is referred to as the model age of the sample where the development of Pb followed the Holmes–Houtermans model, assuming that the common Pb involved remained in a closed system from the time of formation of the lithosphere until it became incorporated in a practically U- and Th-free mineral at the time p. The Pb growth lines in Fig. 4.5 are for three values of μ and constitute trajectories emanating from primaeval lead. The locations of points on such curves indicate times when the relevant leads were removed from their sources to be deposited in the Earth's crust as common Pb minerals such as galena. Many common leads of ore deposits apparently evolved in systems with μ-values between 8 and 10.

The Holmes–Houtermans model assumes that the parent matter had a constant value for the chemical milieu index from time w to time p. Measured α, β and γ-values for a given sample allow its model age (p), its chemical milieu index (μ) and the value x to be calculated. In many samples of common Pb there apears to be a fairly uniform cluster of μ-values of the parent rock. This and other models for the isotopic change with time from the formation of the lithosphere to a time p of mineralization necessitate the hypothesis that the Pb spent the whole time interval ($w - p$) in a chemical milieu characterized by a constant value of μ. Unexpectedly, phenomena such as metamorphism which occurred frequently in Earth history do not seem to have influenced the μ-values of the mother material of Pb minerals to any great extent. Also, there is only a small dispersion in the values of x.

Serious objections to the Holmes–Houtermans model arose from anomalous leads. Common leads giving meaningful model ages are ordinary as opposed to those yielding meaningless values. The number of ore deposits containing such ordinary Pb is smaller, than might be expected because Pb in most such deposits probably underwent a geological history more complex than that envisaged by the model. To conform with this model, the Pb must have had a single-stage history which is confirmable if the model dates accord reasonably well with isotopic ages derived using other minerals from an ore and also if the isotopic ratios of Pb from a particular deposit are constant within experimental error.

The basic assumption in interpreting the development of the isotopic composition of Pb from an ore is the additon of the radiogenic isotopes ^{206}Pb, ^{207}Pb, and ^{208}Pb from the radioactive decay of U and Th with which it is associated through time, these being characterized by the values of μ and $x\mu$, which were constant during the time interval $(w-p)$. The calculated time p often does not coincide with the time of mineralization, i.e. the geological age, of the Pb mineral. Even early measurements on common Pb by Nier showed this. It became clear from fossils in beds containing veins of Pb ores that the model age differed from the palaeontological age given by them. In some cases, absurd results were obtained which led to negative ages so that common Pb ores actually in existence now would have to form in the distant future, say 2 Ga from the present day.

Eberhardt, Geiss and Houtermans (1955) described an anomalous Recent Pb with ^{210}Pb in it found in cotunnite and othe volcanic sublimates from Mount Vesuvius, Italy, considering it to be a new intermediate type designated the Joplin type or J-type. Such J-type leads have model ages lower than their geological ages and the cotunnite Pb, dated by the common Pb method, gave a miximum age of 14 to 17 Ma for the U concentrate, which demonstrated that the dating of extremely young common leads is possible. Its model age was negative, namely -200 Ma. These researchers suggested that this might have resulted from the incorporation of radiogenic Pb when Pb separated from the rock melt. Extrapolating, they proposed that in general J-type leads can be explained either by such incorporation or by an increase in the value of μ in the parent rock. They opined that the existence of large variations in the abundance of Pb isotopes in a Pb deposit may be utilized as a criterion of the anomaly of Pb in such a deposit. J-type leads are also recorded from the Colorado Plateau and extremely anomalous one were detected from Sudbury in Canada. Obviously changes in μ-values invalidate the Homes–Houtermans model, with its requirement that the value of μ remains constant during the time interval $(w-p)$. And of course if common Pb were mixed with a small amount of radiogenic Pb, the resulting isotopic composition would resemble a J-type anomalous Pb. This eventuality could occur in nature at any time during the time interval $(w-p)$ before the U and Th contents of the Pb fell almost to nil. However, a single analysis of an anomalous (J-type) Pb provides practically no information about the date when radiogenic Pb was added during the time interval $(w-p)$ or on the actual isotopic composition, hence the age. To solve this problem, many samples from a region must be analysed. Russell and Farquhar (1960) gave three case histores for three different regions containing Pb ores, two with both common Pb and leads of highly anomalous nature. The regions involved were Sudbury and Thunder Bay, Ontario, and Broken Hill in Australia. These researchers, with others, proposed alternatives to the Holmes–Houtermans model, namely the Russell–Farquhar–Cumming (RFC) and the Russel–Standton–Farquhar (RSF) models, both being based on the work of Alpher and Herman (1951). While the Holmes–Houtermans model assumed a fan of Pb development lines and isochrons commencing from a point (α_w, β_w) corresponding to primaeval Pb (or a similar α,γ diagram), both the RFC and the RSF models assumed a single global source material for all ordinary common leads which may have derived from the upper mantle.

Another anomalous type of Pb is called the Bleiberg type or B-type and this has a model age higher than the geological age. B-type leads were explained as consequences of single or multiple mobilizations of Pb after its original deposition, but without the introducing of noteworthy quantities of young Pb. So, if a Pb mineral formed at a time p, then its model age (p) will give its correct age. However, processes such as pneumatolysis may, through such mobilizations, redistribute Pb minerals in ore bodies or veins and produce a situation in which $g<p$, where g is the geological age. J-type anomalies are named after leads in Joplin (USA) and B-type ones after Bleiberg, Austria.

Houtermans (1953) and Geiss (1954) prepared diagrams for obtaining conventional (model) ages and these were termed Berne graphs, revised sets later being published which showed the changes in the $^{206}Pb/^{204}Pb$, $^{207}Pb/^{204}Pb$ and $^{208}Pb/^{204}Pb$ ratios with time of deposition of Pb ores. It was shown that expressions for the relationship between the isotopic composition of common Pb and time can be derived without using dated samples. The abundance versus time graphs were obtained from old Pb ores and equations given for dating normal common leads from their $^{206}Pb/^{204}Pb$ and $^{208}Pb/^{204}Pb$ ratios. For dating an adequate number of anomalous leads in a given deposit, least-squares analysis gave a value for the $^{207}Pb/^{204}Pb$ ratios which could be used to find the earliest time of addition of radiogenic Pb.

Clearly the conventional age is significant because it is independent of geological changes, changes in crystal structure and chemical alteration. The common Pb method gives the age of the first mineralization. But the conventional ages are reliable only for normal leads. However, if several samples yield the same age in a particular region, this is probably a close approximation to the real age. Reinforcement would be provided if it accords with ages obtained using other dating methods and geological evidence.

A suite of anomalous leads may define an anomalous lead line on $^{207}Pb/^{204}Pb$ versus $^{206}Pb/^{204}Pb$ diagrams and a typical example may have a slope of 0.1 representing the $^{207}Pb/^{206}Pb$ ratio of a radiogenic contaminant which was added in varying quantities to the ordinary Pb to produce a given set of anomalous leads. It is possible to estimate the actual time at which this contaminant was produced by the radioactive decay of U either by calculating the moment in past time when radiogenic Pb with the requisite $^{207}Pb/^{206}Pb$ ratio formed or by assuming that the radiogenic lead accumulated through a time interval terminating when the ore deposit formed. The first approach constitutes an instantaneous growth model and the second a continuous growth one. In the former, the slope of the anomalous Pb line is significant. Presently, the ratio has a value derived from the following:

$$\frac{^{207}\mathrm{Pb}}{^{206}\mathrm{Pb}} = \frac{1}{137.88} \times \frac{9.8485 \times 10^{-10}}{1.551\ 25 \times 10^{-10}},$$

and is therefore 0.04607, a minimum in solar system history because the $^{235}U/^{238}U$ ratio is now lower that it ever was before. At a time t in the past, the value of the ratio may be obtained from

$$r = \left(\frac{^{235}\mathrm{U}}{^{238}\mathrm{U}}\right)_t = \frac{^{235}\mathrm{U}\, e^{\lambda_2 t}}{^{238}\mathrm{U}\, e^{\lambda_1 t}} = \frac{1}{137.88}\frac{e^{\lambda_2 t}}{e^{\lambda_1 t}},$$

and the ratio of radiogenic $^{207}Pb/^{206}Pb$ at the time t is derived as follows:

$$\frac{^{207}\mathrm{Pb}}{^{206}\mathrm{Pb}} = \frac{1}{137.88} \times \frac{e^{(\lambda_2 t)}}{e^{(\lambda_1 t)}} \times \frac{2}{1} = \frac{e^{t(\lambda_2 - \lambda_1)}}{137.88}\frac{\lambda_2}{\lambda_1}.$$

This can be solved applying the relevant slope S thus

$$t = \frac{1}{\lambda_2 - \lambda_1} \ln\left[\frac{137.88\, S \lambda_1}{\lambda_2}\right].$$

However, the second (continuous growth) model is more probable than the first because it is more likely that radiogenic contaminants were generated continuously during the time interval starting with the formation of relevant rocks (such as granites) and ending with the removal of radiogenic Pb from them prior to or in the course of formation of a Pb ore deposit. In that instance, an equation can be used which parallels that of the Holmes–Houtermans model:

$$\frac{^{207}\mathrm{Pb}}{^{206}\mathrm{Pb}} = \frac{1}{137.88}\frac{e^{\lambda_2 t_r} - e^{\lambda_2 t}}{e^{\lambda_1 t_r} - e^{\lambda_1 t}},$$

and, for the analogous Th/U ratio of granite source rocks of a radiogenic contaminant, the $^{208}Pb/^{204}Pb$ ratios may be plotted against the $^{206}Pb/^{204}Pb$ ratios for a suite of anomalous leads and a set of points produced which relate to a straight line with the slope $S = {}^{208}Pb/{}^{206}Pb$, which is the ratio of these isotopes in the radiogenic contaminant. In terms of a continuous growth model, this ratio can be expressed as follows:

$$S = \frac{^{208}\mathrm{Pb}}{^{206}\mathrm{Pb}} = \frac{^{232}\mathrm{Th}}{^{238}\mathrm{U}}\,\frac{e^{\lambda_3 t_r} - e^{\lambda_3 t}}{e^{\lambda_1 t_r} - e^{\lambda_1 t}}\,, \tag{4.7}$$

the values for t_r and t being obtainable from the $^{207}Pb/^{204}Pb$ and $^{206}Pb/^{204}Pb$ ratios of the particular suite being investigated. Hence it is possible to solve equation (4.7) for the ratio $^{232}Th/^{238}U$, from which the Th/U ratio of the granite source rocks can be obtained.

Regarding multistage leads, the general equation of radioactive decay implies that a sample of lead alters its isotopic composition through addition of the isotopes ^{206}Pb, ^{207}Pb and ^{208}Pb from U and Th, with which it has been associated in the mother rock during the time interval $(w - p)$ during which the usual assumptions applied. The following equations may be derived:

$$\alpha - \alpha_w = \mu\epsilon(e^{\lambda_w} - e^{\lambda p})\,,$$

$$\beta - \beta_w = \mu\epsilon(e^{\lambda'_w} - e^{\lambda' p})\,,$$

and

$$\gamma - \gamma_w = \mu\epsilon(e^{\lambda''_w} - e^{\lambda'' p})\,,$$

where ϵ is the ratio $(^{235}\mathrm{U})_{\mathrm{today}}/(^{238}\mathrm{U})_{\mathrm{today}}$.

The first one of these may be cited. Modification of it is necessary where a sample of Pb has undergone transference through say two U-bearing systems characterized by different $^{238}U/^{204}Pb$ ratios designated μ_1 and μ_2. Then the α-value of the sample will be given by

$$\alpha = \alpha_w + \mu_1\,(e^{\lambda_1 \omega}) - (e^{\lambda_1 p_1}) + \mu_2(e^{\lambda_1 p_1}) - (e^{\lambda_1 p_2})\,,$$

Pb may be transferred from one system to the other physically, e.g. through magma formation in the mantle and its later emplacement in the crust. In this way, a Pb sample which was formerly connected with a system in the first enviroment becomes associated with second, the later having a different $^{238}U/^{204}Pb$ ratio. Later, the Pb may recrystallize as galena in an ore deposit. However, the transition from one system to another may occur without any physical removal by changes of the U/Pb ratio of the initial system. Such changes may be induced by loss or gain of U or, alternatively, a fraction of the Pb may be lost. A rarer, but not impossible, possibility is that of Pb gain which will alter the isotopic composition of of the original Pb due to a different isotopic composition of the added Pb. The above considerations refer to a two-stage history, but can be expanded to three or more stages. A two-stage model was discussed by Stacey and Kramers (1975).

The isotopic history of Pb and Sr in basalts and young volcanics from various localities in the Atlantic Ocean initiated this line of investigation (plumbology) in the 1960s. Differences between isotopic ratios of leads in suites of rocks from islands such as Ascension and Gough were paralleled by differences in such ratios in rocks within islands as well. These leads were anomalous and later it was shown that leads, both in continental and oceanic basalts, cannot be explained by a simple single-stage approach. This implied that conformable ore leads do not come from the upper mantle, their uniform Pb growth patterns perhaps stemming from an extensive mixing of leads derived from pelagic sediments subducted under volcanic island-arc systems (Doell and Stacey, 1974). However, volcanic rocks from St Helena plotted far to the right of the geochron and are enriched in ^{206}Pb but depleted in ^{207}Pb. Clearly, such variations in Pb isotope ratios reflect the fact that no common source, but rather diverse ones, were responsible for the lava flows embodying the Pb samples. This suggests heterogeneity in the upper mantle, Pb isotope ratios from oceanic islands forming secondary or higher-order isochrons reflecting multistage histories. Of course, U/Pb and Th/Pb in the upper mantle may have altered either sporadically or continuously through geological time (perhaps in both ways) and rates of change must have been variable too. Another possible explanation for the variability in Pb isotope ratios in volcanic rocks is contamination of single-stage, mantle-derived, leads by varying quantities of Pb coming from granitic continental crustal rocks. But such sialic material is absent from ocean basins, which would rule out this explanation as regards oceanic islands.

Leads from basalts in the Juan de Fuca – Gorda ridge and related seamounts in the north-eastern Pacific Ocean are anomalous and plot along straight lines in Pb development diagrams. These can be interpreted as secondary isochrons, assuming that the leads resided in two systems. Church and Tatsumoto (1975) published isochron dates exceeding 2 Ga for mid-oceanic (MOR) basalts and about 1.75 Ga for seamount samples. They claimed that Pb evolution in the mantle cannot properly be described by a series of closed systems since μ probably varied continuously through geological time. They approximated the history of such leads using five stages entailing a progressive enrichment of the source region in U and Th. This model implies instantaneous changes in the parent/daughter ratios which would be incompatible with the concept of a dynamic Earth suggested by plate tectonics.

Naturally, leads with different isotopic compositions may mix in the mantle, also between the mantle and the crust. Earlier work on the isotopic compositions of Pb and Sr in igneous rocks of Mesozoic and Cenozoic age from the Rocky Mountains, USA, indicated that these orginated from either heterogeneous crustal or mantle sources or were contaminated with Pb and Sr from the granitic crust. Mixing in the mantle or between mantle and crust of leads with different isotopic compositions is always a possibility.

U–Pb systems in young volcanic rocks from the Pacific and the Mid-Atlantic Ridge fit straight line discordia intersecting concordia at ages ranging from 4.5 Ga to 4.6 Ga, paralleling the probable age of the Earth (this being taken as the time interval elapsed since terrestrial Pb had the same isotope ratios as troilite in meteorites). The age of the ridge has been given as 1.23 Ma and the suggestion made that this might accord with a time of change in the U/Pb ratio in the upper mantle. Diversification in this ratio under it may relate to the Late Precambrian, but this assumes that the relevant leads had two-stage histories, and also that the observed ratios resemble those of the source regions of the upper mantle. Where rocks are characterized by U/Pb ratios quite different from those presumed for their sources and in addition where leads involved have passed through a multistage history (i.e. with more than two stages), the concordia would appear to be inapplicable.

Efforts have been made to calculate values for μ_1 from a concordia interpretation of leads in such young volcanic rocks. Where $W = 4.55$ Ga, only slight variations were recorded and an average value of 8.7 ± 0.12 was derived. A second source, μ_2, had a wide range. For the case μ_1, the value of the source region (basalts) was thought to be only slightly lower than that of conformable Pb ore deposits, as shown by their Pb growth curves, which was $\mu = 8.94 \pm 0.13$. Therefore the concordia diagram may provide an alternative means of interpreting the isotopic ratios of lead in young volcanic rocks if applied to two-stage leads only when the U/Pb ratios are close to those in magma sources. This approach gave estimated ages for the age of the Earth which agreed well with those for the age of meteorites.

As regards whole-rock dating, this entails consideration of Pb–Pb isochrons of igneous and metamorphic rocks. If a volume of magma comprising homogeneous material crystallizes, a rock suite is produced which contains different U/ Pb and Th/ Pb ratios. Or perhaps a volcanic rock assemblage may contain Pb isotopically homogenized after a high-grade regional metamorphic episode. Either way, the Pb later evolves along a set of divergent and curved trajectories corresponding to different values of μ in each sample (Fig. 4.5). Where growth continued uninterrupted until the present, the leads lie along an isochron (taking into account the usual assumptions). Recent changes in the concentrations of these elements are permissible if the isotope composition of the Pb remains unaffected. An equation for such an isochron may be derived with a slope given by

$$S = \frac{1}{137.88}\left(\frac{e^{\lambda_2 t} - 1}{e^{\lambda_1 t} - 1}\right),$$

where S is the slope. The age of the sample can be obtained from this using a 207–206 versus Ma graph.

The technique outlined by Sobotovich (1961) was applied to a suite of metamorphic rocks from Lewisian basement in north-west Scotland, this being depleted in U compared with the crustal average. Uranium and Pb concentrations were assessed as 0.24 ppm and 7.9 ppm, respectively, and the average $^{238}U/^{204}Pb$ ratio as 1.76. The isotope ratios of Pb were scattered about a straight line interpreted as a secondary isochron. Instead of using this to obtain the age of the samples, a date was found from the point of intersection of the isochron with the primary lead growth curve (which had a value of μ of 8.68) near to, but slightly lower than, the value for conformable Pb ores. The result was *c.* 2.9 Ga, taken to be the time of variable U loss during pyroxene-granulite metamorphism of the ancestral Lewisian rocks.

Not all linear patterns of isotope ratios on lead evolution diagrams are isochrons because they may be caused by mixing of leads of different isotopic composition in the course of petrogenesis. Moorbath and Welke (1969) illustrated this using the igneous rocks of the Isle of Skye in north-west Scotland. The region is particularly interesting since it includes a wide spectrum of igneous rocks, both intrusive and extrusive, of Tertiary age. Regarding their petrogenesis, controversy arose with regard to the origin of the felsic rocks, which could have formed either by fractional crystallization of basic magma or by partial melting of underlying Lewisian basement. The initial $^{87}Sr/^{86}Sr$ isotope ratio of the felsic and granitic rocks were consistently higher than those of the basement rocks so it was concluded that the granitic rocks resulted, at least partly, from partial melting of the antique Lewisian gneiss (shallow in occurrence here). It was shown later that the isotope ratios of lead extracted from basic and acidic rocks of Skye (after correction for *in situ* decay after formation *c.* 60 Ma ago) fit a straight line on the Pb evolution diagram. This constitutes a chord intersecting a primary Pb growth curve (μ 8.92) at two points which correspond to a mixture of two times, i.e. *c.* 3.1 Ga and 60 Ma ago. The following interpretation was offered. The chord is a mixing line derived from the Lewisian basement complex with single-stage Pb which evolved in the upper mantle until incorporated into basaltic magma approximately 60 Ma ago. This agreed with an earlier interpretation of the Sr data. Consequently, the isotope compositions of both Sr and Pb may well imply that the igneous rocks of Skye contain significant amounts of both these elements derived from the Lewisian basement complex, and also that at least some of the granitic magma may have originated in partial melting of this basement complex.

Turning to Pb in feldspars, igneous and metamorphic rocks contain these minerals enriched in Pb and depleted in U and Th. Concentrations of Pb range up to several hundred ppm compared with (usually) $<$ 1 ppm for U and Th. So the isotopic composition of Pb in feldspars is practically constant for very long time intervals and thus maintains a record of the Pb originally incorporated into the rocks concerned. It is possible to employ the isotope composition of Pb in feldspar to date them by the common Pb method and this will provide information on the petrogenesis of the relevant rocks. Some assumptions are necessary. The Pb must have had a single-stage history and this should include evolution in a closed U, Th–Pb system prior to its incorporation into the feldspar. In addition, the Pb must not be contaminated with other leads of different isotope compositions while the magma is forming, being intruded or crystallizing. Finally, the feldspars must themselves remain closed to Pb and have sufficiently large Pb/U and Pb/Th ratios to ensure that the isotope composition is practically constant.

These requirements are not always met. Sometimes the model Pb dates of feldspars are greater or less than those obtained using other methods and such discrepancies imply the presence of excess radiogenic Pb acquired during a multistage history or through contamination before the feldspars crystallized. Alternatively, radiogenic Pb may become incorporated into feldspars during episodes of thermal metamorphism taking place after crystallization. Doe and Hart (1963) stated that centimetre-sized crystals of K feldspar from the Idaho Springs Formation, Colorado, USA, scavenged radiogenic Pb during contact metamorphism caused by the intrusion of the Eldora stock. The effect was noticeable as far as 300 m from the contact, but there was no alteration in the isotope composition of Sr in the same feldspars at distances exceeding 15 m from the contact. It was inferred that Pb in feldspars from metamorphosed igneous rocks may undergo contamination with varying quantities of radiogenic Pb arising from associated U-bearing minerals. This led to doubt about the reliability of Pb isotope compositions of feldspars from any rock containing zircons or other U-bearing minerals which provide discordant dates as a result of the loss of radiogenic Pb.

4.3 APPLICATIONS

Intersting advances were made during the 1980s and have been discussed in detail in a previous book (Bowen, 1988). Zircons were studied in tonalitic Amitsoq gneiss in the Godthåb district, South West Greenland, and from the Isua supracrustal belt, West Greenland, using

SHRIMP, following early work on them. Several stages of crystal growth were examined and measurements showed that componenets of magmatic zircon from the original protolith are at least 3.822 Ga old. This was from the Amitsoq gneiss, which is apparently significantly older than the felsic volcanic unit of the Isua supracrustal belt as well as older than all ages previously determined by mineral and whole-rock isotope techniques for Amitsoq gneisses at Godthåb and at Isukasia. Overgrowths of younger zircons formed at *c.* 3.63 Ga coinciding with an episode of major Pb loss from older grains. Clearly, the previous whole-rock isochron ages refer to later isotopic redistribution and/or that other phases of Amitsoq gneisses intruded during this later period and dominate the isotope systematics. The Amitsoq zircons were found to be very low in intrinsic common Pb regardless of their U contents. Pre-burning was used in order to minimize the contribution of surface Pb to which low-U spots were particularly sensitive. The worst case was spot 11 − 5 with a $^{206}Pb/^{204}Pb$ ratio of ~ 500 corresponding to 4.2% of the total ^{206}Pb being common. But the work of Baadsgaard *et al.* (1984) on U–Th, Pb dates on zircons from the same region proved to be comparable with the average $^{206}Pb/^{204}Pb$ of the ion probe analyses. The Pb/U ratios were determined using the standard zircon SL and the observed correlation between UO/U and Pb/U in the secondary ion spectrum (now approximated by a quadratic function). Ion probe data revealed a more complex pattern of isotopic compositions than that indicated by orthodox zircon data such as those of Baadsgaard *et al.*

Perhaps the contrasting responses of different grains to the same event relates to their position in the rock fabric. However, Th–Pb isotope data did not evidence any differential movement of Th with respect to U so the present low value of Th and U should be close to the original levels. Hence only minor structural damage resulted from *in situ* decay of actinide elements in the 200 Ma time interval between the formation and recrystallization processes, and also that *c.* 3.6 Ga ago radiogenic Pb was flushed out of the essentially pristine zircon structures. A clue as to how this may have happened may be gleaned from secondary fluid inclusions in the most disturbed grains. They appear to be remnants of fracture planes which, if present when metamorphism occured, would have effectively increased the surface area of the grains and also facilitated the wholesale leaching of radiogenic Pb from the zircons through the metamorphic fluid phase.

Baadsgaard *et al.* (1984) examined the zircon geochronology of the Akilia association and Isua supracrustal belt, both regarded as deriving from a single volcanosedimentary sequence because they occupy the same position in local stratigraphic sequences and also show a similar association of lithologies as well as being intruded by geologically and geochronologically correlated orthogenesis. Zircons from Akilia gave a concordia intersection age of *c.* 3.6 Ga and from the Isua belt of *c.* 3.8 Ga, the latter approaching the original crystallization ages of the zircons. The former was ascribed to the resetting of the zircon date by granulite facies metamorphism about 2.6 Ga ago.

Compston *et al.* (1985) used SHRIMP to investigate zircon xenocrysts from western Australia. Samples were taken from the Kambalda volcanics in order to review age constraints and direct evidence for older continental crust below the Kambalda–Norseman greenstones. The Hanging Wall basalt at Kambalda contains zircons which ion microprobe analysis showed to have a very high content of U and Th together with a wide variety of crystallization ages. Most are probably xenocrysts, but a few may relate to intrusive veinlets. The age of the youngest was given as *c.* 2.7 Ga which, it was claimed, demonstrated that the eruptive age of the basalt cannot exceed 2.743 Ga and confirmed that the apparent Sm–Nd isochron giving 3 Ga for Kimbalda mafic and ultramafic rocks is a mixing line between unrelated components enriched and depleted in light rare earth elements. Mixing may have taken place at depth by erosion of 3.2 Ga to 2.5 Ga old felsic crust from the walls of the Hanging Wall basalt conduits. The zircon xenocrysts' ages were taken to represent the first direct evidence for the presence of very old felsic crust in the eastern Yilgara Block. This implies that the Kalgoorlie–Norseman greenstone sequences formed in a continental rather than in an oceanic environment.

Schärer, Rong-Hua and Allègre (1986) published data on zircon fractions and single grains together with fractions of monazite from Tibet in regard to the ages of Himalayan leucogranites. These occur in the High and Tethys Himalayas and can be distinguished from other Himalayan plutons such as the calc-alkaline Gangdese belt. They have a distinct emplacement context showing a strong relationship with both high-grade metamorphism and large-scale thrusting. The favoured model orginates the granites through crustal melting during intraplate subduction within the northern border of India, reacting to continuing post–collisional northward drift of the subcontinent into Eurasia. The exact ages of some of them are not known because Rb–Sr dating is uncertain since there is a lack of equilibrium in this isotopic system at whole-rock scales. Also, even fairly good linear arrays of whole-rock samples may be fortuitous isochrons defining quite meaningless geological ages. A good example of this is given by the Makalu granite for which precise U–Pb ages of *c.* 24 and 21.9 Ma showed that a 92.7 ± 9.4 Ma isochron from the same granite represented an accidental linear array.

Rb–Sr and K–Ar dating of minerals from various granites and country gneisses in the Himalayas define ages between 17 and 1 Ma, and most have been interpreted as cooling ages, thus fixing mineral-dependent minimum ages for granite emplacement and regional metamorphism. As cooling ages are probably cooling rate-dependent, this could result in age differences of up to several million years for contrasting cooling histories of the same mineral. This constitutes a potential handicap for precise dating of Cenozoic rocks using Rb–Sr and K–Ar analyses on minerals. Hence Schärer, Rong-Hua and Allègre (1986) selected U–Pb dating, asserting that this geochronometric approach better approximates the time of primary rock crystallization due to the much higher blocking temperatures of this isotope system in accessory phases.

Exact rock ages are required in order to establish that the plutonism and related high-grade metamorphism really occurred in post-collisional times *c.* 40 Ma ago, and also to determine time constraints on the thermal regime during the Himalayan orogeny (important in modelling anatectic granite genesis). They are also necessary to facilitate a temporal–spatial reconstruction of the thrust events (break-off) in the Indian margin, i.e. the formation of the Himalayas. Another aim was to investigate the nature and age of the relevant magma sources as traced by inherited radiogenic Pb in zircons, and also to compare data with U–Pb systematics in country gneisses (the possible source rocks of the granites).

The analytical technique used was to dissolve zircons in 50% HF and monazites in 33% HCL in Teflon bombs for about 100 h at 200°C. In the case of Th–U analyses of whole-rock powders, the method used was described by Allègre and Condomines (1976). Manhes, Minster and Allègre (1978) gave the chemical procedure for U–Pb dating. For common Pb isotope measurements in K feldspars, the leaching technique of Gariépy, Allègre and Xu (1985) is applicable. For the leucogranites, isotope compositions and concentrations were determined using Thomson mass spectrometers with thermal ionization sources and mixed spikes of ^{205}Pb–^{235}U for mineral dating and ^{230}Th–^{235}U for Th–U concentrations of rock powders. Except for Pb isotope compositions in K feldspars, all measurements were carried out on secondary electron multipliers. The Pb isotope compositions were obtained using Faraday cup collectors. Correction for incorporated common Pb in zircon and monazite was effected with the isotope compositions of the element in K feldspars from the same sample. The concentration and isotope composition of blank lead were evaluated from a set of *c.* 20 independent measurements giving 30 ± 15 pg of blank Pb with the following isotope composition: ^{206}Pb/^{204}Pb, 17.98 ± 0.01; ^{207}Pb/^{204}Pb, 15.45 ± 0.10; ^{208}Pb/^{204}Pb, 37.28 ± 0.20. For lead determinations in the K feldspar and Th–U in whole rocks, the blanks proved to be negligible.

Schärer, Rong-Hua and Allègre (1986) referred to special problems with U–Pb dating, mentioning two significant constraints which must be taken into account where mineral samples younger than *c.* 100 Ma are involved. First, they indicated that the trajectories of Pb loss are virtually coincident with the concordia curve, which is practically a straight line for the segment zero to *c.* 100 Ma. A result is that all data give concordant ages, thus showing a closed system type of behaviour even for perturbed minerals. Another disadvantage of the coincidence is that it precludes determining ages by interception of the concordia curve with regression lines through discordant data (upper and lower intercept ages). Repeated analyses of different fractions of the same mineral are required to show whether or not it really acted as a closed system with respect to U, Th and Pb. Second, they pointed out that Th-rich minerals (such as monazite) must be corrected for excess ^{206}Pb originating from original excess ^{230}Th. Such an excess arises from Th enrichment in the mineral relative to the Th/U ratio of the magma so that the excess ^{206}Pb does not come from ^{238}U and cause a secular disequilibrium in the system necessitating a correction for the ^{206}Pb/^{238}U ages. The relevant corection is obtained by determination of the Th/U ratio of the rock, the whole-rock powder being assumed to reflect the Th/U ratio of the magma. Of course, it cannot be obtained if U was leached from the rock after the mineral (say monazite) crystallized. Then the ^{207}Pb/^{206}Pb ages remain negative even after adjustment for the under-correction due to whole-rock Th/U ratios, which are too high because of such leaching of U. In a manner similar to ^{230}Th for the ^{238}U radioactive decay series, ^{231}Pa may produce a small excess of ^{207}Pb in the series; the amount of such an initial ^{231}Pa disequilibrium can only be guessed at. Thus direct measurement in rocks older than *c.* 1 Ma is not feasible because of the lack of a stable Pb isotope. In the case of the leucogranite monazites analysed, the possible ^{207}Pb excess was regarded as insignificant because the extent of initial Pa enrichment was taken to be much lower than that for Th.

The zircons selected for analysis were small, needle-like, essentially newly grown crystals having little inheritance, the larger and short prismatic zircons containing on average more inherited radiogenic Pb. They were transparent and colourless, indicating low U concentrations with a lesser quantity of post-crystallization Pb loss and lower contents of common Pb. Highly translucent, dark-coloured, inclusion-bearing or altered grains were avoided. The monazite fractions were chosen solely on the basis of grain size. The zircon and monazite data from the leucogranites were similar. All zircons contained components of inherited radiogenic Pb (including the needle-like ones) with no microscopically distinguishable cores. The inherited components originated from the source rocks and confirm the anatectic source of all the granites and migmatites. The inherited Pb was regarded as showing that the material in the magma source regions is rather old (apparent minimum ages between 400 Ma and 2.2 Ga). As the individual zircon analyses (apart from a single grain) were grain fractions, each point on the diagrams presented by these researchers could in fact represent a mixture of different zircon generations and the minimum ages could be considered as wide estimates for the age of the source rocks. However, the inherited Pb indicates that Precambrian material may be a major component in the magma sources. Crustal residence times of such length for the source materials are consistent with the highly radiogenic isotope composition of Pb in K feldspars revealing $^{207}Pb/^{204}Pb$ ratios up to 16.04. Neodymium model ages from some of these granites suggest mean source ages of the order of 1.1 G.a to 2.3 Ga and accord with ages inferred from inherited Pb in zircons. On the other hand, the monazites showed no inherited components which allowed useful inferences to be made regarding granite crystallization. The ages obtained were 15.1 ± 0.5 Ma for the Lhagoi Kangri granite of the Tethys Himalaya belt and 9.8 ± 0.7 and 9.2 ± 0.9 Ma, respectively, for two varieties of the Maja granite. For the High Himalayan granites, the monazites defined ages of 14.3 ± 0.6 Ma for a granite near Mount Everest and 16.8 ± 0.6 Ma for a migmatitic granite north of Nialam. The age for the Maja granite seems to be compatible with, if slightly older than, preliminary $^{39}Ar/^{40}Ar$ ages obtained from minerals from the same granite. For biotite an age of 7.2 ± 0.6 Ma was obtained and for muscovite an age of 7.2 ± 0.4 Ma was obtained.

From the Th–U data from whole rocks, concentrations of these elements in the relevant powders imply, if compared with the ratio in monazites, that U was leached from the rock, thus falsifying the initial Th/U ratio of the magma from which the minerals crystallized. Since most of the $^{207}Pb/^{206}Pb$ ages of monazites remain negative even after correction for excess ^{206}Pb, it appears that leaching of U is a widespread phenomenon in the Himalayas. Curiously, a migmatitic granite showed an uncommonly high Th concentration (*c.* 37 ppm) implying a late exchange with a very Th-rich fluid which on a priori grounds would seem improbable (U is far more mobile than Th during hydrothermal alteration). Interestingly, a few Th/U ratios may be too low due to the presence of xenotime, a Th-depleted mineral.

Monazite and xenotime are pale green to green transparent yttrium phosphate minerals, the former, a rare earth mineral, being relatively enriched in Th. However, they are hard to distinguish in the course of individual grain hand picking, but monazite can be separated from cogenetic xenotime in the majority of rocks containing these minerals.

A significant implication of the above researches is that, together with previous U–Pb ages of 21.9 ± 0.2 and 24 ± 0.4 Ma for two phases of the High Himalayan Makalu granite, the later ages of 16.8 ± 0.6, 15.1 ± 0.5, 9.8 ± 0.7 and 9.2 ± 0.9 Ma show that the High and Tethys Himalaya leucogranites postdate the collision between the Lhasa Block (Eurasia) and India, which took place approximately 40 Ma ago. The High

Himalaya includes Mt Everest. The inherited radiogenic leads found in all zircons from them substantiate a basically anatectic origin which was confirmed by field observation, e.g. those of Le Fort (1981). Additional evidence also came from isotopic tracers such as Sr, Nd, Pb and O (e.g. Allègre and Ben Othman, 1980; Blattner, Dietrich and Gansser, 1983. This gave crustal signatures and allowed crustal residence times of the magma source materials to be inferred. That sizeable quantities of Precambrian material are included in the source was claimed on the basis of the average minimum ages of the inherited leads in zircons (ranging from *c.* 500 Ma to 2.2 Ga). Also, the U–Pb ages showed that the leucogranites both of the Tethys and High Himalayan belt constitute intrusions which occurred *c.* 14 Ma to 17 Ma ago. This leads to a query, which is why separation of these two is suggested by the actual pattern of outcrops in the relevant regions.

Other matters remain to be clarified, such as whether 22 Ma to 24 Ma old granites, e.g. that at Makalu, were also emplaced in the Tethys Himalayan (Lhagoi Kangri) belt, and whether roughly 9 Ma old granite such as Maja occur also in the High Himalayan range. There is too an open question regarding the precise age relationship between the granite emplacement and high-grade metamorphism. For instance, the *c.* 17 Ma old migmatitic granite north of Nialam (emplaced following the main planar structures of regional deformation) implies that, in some areas, the metamorphism is younger than early (22 Ma to 24 Ma old) granites. But the only 9 Ma old showed that igneous rocks of this type formed after major phases of both tectonic and metamorphic activities took place, e.g. some 13 Ma ago at Kangmar in the east of the Lhagoi Kangri belt. There is field evidence that the alternation of regional tectonometamorphism and granite emplacement is not followed by deformation in the case of the 24 Ma old and 16 Ma old granites in the Tethys Himalaya and High Himalaya belts. The granites cross-cut the dominant Alpine foliations in the country rocks. This implies that the various phases of strong deformation and metamorphism probably occurred in alternation with granite empacements, i.e. before and after discrete pulses of granite formation – otherwise all granites older than the latest deformation would be deformed and metamorphosed in a manner similar to the Palaeozoic Kangmar granite. If Schärer, Rong-Hua and Allègre (1986) were correct, the main phases of regional metamorphism which led to magma formation took place at different times in different segments of the crust brought together by the continuous process of thrusting. It is necessary that granite formation occurs from and within continental crust at depths of less than 20 km at pressures of *c.* 5 to 6 kbar and temperatures of *c.* 650°C, as suggested by mineral paragenesis in country gneisses and granites discussed by Le Fort (1981). The known thickness of the Himalayan units also supports this hypothesis.

Jauspart and Provost (1985) have suggested possible heat sources for anatectic granite formation in the Himalayas. Their model explains melting by heat refraction effects between the buried sediments and the basement–it has the advantage that no heat source apart from *in situ* radioactive decay is necessary.

In summary, the U–Pb ages so far determined suggest pulses of plutonism around 24, 16 and 9 Ma ago and strongly contrasting heat flow measurements and zones with high magnetotelluric conductivity data imply that a fourth phase of plutonism may be active today. It may be that shortening (intracontinental subduction in the Indian margin after collision with Eurasia) occupied four phases separated by time intervals of between 7 ma and 9 Ma. If true, then the large-scale thrustings juxtaposing the various terranes must have been rapid, lasting only a few million years at most between magma generation pulses. Such an episodic alternation of plutonism and thrusting could explain also the means by which granites and high-grade gneisses were brought up to the surface, namely that the initial granite/high-grade gneiss terranes were upthrust during a younger phase which itself promoted metamorphism and granite formation and so on. More precise dating, especially of metamorphism in the country rocks, would facilitate comsolidation of such a model.

Other interesting zircon-based researches have been carried out by many investigators and these are based on the extraordinary properties exhibited by this mineral ($ZrSiO_4$). Thus the zircon U–Pb geochronology of the Zambales and Angat Ophiolites, Luzon, Philippines and evidence for an Eocene arc-back arc pair were discussed by Encarnación, Musaka and Obille (1993). These ophiolitic basement terranes are separated by a *c.* 10 km thick and *c.* 100 km wide sedimentary basin and yielded zircon samples which were examined so as to constrain their origin. Two zircon fractions from tonalite in the Acoje Block of the Zambales ophiolite gave concordant U–Pb ages at 44.2 ± 0.9 Ma. Two zircon fractions from plagio-granite and one fraction from diorite in the Coto Block of the Zambales ophiolite yielded concordant U–Pb ages of 45.1 ± 0.6 Ma. These results showed that this ophiolite is Middle Eocene in age, which agrees with the minimum Late Eocene age given for the overlying Aksitero Formation. No difference in age was observed between the arc-like Acoje Block and the MORB-like Coto Block. Four zircon fractions from two sampling sites in the Angat ophiolite gave concordant ages of 48.1 ± 0.5 Ma, this being much younger than the Late Cretaceous age based on radiolarian fauna obtained from a sheeted dyke-pillow lava sediment sequence south-south-east of the main ophiolite. Such a small difference in age between the Zambales and Angat ophiolites implies a common origin and eliminates the necessity for a major structural discontinuity west of the southern Sierra Madre beneath the central valley of Luzon. The Cretaceous biostratigraphic ages of the ophiolitic rocks south-east of the Eocene Angat ophiolite suggest that there are two ophiolite basements exposed in the southern Sierra Madre. Encarnación, Musaka and Obille (1993) regarded the relationship between the two ophiolites as being constrained by the overlying stratigraphic relations and took this as indicating that an Eocene volcanic arc and associated volcaniclastic apron was constructed on both the Eocene and Cretaceous ophiolite basement. They inferred that the Zambales–Angat ophiolites represent a preserved Eocene back-arc basin that opened behind an Eocene arc which developed within Cretaceous oceanic basement. In their model the Zambales–Angat ophiolites do not comprise allochthonous terranes, but are rather part of a single plate generated *in situ* and forming part of the autochthonous basement of Luzon.

A timescale for producing the geochemical signature of island arc magmas was developed by Gill, Morris and Johnson (1993) using U–Th–Po and Be–B systematics in Recent Papua New Guinea lavas. Twenty samples from 14 Holocene volcanic centres in this region were used and span a number of tectonic settings. Their work was based on the geochemical features of arc magmatism, e.g. abundant andesite or enrichment in Ba and Sr and depletion in Nb and Ta relative to rare earth elements. These features are associated with volcanoes currently underlaid by subducted lithosphere or with areas beneath which subduction took place during the last few million years (Gill, 1981). The above characteristics comprise a geochemical signature of such subduction. Unfortunately, their spatio-temporal scales are inadequately constrained. In addition, enriched mantle relative to the depleted MORB source is believed to exist within the mantle wedge at some convergent plate boundaries quite independent of contemporaneous subduction. Confirmation of this makes it necessary to distinguish between the ages and the processes of enrichment. Gill, Morris and Johnson (1993) examined the fractionation between elements believed to be mobile in subduction zones, of which some have half-lives sufficiently brief to respond to current subduction-related processes. Among these, the ones selected were U-series disequilibrium and $^{230}Th/^{232}Th$, $^{10}Be/^{9}Be$ and B/Be ratios, all of which may be affected by recent subduction. The Holocene lavas erupted mostly in the 19th century in the complex tectonic region of Papua New Guinea, a region including extremely depleted (low K) arc tholeiitic magmas, volcanoes at various distances behind the plate boundary, a time-transgressive arc-continent collision and contrasting geochemistry in lavas on opposite sides of the same subducting plate. Sr and Nd isotope ratios in New Britain volcanic rocks are constant ± 0.0002 and ± 0.00007, respectively, in spite of sizeable differences in the ratios of incompatible trace elements. Gill, Morris and Johnson (1993) claimed that they showed that the isotopic constancy mirrors the insensitivity of Sr and Nd isotopes to recent changes in source composition. Their aproach allowed them to distinguish between effects caused by rapid fluid fluxing in the course of magma genesis against those necessitating storage of the slab signature in the mantle wedge before magma fomation.

The samples from centres of volcanism on and behind the volcanic front in the Eastern Bismarck Arc came from a region of normal subduction. The volcanoes overlie a seismic zone 100 km to 400 km in depth. Samples from the Western Bismarck Arc came from different segments of an eastward-propagating collision zone between Papua New Guinea and the arc. The measured ratios of $^{230}Th/^{232}Th$, $^{10}Be/^{9}Be$ and B/Be in basalts and andesites from all the settings were 1.2–2.9, 2–28 and 8–160, respectively. The Eastern Bismarck Arc rocks are characterized by $\sim$10% excess U and also by positively correlated $^{230}Th/^{232}Th$, $^{10}Be/^{9}Be$ and B/Be ratios. Gill, Morris and Johnson (1993) noted that the positive correlation of $^{10}Be/^{9}Be$ with the Th ratio rather than with $^{238}U/^{230}Th$ ratios, together with the association of high values with larger fluxes of U than Th show that U was added to the arc source and stored there for more than 10^5a before magma genesis. There seems to have been a time lag of hundreds of thousands of years between slab dehydration and melt formation. Lavas from volcanoes up to 230 km above the slab were stated to have the same U-series relationships as those at the volcanic front. Beryllium isotope and B–Be relationships behind the front differ from those at the front but still show slab input. The subduction signature is lost only where the slab is 400 km deep. Lavas from the Western Bismarck Arc have most of the characteristics of those from the Eastern Bismarck volcanic front except that ^{10}Be concentrations decline from low to unmeasurable levels westward in the direction of earlier collision. The ratio of ^{10}Be (half-life 1.5 Ma) to stable B in the lavas is usable to date the collission. Gill, Morris and Johnson (1993) suggested that suturing commenced in the most western segment in the early Pliocene, and within the last 1 Ma at 148°E, which is the location of the present triple junction. Large fluxes of slab-derived elements, including excess ^{238}U which probably developed within the last $\sim$350 Ka, are observed in the collision zone lavas even where the slab now underlying the volcanoes may have had a long residence time (> 3 Ma) in the mantle. The U-series data from the whole Bismarck volcanic front implies $\sim$200 ka between slab dehydration and magma formation and eruption, a marked contrast to a similar set of data from southern Chile where the time elapsed is though to be less than 20 ka.

Cosmogenic ^{10}Be is an effective tracer for sediment recycling contemporaneous with the genesis of magma at volcanic arcs, wheras ^{9}Be in magmas behaves rather like neodymium. Boron is another good tracer for sediment or hydrothermally altered rocks during subduction, mirroring its high concentration in sea water and also its strongly adsorptive behaviour in the course of marine sedimentation. Neither ^{10}Be nor B are fractionated significantly from ^{9}Be during crystal fractionation or partial melting, so that their ratios can be employed to study mantle characteristics caused by the incorporation of a subducted component.

The Pb geochronology of zircon by laser probe-inductively coupled plasma mass spectrometry (LC-ICPMS) was described by Feng, Machado and Ludden (1993). Conventional zircon U–Pb geochronology using chemical separation and TIMS was the accepted technique for obtaining accurate ages for the past 20 years or more in which careful selection of zircon crystals, air abrasion, zircon dissolution, chemical separation of U and Pb were succeeded by analysis of these elements using thermal ionization mass spectrometry (TIMS). With low procedural blanks (if these can be successfully attained), it becomes feasible to analyse submicrogram quantities of zircon, e.g. parts of cores, but the technique necessitates very low level blank values in the chemical work and is very time consuming. Alternatively the SHRIMP has been used, but this is less precise and cannot resolve small time differences. In addition, in spite of its ability to measure U–Th–Pb isotopic compositions in zircon at the 20 to 30 μm scale and to produce much data relatively quickly, it is costly to buy and operate. Kröner and Todt (1988) developed a simple zircon Pb evaporation method using a double filament and TIMS so as to obtain ^{207}Pb/^{206}Pb radiogenic ages. Here there are no chemical procedures and hence no blank contamination. Through stepwise evaporation, zircons can be identified as either concordant (^{207}Pb/^{206}Pb ratios uniform over several steps) or discordant (these ratios increase from rim to core). This approach is both quick and economical, but not as precise, though it is an alternative to conventional U–Pb techniques.

Turning to the laser ablation sampling technique utilized with an inductively coupled plasma mass spectrometer (ICPMS), this is a relatively simple and cheap instrument able to effect direct elemental and isotope ratio analysis of solids (e.g. Denoyer, Kenneth and Hager, 1991). Usually a Nd–YAG laser is used and operates in the near-infrared region at 1064 nm and the pit produced by laser ablating the sample is *c.* 100 to 150 μm in diameter. This kind of laser can be modified by reduction of its divergence and attenuation of the beam energy to produce a focused beam which causes pits of 20 to 30 μm diameter. Such a commercially available modified laser is Fisons-VG LaserProbe; with this it is possible to use ICPMS to analyse U–Th–Pb isotopic composition in single zircon crystals which frequently have grain sizes $< 100\ \mu$m (Fryer, Jackson and Longerich, 1993). From such pits, vaporized materials are conveyed by Ar gas to a Fisons-VG PQII +ICPMS for analysis. Feng, Machado and Ludden (1993) evaluated both the accuracy and the limitation of the method using 21 zircon samples with an age range from 2.7 Ga to 1 Ga which had been dated also by conventional U–Pb TIMS. The LP–ICPMS $^{207}Pb/^{206}Pb$ ages for zircons with grain size > 60 μm and ^{207}Pb concentration > 3 ppm were within 1% of the TIMS ages. Smaller zircons and those with ^{207}Pb concentration < 2 ppm gave inaccurate ages. Operation of the LP-ICPMS under conditions offering a compromise between the ideal spatial resolution and instrumental sensitivity showed that the limits of detection were 0.2ppm for ^{206}Pb, ^{207}Pb and ^{208}Pb. The precision of the $^{207}Pb/^{206}Pb$ ratio was usually 0.5 to 6% (1σ). Feng, Machado and Ludden (1993) averred that their results showed that the LP–ICPMS may be applied to determine $^{207}Pb/^{206}Pb$ ages of zircons and that reliable ages are obtainable from high-quality limpid zircons with a simple Pb loss history. They added that using the approach includes screening zircon populations from different rocks in areas having poorly understood age relationships as well as studying the provenance of detrital zircons from ancient and modern sedimentary sequences. Other uses are the examination of growth zones and of inherited components in complex populations of zircons. When compared with SHRIMP, however, they admitted that the LP-ICPMS technique has the disadvantage that it partially destroys the sample and presently is unable to give accurate U/Pb ratios. On the other hand, it has better sensitivity, similar precision but somewhat worse spatial resolution for measuring $^{207}Pb/^{206}Pb$ relative to the SHRIMP.

Krogh *et al.* (1993) cited evidence supportive of the buried impact crater nature of the Chicxulub structure in Mexico and of its being the likely sorce of ejecta found in Cretaceous–Tertiary (K–T) boundary sections all over the world, adding that a remaining query is whether there may be other K–T-aged craters which also emitted such ejecta. Shocked zircons from a K–T ejecta layer in Colorado were claimed to have preserved a primary source age of 545 ± 5Ma, with variable degrees of isotopic resetting consistent with partial Pb loss at the time of impact (*c.* 65 Ma ago). Krogh *et al.* (1993) obtained zircons from the Beloc site in Haiti and also used a 30 g carbonate-rich polymict breccia sample with 20% to 25% clasts comprising 0.2 to 2.0 mm silicate basement rocks from the Y6 drill hole (sample NL3 from a depth of 1100 m to 1103 m) sited within and 50 km south-west of the centre of the Chicxulub structure. Results obtained from these were compared with those for zircons from K–T boundary distal ejecta from the Raton basin in Colorado. It was asserted that zircons in ejecta from Haiti and Colorado ought to comprise a similar mixture, and that grains from all sites should record the primary age(s) of the target basement together with both physical and isotopic changes induced during the short (10^2 s) time interval between the start of the impact and dispersal above the upper atmosphere. Age variations were anticipated because of the large size of the transient cavity (diameter estimated at both 90 km and 170 km) and the fact that even short-lived crustal events may extend over 50 Ma to 100 Ma. Eighteen analyses from 26 carried out gave a primary, i.e. crystallization, age of 544.5 ± 4.7 Ma and only six yielded inconsistent ages. Shocked zircons represent discrete refractory particles of the actual target and gave the U–Pb primary age of their individual source despite substantial loss of Pb in the course of the impact. Collinearity of data points for three sites which span 2000 km showed that there has been no post-impact disturbance and that each site has an identical 545 Ma primary age component. Rocks of this age are rare in North America and this reinforces the view that the presence of shocked zircons of this age in the 65 Ma old ejecta and in the Chicxulub crater can best be accounted for by an impact–ejection process. However, an extra source age of 418 ± 6 Ma found only in Chicxulub and Haiti implies that the Haiti and Colorado ejecta may have sampled different parts of the target stratigraphy. Krogh *et al.* (1993) concluded that their results accord with Chicxulub being the only significant continental K–T impact crater.

Zircon U–Pb ages concordant with Rb–Sr ages were used by Bingen, Demaiffe and Hertogen (1990) in connection with the evolution of feldspars at the amphibolite–granulite facies transition in augen gneisses in south-western Norway. They stated that the Sveconorwegian Augen Orthogneisses of Rogland–Vest-Agder (south-west Norway) were emplaced as amphibole- and biotite-bearing granodiorites at 1.04 Ga (concordant zircon U–Pb and Rb–Sr ages). They underwent prograde metamorphism which increased from lower amphibolite facies in the western zone, close to the Rogaland anorthosite complex. Potassium-feldspar megacrysts first crystallized as phenocrysts and then chemically equilibrated during the metamorphic episode in the course of which there may have been an increase in temperature. Amphibole and biotite inclusions in the megacrysts also equilibrated in the course of metamorphism. Also, groundmass K feldspar and plagioclase underwent late metamorphic changes during uplift. An internal Rb–Sr mineral isochron (plagioclase, apatite, K feldspar) defined an age of 870 Ma (representing the closure of the Rb–Sr isotopic system in minerals of the augen gneiss) which also represents a K feldspar cooling age in regionally distributed augen gneiss samples. The K feldspar cooling age seemed to be similar to or slightly older than the biotite cooling age.

Miller, Goldstein and Langmuir (1994) discussed Ce/Pb and Pb isotope ratios in arc magmas and the enrichment of Pb in the continents. They indicated that Pb is anomalously enriched in the continental crust and in the magmas which create new continental crust at convergent margins. A detailed examination of volcanic rocks from the Aleutian Island Arc demonstrated that this enrichment is a continuing process and is caused by the efficient non-magmatic transfer of mantle-derived

Pb into the source of convergent margin magmas. This process can account also for the pervasive depletion of Pb in the oceanic mantle through time.

Goldstein *et al.* (1994) studied off-axis volcanism at the East Pacific Rise using U-series dating of basalts. $^{238}U/^{230\rightarrow}Th$ and $^{235}U/^{231}Pa$ ages for basalts were utilized in order to quantify the spatial extent of young volcanism and crustal accretion at 9°31′ north of the rise. Most of the ages are younger than would be expected taking into account the off-axis distance and spreading rate. It was inferred that most of the dated basalts on the crestal plateau were erupted 0.5 km to 2 km outside the axial summit caldera and some volcanism occurred as far as 4 km off-axis. Melts errupted outside the axial summit caldera could have crustal residence times and magmatic supply systems differing from those of axial lavas.

Finally, reference may be made to an interesting single age determination using stepwise direct evaporation on a VG Sector 54 mass spectrometer (mass dispersion 54 cm with 30 cm radius electrostatic filter behind a Faraday multiple collector). In 1992 Fisons Instruments, acknowledging Professor Clark Johnson, University of Wisconsin, Madison, issued an application note on this matter, observing that the VG Sector 54 in conjunction with a Daly Multiplier constitutes an excellent instrument for this purpose. Most precise results come from using a double filament assembly where the zircon grain is contained in one filament and stepwise heated so as to evaporate Pb and Si onto an adjacent (facing) filament. The second filament is then heated so as to yield a stable Pb ion beam. The technique gives detailed data on the $^{207}Pb/^{206}Pb$ ages within individual zircons at relatively low expense. The results of analyses on two zircons from a 1.8 Ga old pluton were obtained. They were *c.* 150 μm in length and embedded in 0.030 to 0.060×0.0010 (0.75 to 1.5×0.04 mm) Re filaments. The evaporation of Pb and Si was effected at temperatures between 1600°C and 1900°C for time intervals of 15 min each. Stable ^{206}Pb ion beam intensities of 2×10^{-13} A (2 V on the Daly Multiplier) were obtained for each step and usually 20 to 50 $^{207}Pb/^{206}Pb$ (measured twice), $^{208}Pb/^{204}Pb$ and $^{206}Pb/^{204}Pb$ ratios being derived for each step utilizing single-collector, peak-hopping analyses. The General Analysis Program permitted determination of lead isotope ratios both in ascending and descending mass order so that the effects of beam stability could be examined. Also, there is provision for on-line printout of all ion beam intensities and ratios together with preliminary $^{207}Pb/^{206}Pb$ ages. For six evaporation steps involving the relevant zircon, internal precision (2SE) of the $^{207}Pb/^{206}Pb$ ratios varied from 0.06 to 0.32% (comparable to that obtainable with common Pb using static multiple collection and Faraday collectors plus ion beams 100 times larger). The inference is that data derived by peak-hopping and the Daly detector are limited in quality solely through the effects of mass fractionation and instabilities of the beam. The clean source and filament design of the VG Sector 54 was illustrated by the fact that the $^{206}Pb/^{207}Pb$ ratios were essentially infinite. The resolvable effects of ancient Pb in the 1.8 Ga old pluton were demonstrated. Zircon no.1 seems to be homogeneous in age and three evaporation steps over a range of 120° C gave the same $^{207}Pb/^{206}Pb$ age of 1.848 Ga $\pm$ 5 Ma. Interestingly, zircon no.2 proved to contain apparent $^{207}Pb/^{206}Pb$ ages spanning a minimum range of 2.678 to 3.072 Ga, increasing with rising evaporation temperature (1600 to 1740°C) and sequential evaporation stage. Excellent 'precision' for the $^{207}Pb/^{206}Pb$ ages was obtained using the Daly detector and was generally superior to error estimates for single-zircon evaporation utilizing an electron multiplier.

5

Uranium series disequilibrium dating

Decay series commencing with the parents ^{238}U and ^{235}U include radioactive isotopes of numerous elements. These daughters can be separated by geological processes such as chemical weathering, adsorption on suspended clay minerals, mineral precipitation from aqueous solutions through biological and inorganic intervention and, during deep-seated magma formation in the Earth, by lava flow crystallization at its surface. Such processes produce disruption in the series through differences in the geochemical properties of the daughters and create geochronological opportunities; for instance, a member separates from its parent and then decays at a rate determined by its own half-life. These effects are utilized in a number of geochronometers applicable to ages extending from several years to 1 Ma or even more. Thus, the geological history of the Earth during the past million years can be investigated, and this plugs a gap between ^{14}C and K–Ar as dating mechanisms.

Uranium and Th decay produces radium isotopes. Radium is a very rare metal which is radioactive with an atomic weight of 226.0254 and an atomic number of 88. It has relative density of 5 and a melting point of 700°C. It has 25 isotopes and belongs to the alkaline-earth group. Its geochemistry mainly relates to the distribution of two of its four naturally occurring isotopes, namely ^{226}Ra and ^{228}Ra. This is because their half-lives are significantly longer than those of ^{223}Ra and ^{224}Ra (Table 5.1)

The approach has been used especially for dating sediments, in particular carbonates, which accumulated in lakes and oceans during the Pleistocene. The geochronometers employed were those involving measurements of the $^{230}Th/^{232}Th$, $^{234}U/^{238}U$, $^{230}Th/^{238}U$ and $^{230}Th/^{234}U$ ratios, all these nuclides being daughters of ^{238}U. Additionally, a daughter of

Table 5.1 Radioisotopes of Ra found naturally

Parent	*Ra isotope*	*Half-life, $T_{1/2}$*
^{235}U	^{223}Ra	11.435 d
^{232}Th	^{224}Ra	3.64 d
^{238}U	^{226}Ra	1620 a
^{232}Th	^{228}Ra	6.7 a

^{235}U, namely ^{231}Pa, constitutes a part of the dating technique which comprises the ratio $^{230}Th/^{231}Pa$. Protactinium has an atomic weight of 231.0359, an atomic number of 91 and 21 isotpes of which ^{231}Pa is the most abundant natural one with a half-life of 3.248×10^4 a, the corresponding decay constant being $\lambda = 2.134 \times 10^{-5}\ a^{-1}$. Finally, daughters of U are valuable in dating snow and ice by the ^{210}Pb technique. This has also been utilized for marine and even lacustrine sediments deposited within the past century or so. Table 5.2 summarizes relevant information.

Uranium daughter abundances are assessed in practice by examining their activities with radiation detectors, so that equations are given in appropriate terms. An activity A of a radionuclide is related to the relevant number of atoms thus:

$$A = c\lambda N$$

where λ is the decay constant and c is the counting efficiency. A recent review of U series disequilibrium dating methodolgy was given in Harmon *et al.* (1979).

5.1 IONIUM DATING OF DEEP-SEA SEDIMENTS

Ionium (^{230}Th) is produced by ^{234}U and produces ^{226}Ra in the decay series of ^{238}U. It can be applied to the dating of deep-sea sediments as a result of the separation of Th isotopes in the oceans from U, a phenomenon caused by differences in relevant chemical properties. In oxidizing conditions on the terrestrial surface U is found as the uranyl ion, UO_2^{2+},

Table 5.2 Data for U daughters of value in geochronometry

Radionuclide	$T_{1/2}$	$\lambda\ (a^{-1})$
$^{234}_{92}U$	2.48×10^5	2.794×10^{-6}
$^{231}_{91}Pa$	3.248×10^4	2.134×10^{-5}
$^{230}_{90}Th$	7.52×10^4	9.217×10^{-6}
$^{226}_{88}Ra$	1.622×10^3	4.272×10^{-4}
$^{210}_{82}Pb$	22.26	3.11×10^{-2}

whereas Th stays tetravalent and is quickly abstracted from sea water by adsorption on solid surfaces and also through incorporation into authigenic minerals, U concentrating in the aqueous phase. Consequently, the residence time of U in oceans is longer by orders of magnitude than that of Th – approximately 0.5 Ma compared with about 300 a, respectively. The residence time is given by the expression $t = A/(\mathrm{d}A/\mathrm{d}t)$, where A is the total quantity of the dissolved element in the oceans and $\mathrm{d}A/\mathrm{d}t$ is the quantity either precipitated from or added to the oceans annually.

Mostly, ^{230}Th in sea water forms by the decay of its parent, the balance (*c.* 25%) arising from river input. The preferential removal of ^{230}Th from sea water separates the radionuclide from its parent and produces an excess of unsupported ionium in the sediments. Of course, this causes activity which diminishes as a function of time, while the ^{230}Th decays to its daughter with a half-life of 7.52×10^4 a. The technique is based upon the assumption that both this ^{230}Th and ^{232}Th (entering the oceans in the river input) are removed together from sea water. The latter has a much longer half-life (1.40×10^{10} a); hence the activity ratio ^{230}Th/^{232}Th decreases with time solely as a consequence of the decay of the unsupported ^{230}Th.

It is possible to date a sample of sediment taken at a give depth below the water–sediment interface from its ^{230}Th/^{232}Th ratio if this has stayed constant for at least several hundred thousand years in the water body next to the sediment in any given ocean basin, and also if there is no migration of the relevant isotopes in the sediment, if their chemical speciation is the same in sea water (with no isotope fractionation between sea water and the mineral phases involving Th in the sediment) and if the ^{230}Th and ^{232}Th found in detrital mineral particles are not included in the analysis. The question of migration of the Th isotopes in the sediments is a thorny one because there is some evidence that ^{234}U, the parent of ^{230}Th, is mobile, and so could contribute to its activity, but this factor is probably of minor importance.

The analysis is effected by treating the sample of sediment with hot hydrochloric acid. This releases Th absorbed on detrital particles or existing in authigenic phases. Since the main detrital component minerals, namely quartz and feldspar, are not dissolved, it is feasible to recover that part of the ^{230}Th and ^{232}Th which went into the sediment from sea water. After they have been purified, the Th isotopes are electroplated onto Pt discs and the alpha activites of ^{230}Th and ^{232}Th measured using solid-state detectors or ionization chambers. It is possible to represent the total activity of ^{230}Th adsorbed on particle surfaces or present in authigenic minerals by the following expression:

$$^{230}\mathrm{Th}_A = {}^{230}\mathrm{Th}_{Ax} + {}^{230}\mathrm{Th}_{As},$$

in which $^{230}\mathrm{Th}_A$ is the total activity of ^{230}Th in the leachable fraction of the sediment at a given depth below the surface in units of disintegrations per unit time per unit weight of the dry material, $^{230}\mathrm{Th}_{Ax}$ is the same parameter for the excess, unsupported fraction of ^{230}Th and $^{230}\mathrm{Th}_{As}$ is the same parameter for the fraction of ^{230}Th which is supported by the radioactive decay of ^{234}U. Considering the second term, the activity of the relevant fraction at a time t is expressible as follows:

$$^{230}\mathrm{Th}_{Ax} = {}^{230}\mathrm{Th}^{0}_{Ax}\, e^{-\lambda_{230} t},$$

in which $^{230}\mathrm{Th}^{0}_{Ax}$ is the activity of the excess, unsupported ^{230}Th in the sediment at a time $t = 0$, i.e. in newly laid down sediment actually at the water–sediment interface, λ_{230} is the decay constant of ^{230}Th ($9.217 \times 10^{-6}\mathrm{a}^{-1}$), and t is the time which has passed since the deposition of the sediment.

If the activity of the very long-lived ^{232}Th is taken as a reference, then the activity of the unsupported ^{230}Th may be described by

$$\left(\frac{^{230}\mathrm{Th}}{^{232}\mathrm{Th}}\right)_{Ax} = \left(\frac{^{230}\mathrm{Th}}{^{232}\mathrm{Th}}\right)^{0}_{Ax} \mathrm{e}^{-\lambda_{230} t} \tag{5.1}$$

Taking into account the fact that the contribution made by U-supported ^{230}Th is probably negligible, then the activity of a sample, expressed as an activity ratio R, may be related to its age t, thus

$$R = R_0 \mathrm{e}^{-\lambda t} \tag{5.2}$$

where R is the $^{230}\mathrm{Th}/^{232}\mathrm{Th}$ activity ratio of the leachable fraction of sediment obtained at a given depth below the top of a core, and R_0 is the same ratio for newly laid down sediment, λ being the decay constant of ^{230}Th and t the time elapsed since deposition took place. R is measured and, if R_o is known, then

$$t = \frac{1}{\lambda} \ln\left(\frac{R_0}{R}\right).$$

Equations (5.1) and (5.2) are used where it is justified to assume that the activity of U-supported activity of ^{230}Th is negligible. The approach has been employed widely to determine the rate of sedimentation in open oceans. If this rate is a constant $a = h/t$, where h is the depth below the water–sediment interface, then

$$R = R_0 \mathrm{e}^{-\lambda h/a},$$

and

$$\ln R = \ln R_o - \frac{\lambda h}{a},$$

the equation of a straight line in coordinates of R and H of which the slope m is $-\lambda/a$. The rate of sedimentation is then

$$a = -\frac{\lambda}{m}.$$

The alternative scenario in which there is a contribution to the activity by U-supported ^{230}Th in the leachable fraction of marine sediment involves a number of U-supported ^{230}Th atoms given by

$$^{230}\mathrm{Th}_s = \frac{\lambda_{234}}{\lambda_{230} - \lambda_{234}}\ ^{234}\mathrm{U}^{0}\, \mathrm{e}^{-\lambda_{234}t} - \mathrm{e}^{-\lambda_{230}t} \tag{5.3}$$

If ^{234}U is in secular equilibrium with ^{238}U, then its activity may be taken as equal to that of the parent, so that

$$^{234}\mathrm{U}^{0}_{234} = {}^{238}\mathrm{U}_{\mathrm{A}}\,.$$

Now, the activity of the parent in question may be regarded as practically constant over time intervals of only a few million years (this is a logical inference from the extremely long half-life of ^{238}U, i.e. 4.51×10^9 a), the activity of ^{234}U being effectively controlled by the decay constant of the parent which is much smaller than that of ^{230}Th, and

$$\lambda_{230} - \lambda_{234} \simeq \lambda_{230}\,,$$

$$\mathrm{e}^{-\lambda_{234}t} \simeq 1\,,$$

from which equation (5.3) can be expressed in the following manner:

$$^{230}\mathrm{Th}_{\mathrm{As}} = {}^{238}\mathrm{U}_{\mathrm{A}}\,(1 - \mathrm{e}^{-\lambda_{230}t}) \tag{5.4}$$

The total activity of the ^{230}Th in the leachable fraction of a sample of sediment of age t is then given by

$$^{230}\mathrm{Th}_{\mathrm{A}} = {}^{230}\mathrm{Th}_{\mathrm{Ax}}\,\mathrm{e}^{-\lambda_{230}t} + {}^{238}\mathrm{U}_{\mathrm{A}}\,(1 - \mathrm{e}^{-\lambda_{230}t})\,,$$

and, including the reference activity of the long-lived ^{232}Th, the following expression is obtained:

$$\left(\frac{^{230}\mathrm{Th}}{^{232}\mathrm{Th}}\right)_{\mathrm{A}} = \left(\frac{^{230}\mathrm{Th}}{^{232}\mathrm{Th}}\right)^{0}_{\mathrm{Ax}} \mathrm{e}^{-\lambda_{230}t} + \left(\frac{^{232}\mathrm{U}}{^{232}\mathrm{Th}}\right)_{\mathrm{A}} (1 - \mathrm{e}^{-\lambda_{230}t})\,.$$

This is a way of describing the variation in the $^{230}Th/^{232}Th$ activity ratio through time due to the decay of excess unsupported ^{230}Th and the growth of U-supported ^{230}Th. The value of the first term diminishes with time and that of the second increases, i.e. the contribution by U-supported ^{230}Th could be almost negligible at the top of a particular core, but may become significant for older sediment towards greater depth in it. Four patterns of variation in this ratio were recognized from sediment cores derived all over the world. In the first, there is a linear decrease of ln R with depth (sediment deposited at a constant rate). In the second, there is a pair of straight-line sections which demonstrates an alteration in the rate of sedimentation. In the third, there is a constant ratio at the top which is succeded by a linear decrease in ln R with depth. This is caused by superficial mixing by burrowing organisms or currents, and the initial activity can be derived by extrapolation of the straight line to the interface between water and sediment. In the fourth, the activity ratio rises with depth and ultimately plateaus at a constant value at a greater depth. The effect results from the growth of ^{230}Th from U in the leachable fraction of the sediment.

Obviously, the rates of sedimentation in the oceans will vary considerably in response to a number of parameters such as amounts of detritus introduced by rivers, the rates of melting of icebergs and the numbers of particles transported through and deposited from the atmosphere. The ^{230}Th/^{232}Th technique provides a range of under 0.5 mm (10^3 a) to 50 mm (10^3 a) on a carbonate- and opal-free basis, the lowest occurring in the South Pacific and the highest in the Atlantic. In an analogous fashion, there is a variation in the ^{230}Th/^{232}Th activity ratios on the surfaces, the highest values being recorded in the Pacific ($R_0 = 148–153$) and the lowest in the Atlantic ($R_0 = 1.5–19$), reflecting a greater input of ^{232}Th by the many large-scale rivers entering the Atlantic. However, this explanation may be affected by work done by Bernat and Allègre (1974) who found that an excess ^{230}Th method could be applied to separate oceanic sediments into fractions which formed isochrons of which the intercepts decreased with depth in the core. The slope of the isochron in their youngest sample was not the expected zero, but indicated instead a disequilibrium among the Th isotopes in the sediment.

5.2 THE ^{234}U–^{238}U GEOCHRONOMETER

The attainment of secular equilibrium in decays series starting with ^{238}U, ^{235}U and ^{232}Th depends upon none of the relevant daughter elements leaving or entering the systems concered. Once the equilibrium is established, the decay rates of the daughters in any particular system equal that of the parent:

$$\lambda_1 N_1 = \lambda_2 N_2 = \lambda_3 N_3 = \cdots = \lambda_n N_n ,$$

in which $\lambda_1 N_1$ is the decay rate of the parent and the other terms represent those of the daughters (Faure, 1986). Therefore, under secular equilibrium conditions, the ratio between the decay rate of any daughter and its parent is equal to one. The production of ^{234}U from its parent goes through several stages thus:

$$^{238}\mathrm{U} \xrightarrow{\alpha} {}^{234}\mathrm{Th} \xrightarrow{\beta^-} {}^{234}\mathrm{Pa} \xrightarrow{\beta^-} {}^{234}\mathrm{U}$$

and, where secular equilibrium obtains, the rate of decay of ^{234}U in a U-bearing mineral equals that of ^{238}U, so that the ratio $^{234}U_A/^{238}U_A$ must be equal to one. Unfortunately, the situation is not quite so simple; in fact, samples of groundwater have demonstrated that such a secular equilibrium did not exist and was also absent from minerals which deposited from it. Later researches confirmed these findings from the former USSR, wide variations being recorded in the activity ratios of $^{234}U/^{238}U$ of numerous samples collected from weathered rocks in the USA by Thurber (1962).

Usually, the $^{234}U/^{238}U$ activity ratio of both water and secondary U-bearing minerals on continents exceeds unity because the isotope ^{234}U is enriched by processes such as the decay of its parent dissolved in ground water. The alpha decay of the parent causes damage to the crystal lattice and allows the daughter to migrate in microcapillaries in the minerals where oxidation converts it into uranyl ions which dissolve in water and remove it into an aqueous phase. Of course, ^{238}U occurs in stable lattice positions. Also, ^{234}Th can be ejected from the surfaces of grains through recoil in the α-decay process of ^{238}U, the radionuclide then decaying swiftly to ^{234}U via the short-lived ^{234}Pa (e.g. Fleischer, 1980). The phenomena listed above account for the fact that the activity ratios of $^{234}U/^{238}U$ in both surface and groundwaters as well as in secondary minerals containing U on continents normally exceed one and indeed may reach values as high as ten. Interestingly, such activity ratios of primary U minerals in rocks which are subject to chemical weathering may be less than one, ^{234}U having been removed preferentially.

If U occurs in solution in groundwater, it may enter the oceans and become isotopically homogenized. This is inferred from measurements which show that the relevant activity ratio, $^{234}U/^{238}U$, has a narrow spread of values with a mean *c.* 1.15. The element can be removed from oceans by incorporation into authigenic minerals (e.g. $CaCO_3$) or by adsorption on grain surfaces under reducing conditions. Once U has been isolated form sea water, excess ^{234}U decays to ^{230}Th until the activity ratio $^{234}U/^{238}U$ nears an equilibrium value. However, it is not feasible to use this parameter as a dating technique because of possible migration of ^{234}U after deposition. Nevertheless, the decay of excess ^{234}U and growth of its daughter ^{230}Th can be so employed for dating both marine and non-marine carbonates of Pleistocene age. If $^{234}U_A$ is the activity of this radioisotope per unit weight of sample now, $^{234}U_{As}$ is the activity of it in secular equilibrium with ^{238}U, and $^{234}U_{Ax}$ is the activity of excess ^{234}U

per unit weight of sample, then

$$^{234}\mathrm{U_A} = {}^{234}\mathrm{U_{As}} + {}^{234}\mathrm{U_{Ax}}\,.$$

The activity of the excess ^{234}U diminishes as a function of time:

$$^{234}\mathrm{U_{Ax}} = {}^{234}\mathrm{U^0_{Ax}}\,\mathrm{e}^{-\lambda 243 t}\,,$$

where the first term is the initial activity of the excess ^{234}U, expressible as

$$^{234}\mathrm{U_{Ax}} = {}^{234}\mathrm{U}^0_A - {}^{234}\mathrm{U_{As}}\,,$$

in which $^{234}\mathrm{U}^0_A$ is the total initial activity of ^{234}U and $^{234}\mathrm{U}_{As}$ is the component supported by ^{238}U. Of course, $^{234}\mathrm{U}_{As} = {}^{238}\mathrm{U}_A$: this leads to a combined equation embodying

$$^{234}\mathrm{U_A} = {}^{238}\mathrm{U_A} + ({}^{234}\mathrm{U^0_A} - {}^{234}\mathrm{U_A})\,\mathrm{e}^{-\lambda 243 t}\,.$$

Division by the constant activity of ^{238}U produces

$$\left(\frac{^{234}\mathrm{U}}{^{234}\mathrm{U}}\right)_\mathrm{A} = 1 + \left(\frac{^{234}\mathrm{U^0_A} - {}^{238}\mathrm{U_A}}{^{238}\mathrm{U_A}}\right)\mathrm{e}^{-\lambda 243 t}\,.$$

If $(^{234}\mathrm{U}/^{238}\mathrm{U})^0_\mathrm{A} = \gamma_0$, then

$$\left(\frac{^{243}\mathrm{U}}{^{238}\mathrm{U}}\right)_\mathrm{A} = 1 + (\gamma_\mathrm{o} - 1)\,\mathrm{e}^{-\lambda 243 t} \qquad (5.5)$$

γ_o being the initial $^{234}\mathrm{U}/^{238}\mathrm{U}$ ratio of the sample.

The geochronometer based upon these considerations involves a determination for $\gamma_0 = 1.15$ and has been applied to dating the $CaCO_3$ of biogenic and inorganic origin deposited in marine and also in non-marine environments, the reliability of ages from the latter being restricted by an uncertain value for γ_0. In non-marine environments, γ_0 shows time-dependent variations, since shells of some non-marine molluscs actually acquire U after their demise.

The fact that groundwater can become enriched in ^{234}U relative to its parent is a useful means of assessing its specific sources as well as of determing mixing between different water masses characterized by their individual $^{234}\mathrm{U}/^{238}\mathrm{U}$ activity rates (e.g. Kronfeld and Rosenthal, 1981).

5.3 ^{230}Th–^{238}U and ^{230}Th– ^{234}U DATING

Practically no Th, but substantial amounts of U are contained in $CaCO_3$ deposited in saline lakes and in the oceans where they sometimes reach 5 ppm. Consequently, there is almost no ^{230}Th initial activity in newly deposited carbonates but, due to the decay of ^{234}U, activity of this radioisotope increases as a function of time. The activity produced by the decay of ^{234}U in secular equilibrium with ^{238}U follows equation (5.4) but the matter is rendered more difficult if excess ^{234}U is present as this contributes to the activity of ^{230}Th until such time as the excess in question has completely decayed. The following expression can be derived:

$$^{230}\mathrm{Th}_{\mathrm{Ax}} = {}^{230}\mathrm{Th}_{\mathrm{x}}\,\lambda_{230} = \frac{\lambda_{238}}{\lambda_{230}-\lambda_{234}}\,{}^{234}\mathrm{U}^{0}_{\mathrm{Ax}}\,\mathrm{e}^{-\lambda_{243}t} - \mathrm{e}^{-\lambda_{230}t}\,.$$

The excess activity of the ^{234}U is given by

$$^{234}\mathrm{U}^{0}_{\mathrm{As}} = {}^{234}\mathrm{U}^{0}_{\mathrm{Z}} - {}^{234}\dot{\mathrm{U}}_{\mathrm{As}}\,,$$

in which $^{234}U^0_A$ is the total initial activity of excess ^{234}U when deposition occurred and $^{234}U_{As}$ the activity of ^{234}U which is in secular equilibrium with ^{238}U. Consequently,

$$^{230}\mathrm{Th}_{\mathrm{Ax}} = \frac{\lambda_{30}}{\lambda_{230}-\lambda_{234}}\left({}^{234}\mathrm{U}^{0}_{\mathrm{A}} - {}^{234}\mathrm{U}_{\mathrm{As}}\right)\mathrm{e}^{-\lambda_{243}t} - \mathrm{e}^{-\lambda_{230}t}\,,$$

and the activity of $^{230}Th_{Ax}$ relative to $^{238}U_{Ax}$ is

$$\left(\frac{^{230}\mathrm{Th}}{^{238}\mathrm{U}}\right)_{\mathrm{Ax}} = \frac{\lambda_{230}}{\lambda_{230}-\lambda_{234}}\left(\frac{^{234}\mathrm{U}^{0}_{\mathrm{A}} - {}^{234}\mathrm{U}^{0}_{\mathrm{As}}}{^{238}\mathrm{U}_{\mathrm{A}}}\right)\mathrm{e}^{-\lambda_{243}t} - \mathrm{e}^{-\lambda_{230}t}\,.$$

Broecker (1963) gave a description of the activity ratio of ^{230}Th to ^{238}U as a function of time; this may be given as

$$\left(\frac{^{230}\mathrm{Th}}{^{238}\mathrm{U}}\right)_{\mathrm{A}} = (1-\mathrm{e}^{-\lambda_{230}t}) + \frac{\lambda_{230}}{\lambda_{230}-\lambda_{234}}(\gamma_0 - 1)\,(\mathrm{e}^{-\lambda_{234}t} - \mathrm{e}^{-\lambda_{230}t}) \qquad (5.6)$$

in which $\gamma_0 = {}^{234}U^0_A/{}^{238}U_A$ and $^{234}U_{As} = {}^{238}U_A$.

Such equations, based on the activity ratio of $^{230}Th/^{238}U$, are applied to the dating of corals and molluscs as well as calcareous oozes if a number of conditions are observed. The first is that the initial activity ratio is near zero, and the second that the sample constituted a closed system regarding U and intermediate nuclides between ^{238}U and ^{230}Th. Finally, the initial activity of $^{234}U/^{238}U$ must be available. However, it is possible to relate the activity of ^{230}Th to that of its immediate parent, ^{234}U, since, if the ^{234}U is in secular equilibrium with ^{238}U, then the $^{230}Th/^{234}U$ activity ratio is related to time thus:

$$\left(\frac{^{230}\mathrm{Th}}{^{234}\mathrm{U}}\right)_{\mathrm{A}} = 1 - \mathrm{e}^{-\lambda_{230}t}\,,$$

an equation which does not take into account the presence of excess ^{234}U. In those instances where $\gamma_0 > 1$, the $^{230}Th/^{238}U$ activity ratio as a function of time is obtainable by a combination of Broecker's equation (5.6) with (5.5). This refinement may well be unnecessary, because it appears probable that the presence of such excess ^{234}U does not make any really appreciable difference to the observed $^{230}Th/^{234}U$ activity ratios of marine carbonates. Use of the inexact equation produces overestimates of *t* of a mere 2% or so for samples which are 130 ka of age. In fact, Broecker *et al.* (1968) utilized this approach in order to date corals from the three lowest terraces of Barbados, an island which seems to be in the process of rising out of the sea at a uniform rate and possesses at least 18 major terraces. Usually, such terraces represent coral reefs, the ages of which increase with increasing elevation, recording different sea-level stands in the past. Data obtained on the $^{230}Th/^{234}U$ activity ratios of relevant corals gave dates of 122 ka, 103 and 82 ka. Sample 1064-C, a *Montastria annularis* from the lowest terrace, gave a value for the $^{230}Th/^{234}U$ activity ratio of 0.53 ± 0.02, from which an age of 81 916 a can be deduced. It was determined that the initial activity ratio of $^{243}U/^{238}U$ in this coral equalled that of sea water. These researchers also indicated that the ages of the corals from these three lowest terraces in Barbados coincide with times of elevated summer radiation at 45°N latitude which were predicted by Milankovitch on the basis of celestial mechanics.

TIMS (thermal ionization mass spectrometry) U-series dating was used in Papua New Guinea by other researchers. In this region, emergent coral reef terraces extend about 80 km parallel to the coast along the Huon Peninsula and comprise a geological record of sea-level changes and tectonic movements there over at least the past 250 ka (Chappell, 1974). Their elevation increases south-easterly reflecting increasing rates of uplift. Theses vary from *c.* 0.5 m ka^{-1} to almost 4 m ka^{-1} where the elevation reaches *c.* 700 m above sea level.

Stein *et al.* (1993) were among investigators who utilized U-series dating as well as stable isotope investigations of the last interglacial event there. In 1988 an international expedition was made to the Huon Peninsula where uplifted reef terraces along the Vitiaz strait on its northern flank were targeted, especially regarding appropriate samples for U-series dating in a reef complex designated VII which is correlated with the last interglacial episode and high sea-level stand. The complex comprises a barrier reef (VIIb), a lagoon and a fringing reef (VIIa). Twelve corals from the terraces and two from the older reef complex VIII were chosen for analysis. The petrography, O and C isotope compositions plus the Mg and S concentrations were determined together with the concentrations and isotope compositions of U and Th. $^{230}Th/^{234}U$ ages of the corals with >99% aragonite having primary textures and showing U/Sr ratios of *c.* 0.4×10^{-3} and initial $\delta^{234}U$ values close to that of modern sea water were thought to be reliable. The 'most reliable' (sic) ages from complex VII corals fell within two groups centred at 118 ka and 134 ka. Corals with $\delta^{234}U(T)$ values exceeding 160 and U/Sr ratios much lower than sea water were taken to imply diagenetic alteration (seemingly a common phenomenon in the region). Stein *et al.* (1993) suggested the simplest model for sea-level height for terrace VII to be a continuous rise between 134 ka and 118 ka, or there may have been two periods of rapid sea-level rise. However, they pointed out that in the Bahamas there is evidence that sea level stayed fairly constant during this time interval. Lack of ages between 132 ka and 120 ka in the Papua New Guinea locality might have been caused by alterations in local tectonic uplift rate through this period of time. Also, they might be the result of erosion disrupting the continuous record. In any case, Stein *et al.* (1993) found no basis for acccepting a single short time for the age of the last interglacial and using this age as a precise chronometer for global correlation or for testing climatic models. Their older ages preceded the Milankovitch solar insolation peak at 128 ka and the younger ages are ~ 10 ka after this peak. If the present high-precision data turn out to be correct, it will mean that the validity of the Milankovitch theory of climate changes will require reassessment. A fundamental issue requiring resolution is how to identify coral samples which have not been disturbed by diagenetic processes.

5.4 ASSESSING $^{230}Th/U$ AGES IN IMPURE CARBONATES

This relationship in authigenic carbonates is a function of the time elapsed since their deposition and, as noted earlier, pure carbonates such as corals and numerous speleothems can yield reliable dates from their ratios. A different situation arises where geochronological information is required from impure carbonates such as spring travertines and soil calcretes. To deal with complexities encountered in them, correction methods have to be applied and a number of basically different ones have been proposed. Most have suffered from a lack of independent knowledge of the real ages of the samples examined, but Kaufman, Yecheili and Gardosh (1992) presented evidence that the ages of a suite of lacustrine sediments from the Dead Sea Basin as observed by one of the methods is in fact valid. Kaufman (1993) used this result to evaluate how reasonable are the ages derived by applying three other correction methods to the same samples. The methods in question are schemes I, II and SL and that giving the valid age is scheme L/L, which involved 30 sample sets comprising 170 samples of the impure carbonates. Scheme L/L effected the necessary isotopic correction by comparing the ratios found in the leachates of several coeval samples, and the validity of two of the others, namely I and II, was tested by comparing their results with the scheme L/L data from eight of the same 30 sample sets. Schemes I and II made corrections by comparing the isotope ratios in the leachate and residue of one impure carbonate, the two schemes differing in their assumptions about the leachate–residue relationship and so in their mathematical treatments of the isotopic ratios. A comparison of scheme L/L with I and II demonstrated that the latter are invalid in more than half the cases. The fourth correction scheme, scheme SL, performs the corrections on the isotopic ratios of a single leachate by assuming that the correction factor, R_0 (the original detrital $^{230}Th/^{232}Th$ ratio) has a more or less constant value worldwide, this being 1.5 to 1.7.

Kaufman's study indicated that R_0 is about 3.3 for the Dead Sea Basin and appraised the errors incurred if the incorrect R_0 is used. An effort was made to find the best worldwide estimate of R_0 and its uncertainty for any new area to be investigated. A graph was given for use in estimating the consequent age uncertainty.

5.5 ^{230}Th–^{231}Pa DATING

Protactinium is a radioactive element with an atomic weight of 231.0359 and an atomic number of 91; it has 21 isotopes of which the most abundant is ^{231}Pa with a half-life of 3.248×10^4 a and a corresponding decay constant of $2.134 \times 10^{-5} a^{-1}$. It is a daughter of ^{238}U which arises thus:

$$^{238}U \xrightarrow{\alpha} {}^{231}Th \xrightarrow{\beta^-} {}^{231}Pa \, .$$

Geochemically, in oceans, its properties parallel those of Th since the isotopes of both of these elements are removed from sea water either by adsorption on mineral grains or by incorporation into authigenic minerals. Consequently, sediment which is deposited in oceans contains not just the excess of unsupported ^{230}Th mentioned earlier, but also unsupported ^{231}Pa. If both are removed equally efficiently from the sea waters, then their activity ratios alter with time in a fashion controlled by their decay constants:

$$\left(\frac{^{230}Th}{^{231}Pa}\right)_{Ax} = \left(\frac{^{230}Th}{^{231}Pa}\right)^0_{Ax} \frac{e^{-\lambda_{230}t}}{e^{-\lambda_{231}t}} \tag{5.7}$$

in which the left-hand side represents the activity ratio of the unsupported nuclides per unit weight of sediment at a time t, the second ratio relates to the time of deposition, and the decay constants refer to the two nuclides. This expression assumes that the U concentration in the leachable fraction can be neglected. If this is valid, then the effective decay constant of the unsupported $^{230}Th/^{231}Pa$ activity ratio is given by

$$\left(\frac{^{230}Th}{^{231}Pa}\right)_{Ax} = \left(\frac{^{230}Th}{^{231}Pa}\right)^0_{Ax} e^{-\lambda' t} \, ,$$

in which $\lambda' = \lambda_{231} - \lambda_{230} = 12.123 \times 10^{-6} a^{-1}$ and the effective half-life is 57 100 a. From this it is apparent that the ratio of the activities of the unsupported nuclides in question increases with depth in a core at a rate proportional to λ'. If the sediment deposited at a constant rate, then

$$\ln\left(\frac{^{230}\mathrm{Th}}{^{231}\mathrm{Pa}}\right)_{\mathrm{Ax}} = \ln\left(\frac{^{230}\mathrm{Th}}{^{231}\mathrm{Pa}}\right)^{0}_{\mathrm{Ax}} + \frac{\lambda' h}{a},$$

where h is the depth in the core and a the constant rate of sedimentation (h/t). This is a straight-line equation in coordinates of h and ln $(^{230}\mathrm{Th}/^{231}\mathrm{Pa})_{Ax}$. Its slope is inversely proportional to the sedimentation rate and its intercept equals the initial value of the activity ratio of the relevant unsupported nuclides in the sediment. This was used by Ku and Broecker (1967) in order to determine a for a core (T3-63-1) from the Arctic Ocean and a value of 0.2 cm (10^3 a) was obtained. The initial activity ratio at the top of this particular core turned out to be 26.

Rosholt *et al.* used a technique which alowed dating of one sediment sample, i.e. analysis of a suite was not required. If t is the time interval between ^{231}Pa production and its incorporation into a sediment, then it is possible to write

$$^{231}\mathrm{Pa}_{\mathrm{A}} = {}^{235}\mathrm{U}_{\mathrm{A}}\,(1 - \mathrm{e}^{-231t})\,.$$

Growth of ^{230}Th from ^{234}U in sea water is provided by a similar equation. Consequently, the activity ratio of U-supported ^{230}Th to ^{231}Pa in sea water is

$$\left(\frac{^{230}\mathrm{Th}}{^{231}\mathrm{Pa}}\right)_{\mathrm{A}} = \left(\frac{^{234}\mathrm{U}}{^{235}\mathrm{U}}\right)_{\mathrm{A}} \frac{1 - \mathrm{e}^{-\lambda_{230}t}}{1 - \mathrm{e}^{-\lambda_{231}t}},$$

If the relative abundances of the U isotopes in the sea water are normal, namely if the atomic abundance ratio $^{238}\mathrm{U}/^{235}\mathrm{U} = 137.88$ and the activity ratio $^{234}\mathrm{U}/^{238}\mathrm{U} = 1.15$, then

$$\left(\frac{^{234}\mathrm{U}}{^{235}\mathrm{U}}\right)_{\mathrm{A}} = 1.15 \times \frac{238}{235} \times \frac{^{234}\mathrm{U}}{^{235}\mathrm{U}} = 24.97\,.$$

If residence times for ^{230}Th and ^{231}Pa in oceans are short and about equal (*c.* 300 a), then

$$\frac{1 - \mathrm{e}^{-\lambda_{230}t}}{1 - \mathrm{e}^{-\lambda_{231}t}} = \frac{\lambda_{230t}}{\lambda_{231t}} = \frac{9.217 \times 10^{-6}}{2.134 \times 10^{-5}} = 0.4319\,.$$

and the activity ratio of U-supported ^{230}Th to ^{231}Pa in sea water is

$$\left(\frac{^{230}\mathrm{Th}}{^{231}\mathrm{Pa}}\right)_{\mathrm{A}} = 24.97 \times 0.4319 = 10.78\,.$$

Naturally, if both of the isotopes are removed with the same efficiency from the sea water, then the initial activity ratio of a newly deposited sample of sediment ought also to possess this same value.

Current activities of ^{230}Th and ^{231}Pa in a sediment sample laid down *t* years BP has to be adjusted for the activities supported by the decay of U after its deposition, and

$$\left(\frac{^{230}\mathrm{Th}}{^{231}\mathrm{Pa}}\right)_{\mathrm{Ax}} = \left(\frac{^{230}\mathrm{Th_t} - {}^{230}\mathrm{Th_s}}{^{231}\mathrm{Pa_t} - {}^{231}\mathrm{Pa_s}}\right)_{\mathrm{A}},$$

in which the x represents excess unsupported material, t is the total and s the supported. After substitution into equation (5.7), the following expression is obtained:

$$\left(\frac{^{230}\mathrm{Th}}{^{231}\mathrm{Pa}}\right)_{\mathrm{Ax}} = \left(\frac{^{231}\mathrm{Th_t} - {}^{230}\mathrm{Th_s}}{^{231}\mathrm{Pa_t} - {}^{231}\mathrm{Pa_s}}\right)_{\mathrm{A}} = 10.78\,\frac{e^{-\lambda_{230}t}}{e^{-\lambda_{231}t}},$$

and, solving for *t*,

$$t = \frac{1}{\lambda_{231} - \lambda_{230}} \ln \left[\frac{\left(\dfrac{^{230}\mathrm{Th_t} - {}^{230}\mathrm{Th_s}}{^{231}\mathrm{Pa_t} - {}^{231}\mathrm{Pa_s}}\right)_{\mathrm{A}}}{10{\cdot}78} \right],$$

the assumption being that the U-supported activities of ^{230}Th and ^{231}Pa are in secular equilibrium with their parents ($^{231}\mathrm{Pa_{As}} = {}^{235}\mathrm{U_A} = \lambda_{235}\,{}^{235}\mathrm{U}$ and $^{230}\mathrm{Th_{As}} = {}^{234}\mathrm{U_A} = \lambda_{234}\,{}^{234}\mathrm{U}$, where ^{235}U and ^{234}U are the numbers of atoms of these radioisotopes per unit weight of sample).

Although ^{230}Th–^{231}Pa dates often concord with radiocarbon dates, some cores have yielded erroneous ones which may be due to fractionation of the relevant radioisotopes during adsorption and deposition with the sediment (Osmond, 1979).

5.6 ^{210}Pb DATING

In the decay series commencing with parental ^{238}U occurs ^{222}Rn which enters the atmosphere at a rate of some 42 atoms min^{-1} cm^{-2} of land surface and later decays through several short-lived daughters to ^{210}Pb. This is then removed in precipitation, having a residence time of only some 10 d. Subsequently, the radioisotope is deposited in the snow and ice in glaciers as well as in lakes and coastal oceans, the activity of unsupported ^{210}Pb thereafter diminishing as a function of time at a rate depending upon the relevant half-life, which is 22.26 years.

^{210}Pb is therefore useful as a dating tool in connection with materials containing it which arose within the past century. One of its first applications entailing the measurement of depositional rates of snow was on samples from Greenland and Antarctica, and later work involved Alpine glaciers. More recently, lake sediments and coastal sea deposits have been dated too. Nowadays, it is common to obtain primary timescales for lake sediments through the same method, which is based upon the defined decay kinetics of this sedimentary radionuclide. In conjunction with ^{210}Pb it is also possible to utilize ^{137}Cs, from which an artificial marker horizon was produced by nuclear weapons tests in the early 1960s. More details regarding the technique are available in papers by Robbins (1978), Robbins, Edgington and Kemp (1978) and Smith and Walton (1980).

The basic theory of the dual origins of sedimentary ^{210}Pb was discussed by Robbins (1978) and may be recapitulated here. Initially, it is found in nature in sediment minerals in a supported form, i.e. at equal activity to its parent nuclides, by *in situ* production from, decay of ^{238}U or ^{226}Ra, this component being controlled by sediment mineralogy and therefore giving no timescale information; indeed, the ^{238}U or ^{226}Ra is assumed to be constant down a core profile. As noted above, the dating potential springs from the second fraction namely the unsupported (excess) ^{210}Pb of atmospheric origin, the result of ^{222}Rn decay.

Unsupported ^{210}Pb is conventionally quantified by subtracting the measured activity of supported ^{210}Pb from the total ^{210}Pb inventory at each depth. The former may be determined by measuring the ^{222}Rn activity emanating from dissolved sediment and the latter by analysing ^{210}Pb decay products ^{210}Bi or ^{210}Po. In all cases, secular equilibrium assumptions are made for the parent–daughter pairs. Thus the conventional dating method is indirect and specialized as regards both nuclear counting equipment and radiochemical expertise. This means that it is expensive as well as time consuming. Recovery of the radioisotope from solutions is effected by anion-exchange chromatography using a few milligrams of ordinary Pb (^{210}Pb-free) as carrier. The end-concentrate is either evaporated directly into a planchet for counting or precipitated as a sulphate or chromate. The decay of ^{210}Pb to ^{210}Bi is by β-emission and the maximum energy is only 0.02 MeV. Hence, the activity of ^{210}Pb is assessed by counting ^{210}Bi ($E_{max} = 1.2$ MeV), half-life $T_{1/2} \simeq 5$ d. In the alternative scenario, i.e. using ^{210}Po, this forms by β^- decay of ^{210}Bi, its α-particles being countable.

Baxter *et al.* (1981) investigated a particle track method which depends upon optical counting of radiation damage tracks produced in selected plastic films by both α-particles and fission products emitted from essentially unprocessed dried sediment. Supported ^{210}Pb is measured indirectly by ^{235}U analysis. Because the ^{235}U/^{238}U ratio does not vary significantly in nature, the ^{238}U content of each sample can be calculated and, again, secular equilibrium with ^{210}Pb invoked. The supported ^{210}Pb profile is then subtracted from the total-α record and the major assumption made that the resultant excess-α plot is dominated by the decay of unsupported ^{210}Pb (more precisely of ^{210}Po). Successful application of the method is dependent upon the idealized non-migratory behaviour of all other U-series α-emitters and on total constancy with depth of ^{232}Th and its decay products. Of course, such assumptions are broader than those entailed in the conventional method, and the practical advantages are considerable: the approach is fast, simple and cheap, using not much more than an optical microscope. It is quite possible that it may lead to a wider application of ^{210}Pb geochronology. The sediment core analysed came from Loch Lomond in Scotland and was collected using a 1 m mini-Mackereth corer in 28 m depth of water 4 km from the mouth of the Endrick Water. Its length was 96 cm and its number LLRPM1.

The principles of the particle track method resemble those given for excess-Th dating by Fisher (1977). The premises are that the main trend in α-track density with depth results from decay of unsupported ^{210}Pb and that the supported ^{210}Pb component can be measured through U assay. Weighed dried samples are first taken in weights of *c.* 100 mg and pressed into a 13 mm diameter disc using cellulose powder as binder. For total-α determination, each such disc is placed in contact with a sheet of cellulose nitrate plastic and the assembly stored long enough to generate a reasonable densityof alpha tracks; this took about 8 weeks in the case of the Loch Lomond samples and yielded a track density of 800–1000 tracks cm^{-2}. Of course, damage trails are on an atomic scale and, to render tracks visible after irradiation, etching of the plastic is carried out in 2.5 M NaOH at 60°C for 1.5 h followed by washing in hot distilled water. After drying, the plastic in question is mounted on a glass slide and the track density assessed by counting under an optical microscope or , automatically, employing a Quantimet Image Computer. Usually, the relevant magnification is within the range $\times$ 400 to $\times$ 600. Each individual sample is counted over many random fields of view until at least 10^3 tracks have accumulated. After this , the rate R at which α-particle tracks are produced per square centimetre per minute is calculated from the following expression:

$$R = C/T\,,$$

where C is the density of tracks per square centimetre and T the exposure time in minutes. The procedure permits determination of the total α-particle activity profile as well as a subsequent examination of any depth-related trends in a sediment core.

The supported ^{210}Pb component is measured through U assay by means of the fission track technique described in chapter 12. The sample pellet is attached to a sheet of Lexan polycarbonate and irradiated, together with standard glasses, in the thermal-neutron column of a nuclear reactor for 5 min to accumulate a total flux of 10^{16} neutrons cm^{-2}. This irradiation produces fission in the ^{235}U in the sample and the resultant fission fragments cause damage trails in the Lexan detector. After irradiation and an appropriate cooling period, the latter is etched in 6 M

NaOH at 60°C for 20 min and the track density determined. The U concentration is then obtained from

$$U_s = \frac{D_x}{D_s} U_x \frac{I_s}{I_x},$$

in which U_x, U_s are the U contents in the sample and standard respectively, D_x and D_s are the track densities of each, I_s and I_x being the respective isotope abundance ratios. The total -α activities and U concentrations can be measured routinely by this approach, usually to within $\pm 5\%$. The detection efficiencies attained were $>95\%$.

Radiocaesium analyses were carried out, weighed dried sediment samples being packed into standard geometrical countings vials which fitted reproducibly into a plastic holder over the face of a 100 cm^3 Ge(Li) γ-detector. A maximum replicate error of 6.6% applied to the ^{137}Cs analyses.

For metal analysis, the chemical procedure that was adopted comprised overnight leaching of 0.5-1.0 g dried ground sediment at 60°C with 100 ml 25% CH_3CO_2H solution, 0.25 M in NH_2OH-HCl, leaching of residual sediment with hot 30% H_2O_2 and final dissolution of the resistant residue using $HCl/HNO_3/HF/H_3BO_3$. After the various solutions were processed, the elemental analyses were effected by atomic absorption spectrometry using an IL151 or PE306 spectrophotometer in the appropriate flame (air/C_2H_2, N_2O/C_2H_2) or flameless (HGA-74) mode. The sediment analyses were performed in duplicate with overall analytical precision in the range ± 5–10% ($\pm 1\sigma$). Arsenic was determined in whole sediment by instrument neutron activation analysis with a relative standard deviation of the mean of typically two or three replicate determinations of $\pm 15\%$.

The radionuclide date obtained showed an exponential decrease in the ^{210}Pb profile from ~ 22 dpm g^{-1} at the surface to ~ 1.7 dpm g^{-1} at depths exceeding 15 cm. The ^{226}Ra concentrations showed a small, but steady, decline with depth. As there is no parallel evidence of a slowly altering sediment composition, it is believed that the ^{226}Ra trend reflects the association of Ra with siliceous plankton, the remains of which are enriched towards the surface, before dissolution at depth and upwards diffusion. To some extent, this confuses the estimating of supported ^{210}Pb values for each sample. This is because, in the upper 12 cm, it is not possible to define the amount of secular equilibrium existing between ^{226}Ra and ^{210}Pb. Luckily, as unsupported ^{210}Pb predominates over supported ^{210}Pb in both magnitude and trend, this uncertainty becomes insignificant.

When the data are compared with palaeomagnetic and radiocarbon dating, the sedimentation rates derived are in good agreement at ~ 22 mg cm^{-1} a^{-1}.

Turning now to the question of snow and ice chronology, the time which passed since a sample of snow was deposited at a depth h below the surface can be calculated from the activity of ^{210}Pb if the initial activity of this radioisotope has remained constant, so that

$$^{210}\mathrm{Pb}_A = {}^{210}\mathrm{Pb}^0_{\mathrm{A}}\, \mathrm{e}^{-\lambda t},$$

in which the left-hand side expresses the activity of the radioisotope per unit weight of sample at depth h, $^{210}\mathrm{Pb}^0_{\mathrm{A}}$ is its activity at the surface (depth $h = 0$), λ is the decay constant of ^{210}Pb ($3.11 \times 10^{-2}\, \mathrm{a}^{-1}$) and t is the age of the snow sample. Therefore

$$t = \frac{1}{\lambda} \ln \left(\frac{^{210}\mathrm{Pb}^0}{^{210}\mathrm{Pb}} \right)_{\mathrm{A}}.$$

If the rate of annual water accumulation (a) is constant and $a = h/t$, then it may be stated that

$$\ln\,(^{210}\mathrm{Pb}_{\mathrm{A}}) = (^{210}\mathrm{Pb}^0_{\mathrm{A}}) - \frac{\lambda h}{a},$$

this being the equation of a straight line with a slope $m = -\lambda/a$ and an intercept on the ordinate in $^{210}\mathrm{Pb}^0_{\mathrm{A}}$. That is to say that, if the initial activity of the radioisotope and the rate of accumulation of water have remained constant at a particular site, then the ^{210}Pb activities of snow samples collected at different depths below the surface will approach a straight line in coordinates of $^{210}\mathrm{Pb}^0_{\mathrm{A}}$ and h. The initial activity $^{210}\mathrm{Pb}^0_{\mathrm{A}}$ may be obtained by extrapolation of the line to its point of intersection with the ordinate at $h = 0$.

Finally, it is necessary to examine the U-series disequilibrium in volcanics. This results, since daughters of ^{238}U are not in radioactive equilibrium with their parent, from chemical fractionation during the formation of magmas by partial melting in the upper mantle (assuming that secular equilibrium was present in these rocks prior to melting). The phenomenon is detectable from the activity ratios of $^{230}\mathrm{Th}/^{238}\mathrm{U}$, $^{226}\mathrm{Ra}/^{238}\mathrm{U}$ and $^{210}\mathrm{Pb}/^{238}\mathrm{U}$, all possibly exceeding unity. Since the isotopes of U are not subject to fractionation during the process, no disequilibrium has been noted between ^{234}U and ^{238}U.

Whole-rock samples of volcanics from Tristan da Cunha, Faial in the Azores and Mt Vesuvius were studied by Oversby and Gast (1968); their ages ranged from *c.* 500 to only a few years for lava flows which were extruded in 1961. The results obtained are listed in Table 5.3. All the activity ratios exceed 1, but there are important regional variations. However, all contain varying quantities of excess unsupported ^{230}Th, ^{226}Ra and ^{210}Pb. The excess of this latter radioisotope could be, at least in part, a consequence of the decay of excess ^{226}Ra. The sample from Mt

Table 5.3 Activity ratios of ^{230}Th, ^{226}Ra and ^{210}Pb and ^{210}Pb/ ^{230}U relative to ^{238}U in recently formed volcanic rocks

Location	*Number of Samples*	$\left(\frac{^{230}Th}{^{238}U}\right)_A$	$\left(\frac{^{236}Ra}{^{238}U}\right)_A$	$\left(\frac{^{210}Pb}{^{238}U}\right)_A$
Tristan da Cunha	5	1.14	1.41	1.04
Faial	4	1.45	2.52	1.90[a]
Vesuvius	1	1.05	10.8	8.40

Vesuvius is unusually interesting since it possesses an extremely large enrichment in ^{226}Ra as shown by its ^{226}Ra/^{238}U activity ratio of 10.8. The enrichment in ^{210}Pb can be explained because of the decay of ^{226}Ra, and indeed it is well established that there is high ^{210}Pb activity in the mineral cotunnite, $PbCl_2$, occurring as a sublimate in fumaroles round the volcano. Possibly the Pb in this mineral derives from volcanization from basaltic magmas. Recently, magma formation has been studied using the radioactive disequilibrium among the daughters of ^{238}U in such young volcanic rocks. In addition, a geochronometer useful in dating volcanic rocks is based upon the radioactive disequilibrium between ^{230}Th and ^{238}U in them (Allégre and Condomines).

It must be remembered that radioactive disequilibrium among the daughters of ^{238}U in such young volcanic rocks may arise not only from processes occurring during magma formation, but also from those taking place during subsequent movement surfacewards. Processes which may disturb the radioactive equilibria include the partitioning of trace elements between solids and the melt during partial fusion with later phenocryst formation as well as vapour fractionation and the contamination of magma through the assimilation of crustal rocks or even as a result of interaction with sea water. Some also promote systematic variations in the isotope compositions of Sr, Nd, Pb and O in volcanic rocks and indeed correlations between $(^{230}Th/^{232}Th/)_A$ and $(^{87}Sr/^{86}Sr)_A$ as well as $\delta^{18}O$ have been established, for instance by Newman, Finkel and MacDougall (1983).

Nevertheless, secular equilibrium has been shown to exist as regards the activities of ^{230}Th and ^{238}U in recent lava flows on volcanoes which are situated on plate margins. e.g. Mt St Helens in Washington, USA, by Krishnaswami, Turekian and Bennett (1984). Similar results have been obtained from lavas erupted by Kilauea and Mauna Loa on Hawaii, the ^{230}Th/^{238}U activity ratios only slightly exceeding unity (although there is ^{230}Th enrichment in prehistoric lava flows on the latter volcano in which $(^{230}Th/^{238}U)_A = 1.067 \pm 0.049$ (1σ) based on four measurements. Of course, this could also be due to ^{238}U-depletion by hot groundwaters after eruption. Some interesting data made it apparent that, despite the enrichment of some samples in ^{226}Ra, there is no appreciable fractionation in the relevant lavas. The enrichment in question may be due to processes taking place after magma evolution.

The activity of ^{210}Pb in volcanic rocks increases through the decay of excess ^{222}Rn escaping from them after the alpha decay of ^{226}Ra. Thus, those deviations noted in the $^{210}Pb/^{238}U$ activity ratio should not unequivocally be attributed to elemental fractionation effects in the magma. This caveat is applicable to samples with elevated values for this ratio, but normal $^{226}Ra/^{238}U$ activity ratios. In the case of the Galapagos, data include an anomalous sample (one of four), but seem to be in equilibrium for the series $^{228}U-^{234}U-^{230}Th-^{226}Ra$. Excess ^{210}Pb is believed to have been incorporated through the scavenging of this radioisotope from sea water.

Recent advances include an interesting paper by Nozaki (1986) dealing with $^{226}Ra-^{222}Rn-^{210}Pb$ systematics in sea water near the bottom of the ocean. In the preceding years, the distribution of ^{210}Pb was studied through the GEOSECS Program, particularly in the Pacific Ocean. Ocean boundaries were envisaged as potential sinks in regard to reactive metals in sea water. The ^{210}Pb deficiency relative to its ancestral ^{226}Ra was found usually to increase from mid-oceanic interiors towards the continental edges and bottom sea floor where it is more intensively scavenged. Within *c.* 100 m above the latter, there is a large quantity of excess ^{222}Rn relative to ^{226}Ra as a result of its supply by sediments through porewater diffusion. This ^{222}Rn decays to ^{210}Pb and therefore its effect on the distribution of the latter in the bottom water arises. Actually, four vertical profiles in the deep west North Pacific Ocean showed no significant variations with depth, irrespective of this excess ^{222}Rn. Consequently, the apparent box model residence time based on $^{210}Pb/^{222}Rn$ activity ratios decreases from ~ 100 a in ocean interiors to 10 a near the bottom. However, the ^{210}Pb data are explicable using a vertical-mixing and first-order scavenging model without enhanced scavenging near the sediment–water interface. This model calculation indicated that bottom excess ^{222}Rn can hardly influence ^{210}Pb in the water and the short residence time derived from the box model is an artefact which does not reflect the chemical reactivity of Pb. This implies that the deficiency of ^{210}Pb as a result of removal by scavenging is best estimated by using ^{226}Ra and not ^{222}Rn. The ^{210}Pb scavenging is best estimated by using ^{226}Ra and not ^{222}Rn. The ^{210}Pb scavenging at the sediment–water interface inferred from the oceanic distributions of ^{210}Pb appears difficult to detect from bottom ^{210}Pb profile measurements.

5.7 ACTINIDE-SERIES DISEQUILIBRIUM USED TO DETERMINE THE CHRONOLOGY OF DEEP-SEA HYDROTHERMAL ACTIVITY

Lalou, Reyss and Brichet (1993) investigated this subject, describing the various radiochronological techniques applicable for dating geologically recent deep-sea hydrothermal deposits less than 400 ka old and the fundamental conditions which must be met in order to obtain reliable dates. They indicated that existing approaches had to be either further developed or modified so as to be able to deal with materials from this environment, one discovered only in 1979. In the oceans the behaviour of the elements is dependent upon their geochemical properties in those physicochemical conditions imposed by sea water. Differing behaviour patterns disturb secular equilibrium in the three natural actinide decay series, ^{238}U, ^{235}U and ^{232}Th, and some of the disequilibria are in geochronological usage. However, in each instance, the choice of nuclide selected has to fulfil three constraints: (1) its half-life must accord with the timescale of the event to be dated, (2) initial conditions must be known, i.e. the activity ratio of the radionuclides employed, and (3) the sample being dated must have been a closed system from the time of formation to the time of measurement. Because of the complex nature of hydrothermal activity, the last two requirements must be verified for each deposit examined. Lalou, Reyss and Brichet (1993) presented two timescales, one from $\approx$ 0 to $\approx$ 200 a and a second from $\approx$200 to 400 000a. Three radionuclides were considered, namely ^{228}Th ($T_{1/2} = 1.913$ a), ^{228}Ra ($T_{1/2} = 5.75$ a) and ^{210}Pb ($T_{1/2} = 22$ a). Sulphides and barite-rich samples were analysed. Previous work showed that ^{232}Th cannot be detected in sulphides when they precipitate so that, if ^{228}Ra occurs, ^{228}Th will ingrow into transient equilibrium (^{228}Th/^{228}Ra ≈ 1.5) after *c.* 10 a. In the ^{238}U series, ^{210}Pb was considered appropriate for dating older samples up to $\approx$200 a if its origin and activity are known. But since sulphides are of variable chemical composition, the ^{210}Pb activity has to be normalized to stable Pb to be representative. Two origins were considered for Pb which have completely different specific activities: sea water (2.10^8 dpm g^{-1} Pb) and alteration of basalt (5.10^4 dpm g^{-1} Pb). The ^{210}Pb/Pb ratio (even where the Pb comes from basalts) can change after a long residence time of the fluid in the hydrothermal circuit either by a decrease if the ^{210}Pb is unsupported or by an increase of ^{210}Pb by growth *in situ* from ^{222}Rn. These researchers overcame the problem by using the ^{210}Pb/Pb ratio of that particulate matter formed in the plume as the initial ratio. They measured the ratio in particulates to look into the spatial and temporal variability of the various fields and, where no particles exist, the initial ratio was estimated from anhydrite samples. Anhydrite precipitates when a chimney is forming because of heating of ambient sea water by the high-temperature fluid, and it dissolves after a few years. Alternatively, it is possible to cross-check the method with the $^{228}Th/^{228}Ra$ one and calculate the initial $^{210}Pb/Pb$ ratio. The ratio was found to be constant for one field for at least 5a.

As regards barite-rich samples, different problems arose. Mixtures with sulphides may be used and the $^{228}Th/^{228}Ra$ approach was used with barite-rich chimneys from the Mariana Trough and the Juan de Fuca Ridge by Stakes and Moore (1991). A bulk sample was measured by γ spectrometry and it was assumed that the sulphide and barite precipitated simultaneously. Enrichment in Ra is connected with enrichment in Ba since the two have similar properties deriving from their similar ionic radii (0.142nm for Ba, 0.148nm for Ra). Radium can be normalized to Ba and, if the ratio $^{228}Ra/Ba$ at time t_0 (time of initial formation) is known, the decrease in the ratio can be employed in dating for *c*. 30 a with a resolution of several years. Also if ^{228}Ra is present, then ^{226}Ra must be present too at time t_0. Consequently, the ratio ($^{228}Ra/^{226}Ra$) is usable as a geochronometer if the initial ratio is known. The half-life of ^{226}Ra (1622 a) is long enough for it to be stable during the time pertinent to the method, i.e. $\approx$ 30 a. One value for the initial ratio was given as 0.207 ± 0.008.

The main drawback in using the $^{210}Pb/Pb$ ratio in dating barite-rich chimneys is greater than that relevant to sulphide samples because barite concentrates Ra and, if sulphide and barite are coprecipitated, both origins must be considered and the ratio corrected from the ingrowth of ^{210}Pb from ^{226}Ra. If barite and sulphide are not coprecipitated, the radionuclides must be measured in the two phases. In the sulphides ^{210}Pb will decrease since it is not supported by ^{226}Ra. In the barite, which is both Ra-rich and devoid of Pb, ^{210}Pb grows from ^{226}Ra. The sulphide fraction can be dated by the $^{210}Pb/Pb$ ratio and the barite phase by the $^{210}Pb/^{226}Ra$ ratio. As regards the 200 to 400 000 a timescale, it is possible to apply the $^{230}Th/^{234}U$ and $^{231}Pa/^{235}U$ methods by α-spectrometry, and the conditions at the time t_0 were verified on active chimneys. In young sulphide samples there are variable amounts of U, but no detectable ^{230}Th, ^{232}Th or ^{231}Pa. In samples older than 2000 a, the assumption of a closed system is verifiable if there is no ^{232}Th, when both methods give the same age or if two pieces of the same chimney with very different U concentrations yield the same age. These requirements are usually fulfilled if the oxidized part of the sample is avoided. On the timescale of 200 to 8000 a, the ratio $^{226}Ra/Ba$ is useful for dating, although the initial value of this varies between fields. Determination of the ratio at t_0 is best achieved by measuring it in the particulate matter of fluids escaping from the chimneys of a studied field. The $^{230}Th/^{234}U$ method can be applied also to dating hydrothermal Mn oxides if either samples lacking ^{232}Th are found or a correction for initial ^{230}Th can be made.

Turning to conclusions obtained from these researches, obtaining a synthesis of geochronological data from all of the submarine hydrothermal systems examined was limited by a number of factors. Perhaps the most important lies in the difficulty of obtaining representative samples and the best approach to this seems to be using deep submergence vehicles. There are problems of how to be sure that all relevant fossil events have been sampled in one field and how to avoid altered specimens. Also, some may be unobtainable due to burial by sediments. Finally, of the 66 000 km long active ridges known, only *c.* 1% has been explored.

However, some general information may be presented. Usually chimneys form rapidly at the spreading axis and may stop emitting fluid thereafter or they may remain active without growing. In addition, they may be reactivated after stopping to produce adventive chimneys. It appears that the entire life of an individual chimney is normally no more than a century or so. On the East Pacific Rise, fossil deposits are separated from active ones and usually found outside the rift valley. The ages of the former are frequently less than the 'magnetic age' of the basement. Should the high-temperature hydrothermal output be due to upwelling of magma under the rift valley of a given segment of the mid-ocean ridge system (itself the cause of filling of gaps between spreading plates), there exists a new possible means of acquiring more exact data about the spreading rate in a given segment. Lalou, Reyss and Brichet (1993) calculated a half-spreading rate for the past 4000 a (the age of the fossil deposit 700 m west from the axis) of $17 cm^{-1} a^{-1}$ instead of $3 cm\ a^{-1}$ given by others. The assumption was made that high-temperature events can only occur in the central valley, i.e. on a basement of zero age. They asserted that their work showed that hydrothermal activity does not represent a regular input of matter to the ocean so that its pulsed nature must be considered in all attempts at modelling (chemical, biological or tectonic).

6

Uranium–xenon and uranium–krypton dating

6.1 THEORY

Uranium undergoes spontaneous fission as well as radioactive α-decay and this produces fissiogenic Xe and Kr which accumulate in U-bearing minerals through geological time. The proposal to use these for determining the age of such minerals was made as early as 1947 when Khlopin, Gerling and Baronovskaya discovered Xe produced by the spontaneous fission of ^{238}U in pitchblende. MacNamara and Thode (1950) were the first to give the isotopic compositions of Xe_{sf} and Kr_{sf}. A U–Xe or U–Kr age can be calculated as follows:

$$t_{(U-Xe)} = \frac{1}{\lambda_a + \lambda_{sf}} \ln\left[\frac{^{i}Xe_{sf}(\lambda_a + \lambda_{sf})}{^{238}U\,^{i}Y_{sf}\,\lambda_{sf}} + 1\right],$$

where λ_a is the α-decay constant of ^{238}U, λ_{sf} is the ^{238}U spontaneous fission decay constant and $^{i}Xe_{sf}$ and $^{i}Y_{sf}$ are concentration and mass yield, respectively, for a Xe or Kr isotope with an atomic mass i resulting from spontaneous fission. Eikenberg, Signer and Wieler (1993) noted that, since $\lambda_a \gg \lambda_{sf}$, the above equation can be reduced to

$$t_{(U-Xe)} = \frac{1}{\lambda_a} \ln\left[\frac{^{i}Xe_{sf}\,\lambda_a}{^{238}U\,^{i}Y_{sf}\lambda_{sf}} + 1\right].$$

Determining the relevant ages necessitates knowledge of the product of $\lambda_{si} \times {}^{i}Y_{sf}$, but values published for $\lambda_{sf} \times {}^{136}Y_{sf}$ scatter between 7.4×10^{-18} a^{-1} and 5.4×10^{-18} a^{-1} with most of the published λ_{sf}-values clustered around 6 and 9.10^{-17} a^{-1}. The yield of the main ^{238}U spontaneous fission isotope ^{136}Xe is roughly 6%. Because the rate of production ratio $^{136}Xe_{sf}$

$^{206}Pb_{rad}$ is of the order of 4×10^{-8}, the U–Xe method was applied initially to U-rich minerals such as pitchblende, but later U-poorer accessory minerals such as monazites and zircons have been investigated as well. Eikenberg, Signer and Wieler (1993) mentioned that pitchblendes are liable to be affected by geological events causing alteration or dissolution and subsequent redistribtion of U and its daughters so that it is the case that the reported U–Xe ages of pitchblendes are frequently below their formation ages. Earlier, this disparity was ascribed to gass losses by diffusion from which it could be inferred that pitchblende is unsuited for dating by the U–Xe approach. Later work using the U–Xe, U–Kr and U–Pb systems together with electron microprope analysis demonstrated that the fission gases in pitchblendes could well have survived even greenschist facies metamorphism so that studying them would aid in the interpretation of discordant U–Pb isotope ages. Obviously if geologically undisturbed pitchblendes gave concordant U–Pb ages and underwent no losses of fission gases, the U–Xe and U–Kr ages should both be identical and agree with the U–Pb ages if appropriate values for λ_{sf}, ${}^{i}Y_{sf}$ and $(^{136}Xe/^{86}Kr)_{sf}$ are employed. Where gas losses did occur, however, Kr seems always to be lost preferentially with respect to Xe and both fissiogenic Xe and Pb losses ran almost parallel. Discordant fission gas ages, that is to say $^{36}Xe_{sf}/^{86}Kr_{sf} \geqslant 6.1 \pm 0.1$, clearly indicate perturbation of the U–Pb system. Also, extensive fractionation between Xe and Kr in pitchblendes has been observed. Eikenberg, Signer and Wieler (1993) attempted to deduce a value for $\lambda_{sf} \times {}^{136}Y_{sf}$ as well as the isotopic composition of spontaneous fission Xe and Kr together with the spontaneous fission $^{136}Xe/^{86}Kr$ ratio. They analysed two types of pitchblende, coarse-grained and finely disseminated, applying U–Pb analysis, noble gas analysis, and electron microprobe and atomic absorption analysis, obtaining important data. Some samples gave discordant U–Pb, but concordant U–Xe and U–Kr, ages coinciding with lower U–Pb discordia intercepts and U–Pb microprobe ages from single grains. They contained pitchblende as fine-grained ore disseminations probably representing recrystallized minerals. In the course of dissolution and precipitation of U, fission gases were released quantitatively from the system which produced a resetting of the U–Xe–Kr clock. Part of the radiogenic Pb still occurred in minerals adjacent to pitchblende.

In the samples of Eikenberg, Signer and Wieler (1993), Xe and Kr constituted mixtures from four sources, namely (1) the spontaneous fission of ^{238}U, (2) the slow neutron-induced fission of ^{235}U, (3) the fast neutron-induced fission of ^{238}U and (4) atmospheric gases. Atmospheric Xe and Kr impurities were removed by the assumption that ^{130}Xe and ^{82}Kr are wholly atmospheric. This was justified on the basis that these isotopes are shielded by the stable isobars ^{130}Te and ^{82}Se, respectively, and also because the direct fission yields of both are below 0.01%. As the measured $^{130}Xe/^{136}Xe$ ratio was usually less than 10^{-2} and in most cases below 10^{-3}, probably very little contamination with atmospheric Xe takes place. However, contamination of samples with atmospheric Kr was regarded as more serious than Xe since the atmosphere contains approximately an order of magnitude more Kr than Xe. In addition, the fission yields of ^{84}Kr and ^{83}Kr are *c.* 40 and 150 times lower than those of ^{136}Xe. Luckily, most samples were only very slightly contaminated with atmospheric Kr, as demonstrated by their low $^{82}Kr/^{86}Kr$ ratios.

To use the U–Xe–Kr dating method, the product $\lambda_{sf} \times {}^{i}Y_{sf}$ must be known. Additionally, so should the concentrations of spontaneous fission Xe and Kr, the relative Xe and Kr spontaneous fission yields and the $(^{136}Xe/^{86}Kr)_{sf}$ ratio. Any contribution from ^{238}U which underwent fast neutron-induced fission can be neglected. Eikenberg, Signer and Wieler (1993) commenced their iteration with the following values; for ^{136}Y from the slow neutron-induced fission of ^{235}U, 6.39% and for $^{136}Y_{sf}$, 6.3%, both taken from earlier researchers. Other germane values were adopted from them as well. From isotope pairs 136–131, 134–131 and 132–131 for which the differences between induced and spontaneous composition are large, a mean value for a factor R was calculated for each sample, R being the ratio of induced to spontaneous fission events. Then U–^{i}Xe and U–^{i}Kr ages for each sample were calculated. While the U–^{i}Xe ages were only slightly different, there were significant differences between the U^{136}–Xe and U–^{86}Kr ages of geologically unaffected pitchblendes, a matter regarded as unsurprising by Eikenberg, Signer and Wieler (1993) on the grounds that varying compositions of fission Kr from $^{238}U_{sf}$ have been reported. Keeping $^{136}Y_{sf}$ fixed, they changed the $^{i,j}Y_{sf}$ yields of the deviational U–^{i}Xe and U–^{i}Kr systems to obtain agreement between all ages ($^{i,j}Y_{sf}$ being the mass yields of the isotopes i and j from ^{238}U spontaneous and ^{235}U slow neutron fission, respectively). The new yields were inserted into suitable equations so as to calculate a new set of ages, a procedure iterated until all ages agreed satisfactorily within analytical error. Most samples gave $(^{136}Xe/^{86}Kr)_{sf}$ values around 6.1. And back-up data showed that they were not geologically disturbed. However, some samples had undergone Xe, Kr and Pb losses of up to 50% –but they were still suitable to determine the relative $^{i}Xe_{sf}$ and $^{i}Kr_{sf}$ isotopic ratios. This is because minute losses of noble gases have an undetectable effect on the isotopic composition of Xe and Kr.

As regards such geologically undisturbed samples, the product $\lambda_{sf} \times {}^{136}Y_{sf}$ was obtained and required that the minerals analysed remained closed both to U and the fission gases. Only samples with concordant U–Pb ages and the recrystallized pitchblendes (U–Pb discordant, but U–Xe–Kr concordant) were examined, the latter having formation ages determinable from the intercept of the U–Pb discordia with the concordia and also from U–Pb microprobe analysis. All values of $\lambda_{sf} \times {}^{136}Y_{sf}$ were around $5.7 \times 10^{-18}\,a^{-1}$ and agreed within error. No fractionation between the heavier Xe and the lighter Kr was observed, from which it was concluded that the selected minerals retained their fission Xe quantitatively. Comparing the value for $\lambda_{sf} \times {}^{136}Y_{sf}$ obtained with values given using other techniques such as radiochemical isotope counting and fission track methods gave interesting results. For λ_{sf}, there are two widely used values, one derived from direct fission counting, which is $8.5 \times 10^{-17}\,a^{-1}$ and a second based on fission track results, which is $6.4 \times 10^{-17}\,a^{-1}$. Refining the value was achieved by fitting spontaneous fission mass yield data from various sources and this produced $(8.6 \pm 0.2) \times 10^{-17}\,a^{-1}$, the value published by von Gunten (1969). Considering his figure and fission yields given by Kuroda, Edwards and Ashizawa (1956), a value for $\lambda_{sf} \times {}^{136}Y_{sf}$ of $5.4 \times 10^{-18}\,a^{-1}$ can be obtained and this compares well with the Eikenberg, Signer and Wieler (1993) value of $(5.7 \pm 0.4) \times 10^{-18}\,a^{-1}$; these researchers noted that using a lower value of $\lambda_{sf} = 6.4 \times 10^{-17}\,a^{-1}$ would require ${}^{136}Y_{sf} = 9\%$ to accord with their result. They added that such a large ^{238}U fission yield at mass 136 is impossible because it would entail a total mass yield beyond 100%, taking the relative yields from Kuroda, Edwards and Ashizawa (1956) or von Gunten (1969). They concluded that their work and earlier studies on U-bearing minerals demonstrated that the high value of $8.6 \times 10^{-17}\,a^{-1}$ for λ_{sf} is correct, but a discrepancy of $\sim 20\%$ remains between the $\lambda_{sf} \times {}^{i}Y_{sf}$ values found from U minerals such as pitchblende or uraninite on the one hand and U-bearing accessory minerals such as zircon or monazite on the other. Taking the most commonly used value of 6.3% for ${}^{136}Y_{sf}$ and the high value of $7.4 \times 10^{-18}\,a^{-1}$ for $\lambda_{sf} \times {}^{136}Y_{sf}$, a value of $11.7 \times 10^{-17}\,a^{-1}$ would result for λ_{sf}, which is significantly greater than either the fission decay constants proposed by von Gunten (1969) or Fleischer and Price (1964a). Such a result would not accord with almost all values for λ_{sf} obtained from various methods.

Eikenberg, Signer and Wieler (1993) investigated the isotopic composition of spontaneous fission Xe and Kr, noting that their isotopic abundances of ^{238}U-derived Xe_{sf} were close to previously published values with a small depletion of the lighter fission Xe isotopes relative to ${}^{136}Xe_{sf}$. They noted that Shukolyukov (1970) measured Xe in samarskite and recorded a fissiogenic Xe and Kr composition close to theirs, adding that this mineral has high REE concentrations and apparently the high neutron absorption cross-sections of several REE isotopes favour the mineral in studying the spontaneous fission Xe and Kr isotope spectrum.

Leaving aside the samarskite case, the isotope ratios for spontaneous fission Kr presented disagree with values reported earlier. However, all data defined a linear array extrapolating through a calculated spontaneous fission point as well as a ^{235}U slow neutron-induced fission point given by Crough (1977). Eikenberg, Signer and Wieler (1993) took this as indicative that, like Xe, fission Kr in their samples comprises two components, i.e. spontaneous fission of ^{238}U and slow neutron-induced fission of ^{235}U. Results showed strong support for the validity of the isotopic composition of fissiogenic Kr from $^{238}U_{sf}$.

Regarding the $^{138}Xe/^{86}Kr$ spontaneous fission ratio, these researchers noted that nine undistributed samples having minor contribution of induced fission krypton ($R \leqslant 0.05$) gave $(^{136}Xe/^{86}Kr)_{sf}$ ratios which ranged from 5.9 to 6.3 with a mean value of 6.1 ± 0.1. This agreed well with a zircon-derived value of 6.0 ± 0.4 given by Hebeda, Freudel and Schultz (1987). Some samples lost fissiogenic daughter products which caused fractionation between Xe and Kr, thus raising $(^{136}Xe/^{86}Kr)_{sf}$ ratios. The U–Pb concordant uraninite B-1 gave a surprisingly high $(^{136}Xe/^{86}Kr)_{sf}$ ratio of 6.5. As this particular sample was much the oldest U-rich sample at *c.* 1.5 Ga, probably the long self-irradiation gave rise to lattice deformations resulting in a small Kr loss of 6%. Eikenberg, Signer and Wieler (1993) suggested that U-bearing minerals with $(^{136}Xe/^{86}Kr)_{sf}$ 6.1 ± 0.1 show concordant U–Xe and U–Kr ages.

6.2 FISSIOGENIC RARE GASES IN THE ATMOSPHERE

Eikenberg, Signer and Wieler (1993) estimated a maximum contribution of ^{136}Xe from ^{238}U spontaneous fission of 1.6% to the total inventory of ^{136}Xe in the atmosphere, which agrees with estimates given by Pepin and Phinney (1976) and Bernatowicz and Podosek (1978). The source for ^{129}Xe is not yet established, but an excess of it has been recorded in terrestrial gases and minerals, for instance by Staudacher and Allègre (1982). Such excesses may be due to the production of now-extinct ^{129}I ($T_{1/2} = 16$ Ma) during nucleosynthesis or perhaps arose from a non-primordial origin. However, Eikenberg, Signer and Wieler (1993) ruled out significant contribution of ^{129}Xe from $^{238}U_{sf}$ to the Earth's atmosphere on the basis of their $^{129}Xe/^{136}Xe$ spontaneous fission ratio of $\leqslant 0.001$.

7

Rubidium–strontium dating

7.1 GEOCHEMISTRY

Rubidium (Latin *rubidus* meaning deep red) is an element with relative atomic mass of 85.4678 and atomic number 37. A soft and highly reactive white metal, it resembles Na. Its relative density is 1.53, its melting point is 38.9°C and its boiling point is 705°C. It does not form any minerals, but occurs dispersed in common K-bearing ones such as the micas, K feldspar, some clay minerals and evaporites. The reason is that its ionic radius is 0.148 nm, close to that of K (0.133 nm), for which it can substitute in all K-bearing minerals. It is found in lepidolite to the extent of *c.* 1.5% so that this mica constitutes a commercial source. It is also present in the simple feldspathoid leucite, a KAl silicate present as disseminated grains and crystals in some basaltic rocks and mined in Italy as a fertilizer meterial. Other commercial sources include brines in Michigan and K minerals found at Searles Lake in California. With Li, Na, K, Cs and Fr, Rb is assigned to Group IA (alkali metals) in the Mendeleev and other 'Short' forms of the Periodic Table. In different nomenclatures, it is included among typical, main group, non-transition or s and p block elements, and has an electronic configuration of [Kr] $5s^1$. In a new set of designations which was recommended by the International Union of Pure and Applied Chemistry, Rb is in category 1.

Rubidium has 27 isotopes of which only two are found in nature, namely the stable ^{85}Rb and the radioisotope ^{87}Rb, of which the isotopic abundances are 72.17% and 27.83%, respectively. All its other isotopes are radioactive and short-lived, but ^{87}Rb is long-lived and can be used in Rb–Sr dating. An exact knowledge of its half-life is fundamental and initially greatly varying values were presented, resulting in confusion as regards adopting and rejecting ages so determined. Now a value of

4.8×10^{10} a is accepted and the following expression applies to its decay:

$$^{87}Rb \rightarrow {}^{87}Sr + \beta^- + \bar{v} + Q$$

where β^- is a beta particle (negatron), $\bar{v}$ is an antineutrino and Q is the decay energy (0.275 MeV).

Strontium (from Strontian, a Scottish town) is an element with relative atomic mass of 87.62 and atomic number 38. A reactive metal, it resembles Ca. Its relative density is 2.583, its melting point is 770°C, much higher than that of Rb, which can be liquid at room temperature. Its boiling point is 1390°C. It occurs as celestine ($SrSO_4$) and strontianite ($SrCO_3$). As its ionic radius is 0.113 nm and that of Ca somewhat smaller (0.099 nm), it is able to replace Ca in many minerals despite the difference in size. Consequently, like Rb, it too is found dispersed, especially in Ca-bearing minerals such as plagioclase, apatite and aragonite. However, the size difference in the ionic radii of the two elements does impose constraints on such replacement because Sr ions favour eightfold coordinated sites whereas Ca ions are small enough to fill both into these and also into sixfold ones. In addition, Sr^{2+} ions are subject to capture by K feldspar in place of K^{1+} ions. But the replacement of the latter by the former must be associated with the replacement of Si^{4+} by Al^{3+} in order to maintain electrical neutrality.

Strontium has 23 isotopes of which four occur naturally. They are stable and are ^{84}Sr, ^{86}Sr, ^{87}Sr and ^{88}Sr; they have abundances of 0.56% 9.86% 7% and 82.85%, respectively. One of them, namely ^{87}Sr, has a metastable state with a γ-energy of 388 keV for the isomeric transition and decays by electron capture over a time interval governed by the half-life which is 2.81 h. Clearly the abundance of ^{87}Sr is variable, reflecting the formation of radiogenic ^{87}Sr through the decay of the naturally occurring ^{87}Rb. Hence the exact isotopic composition of strontium in a mineral or rock which contains Rb will depend upon the age and the Rb/Sr ratio of that particular mineral or rock.

Average concentrations of Rb, K, Sr and Ca in various types of igneous and sedimentary rocks were given by Turekian and Wedpohl (1961) and are listed in Table 7.1. These data show the geochemical coherence of the relevant four elements and most common rocks contain measurable concentrations both of Rb and Sr in amounts up to several hundred ppm. Rb/Sr ratios in common igneous rocks are variable and range from 0.06 in basaltic ones to more than 1.7 in highly differentiated granites with low Ca contents. It appears likely that Sr concentrates in early-formed calcic plagioclase when magma fractionally crystallized while Rb remains in the liquid phase. The Rb/Sr ratio of the remaining magma may increase gradually during ongoing crystallization so that suites of differentiated igneous rocks usually have Rb/Sr ratios which increase as a function of increasing differentiation.

Table 7.1 Average concentrations (ppm) of Rb, K, Sr and Ca in various type of igneous and sedimentary rocks (from Turekian and Wedepohl, 1961)

Rock type	*Rb*	*K*	*Sr*	*Ca*
Ultrabasic	0.2	40	1	25 000
Basaltic	30	8 300	465	76 000
High Ca granite	110	25 000	440	25 300
Low Ca granite	170	42 000	100	5 100
Syenite	110	48 000	200	18 000
Shale	140	26 600	300	22 100
Sandstone	60	10 700	20	39 100
Carbonate	3	2 700	2 000	312 400
Deep–sea carbonate	10	2 900	2 000	312 400
Deep–sea clay	110	25 000	180	29 000

7.2 METHODOLOGY

7.2.1 Theory

The growth of radiogenic ^{87}Sr in a Rb-enriched mineral over a time interval t may be expressed thus:

$$^{87}\text{Sr} = {}^{87}\text{Sr}_\text{i} = {}^{87}\text{Rb}\,(e^{\lambda t} - 1)\,, \tag{7.1}$$

where ^{87}Sr and ^{87}Rb are the total numbers of atoms of the respective isotopes in a unit weight of the mineral now and $^{87}\text{Sr}_i$ is the amount of ^{87}Sr incorporated into it when it formed, λ is the decay constant in reciprocal years of ^{87}Rb and t is the age of the mineral sample in years. If each term is divided by the number of ^{86}Sr atoms found (a constant because this is a stable isotope), then

$$\frac{^{87}\text{Sr}}{^{86}\text{Sr}} = \left(\frac{^{87}\text{Sr}}{^{86}\text{Sr}}\right)_\text{i} + \frac{^{87}\text{Rb}}{^{86}\text{Sr}}\,(e^{\lambda t} - 1)\,, \tag{7.2}$$

and this constitutes the equation for age determination using the method. A requirement is that no loss of Rb or Sr or both has taken place since the mineral formed, i.e. that the system may be regarded as closed if neither element has been added to the system through t. λ is the decay constant and values for this vary according to the half-life of ^{87}Rb accepted. As noted earlier, 4.8×10^{10} a is frequently used, but that of Steiger and Jäger (1977) may be quoted. It is 4.88×10^{10}a and their decay constant value is $1.42 \times 10^{-11}\text{a}^{-1}$.

Measuring Rb and Sr concentrations in a Rb-bearing mineral sample may be done using standard chemical and mass spectrometric procedures described by Papanastassiou and Wasserburg (1973). X-ray fluorescence and isotope dilution techniques are also feasible. In the mass spectrometer, the ^{87}Sr/^{86}Sr ratio is measured using an appropriate pure

Sr salt produced by dissolving the mineral sample in acid and separating Sr by means of cation exchange chromatography.

The ratio of concentrations of Rb and Sr is converted into an ^{87}Rb/^{87}Sr ratio thus:

$$\frac{^{87}\mathrm{Rb}}{^{86}\mathrm{Sr}} = \left(\frac{\mathrm{Rb}}{\mathrm{Sr}}\right)_{\mathrm{c}} \times \frac{\mathrm{Ab}(^{87}\mathrm{Rb}) \times \mathrm{W}(\mathrm{Sr})}{\mathrm{Ab}(^{86}\mathrm{Sr}) \times \mathrm{W}(\mathrm{Rb})},$$

where the derived ratio is that between the isotopes involved given in terms of the numbers of atoms present in a unit weight of the mineral sample at present, $(\mathrm{Rb/Sr})_{\mathrm{c}}$ is the ratio of the concentrations of the elements, Ab refers to the respective isotope abundances and W(Sr), W(Rb) are the respective atomic weights. The abundance of ^{86}Sr relates to the abundance of ^{87}Sr and varies from sample to sample. The abundance and atomic weight of ^{87}Rb probably changed continuously since nucleosynthesis, but now all samples of terrestrial, meteoritic and lunar Rb have the same isotopic composition and atomic weight.

In obtaining an age, a suitable value must be used for the ^{87}Sr/^{86}Sr ratio of that Sr which was incorporated into the mineral sample when it formed. But when this mineral is highly enriched in radiogenic ^{87}Sr, data calculated from an appropriate age equation may be regarded as independent of this ratio and it is permissible to select an initial ratio of, e.g. 0.706, as being representative of a large mumber of recent basic volcanics containing strontium probably derived from the Earth's upper mantle.

After measurement of the rubidium and strontium concentrations in a Rb-bearing mineral and employing a suitable value for the initial ^{87}Sr/^{86}Sr ratio, a value for t may be obtained as follows:

$$t = \frac{2.303}{\lambda} \ln \left[\frac{\dfrac{^{87}\mathrm{Sr}}{^{86}\mathrm{Sr}} - \left(\dfrac{^{87}\mathrm{Sr}}{^{86}\mathrm{Sr}}\right)_{\mathrm{i}}}{\dfrac{^{87}\mathrm{Rb}}{^{86}\mathrm{Sr}}} + 1 \right], \tag{7.3}$$

and the validity of t depends both on the mineral having remained a closed system with respect to Rb and Sr during its geological history, and on having a suitable assumed value for the initial ^{87}Sr/^{86}Sr ratio. These requirements mean that the calculated value of t becomes a model date for the mineral sample analysed and obviously the degree to which the model represents the real geological history is critical to the validity of t.

Igneous acidic rocks such as granites are suitable for the method since they contain both micas and K feldspar. As with U–Th–Pb dating, both concordant and discordant data may be obtained, the latter resulting from gain or loss of radiogenic ^{87}Sr through such geological events as reheating due to regional metamorphism. Suites of comagmatic igneous rocks with different chemical compositions are the product of fractional crystallization of magma with subsequent separation of crystals from the remaining liquid. If the Sr in such a magma stayed isotopically homogeneous during cooling, all such suites had the same initial $^{87}Sr/^{86}Sr$ ratio. In addition, they will have about the same age because the time taken for crystallization is negligible compared with the date of the event. Hence the age equation may be regarded as representing a group of straight lines in a slope intercept form such that $y = b + mx$ and all rock samples from a comagmatic suite as points on a straight line in coordinates of $^{87}Sr/^{86}Sr$ (y) and $^{87}Rb/^{86}Sr$ (x), this constituting an isochron. All points on this represent systems with the same initial $^{87}Sr/^{86}Sr$ ratio and its slope is related to t in the following manner:

$$m = e^{\lambda t} - 1\,, \tag{7.4}$$

with the value of the initial $^{87}Sr/^{86}Sr$ ratio (b) being provided by the y-intercept. Recapitulating, any suite of comagmatic rocks with an age t defines an isochron if its members all had the same initial $^{87}Sr/^{86}Sr$ ratio and if it was closed to both Rb and Sr since crystallization occurred and neither element was introduced from any extraneous source.

Optimally, the whole-rock isochron approach involves a suite of which the members cover as broad a range of Rb/Sr ratios as possible so as to obtain a well-defined slope for the relevant isochron. A straight-line fit by regression analysis may be effected and both the slope and the intercept of the isochron determined. The age of the suite can be derived from the slope by solving equation (7.4). This age is taken as showing the time elapsed since all the rocks in the suite had the same initial $^{87}Sr/^{86}Sr$ ratio, which should relate to the time of crystallization. In this way, whole-rock Rb–Sr isochrons can produce rather reliable age data for rock samples.

Figure 7.1 shows the isotopic evolution of Sr in a suite comprising two igneous rocks derived from a common magma, but with different Rb/Sr ratios. When crystallized, both plot as points on a straight line of zero slope because they have the same $^{87}Sr/^{86}Sr$ ratio. They cooled until becoming closed systems after which their $^{87}Sr/^{86}Sr$ ratios increased through the decay of ^{87}Rb to ^{87}Sr. Each decay process decreased the $^{87}Rb/^{86}Sr$ ratio and increased the $^{87}Sr/^{86}Sr$ ratio in the same proportion. Therefore the ratios altered along straight lines so that the rock samples stay on the isochron as its slope increases as a function of time. The value of the y-intercept remains constant and demonstrates the initial $^{87}Sr/^{86}Sr$ ratio of the suite.

7.2.2 Isochrons and Fictitious Isochrons

When $^{87}Sr/^{86}Sr$ and $^{87}Rb/^{86}Sr$ ratios have been determined and placed on isochron diagrams, best straight lines must be fitted to the relevant points. This cannot be done perfectly because of analytical errors so that

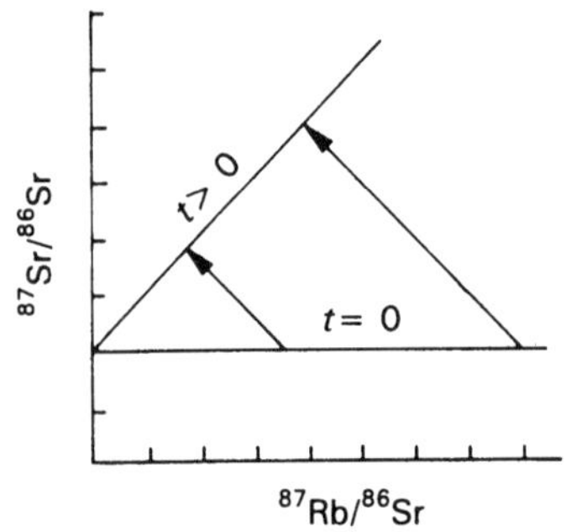

Fig. 7.1 This shows a Rb–Sr isochron diagram illustrating the time-dependent isotopic evolution of rock systems after crystallization from a homogeneous magma.

there will always be an uncertainty in the slope estimate, hence the inferred age. Also, the value of the initial $^{87}Sr/^{86}Sr$ ratio will be uncertain.

Analytical errors relate to random and systematic mistakes. The former have to do with the precision of measurement and the latter restrict accuracy. Fortunately, although they cannot be eliminated altogether, they can be limited. To fit the data points to the isochron, it is necessary to determine how to use the analytical errors of the coordinates in order to identify the best slope and intercept. In addition, it must be decided whether a particular point fits within the analytical error or not. If the point is not acceptable, i.e. does not fit within the analytical error, then the appropriate data coordinates for that point may have systematic analytical error or the sample involved may not fulfil the requisite assumptions for the isochron approach and could either have an initial $^{87}Sr/^{86}Sr$ ratio different from others in the suite or not be a closed system.

While isochrons can be fitted visually to data points, a more objective technique is to apply least-squares regression procedure. With this, deviations of either the x- or the y-coordinate from the best line in the slope-intercept form must be minimized. The following equations can be employed in order to calculate the best slope m and intercept b of a straight line:

$$m = \frac{\Sigma XY - ((\Sigma X)(\Sigma Y))/N}{\Sigma X^2 - (\Sigma X)^2/N},$$

$$\mathrm{b} = \frac{(\Sigma X)(\Sigma XY) - \Sigma Y(\Sigma X^2)}{(\Sigma X)^2 - N(\Sigma X^2)},$$

where Y refers to the $^{87}Sr/^{86}Sr$ ratios, X to the $^{87}Rb/^{86}Sr$ ratios and N to the number of data points.

A simple least-squares regression assumes that deviations of the data points from the best straight line result only from errors in the Y-coordinates, while ignoring such errors in the x-coordinates which certainly exist. Consequently, the fitting of isochrons in this way is not entirely satisfactory. Cameron *et al.* (1981) discussed the statistical analysis and interpretation of imperfectly fitted Rb–Sr isochrons from polymetamorphic terranes.

In cases where mixtures occur, a frequent happening in nature, e.g. if mantle magma intrudes the crust and interacts with country rock, materials of quite disparate chemical and isotopic compositions associate. Mixing models can be used to describe this situation and, where two components are involved, a mixing parameter f can be defined thus:

$$f = A/(A+B)$$

where A and B are the weights of these in a sample of the mixture. The concentration of any element X in this, X_M, is given by

$$X_M = X_A f + X_B(1-f)$$

where X_A and X_B are the concentrations of X in the two components A and B, respectively, given in units of weight. These concentrations will be constant for a suite of samples formed by mixing A and B in differing proportions. Therefore X_M is a linear function of f and $X_M = f(X_A - X_B) + X_B$ and the value of f can be calculated from the concentrations of X in a two-component mixture if its concentrations in the end members X_A and X_B are known. By extrapolation, it is possible to consider the concentrations of two elements, say X and Y, in such a mixture of A and B with values X_A, Y_A and X_B, Y_B, respectively. If f is eliminated the following expression can be derived:

$$Y_M = X_M \frac{(Y_A - Y_B)}{(X_A - X_B)} + \frac{(Y_B X_A - Y_A X_B)}{(X_A - X_B)}.$$

This is the equation of a straight line in the coordinates X_M and Y_M, i.e. the locus for all data points representing mixtures of the two components A and B in varying proportions, including the pure ones. The significance of such mixing relationships lies in their ability to form linear correlations between concentrations of pairs of elements. Straight lines can be fitted to arrays of such points which represent two-component mixtures in differing proportions. If the concentration of one component in the end members is known, the equation can be used to ascertain that of the other. Of course, the value of the mixing parameter is determinable from the penultimate equation and applies to any particular sample in a suite of two-component mixtures from the observed concentration of an element in the mixture in question.

Where two-component mixtures have different $^{87}Sr/^{86}Sr$ ratios, the total number of ^{87}Sr atoms in a unit weight of such a mixture is given by

$$^{87}Sr_M = \frac{Sr_A Ab_A{}^{87} N_A f}{W_A} + \frac{Sr_B Ab_B{}^{87} N_A (1-f)}{W_B},$$

where Sr_A and Sr_B are the concentrations of Sr, Ab_A^{87} and Ab_B^{87} are the isotopic concentrations of ^{87}Sr, W_A and W_B are the atomic weights of Sr in components A and B, respectively, N_A is the Avogadro number (6.0221367×10^{23} mol^{-1}) and f is the mixing parameter. A similar equation can be presented for the total number of ^{86}Sr atoms and formation of the $^{87}Sr/^{86}Sr$ ratio eliminates the Avogadro number to produce

$$\left(\frac{^{87}Sr}{^{86}Sr}\right)_M = \frac{Sr_A Ab_A{}^{87} f w_B + Sr_B Ab^{87}(1-f)\,W_A}{Sr_B Ab_A{}^{87} f w_B + Sr_B Ab^{86}(1-f)\,W_A}$$

For all practical purposes, the atomic weights of Sr and the abundances of ^{86}Sr in the two components $Ab_A{}^{86}$ and $Ab_B{}^{86}$ may be regarded as equal so that the previous equation can be simplified to

$$\left(\frac{^{87}Sr}{^{86}Sr}\right)_M = \frac{Sr_A Ab_A{}^{87} f + Sr_B Ab_B{}^{87}(1-f) W_A}{Ab^{86}(Sr_A f + Sr_B (1-f) W_A},$$

and further modified to

$$\left(\frac{^{87}Sr}{^{86}Sr}\right)_M = \frac{a}{Sr_M} + b,$$

where a and b are constants governed by the relevant concentrations and $^{87}Sr/^{86}Sr$ ratios of the Sr in the components A and B. Such a mixing hyperbola is transformable into a straight line by plotting $(^{87}Sr/^{86}Sr)_M$ versus $1/Sr_M$. This facilitates fitting a straight line to data points in co-ordinates of $^{87}Sr/^{86}Sr$ and $1/Sr_M$ and obtaining the mixing equation through measurement of the parameters involved from samples comprising two components. The fit of the data points to a straight line verifies the validity of the mixing hypothesis taken from Faure (1986). Of course, it assumes that no alteration in the $^{87}Sr/^{86}Sr$ ratios or Sr concentration took place after mixing.

Faure (1986) examined the case where Sr and another element, say Z, occur in a two-component mixture, pointing out that their relationship is linear and

$$Sr_M = Z_M \frac{(Sr_A - Sr_B)}{(Z_A - Z_B)} + \frac{(Sr_B Z_A - Sr_A Z_B)}{(Z_A - Z_B)},$$

where Sr_M and Z_M are the concentrations of Sr and element Z in the mixture. The further expression can be derived

$$Sr_M = cZ_M + d\,,$$

where c and d are constants governed by the concentrations of Sr and element Z in the end members. From this, it is possible to derive

$$\left(\frac{^{87}Sr}{^{86}Sr}\right)_M = \frac{a}{cZ_M + d} + b\,,$$

which is the equation of a hyperbola. However, this cannot be transformed into a straight line because it has an extra term containing the product of the *x*- and *y*-coordinates. Consequently, it is not possible to derive the mixing equations linking the $^{87}Sr/^{86}Sr$ ratio of two-component mixtures to the concentrations of other elements by fitting straight lines to data points in coordinates of $^{87}Sr/^{86}Sr$ and $1/z$. Rather, it is necessary to fit such lines to data points in coordinates of Sr_M and Z_M, thereafter deriving equations of a form following equations which can be solved so as to obtain compatible values of Z_A and Z_B for Sr_A and Sr_B.

Apropos fictitious isochrons, these have slopes devoid of geochronological significance. If a set of samples resulted from the mixing of two components with different $^{87}Rb/^{86}Sr$ and $^{87}Sr/^{86}Sr$ ratios, they will fall on a straight line which links the end members in coordinates of the $^{87}Rb/^{86}Sr$ and $^{87}Sr/^{86}Sr$ ratios. Such a straight line constitutes a fictitious isochron and its slope can be either positive or negative according to the relative values of the $^{87}Rb/^{86}Sr$ and $^{87}Sr/^{86}Sr$ ratios of the end members. The coordinates of points representing mixtures of two components with different concentrations of Rb and Sr plus different $^{87}Sr/^{86}Sr$ ratios can be calculated. The equation representing two-component mixtures in these coordinates is as follows:

$$\left(\frac{^{87}Sr}{^{86}Sr}\right)_M = g\left(\frac{^{87}Rb}{^{86}Sr}\right)_M + e\,,$$

where

$$g = \frac{Sr_A\,Sr_B\left[\left(\frac{^{87}Sr}{^{86}Sr}\right)_B - \left(\frac{^{87}Sr}{^{86}Sr}\right)_A\right]}{k(Rb_B Sr_A - Rb_A Sr_B)}$$

and

$$e = \frac{Rb_B Sr_A\left(\frac{^{87}Sr}{^{86}Sr}\right)_A - Rb_A Sr_B\left(\frac{^{87}Sr}{^{86}Sr}\right)_B}{(Rb_B Sr_A - Rb_A Sr_B)}$$

and k is a constant of which the value is governed by the abundance ^{86}Sr and the atomic weight of that element, i.e.

$$k = Ab_{(^{87}Rb)} W_{Sr} Ab_{(^{86}Sr)} W_{Rb} - 2.90 \pm 0.007 ,$$

for Sr having $^{87}Sr/^{86}Sr$ ratios between 0.700 and 0.750. Both the slope and intercept of straight lines representing two-component mixtures are calculable from the appropriate equations given earlier. In the case of two such components, A and B, where $Sr_A = 200$ ppm, $Rb_A = 460$ ppm, $(^{87}Sr/^{86}Sr)_A = 0.725$, $Sr_B = 450$ ppm, $Rb_B = 40$ ppm and $(^{87}Sr/^{86}Sr)_B = 0.704$, the relevant slope is 0.0038 which gives a fictitious age of 267 Ma.

An instance of correlated $^{87}Sr/^{86}Sr$ and $^{87}Rb/^{86}Sr$ ratios is K-rich lava flows from the Birunga and Toro-Ankole fields along the mutual borders of Uganda, Zaire and Rwanda in East Africa cited by Bell and Powell (1969). These volcanic centres lie along the western rift north of Lake Tanzania and west of Lake Victoria and are about 160 km apart. They have been identified as Pliocene to Recent in age and volcanic activity in the region is believed to have lasted into historic times. Thus there is no doubt that the rocks are young and samples from them should give an isochron with almost zero slope. Nonetheless, there is a significant positive correlation between the average $^{87}Sr/^{86}Sr$ and the $^{87}Rb/^{86}Sr$ ratios of the various rock types, and the straight line which can be drawn through the data points is not an isochron, rather a mixing line. Its slope gave a fictitious age of 773 Ma because the $^{87}Sr/^{86}Sr$ ratios of the rocks resulted from the mixing process and do not relate to the radioactive decay of ^{87}Rb in them after crystallization.

Brooks, James and Hart (1976) listed 30 occurrences both of volcanic and plutonic igneous rocks which gave fictitious isochrons when their initial $^{87}Sr/^{86}Sr$ ratios were plotted against their $^{87}Rb/^{86}Sr$ ratios. Contamination was ruled out as an explanation. Instead, it was suggested that they relate to the age of the mantle from which the relevant magma derived. Others believed that sympathetic variations of the isotopes of Sr or Nd with the isotope composition of oxygen accord better with magma mixing or contamination than with the mantle isochron concept.

Even when isochrons are not fictitious, care must be exercised. An example comes from investigations of a suite of rocks from the micropegmatite at Sudbury, Ontario, Canada. It was assumed that they had the same initial $^{87}Sr/^{86}Sr$ ratio and, that all members crystallized in a relatively short time interval (compared with the common age of the rocks). The data points fitted well, which implied that these hypotheses were correct. The relevant slope gave an age of 1.704 Ga $\pm$ 10 Ma as that of crystallization. However, subsequent recalculation of the data by other researchers using the value of Steiger and Jäger (1977) for the decay constant of ^{87}Rb increased this to 1.843Ga $\pm$ 133 Ma. Hence, although the isochron method of dating such suites of cogenetic igneous rocks is preferable to calculating model ages, caution is necessary.

Isochrons can provide both the age of a suite of rocks and its initial $^{87}Sr/^{86}Sr$ ratio, as well as showing through the good fit of the data points that the samples have remained closed to both Rb and Sr throughout their geological histories, and that neither element has entered the system from extraneous sources. The minerals in igneous rocks should lie on the same isochron as the whole rock but, when heating occurred during metamorphism, a different state of affairs arises. Such an event causes the minerals to deviate from the whole-rock isochron, each mineral having its own individual isochron. This complicates the dating of such metamorphic rocks which, during metamorphism, have undergone temperature and pressure increases accompanied by recrystallization of existing minerals and formation of new ones. Clearly, the situation is very mobile and involves ionic diffusion and perhaps an aqueous phase as well. Metasomatism may occur and hence induce a great effect on the parent–daughter relationships of all the naturally occurring radioisotopes within a particular rock. Even small temperature rises have an influence on these and this may not be reflected in the mineralogy. Probably the sensitivity of isotope systems in rocks to the phenomenon relates to the fact that ionic diffusion rates through crystal lattices and grain boundaries are functions of temperature. Naturally the daughters resulting from radioactive decay comprise isotopes of different elements and have different ionic charges and radii from those of their parents, energy released during the process promoting disruption and perhaps destruction of crystal lattices, at least locally, so that the escape of radiogenic daughters is thereby facilitated.

When equation (7.1) is modified by division of each term by the number of ^{86}Sr atoms found (a constant), it gives the equation (7.3) for age determination. It may be rewritten to a reasonable approximation as follows:

$$\frac{^{87}\mathrm{Sr}}{^{86}\mathrm{Sr}} = \left(\frac{^{87}\mathrm{Sr}}{^{86}\mathrm{Sr}}\right) + \frac{^{87}\mathrm{Rb}}{^{86}\mathrm{Sr}}\,\lambda\,t\,. \tag{7.4}$$

This is the equation of a straight line in slope–intercept form in coordinates of $^{87}Sr/^{86}Sr$ and t. It describes the time-dependent increase of the $^{87}Sr/^{86}Sr$ ratios of Rb-bearing systems as the Sr evolution lines are straight with slopes $m = (^{87}Rb/^{86}Sr)\,\lambda$. A set of lines for various systems having different values for the latter is shown in Fig. 7.2. Regarding metamorphic rocks, many measurements have been made, e.g. samples from the Amitsoq gneiss from the Godthåb district of southern West Greenland were dated by the whole rock Rb–Sr method and the isochron gave an age of 3.66Ga $\pm$ 99Ma. This may be compared with a value of 3.822Ga $\pm$ 5Ma from SHRIMP assessments of the U–Th–Pb isotopic

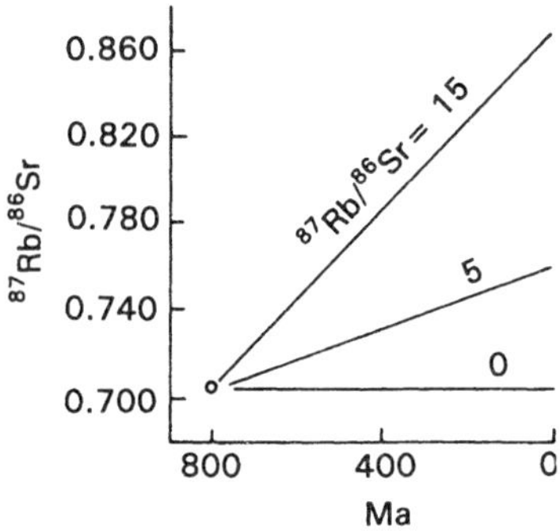

Fig. 7.2 Strontium development diagram for a suite of three rock systems with various $^{87}Rb/^{86}Sr$ ratios. All came into being 800 Ma ago when each had the same $^{87}Sr/^{86}Sr$ ratio of 0.7040. Subsequently, the $^{87}Sr/^{86}Sr$ ratios of the systems evolved along a set of divergent straight lines of which each slope is dependent upon the corresponding $^{87}Rb/^{86}Sr$ ratio. In the case of the rock system with zero slope, the $^{87}Sr/^{86}Sr$ ratio is a constant equal to the initial value for all the systems.

composition of zircons from a tonalitic Amitsoq gneiss. The initial $^{87}Sr/^{86}Sr$ ratio was 0.7009 ± 0.0011.

Metamorphic rocks may derive from pre-existing sedimentary ones which contain minerals that, rendered unstable under such conditions of increased heat and pressure, recrystallize. Rubidium-bearing minerals in metamorphosed rocks include micas and potassium feldspars datable using the isochron approach or by calculating model ages. These represent the time elapsed since termination of the relevant metamorphic episode. Also, whole-rock isochrons can be obtained which provide ages of metamorphic episodes causing recrystallization of pre-existing sedimentary rocks.

Dating unmetamorphosed sedimentary rocks is feasible since many contain minerals having Rb/Sr ratios high enough to produce measurable enrichments of their strontium in radiogenic ^{87}Sr. Some such minerals may be authigenic, for instance glauconite, K feldspar and some evaporite minerals such as sylvite. Others may be allogenic and detrital, e.g. micas, clay minerals and some K feldspars. As authigenic minerals form in the depositional environments of sedimentary rocks, they contain the key to determining the date at which sedimentation took place and thus the time elapsed since it ceased, if the relevant minerals constitute closed systems. With evaporite minerals, caution is necessary because they are easily altered by recrystallization, a process sometimes proceeding continuously in equilibrium with connate brines. This would reduce the value of these minerals as geochronometers. As regards shales, these are detrital and include mineral particles which could have had complex previous histories prior to the deposition of these unmetamorphosed and fine-grained sediments. Often linear arrays on isochron diagrams can be obtained, but these are hard to interpret because they may relate to diagenesis, deformation or recrystallization rather than the time of deposition. As for glauconites, these cover, under the name 'glaucony', a mineral family ranging from glauconitic mica to glauconitic smectite. However, this has little bearing on Rb–Sr dating since both the potassic and ferruginous clay minerals in the glaucony facies can be used to determine the time of deposition of the sediments, if not recrystallized through tectonic considerations or chemical exchange with highly reactive groundwaters.

Clauer (1982) surveyed the relevant factors and, with others, showed that uncertainties could be avoided by analysing only concentrates of illitic clay minerals with grain diameters under 2 μm. These may contain clays which were either authigenic or altered during deposition and diagenesis, in contrast to coarser fractions mainly composed of detrital clays (these lead to overestimation of the time of deposition). Authigenic or transformed illites show broad 001 X-ray diffraction peaks, detrital ones having a narrow peak in the 1 nm position. The width of the (100) peak is expressed by a 'crystallinity index', i.e. the width in $\Delta 2\theta$ of the (001) diffraction peak at 50% height which is derived from analysis of an orientated clay sample. Illite occurs in several polymorphic forms called 1M, 1Md and 2M. The first two form at low temperature and hence are authigenic or diagenetic in origin. The last one forms at temperatures above 250°C, so its presence means that the rock underwent metamorphism (or the illite is detrital). Thus 2M illite polymorphs cannot give data shedding light on the time of deposition of a sedimentary rock.

A thorough mineralogical study of clay-rich sedimentary rocks is necessary before dating is undertaken so that unsuitable material can be rejected. Where appropriate samples are examined, K–Ar ages of clay minerals, whether authigenic or transformed, usually tend to be younger than Rb–Sr dates due to the fact that the former represent culmination of diagenesis and the latter its onset. Consequently, Rb–Sr ages better indicate the time of deposition of the rock. In addition, to be valid, clay mineral ages must decrease upwards in a stratigraphic section of sedimentary rocks.

Some shales comprise detrital muscovite and illite which, through their coarser grain sizes, can be separated from authigenic clay minerals. Morton (1985) derived relevant data from Late Devonian black shale of west Texas, USA. He showed that clay minerals in the 1 to 2 μm fraction of the Woodford shale contain the 2M illite polymorph. After removal of exchangeable Rb and Sr, the coarse clay fractions constituted an isochron corresponding to an age of 539 ± 18 Ma. This exceeds that believed to be the depositional age (between 360 and 374 Ma) and probably indicates the provenance of the coarse illite. This is most likely to be southern hemisphere continents imprinted by the Brazilian or Pan-African events which occurred some 550 Ma ago. The finer clay fractions (< 0.2 μm) in this shale contain the 1Md polymorph and constitute a second isochron which gave an age of 301 ± 4 Ma. Possibly this represents a diagenetic episode which could have been connected with the Ouachita orogeny of the Middle Pennsylvanian when Sr in the fine-grained authigenic clay isotopically re-equilibrated with the porewater Sr. This would agree with the initial $^{87}Sr/^{86}Sr$ ratios of the two isochrons. The authigenic clay fractions yielded an initial value of 0.7086 ± 0.0004, close to that of marine limestone of Palaeozoic age. It is feasible that the pore water in the Woodford shale had in it Sr from underlying Palaeozoic marine carbonates at the time of diagenesis. But the detrital clays have a lower initial ratio of 0.7037 ± 0.0008 which resembles that of young volcanics, implying that they were young in the Late Devonian when the Woodford shale was presumably being deposited.

Separate dating of detrital minerals appears to facilitate identification of their sources, therefore, although minerals resistant to erosion and weathering are required (among these are muscovite, illite, K feldspar and zircon). All can be separated from their host sedimentary rocks and hence can be dated in order to assess the ages of these rocks. Such provenance dates are significant in determining specific sources of sediment for the palaeogeographical reconstruction of depositional basins. The approach was used to reveal the provenance of the Eocene Tyee Sandstone in Oregon, USA, from isotopic dates of muscovite and K feldspar. However, where a particular detrital mineral originated from several source rocks with different ages, mixtures of several grains of it will give a mixed age without geochronological significance. The only information from them pertains to the various proportions of grains coming from the diverse source rocks.

The Rb–Sr method has been applied to unmetamorphosed rocks as far back as the Precambrian, e.g. in the West African craton (Clauer *et al.*, 1982) and in the Carpentarian McArthur Basin, Northern Territories, Australia (Kralik, 1982). In such age determinations, three assumptions are made. The mineral grains involved must have remained closed to Rb and Sr during both weathering and transportation, diagenetic alterations must not have occurred and the initial $^{87}Sr/^{86}Sr$ ratios of minerals coming from sources of different ages must be similar. Naturally, it is very improbable that all three conditions can be satisfied in any one sample. Nevertheless, it has proved possible to assemble valuable geoligical information by dating feldspar in glacial sediments both in North America and in the Transantarctic Mountains. Dating detrital feldspar in till is simplified because of systematic variations in the K feldspar/plagioclase feldspar ratios found in different grain-size fractions. Sometimes, the Rb/Sr ratios in feldspar concentrates increase with grain size because the K feldspar/plagioclase feldspar ratios of the coarse sand fractions in till are higher than those of the fine sand or silt fractions. Whether quartz is present or not is irrelevant since this mineral has such low concentrations of Rb and Sr that no practical change in the Rb/Sr ratio of the feldspar is produced if it is present.

7.3 APPLICATIONS

7.3.1 Sphalerite Dating and Genesis of MVT Ores

An important case of Rb–Sr dating may be considered because it involved tackling the difficult problem of isotopic dating when there is a lack of suitable minerals with high or variable parent/daughter ratios. Nakai *et al.* (1990) described it with reference to Mississippi Valley Type (MVT) ore deposits which commonly comprise some combination of Pb, Zn and Fe sulphides together with barite, fluorite, dolomite and calcite. Clearly, indirect efforts were necessary and one approach previously employed entailed dating apparent resetting from hydrothermal fluids, e.g. the Rb–Sr age of glauconites was used to estimate the age of mineralization in the Viburnum Trend, a 75 km long belt of MVT mineralization in Missouri, based on assuming that the formation of this deposit was the last process affecting the isotope system in the glauconites. Another approach is that using model ages, e.g. the large fractionation in Sm/Nd ratios in fluorites from the North Pennine Orefield was used to impose constraints on the timing of mineralization. A similar exercise gave an estimate for the time of mineralization in the Mascot–Jefferson City district, East Tennessee, as younger than 395 Ma using a crude basin-evolution model. However, Nakai *et al.* (1990) emphasized that, as with model lead ages, interpretations depend critically on the inferred isotope evolution of the source of the components and this is rarely well defined.

The MVT deposits of the central and eastern USA are thought to have formed as a consequence of large-scale lateral fluid flow driven by the tectonic squeezing of sedimentary successions in the course of the Late Palaeozoic Alleghenian orogeny (330 to 250 Ma ago). Those of the Mascot–Jefferson City district are hosted by the Lower Ordovician Knox group of which the top is marked by an extensive Middle Ordovician unconformity, the underlying ores occurring in collapse breccias created through karstification. Large-scale fluid migration in the Late Palaeozoic may be inferred from palaeomagnetic measurements and K–Ar age determinations on clays and K feldspars in the area. This led to the opinion that the Appalachian MVT deposits in East Tennessee resulted from massive movements of fluid expelled from sedimentary basins during the Alleghenian orogeny.

Nakai *et al.* (1990) presented data for eight leached samples of sphalerite (ZnS) together with seven of their (fluid inclusion) leachates.

These sphalerites had low Sr concentrations (< 2 ppm), but variable $^{87}Rb/^{86}Sr$ ratios. Isotope data were plotted on a conventional isochron diagram and, with the exception of one sample, all showed a good correlation between $^{87}Rd/^{86}Sr$ ratios and $^{87}Sr/^{86}Sr$ ratios. An observed scatter may well reflect heterogeneity in the initial Sr isotope compositions of the sphalerites and indeed such isotope variability for mineralizing solutions is well established. Also, the introduction of secondary fluids and the deformation of the ore constituents could have disturbed the Rb–Sr system. The exceptional sample alluded to above was more deformed than its fellows but, neglecting this, the data for the sphalerites define an apparent isochron age of 377 ± 29Ma (2σ) with an inital $^{87}Sr/^{86}Sr$ ratio of 0.7107(3) (2σ uncertainty).

The leachates have lower $^{87}Rb/^{86}Sr$ ratios (< 0.1) than the sphalerites and are clustered round the lower end of the isochron. The calculated apparent Rb–Sr ages for leach–sphalerite pairs were variable and ranged from 150 to 440 Ma. The data indicated that in the relatively large samples necessary for the study, several kinds of inclusion fluids known to exist were in fact sampled, including secondary fluids not directly related to sphalerite precipitation.

A chondrite-normalized rare earth element (REE) pattern for one sample was constructed after removal of inclusion fluids and Nakai *et al.* (1990) were unaware of any high-precision REE analyses of sphalerites before this. Although concentrations were low, this pattern is similar to that of average shales. If the REEs in the sphalerite came from hydrothermal fluids, no major fractionation can have taken place during mineralization.

Exactly where the Rb, Sr and REEs are located in the sphalerites remained uncertain. The Rb/Sr ratio is too high for the components to be derived from carbonate, fluorite or barite inclusions. From the Rb/Sr ratio (0.12) and Sr/Nd ratio (90) of one or the sampes (C16D), clay or feldspar were considered to be the likeliest host phases and the K feldspar appeared to predate sphalerite in all the samples examined. Both are distributed together in silica-rich alteration zones believed to be coeval with mineralization. Scanning electron microscope data showed that feldspar inclusions may comprise $\ll 0.1\%$ of the handpicked sphalerite, but at least this value is required in order to explain the Rb and Sr concentrations found in sample C16D. It was concluded that the Rb, Sr and REEs are probably hosted in the sphalerite or in K feldspar, these phases being regarded as most likely cogenetic so that the Rb–Sr data ought to define the age of sphalerite mineralization. However, Nakai *et al.* (1990) added that, if they are not held in inclusions, but were separated from hydrothermal fluids in minute quantities by the sphalerite, it is unclear how the range in Rb/Sr could have been produced. It is possible that the ore fluid composition evolved with deposition of gangue minerals as a consequence of different bulk partition coefficients for Rb and Sr. Alternatively, the variations may well reflect incomplete mixing of fluids. However, they pointed out that both of these potential explanations seem improbable since the range of Rb/Sr in the analysed fluid inclusion contents is relatively restricted.

The isochron age is geologically rational and implies that the range in initial Sr isotopic composition in the ore fluids was small. The initial $^{87}Sr/^{86}Sr$ ratio from sphalerites from the isochron intercept in the study of Nakai *et al.* (1990) was 0.7107 and this accords with an intermediate situation between the Sr isotopic ratio for pre-ore and post-ore dolomites. Thus the isotope data are compatible with the concept that the Sr in the sphalerite was sequestered from the ore fluids and not inherited from unrelated and older material in the ore. The combined date showed a relatively small range in Sr isotopic compostion of the ore-forming fluids at the time of mineralization.

Data from Nakai *et al.* (1990) suggest that the mineralization predated the Alleghenian orogeny (330 to 250 Ma) by $\sim$50 Ma and this is inconsistent with the idea that basin brines which were tectonically expelled in the course of that event were significant in ore genesis. In addition, the alkali feldspar intergrown with sphalerite in the deposits studied could not therefore have grown during Alleghenian fluid movements as has been argued. It is believed that the thermal effects of the Acadian orogeny (380 to 350 Ma) were restricted to areas further east than the Mascot–Jefferson City district, East Tennessee. However, there was probably major thrusting just south-east of the district at the time. The scenario presented was that uplift and deformation in the early Devonian caused fluid expulsion and the creation of MVT mineralization to the west. Such deposits found along the Appalachian Mountains from Georgia to Newfoundland have a number of common features, e.g. high Zn/(Zn + Pb) ratios and relatively homogeneous Pb and stable isotope compositions. Among these, mineralization in the St George Group, Newfoundland, resembles in stratigraphic setting that in East Tennessee. $^{40}Ar/^{39}Ar$ ages of alkali feldspar from MVT deposits in western Newfoundland were taken as indicative of thermal events at 370 to 350 Ma and 210 Ma ago, the older one being interpreted as the time of sphalerite mineralization. All this is accordant with the view that Appalachian MVT mineralization was linked regionally to the Acadian orogeny.

Brannon, Podosek and McLimans (1992) noted that Nakai *et al.* (1990) inferred a Rb–Sr isochron of Acadian age for sphalerites from eastern Tennessee, and also that the latter researchers extended this to other widely separated North American MVT deposits, these results suggesting a prevailing Acadian association for North American MVT mineralization. With regard to a suitable genetic model for these deposits, Brannon, Podosek and McLimans (1992) stated that this must necessarily entail an appropriate temporal relationship between cause and effect. Hence the hypothesis that MVT deposits were caused by large-scale migration of basinal brines brought about by tectonic orogeny can be examined by comparing the time of ore formation with the time of associated orogeny. The importance of such chronological constraints has been recognized and led to a large volume of experimental work, but presently available data were taken to be inconclusive by Brannon, Podosek and McLimans (1992). They pointed out that the palaeomagnetic and various isotopic systems may have been overprinted by different geological events which, even when mediated by fluid–rock interactions, do not assist in determining which, if any, of the fluids in question were responsible for the ore minerals. However, fluid inclusion studies have shown that there is a link between MVT ores and hydrothermal dolomites deposited throughout the region. This association strongly suggests that long-distance fluid migration took place. But the regionally distributed minerals thought to be conceivably connected with MVT ores on the basis of similarity of fluid inclusion compositions are not all dated, and the palaeomagnetically and radiometrically dated minerals which have been dated are so far not definitively related to MVT mineralization. Clearly, the best approach would be to derive ages for all major ore minerals, but reliable data are rare for technical reasons. This refers to the fact that, with the trace element systems to be utilized, either the parent/daughter ratio is so low that the approach is inapplicable or the absolute abundances of the relevant elements are insufficient for reliable analyses to be effected. Brannon, Podosek and McLimans (1992) indicated another, inadequately explored problem, which is that variations in the isotopic composition of the daughter element in the ore-forming fluids may mask *in situ* accumulation of the radiogenic daughter isotope even when the isotopic analysis necessary is feasible.

Brannon, Podosek and McLimans (1992) reported Rb–Sr dates for sphalerites from the MVT 'type sections', i.e. the Upper Mississippi Valley Zn–Pb district. Rb/Sr data from two distinct sphalerite bands from the West Hayden ore body gave two independent but indistinguishable isochrons of Permian age (269 ± 6 Ma and 270 ± 4 Ma) in accordance with an episode of brine migration initiated by the Alleghenian/Ouachita orogeny. Taking into account previous work showing a connection with the Acadian orogeny, Brannon, Podosek and McLimans (1992) regarded their results as indicating that North American MVT deposits formed in at least two distince tectonic episodes.

7.3.2 Rb–Sr Data from a Volcanoplutonic Complex (Late Hercynian), Italy

In regard to calc-alkaline magma generation and evolution, numerous models have been proposed and include closed-system fractional crystallization, crustal anatexis, AFC (combined assimilation–fractional crystallization), liquid-state thermogravitational diffusion and MASH (melting–assimilation–storage–homogenization). Barth *et al.* (1993) investigated the evolution of a calc-alkaline basic to silicic magma system, the Late Hercynian Atesina–Clima d'Asta volcanic–plutonic complex of northern Italy, this being the largest such structure in the Late Hercynian of the South Alpine domain. The Atesina volcanic complex is associated with the Clima d'Asta pluton. The composition of the former extends from basaltic andesite to rhyolite and that of the latter ranges from quartz norite to monzogranite. Geochemical information suggested that the basaltic andesites are derived from the mantle and have no significant crustal contribution while the andesite and rhyodacites–rhyolites originate from two different magma sources, the latter being wholly anatectic. Isotopic ivestigations have implied rather more complex lower crustal processes involving a MASH-type model. It is thought that hydrothermal fluid circulation during the Triassic may have been connected to incipient rifting of the continental crust and upwelling of alkaline mantle-derived mantle. The phenomenon may have produced variably rejuvenated Rb–Sr and K–Ar ages and induced sizeable secondary shifts in some geochemical and O isotopic signatures. Of course, the Hercynian orogeny (alternatively known as the Variscan orogeny) extended across southern Europe and relates to the American Alleghenian (or Appalachian) orogeny (Late Carboniferous to Permian), both being much younger than the Acadian orogeny of Devonian age.

Barth *et al.* (1993) presented major trace element, REE, Rb–Sr, Sm–Nd and $^{18}O/^{16}O$ isotopic data for the main volcanic complex and pluton lithologies so as to clarify the main mechanisms of Late Hercynian magma genesis and also in order to obtain constraints on the sensitivity of the geochemical and isotopic systems to medium- to low-temperature water–rock interaction. Fluid infiltration may have mobilized both major and trace elements, induced open-system behaviour in total-rock Rb–Sr and shifted $\partial^{18}O$ towards elevated values (total rock up to 16.6‰ and volcanic matrix up to 17.8‰. Oxygen isotope disequilibrium between quartz–feldspar pairs was taken as implying water–rock interaction at medium to low temperatures. However, the $\partial^{18}O$ values of quartz, the REE characterized by regular LREE enrichment and HREE depletion together with the Sm–Nd isotopic signatures remained almost unaffected by secondary processes. Initial ϵ_{Nd} values at 270 Ma of the Atesina volcanic complex and Cima d'Asta pluton were restricted to overlapping ranges of -3.6 to -6.5 and of 2.7 to -6.5, respectively, taken to indicate a significant crustal contribution. Both these values and associated model ages of 1.1 to 1.6 Ga accorded with those of typical South Alpine lower crustal magmatites. Neither volcanic nor plutonic rocks pursued the 'normal' trend of increasingly crustal Nd isotopic signatures with the progressive extent of magma evolution which could be anticipated for a single-stage process. Rather they show an inversion of the relationship. Barth *et al.* (1993) regarded geochemical and isotopic constraints as suggesting a model of a large-scale MASH-type melting and mixing zone either at or near to the base of the continental crust. In addition, distinct elemental enrichment/depletion and REE crossover patterns shown by high-silica as compared with less silicic Atesina volcanic complex rhyolites imply later magma evolution in a shallow level, compositionally zoned chamber.

7.3.3 Age of the Hercynian Orogeny by Rb–Sr Dating

Mrini *et al.* (1992) effected a Rb–Sr whole-rock isochron examination of Hercynian granitoids from the Meseta domain and the Atlas domain in Morocco which demonstrated that the Hercynian orogeny extended between 330 Ma and 250 Ma. Two main magmatic events were identified, namely 'orogenic' granites with biotite and sometimes cordierite of mixed origin plus leucogranites of crustal origin, both emplaced between 330 Ma and 280 to 275 Ma and showing calcalkaline affinities, together with 'late-orogenic' granites which range from 275 Ma to 250 Ma in age, these being essentially alkaline.

7.3.4 Disturbance of Rb–Sr System in Schwarzwald (Variscan) Granites

Simon (1990) examined Hercynian S-type granites from the south-eastern Schwarzwald granite series which represent cogenetic biotite and two-mica granites. Oxygen and H isotope data demonstrated that hydrothermal alteration involving isotopically light surface waters caused a dramatic reduction in $\delta^{18}O$ and δD together with a marked disequilibrium between the minerals. Effective water–rock ratios were calculated as high (*c.* 0.8 vol. units). A shift in the $^{18}O/^{16}O$ isotope ratios and the chemical composition of the fluid due to water–rock interaction was continuously traceable from pure water with meteoric isotopic character in the deep-seated biotite granites to slightly saline water with rock-equilibrated isotopic composition in the shallower two-mica granites. Simon (1990) alluded to substantial retrograde hydrometamorphism within the temperature range 500°C to 200°C as having mineralogical results, including high-temperature chloritization of biotite and low temperature muscovitization as well as feldspar alteration, respectively. Another result of the re-equilibration of cations is a strong disturbance of the Rb–Sr system which affects measured ages and initial $^{87}Sr/^{86}Sr$ values. Hydrothermal differentiation and alteration most probably overlap the magmatic differentiation process to a great extent.

Simon (1990) pointed out that interactions between hydrosphere and lithosphere have commanded great attention in the Earth sciences ever since the fundamental work of Taylor (1968, 1977, 1978) in stable isotope geochemistry validated this as significant in evaluating them. The sizeable isotopic contrast between rock and meteoric water facilitates monitoring of the isotopic exchange processes involving them. In fact, the O isotope ratios of minerals with differing kinetic exchange behavioural patterns can be employed to assess the possible primary isotopic compositions of rocks and water as well as the water–rock isotopic ratios, fluid fluxes and the duration of isotope exchange. Oxygen isotope investigations by Hoefs and Emmermann (1983) showed that the Variscan granites of the southern Schwarzwald were greatly affected by hydrothermal alteration by isotopically light meteoric waters.

Turning to the subject of Rb–Sr analyses, hydrothermal convection presumably began through the intrusion of epizonal plutons, but the matter of the age of this event remained to be solved. If a temperature range of 600°C to 200°C in which fluid circulation took place is accepted, computer simulations imply that this phenomenon occurred very soon after the actual intrusion. However, the actual temporal sequence of the intrusions was a matter for discussion. The two-mica granites have been described as being intruded simultaneously and also as having been intruded in two separate events several million years apart. Leutwin and Sonet (1974) gave Rb–Sr ages for the encompassing gneiss basement of 441 ± 10Ma and for the Albtal biotite granite in the range 336 Ma to 354 Ma. Simon (1990) thought that it remained questionable whether the granite series was generated by magmatic pulses over a long time interval, say of *c.* 50 Ma, or whether the differing chemical and isotopic characteristics of the plutons relate to differentiation and/or hydrothermal alteration in a single event.

From the Bärhalde two-mica granite, Rb–Sr analyses gave data from which $^{87}Rb/^{86}Sr$ against $^{87}Sr/^{86}Sr$ values of the feldspars were plotted and resulted in what Simon (1990) termed an 'errochron' with a slope corresponding to 320 ± 13 Ma and an initial $^{87}Sr/^{86}Sr$ ratio of 0.704 ± 0.010. The subdivision of the feldspars into two groups reflecting their $\partial^{18}O$ values is shown in Table 7.2. However, the correlation of oxygen isotope and Sr isotope ratios became clearer if calculated model

Table 7.2 Subdivision of feldspars from Bärhalde two-mica granite, based on $\delta^{18}O$ values.

$\delta^{18}O$ (‰)	*Age (Ma)*	*Initial $^{87}Sr/^{86}Sr$*
< 9.5	317 ± 8	0.7135 ± 0.0086
> 9.5	300 ± 38	0.7153 ± 0.0026

ages using an initial $^{87}Sr/^{86}Sr$ of 0.716 are plotted against the $\partial^{18}O$ value of the feldspars and decreasing model ages. However, no correlation was noted between the O isotope ratios and and model ages for biotite and muscovite since there is no spread in the $\partial^{18}O$ values, though model ages also vary in the range from 267 Ma to 324 Ma that Simon (1990) regarded as suggesting a strongly disturbed Rb–Sr system due to the hydrothermal alteration.

7.3.5 Magmatic and Hydrothermal REE Fractionation in Chinese Granites

Maruéjol, Cuney and Turpin (1990) investigated the Xihuashan stock, South Jiangxi, China, which is composed of cogenetic granite units emplaced during the Yanshanian orogeny (153 ± 0.2 Ma ago). Two feldspars were involved together with Fe-rich biotite and garnet plus slightly peraluminous granites as well as numerous primary accessory minerals. The granites were categorized as Xe, γa, γc, γd and γb. Petrographic observations, major element, REE, Y and Rb–Sr isotopic data indicate a magmatic suite (granites Xe and γa$\rightarrow$granites γc and γd$\rightarrow$granite γb) distinct from hydrothermal Na or K alteration of γb. From granite Xe to granite γb, LREE, Eu, Th and Zr contents were strongly depleted, while HREE, Y and U contents increased. During the K alteration of γb, such variations are not important. However, evidence from major and accessory minerals as well as from geochemical and fluid inclusion sources suggest two successive alteration fluids interacting on γb. One was a late magmatic F^- and CO_2 rich fluid and the second a postmagmatic, aqueous and slightly saline fluid. Depletion of LREE and Th contents and the increase in HREE, Y and U contents correspond in the magmatic suite to the early fractionation of monazite in the granites where no hydrothermal alteration took place, i.e. granites Xe and γc and to the hydrothermal change of monazite to parisite and secondary apatite, intensive new formation of yttroparisite, Y enrichment and U loss in the uranothorite and late crystallization of uraninite in the granites γc and γb. The simulated crystallization of monazite and temperature of monazite saturation show early fractionation of monazite from the magma in the less evolved granites (Xe and γe) and prevailing hydrothermal leaching of monazite in the most evolved ones (γc, γd and γb) related to a late magmatic event. The slight variations of REE, Y, Th and U content in the K-altered granites compared with granite γb emphasizes the distinct chemical nature of the successive hydrothermal fluids. Both Rb–Sr and Sm–Nd data point to a 30 Ma time interval between the late magmatic and the postmagmatic fluid circulation.

Rb–Sr isotopic measurements on granites Xe, γa, γc and γb fit an isochron yielding an age of 153 ± 0.2 Ma and an initial $^{87}Sr/^{86}Sr$ ratio of

0.7137 ± 0.0002. These data indicated that the different granitic units are cogenetic and that emplacement of the Xihuashan stock must have taken place over a relatively short time interval. The initial $^{87}Sr/^{86}Sr$ ratio necessitates a crustal source for the different units. The data points of granite γb (hydrothermal REE minerals crystallization, leaching of LREE and enrichment in HREE and Y) lie on this isochron and suggest fluid circulation and alteration at a magmatic stage as was proposed earlier by Le Bel *et al.* (1984). Potassium-altered granites from granite γb were enriched in Rb and Sr and define an alignment which gave 128 ± 0.1 Ma and and $^{87}Sr/^{86}Sr$ ratio of 0.760 ± 0.003. Maruéjol, Cuney and Turpin (1990) noted that, even if the points do not suffice to define precisely an isochron age, the apparent age is nevertheless significant. The data imply that the fluid connected with the removal of quartz and K-alteration of granite γb was not equilibrated with the melt and was, in fact, external to the granite. But it is not possible to exclude the likelihood that the shift in the Sr isotopic composition is related to the interaction with the late stage fluid observed in vugs of the K-altered granites.

Some of the conclusions drawn by Maruéjol, Cuney and Turpin (1990) may be cited. The Rb–Sr and some trace element data validate the inference that the different granitic units of the Xihuashan stock may well be cogenetic with a crustal signature. On the other hand, the lack of xenotime in one of the units, namely the granite Xe, the biotite chemistry and evolution of plagioclase An content and of Nb–Ta concentrations all suggest non-cogenetic relationships between some of the granitic units. The apparent succession is granites Xe and $\gamma a \rightarrow$ granites γc and $\gamma d \rightarrow$ granite γb. These leucocratic granites are slightly peraluminous compared with Hercynian or Himalayan two-mica leucogranites. They offered Maruéjol, Cuney and Turpin (1990) a unique chance to decipher in a single granitic body both magmatic and hydrothermal effects on both major and trace elements behaviour. Indeed, these researchers were able to delineate in thin sections successive stages of the hydrothermal alteration of a complex accessory minerals association (especially of monazite) as well as producing detailed petrographical and mineralogical analyses of these minerals and their alteration products together with an invaluable examination of whole-rock major element, trace element and Sr isotopic geochemistry. In addition, REE modelling of the various granitic units and alteration minerals allowed an evaluation to be made of the contributions of magmatic and hydrothermal fractionation.

7.3.6 Strontium Through Time

7.3.6(a) Proto-Earth

The isotopic composition of Sr altered continuously after nucleosynthesis and the rate of change of the $^{87}Sr/^{86}Sr$ ratio can be expressed as

($^{87}Sr/^{86}Sr$) λ, which takes into account its proportional relationship to the Rb/Sr ratio of the host system. Obviously the rocks and minerals on Earth reached different values for the ratio in accordance with their varying Rb/Sr ratios and their diverse ages. At first, however, the material of the proto-Earth accreted from the solar nebula and must have had very similar $^{87}Sr/^{86}Sr$ ratios grouped around a primordial value. Later diversity was due to the general geochemical differentiation of the planet, but this has been obscured by geological processes which continouously modified and destroyed rocks through geological time and entailed recombination of material to create new ones in which the Sr originated from preceding rocks. There is evidence in the 'initial' $^{87}Sr/^{86}Sr$ ratios of igneous rocks which mirror the history of Sr prior to its incorporation in the present host rock. Hence the study of Rb–Sr decay schemes in such rocks is useful not just in revealing ages, but also in clarifying geochemical history.

7.3.6(b) Meteorites

These bodies may have formed in the solar system along with the planets, satellites, etc., representing fragments of larger parents which underwent chemical differentiation through relatively short time intervals of tens to hundreds of Ma. Melting will have occured with segregation of metallic Fe–Ni and silicate liquids, the latter subsequently cooling to form a group of solids with different Rb/Sr ratios. Later frequent collisions yielded breccias comprising fragments of very variable origins and compositions. Meteorites impacting the Earth, whatever their source, represent solar system matter which preserves a record of the isotopic composition of Sr in the above-mentioned parents when they became inactive through rapid heat loss.

The Rb–Sr method has been applied to chondritic meteorites from which an average age of 4.49 Ga (based on $\lambda = 1.42 \times 10^{-11} a^{-1}$ for the decay constant of ^{87}Rb) resulted. Achondrites gave an average figure of 4.36 Ga with a slightly greater uncertainty. So it seems probable that stony meteorites formed about 4.5 Ga ago. The Rb–Sr isochrons of meteorites suggest that they crystallized rapidly and they appear to have very similar initial $^{87}Sr/^{86}Sr$ ratios. Still, there are important differences in this ratio in some stones, giving information about events which happened at the beginning of the solar system. It has been suggested that the solar photosphere has a Rb/Sr ratio of *c.* 0.65. If this is so and if it resembles that of the solar nebula, then it may be inferred that the $^{87}Sr/^{86}Sr$ ratio increased by 2.6×10^{-5} or $10^{-6} a^{-1}$. Therefore the isotopic composition of Sr in a Rb-free phase coming from the solar nebula depends on the time of its formation.

Stony meteorites are remnants of parents and cannot have had the same initial $^{87}Sr/^{86}Sr$ ratios. Nevertheless, many fit well on isochrons because they formed in such a relatively short time interval. To determine the primordial $^{87}Sr/^{86}Sr$, however, would seem to be a task fraught with problems. It might be feasible to obtain a value representative of the ratio in the solar nebula when the parents formed. Basaltic achondrites may be optimal for this purpose because they probably crystallized from silicate liquids and have very low Rb/Sr ratios ($\simeq 0.002$), implying that their $^{87}Sr/^{86}Sr$ ratios have changed little since crystallization. Papanastassiou and Wasserburg (1969) examined seven and obtained an initial ratio of $0.698\,990 \pm 0.000\,047$ which was termed BABI, basaltic achondrite best initial. A similar value for the achondrite Juvinas was termed JUSI and it was verified by Allègre *et al.* (1975) as $0.698\,976 \pm 0.000\,055$. BABI was later adjusted using an interlaboratory standard NBS No. SRM-987 to give a value of $0.698\,97 \pm 0.00003$. BABI may represent the $^{87}Sr/^{86}Sr$ ratio of a group of parents when they cooled and crystallized to form basaltic achondritic meteorites. Since the Rb/Sr ratios of the parents was low, not much isotopic evolution can have taken place during their internal differentiation, so that it is justifiable to make this assertion. But basaltic achondrites must not be taken to be the earliest objects to form in the solar system. The augite achondrite Angra dos Reis and a basaltic clast of the polymict breccia meteorite Kapoeta both have lower initial $^{87}Sr/^{86}Sr$ ratios than BABI, hence their parents must have separated from the solar nebula before other planetary bodies. The initial $^{87}Sr/^{86}Sr$ of the former (termed ADOR) is $0.698\,84 \pm 0.000\,03$ and that of fragment C in Kapoeta is $0.698\,85 \pm 0.000\,04$. Some relevant ages are given in Table 7.3.

Nakhla seems to be a young meteorite which formed on a planet still active about 1.3 Ga ago. Shergotty was described by Nyquist (1982). Its feldspar has been practically completely converted into a glass (maskelynite) through shock metamorphism *c.* 165 ± 11 Ma ago. This was shown by an internal Rb–Sr isochron. Sm–Nd investigations showed that it and Nakhla belong to the SNC group of meteorites which crystallized *c.* 1.3 Ga ago. Data from high $^{40}Ar/^{36}Ar$ ratios (*c.* 1750) and the

Table 7.3 Mineral isochrons from 'young' meteorites ($\lambda^{87}Rb = 1.42 \times 10^{-11}\,a^{-1}$).

Meteorites (Acondrite)	*Date (Ga)*	$(^{87}Sr/^{86}Sr)_i$
Sioux County	4.09 ± 0.14	0.69897 ± 0.00008
Stannern	3.2 ± 0.5	0.6993 ± 0.0003
Pasamonte	2.4 ± 2.9	–
Nakhla	1.34 ± 0.02	0.70232 ± 0.00006
Nakhla	1.21 ± 0.01	0.70254 ± 0.00003
Shergotty	0.165 ± 0.0011	0.72260 ± 0.00012

abundance of noble gases imply that the planet from which Nakhla originated may well have been Mars. This is large enough to have stayed active until 1.3 Ga ago and has huge volcanoes, including Olympus Mons. Shock mechanism may have operated during the ejection process consequent on meteorite impacts from space and would have reset the Rb–Sr isochron clocks.

Isotopic anomalies occur in meteorites, i.e. their isotopic compositions of some constituent elements differ from those in terrestrial samples. For instance, excess ^{129}Xe in meteorites is caused by the beta decay of ^{129}I ($T_{1/2} = 1.57 \times 10^7$a) and this can be used geochronometrically since the ratio of ^{129}I to stable ^{127}I in the solar nebula decreased as a function of time. This I–Xe dating method indicates the time elapsed between nucleosynthesis and the condensation of a solid object from this nebula. Niemeyer (1979) recorded I–Xe dating of silicate and troilite from 1Ab iron meteorites. There are numerous categories of irons, 1AB referring to the first category. AB refers to their chemical signature, and they comprise Fe alloyed with Ni, are rather primitive, contain silicate inclusions (chondrites) and hence may have had a two-phase origin. The Fe part is *c.* 4.5 Ga old. Not all isotopic anomalies relate to decay products and elements such as O, Si, S, Ti, Cd and the noble gases have been found in the carbonaceous chondrite Allende. Its anomalies show that the solar nebula was not isotopically homogeneous. Some anomalies are associated with FUN, i.e. fractionation and unknown effects, cf. Papanastassiou and Wasserburg (1978). Isotopic fractionation may have resulted from the selective partitioning of isotopes between gas and solid phases condensing out of the nebula. Anomalous isotopic compositions of Sr not attributable to the decay of ^{87}Rb to ^{87}Sr have been described form Ca–Al-rich inclusions and chondrules in Allende. This implies that the Rb–Sr chronometer in meteorites is unreliable and would preclude the use of initial ^{87}Sr/^{86}Sr ratios in determining short time intervals during the formation of planetary bodies in the solar system. Patchett (1980) worked on Ca–Al inclusions in Allende and used a double spike approach in order to show that both the ^{87}Sr/^{86}Sr and ^{84}Sr/^{86}Sr ratios of many of its inclusions and several chondrules are negatively correlated. This showed that they have variable enrichment in ^{84}Sr compared with terrestrial samples, bulk meteorites and the matrix in which the Allende inclusions occur. Correlation of the isotope ratios of Sr in the inclusions accords with the anticipated results of mass fractionation, but does not wholly exclude unknown nuclear effects. It is most likely this meteorite comprises particles formed in different parts of the solar nebula. Maybe they accreted into a small planetary body in which isotopic homogenization took place, either through diffusion during recrystallization or (possibly with) partial or total melting.

7.3.6(c) The Moon

Rb–Sr dating of lunar rocks led to the view that the Moon formed *c.* 4.5 Ga ago from the solar nebula and had an initial $^{87}Sr/^{86}Sr$ ratio termed LUNI which is close to BABI. A mean value for the ratio is $0.0698\,92 \pm 0.000\,09$, cf. BABI $= 0.698\,97 \pm 0.000\,03$ (Nyquist, 1977). The inference is that the Moon formed around the same time as the meteorites, but was bombarded later by planetary fragments that created the huge basins and most impact craters now visible in the highland regions of its surface.

7.3.6(d) Terrestrial Crust, Mantle and the Seas

The initial $^{87}Sr/^{86}Sr$ ratios of very old granitic rocks are low (0.700 to 0.702), but exceed BABI. Such rocks, enriched in Rb, started to form early in Earth history and played a significant role in the later isotopic evolution of this element. Later enrichment of the continental crust in radiogenic ^{87}Sr is different from the smaller increase in the $^{87}Sr/^{86}Sr$ ratio in the upper mantle. This is important because it facilitates distinguishing Sr in the continental crust from Sr coming from the upper mantle. Obviously this is a means of identifying sources of igneous rocks. The isotopic composition of Sr in the mantle is inferred from basalts and large gabbroic intrusives believed to have originated there and to have remained free of any significant contamination by crustal Sr. While it was believed that existing $^{87}Sr/^{86}Sr$ ratios in the mantle fall into a narrow range *c.* 0.704 ± 0.002, differences have been recorded from Recent basaltic rocks. Indeed, these proved systematic enough for them to be assigned to four groups, namely oceanic floor, oceanic islands, island arcs and continents. The relevant ratio values for these are 0.702 80, 0.703 86, 0.704 37 and 0.705 77, respectively. Skewing of relevant histograms towards greater values of the ratio might imply a contamination of the magma with radiogenic ^{87}Sr arising from old sial. Alternatively, this may be caused by terrigenous sediments being introduced from adjacent continents, or by interactions with seawater.

The fact that ore deposits may form in subduction zones led to consideration of the isotopic composition of Sr in relevant volcanic and plutonic igneous rocks. Two subductive processes are relevant, namely that entailing oceanic crust underriding oceanic crust to produce volcanic island arcs and that involving continental crust to produce igneous activity, e.g. in the Andes. With both, chemical compositions and ages of the igneous rocks vary systematically with distance from the trench, and hence with depth to the Benioff zone. Therefore basalt adjaccent to the trench transforms into andesite and dacite at greater distance. It has been claimed that Sr concentrations increase as a function of depth to this zone in island arc systems in the Pacific Ocean (Hart *et al.*, 1970). But this is not always so, perhaps because expected overall similarities are swamped by local differences both in magma generation and its later developments in different island arcs. For instance, Whitford (1975) found that the $^{87}Sr/^{86}Sr$ ratios in volcanics in the Sunda arc, Indonesia, increase to some extent with inferred depth to the Benioff zone, but the correlation of the parameters was poor. This reflects the increase in these ratios from calcalkaline rocks about 170 km to 200 km above the zone along the strike of the arc from east to west. This may relate to a westward thickening of crust overlying the subduction zone. If this is the case, then Sumatra may be underlaid by continental crust and Java, to the east, by thinner crust intermediate in type between continental and oceanic.

A contrary geographical phenomenon was recorded in the igneous rocks of the central Andes by McNutt *et al.* (1975). They noted that the initial $^{87}Sr/^{86}Sr$ ratios in these increase more or less systematically from west to east. As the coastal rocks are of early Jurassic age and those of the interior Quaternary, it is possible that there has been an easterly shift in the igneous activity. From the Middle Cretaceous, these ratios diminish with increasing distance from the coast and also with decreasing age from a value of 0.7077 to 0.7022. In the Jurassic plutons the initial ratios range from 0.7043 to 0.7059 and do not correlate with age.

The isotopic evolution of Sr in the mantle has been examined by plotting initial $^{87}Sr/^{86}Sr$ ratios of basalts and gabbroic rocks from various sources versus their Rb–Sr method-determined ages. Results strongly suggested that Rb/Sr ratios vary greatly in the mantle. Thus Sr evolution must be represented by divergent lines in a Sr development diagram. It must be borne in mind also that the Rb/Sr ratio may have decreased through geological time if Rb was removed preferentially into the crust. Then the slopes of the development lines would decrease as a function of time. It is probable that there was a nonlinear evolution of Sr in the mantle. A time-integrated average Rb/Sr ratio for the mantle has been given from the present $^{87}Sr/^{86}Sr$ ratio of 0.704 ± 0.002 relative to BABI 4.5 Ga ago as 0.027 ± 0.011.

The upper mantle is most likely both chemically and isotopically heterogeneous. Volcanics from oceanic islands and mid-oceanic ridges are less likely to be contaminated with old sial than those on continents or at subduction zones. The average $^{87}Sr/^{86}Sr$ ratio of oceanic island volcanics is 0.704 37, greater than the values obtained from sea floor and ridge basalts which gave 0.70280. This implies that the latter pair formed by melting of mantle material with a significantly lower Rb/Sr ratio than that of the source of volcanics in oceanic islands. It is thought that this difference must have existed for at least 2 Ga for it to be reflected in the Sr isotope ratios. From the above, it seems likely that Rb depletion took place in some parts of the mantle quite early in the history of the planet and oceanic tholeiites forming along spreading centres in the oceans may originate from the melting of such depleted mantle material. Oceanic islands volcanics probably derived from primitive, non-depleted, material of the mantle. This relates well with the concept of hot spots in the mantle which produce localized elevated concentrations of U, Th and K, which are radioactive and thus a primary heat source. Such hot spots would indicted mantle heterogeneity.

It is difficult to describe the isotopic evolution of Sr in the continental crust using simple models because it is so heterogeneous in the ages and Rb/Sr ratios of its rocks. Also a sample of Sr in it may have traversed an indeterminate number of systems with different Rb/Sr ratios after its initial separation from the upper mantle early in Earth history. In addition, the $^{87}Sr/^{86}Sr$ ratio may have been altered by a number of geological processes such as metamorphism, weathering, transportation, deposition, diagenesis, etc. Nevertheless, this ratio should be significantly greater in the continental crust than in the upper mantle. In fact, this is the basis for interpreting the initial ratios of both igneous and sedimentary rocks. A high initial $^{87}Sr/^{86}Sr$ ratio indicates crustal Sr and implies that the relevant rock originated, at least partly, from crustal matter. Hence investigation of such ratios provides data useful in delineating the sources of granitic rocks in continental crust. Such crust probably formed early in planetary history and so many granites should have crustal ancestry.

Various criteria may be applied in order to distinguish between crustally derived granites and those which originated from the mantle. If the initial $^{87}Sr/^{86}Sr$ ratio of a sample falls below the range from BABI to 0.704, its Sr probably derived from the mantle directly or indirectly without any appreciable contamination from another source. But if the initial Sr ratio is above this level, the Sr may have derived entirely from sialic rocks which were earlier enriched in ^{87}Sr because of their age or elevated Rb/Sr ratios. Alternatively, the sample may contain mantle-derived Sr which was modified by the later addition of Sr enriched in ^{87}Sr from another source. Either way, a high initial $^{87}Sr/^{86}Sr$ ratio is evidence that pre-existing sialic material was significantly involved in granite formation. Therefore, such rocks with high initial Sr ratios represent 'granitized' crystalline matter or are hybridized rocks containing important quantities of sialic material which was assimilated by magma originating in the upper mantle. Faure and Powell (1972) listed relevant initial Sr isotope ratios for about 130 granitic rocks and found that about half of them had sialic ancestors. That many granitic rocks may have crustal ancestry seems to be likely, for instance with late Precambrian and Phanerozoic granitic gneisses together with associated plutonites perhaps representing the crystallization products of locally derived magmas. Even so, the mantle must have acted as an ongoing source for vast volumes of basaltic and ultramafic rocks.

It must be noted that initial Sr isotope ratios of large granitic batholiths in North America exceed those of the mantle only slightly. Such relatively low $^{87}Sr/^{86}Sr$ ratios may be the result of melting of lower continental crust with low Rb/Sr ratios due to Rb depletion by granulite facies metamorphism, the phenomenon involving the mantle. Later, the magma may have been contaminated by variable amounts of ^{87}Sr from rocks of the upper continental crust. Alternatively, magma may have been generated by the melting of a mixture of deep sea sediment, sea floor basalt and upper mantle in a descending lithospheric plate in a subduction zone overridden by the North American plate.

These hypotheses were examined by Kistler and Peterman (1973, 1978). An overall south-easterly increase occurred in initial $^{87}Sr/^{86}Sr$ ratios in rocks from Californian batholiths north of the Garlock fault and east of the San Andreas fault, the relevant range of values being 0.7031 to 0.7082 except for one value of 0.7094. These researchers regarded the initial Sr isotope ratios as independent of the age of the rocks and controlled by geographical factors. A contour map for the region showing ratios at intervals of 0.001 indicated that the 0.706 and 0.704 contours seem to delimit areas with different crustal histories reflected in the distribution of Palaeozoic sedimentary rocks and ultramafics, ore deposits, topographical elevations and gravity anomalies. The 0.706 contour constituted the border between carbonates, shales and sandstones in the south and east on the one hand and volcanics, cherts and greywackes to the north and west. The 0.704 contour marked the eastern border of Mesozoic granitic rocks of which the trace element concentrations resemble those of oceanic tholeiites and alkali basalts. It also marked the eastern margin of the main exposures of ultramafic rocks.

Such correlations imply that the initial Sr isotope ratios of the granite-type rocks in the batholiths in question reflect compositional variations in the crust of the region which arose in the late Precambrian and early Palaeozoic. Correlations of initial $^{87}Sr/^{86}Sr$ ratios and chemical compositions of Mesozoic granite plutons with those of Late Precambrian to Triassic sedimentary rocks where these are intruded implies a strong crustal influence on the granitic magmas. Objections were made to any anatexis of sedimentary rocks since magmas so produced would have had higher initial Sr isotope ratios than those actually observed. Another model envisaged quartz diorite magmas forming by upper mantle melting near the continental margin and granodiorite with quartz monzonite magmas deriving from lower crustal melting further inland and east of the margin. This would accord with the observation that the granitic rocks are predominantly made of quartz diorite and trondhjemite in which the initial $^{87}Sr/^{86}Sr$ ratio is less than 0.704 and also with the fact that quartz diorites and granodiorites occur where the ratio is between 0.7040 and 0.7060. In addition, the rocks are mostly granodiorites and quartz monzonites in which the ratio exceeds 0.7060. The chemical and isotopic compositions of granitic rocks with ratios less than 0.7040 imply source magmas coming from the upper mantle or perhaps through melting and assimilation or young volcanics and volcanogenic sediments. The intrusives with ratios exceeding 0.7060 may have resulted from melting of lower crustal rocks with high Rb/Sr ratios and long residence times. Kistler and Peterman (1978) used this contour to map the periphery of the Precambrian continental crust in the western USA. They showed that several rifting episodes affected the crust during the late Precambrian and also during the Phanerozoic prior to Mesozoic activity.

Magma forms locally in subduction zones in the upper parts of a descending lithospheric plate and in the mantle above it, the oceanic crust above the plate being thin (consisting of *c*. 5 km of basalt with an overlying layer of pelagic sediment which is much thinner and accumulates during movement of crust from ridge to the subduction zone). In the descending plate, temperature and pressure conditions convert basalt through amphibolite to eclogite which later partly melts to form magma of calc-alkaline composition. Water from pelagic sediments and hydrated basalt in the oceanic crust acts as a flux to promote melting. Clearly the chemical compositions of the various magma involved will reflect that of the parental material and also the degree of melting which was involved plus the extent to which interaction with country rock through which migration occurred took place on the way to the surface. Thus volcanics in island arcs appear to have $^{87}Sr/^{86}Sr$ ratios dominated by Sr typical of the upper mantle, but these may range upwards in response to how much Sr derived from pelagic sediments in the oceanic crust or was afterwards contributed by the interaction of magmas and country rocks. The ratios in magmas generated in subduction zones may be raised becaused the Sr in oceanic curst basalts increased through reaction with sea water, as Hart, Erlank and Kable (1974) indicated. It is also possible that Rb-rich phases such as phlogopite may melt preferentially and so form a magma with an $^{87}Sr/^{86}Sr$ ratio higher than that of the bulk of the parental source material. In view of the number of variables involved, it appears unlikely that $^{87}Sr/^{86}Sr$ ratios are determined by a single factor e.g. depth to the Benioff zone or the composition of melting crust. However, magmatic activity along subduction zones most probably is a main mechanism of interaction between Benioff zones and the continental crust. This latter may have been augmented by ascending mantle material and this might imply that the crust expanded in volume through geological time. Indeed, Hurley and Rand (1969) proposed this on the basis of the distribution of orogenic belts on the continents and also because granitic rocks tend to have low initial $^{87}Sr/^{86}Sr$ ratios which indicate that either they came from the upper mantle or from rocks with short residence times in the crust.

Variability in Sr isotope ratios occurs in basalts formed in oceanic basins, among oceanic islands and at mid-oceanic ridges due to alteration through interaction with sea water or hot brines during the extrusion and cooling of pillow basalt (Degens and Ross, 1969). Also, there is potential isotopic exchange between oceanic crustal rocks and basaltic magma as well as possible growth of variable amounts of ^{87}Sr in basaltic magmas in the course of fractional crystallization. Yet another cause for this variability is differences in the Rb/Sr ratios of source regions produced through vertical and/or lateral variations in the parameter in the mantle under mid-oceanic ridges and oceanic islands. This reinforces the concept of a heterogeneous mantle, also shown by concentrations of alkali metals generally. Peterman and Hedge (1971) noted a positive correlation between Sr isotope ratios and $K_2O/(K_2O+Na_2O)$ ratios of oceanic basalts. This implies that such basalts with high $^{87}Sr/^{86}Sr$ ratios may have derived from sources with higher K concentrations than those with low Sr ratios.

Clemens, Gromet and Farrell (1993) discussed synchronous changes in sea water Sr isotope composition and global climate and noted previous work showing that the $^{87}Sr/^{86}Sr$ ratio in sea water increased gradually over the past 40 Ma, which suggested that there has been a concomitant increase in rates of chemical weathering. They referred to a recent analysis of a 250 ka $^{87}Sr/^{86}Sr$ record on which was superimposed gradual increasing higher frequency Sr isotope ratio variations appearing to follow a 100 ka cycle. This periodicity was taken as corresponding with one of the prominent cycles in the Earth's orbital parameters which have been shown to modulate the patterns of solar insolation and therefore climate. However, the resolution of this record was not sufficient to establish the phase relationship between the $^{87}Sr/^{86}Sr$ variations and the global climatic cycles. Clemens, Gromet and Farrell (1993) presented a high-resolution sea water $^{87}Sr/^{86}Sr$ record spanning the past 450 ka and found that maxima and minima in these ratios coincided with minima and maxima, respectively, in continental ice volume (from the SPEC-MAP oxygen isotope record). It was inferred that maybe there was less chemical weathering in arid glacial periods than in the more humid interglacials. During glacial–interglacial transitions, sea water $^{87}Sr/^{86}Sr$ changes at a rate of $\sim 1\,\mathrm{ppm\,ka^{-1}}$, this being about three times a prevous value arrived at by other researchers. Mass balance calculations showed that simple changes in modern chemical weathering regimes are not able to completely account for such rapid changes and this suggests that revision of current ideas about Sr reservoirs and the mechanisms of exchange between them is necessary.

Clement, Gromet and Farrell (1993) worked with single species of planktonic foraminifera from Indian Ocean Drilling Project (ODP) Site 758. The variability in $^{87}Sr/^{86}Sr$ was at or near analytical resolution at the $\pm 1\sigma$ level, but the persistence of the signal over several 100 ka cycles, a strong in-phase covariation with global ice volume and structure similar to that reported for Pacific core V28-238 supports a possible interpretation as an environmental signal connected with glacial–interglacial climatic change. In order to test the reproducibility and potential character of this apparent signal, Clemens, Gromet and Farrell (1995) generated a high resolution record for the North Atlantic using the benthic foraminifer *Cibicides wuellerstorfi* from core Chain 82-24. This Sr isotope record did not demonstrate resolvable evidence of glacial–interglacial cyclicity. However, during the analyses a systematic and positive relationship between mass fractionation and fractionation-corrected $^{87}Sr/^{86}Sr$ values was observed and corrected. The fractionation bias was caused by a failure of the linear law to correct totally for fractionation effects. Without this extra correction, the Chain 82-24 record showed a 100 ka Sr isotope signal and an in-phase relationship with ice volume. Hence an apparent (uncorrected) record of 100 ka variability in the Sr isotopic composition of sea water for this site constitutes an analytical artefact seemingly linked to environmental parameters which vary on glacial–interglacial timescales. The resolution of this connection is to come, but nevertheless it must be concluded that a glacial–interglacial Sr isotopic signal does not occur in Ch-ain 82-24. Similar high precision attempts to replicate the V28-238 signal by others have also proved unsuccessful.

The revealing of an analytical artefact affecting data from Chain induced rechecking of ODP site 758 results for similar effects by Clemens, Gromet and Farrell (1995). The data were collected in static mode with an exponential fractionation correction and a fractionation bias was not detectable in either standards or samples of the data set. There is thus no reason to assume that the glacial–interglacial signal from Site 758 is an analytical artefact related to fractionation bias. However, the high-precision results from the Chain data comprise a strong argument against the existence of global change in sea water $^{87}Sr/^{86}Sr$ (at the measurement precision level) on glacial–interglacial timescales.

8

Potassium–argon and argon–argon dating

K–Ar DATING

8.1 GEOCHEMISTRY

Potassium (Latin *kalium* from the Arabic *gali* meaning alkali) is an element with relative atomic mass of 39.0983 and atomic number 19. A silvery white and soft, very reactive alkali metal, it resembles Na. Its relative density is 0.86, its melting point is 63.5°C and its boiling point is 770°C. It is widely distributed in sea water and occurs in a number of salts such as carnallite (natural K, Mg chloride) and sylvite (natural KCl). It is essential for life and present in all living matter. Its salts are employed as fertilizers and for many other purposes.

With Li, Na, Rb, Cs and Fr, K is assigned to Group IA (alkali metals) in the Mendeléev and other 'short' forms of the Periodic Table. In different nomenclatures, it is included as a typical, main group of non-transition element and may be allocated to the s block elements with its electronic configuration of $[Ar]4s^1$. In a new set of designations which have been recommended by the International Union of Pure and Applied Chemistry, K is in category 1.

Potassium has 18 isotopes of which the majority are unstable. The isotopic composition of the element was studied by F.W. Aston who, in 1921, reported detecting ^{39}K and ^{41}K. However, the naturally occuring radioisotope ^{40}K was not identified until 1935. Two years later, this was shown to undergo branched decay to ^{40}Ca and ^{40}Ar after it was observed that the Ar abundance in the Earth's atmosphere is orders of magnitude greater than that to be expected in comparison with the cosmic abundances of the other inert gases. From this, it was inferred that excess ^{40}Ar should be present in ancient K-bearing minerals. By 1948 this had been confirmed in suitably old orthoclase, microcline and other minerals so that the K–Ar dating method became practicable once the half-life of ^{40}K was established as 1.28×10^9 a. All that is necessary is to estimate the ratios of the two nuclides ^{40}K and ^{40}Ar in a specimen in order to obtain an indication of its age.

Argon (Greek *argos* meaning inactive) is an element with relative atomic mass of 39.948 and atomic number 18. It is a noble gas with a melting point of −189°C and a boiling point of −185°C. It is a constituent of the Earth's atmosphere (0.9%) used in large amounts for filling electric lamps and in flourescent tubes at a pressure of 400 $N\,m^{-2}$.

With He, Ne, Kr, Xe and Rn, Ar comprises the noble gases. These are sometimes termed inert or rare gases, but neither name is appropriate because, as mentioned above, Ar is far from rare and both Kr and Xe are known to form compounds, for instance KrF_2 and $XePtF_6$. In the Mendeléev Periodic Table, the noble gases make up Group 0 or 8. In other nomenclatures, Ar is a typical, main group, non-transition element falling into the s and p block elements group with its electronic configuration of [Ne] $3s^2p^6$. Among the new designations of the International Union of Pure and Applied Chemistry, Ar is in category 18.

Argon has 15 isotopes, all but three radioactive and of these all but one (^{39}Ar; $T_{1/2} = 269$ a) extremely short-lived (milliseconds to days).

The isotopic abundances of naturally occurring isotopes of K, Ar and Ca are summarized in Table 8.1 While the K–Ar dating method has developed into a widely used and significant approach to measuring the

Table 8.1 Isotopic abundances of naturally occurring isotopes of K, Ar and Ca

Isotope	*Abundance* (%)
Potassium	
^{39}K	93.2581 ± 0.0029
^{40}K	0.1167 ± 0.00004
^{41}K	6.7302 ± 0.0029
Argon (in the terrestrial atmosphere)	
^{36}Ar	0.337
^{38}Ar	0.063
^{40}Ar	99.600
Calcium	
^{40}Ca	96.941
^{42}Ca	0.647
^{43}Ca	0.135
^{44}Ca	2.086
^{46}Ca	0.004
^{48}Ca	0.187

ages of K-bearing rocks and minerals, ^{40}K also decays to ^{40}Ca and this might be considered a feasible alternative, but it has not been much used. This is partly because ^{40}Ca is the commonest, naturally occurring stable isotope of Ca and formation of radiogenic ^{40}Ca in a rock or mineral increases its overall abundance only minimally. Another reason is that Ca isotopoes are fractionated by natural and industrial processes. However, K–Ca dating will be considered in Chapter 11.

The radioactive decay of naturally occurring ^{40}K to the stable isotope ^{40}Ar takes place by electron capture as well as by negatron and positron decay with emission of γ-quanta and energy of 1.51 MeV. About 11% of the ^{40}K atoms decay by electron capture to an excited state of ^{40}Ar which de-excites by emitting a gamma ray (thus going indirectly to the ground state). About 0.16% of the decays take the more direct route by electron capture directly to the ground state of ^{40}Ar. Positron decay is very rare, occurring only about 0.001% of the time with an E_{max} of 0.49 MeV. It is followed by two annihilation gamma rays with a combined energy of 1.02 Mev. Decay to the stable ^{40}Ca involves most of the ^{40}K atoms and takes place by negatron emission directly to the ground state with a decay energy of 1.32 MeV. The result of all these processes is that only 11% or so of the ^{40}K atoms decay to ^{40}Ar.

8.2 METHODOLOGY

8.2.1 Theory

In a K-bearing system which has been closed to K, Ar and Ca during its geological history, the growth of radiogenic ^{40}Ar and ^{40}Ca can be described as follows:

$$^{40}Ar^{+} + {}^{40}Ca^{+} = {}^{40}K\,(e^{\lambda t} - 1)\,,$$

where t is time and λ the total decay constant of ^{40}K, i.e. the sum of the decay constants of the two branches of the decay mode. Various values for these have been given in the literature. Thus, for the $^{40}K \rightarrow {}^{40}Ar^{+}$ branch, two such values are $0.585 \times 10^{-10}\,a^{-1}$ and $0.566 \times 10^{-10}\,a^{-1}$. A widely used value for the $^{40}K \rightarrow {}^{40}Ca^{+}$ branch is $4.72 \times 10^{-10}\,a^{-1}$. The decay constant for electron capture by ^{40}K direct to the ground state of $^{40}Ar^{+}$ is often stated as $(8.67 \pm 1.74) \times 10^{-13}\,a^{-1}$. Values generally accepted are $\lambda_e = 0.581 \times 10^{-10}\,a^{-1}$ and $\lambda_\beta = 4.962 \times 10^{-10}\,a^{-1}$, where λ_e refers to the decay of ^{40}K to ^{40}Ar and λ_β refers to the decay of ^{40}K to $^{40}Ca^{+}$. Hence the total decay constant of ^{40}K is the sum of these, namely $5.543 \times 10^{-10}a^{-1}$, the corresponding half-life being given by the expression $T_{1/2} = 0.693/5.543 \times 10^{-10} = 1.250 \times 10^{-9}$ a, where $T_{1/2}$ is the half-life. The branching ratio R is defined as λ_e/λ_β and it has a value of 0.117. The fraction of ^{40}K atoms decaying to $^{40}Ar^{+}$ is given by $(\lambda_e/\lambda_\beta)\,^{40}K$.

The growth of radiogenic ^{40}Ar atoms in a K-bearing mineral or rock can be shown in this manner:

$$^{40}\mathrm{Ar}^{+} = \lambda_e/\lambda\,^{40}\mathrm{K}\,(e^{\lambda t} - 1)\,,$$

and the total number of $^{40}Ar^{+}$ atoms is

$$^{40}\mathrm{Ar} = {}^{40}\mathrm{Ar}_i + {}^{40}\mathrm{Ar}^{+},$$

this being the sum of any present when the mineral formed ($^{40}Ar_i$) and the radiogenic ones ($^{40}Ar_i$).

A basic assumption is that there was no ^{40}Ar present when the mineral or rock formed, i.e. that $^{40}Ar_i = 0$. This simplifies matters so that, if the K concentration and quantity of accumulated radiogenic $^{40}Ar^{+}$ in a mineral sample are known, the penultimate equation can be solved for t thus:

$$t = \frac{1}{\lambda} \ln\left[\frac{^{40}\mathrm{Ar}^{+}}{^{40}\mathrm{K}}\left(\frac{\lambda}{\lambda_e}\right) + 1\right].$$

Another assumption must be made. This is that no radiogenic ^{40}Ar arising from the decay of ^{40}K in the mineral during its geological history has escaped by diffusion. Of course this entails the hypothesis that the mineral closed to ^{40}Ar very soon after it crystallized, in other words that rapid cooling took place. In fact, it is implicit in the method that no excess ^{40}Ar became incorporated into the mineral sample when it crystallized or during any subsequent episodes of metamorphism.

In addition, the mineral must have remained closed to K throughout its geological history and the isotopic composition of the K in it must have changed solely through the radioactive decay of the ^{40}K; therefore fractionation did not occur.

A correction must also be made for the presence of atmospheric ^{40}Ar. In dating a K-bearing mineral or rock, both the concentration of K and the quantity of radiogenic ^{40}Ar have to be measured. Suitable techniques include atomic absorption spectrometry, isotope dilution, neutron activation and flame photometry.

Metamorphic effects are adverse since they often induce Ar loss, a phenomenon reflecting the inability of the gas to bond with other atoms in a crystal lattice. Actually, minerals have difficulty in retaining Ar even at quite low temperature and at atmospheric pressure. This would also be the case even where argon over-pressure in basement rocks triggers the release of radiogenic Ar during metamorphism. But Ar over-pressure involving either radiogenic or atmospheric Ar does not always take place, as was shown in Alpine muscovites in granites from the Helvetic basement (Aar Massif) in the central Swiss Alps.

In view of the above considerations, it is not surprising that the concept of closure temperatures, i.e. temperatures below which, when minerals form, crystallization ages are recorded, has come under increasing criticism. Also, studies in the western Alps and in south-west Norway demonstrated that old ages had been retained despite later metamorphic temperatures estimated as exceeding proposed isotopic closure temperatures. The implication is that other factors may control mineral age relations to a greater or lesser extent.

Other possible causes of Ar loss include rises in temperature due to deep burial, chemical weathering and alteration by aqueous fluids, deformation, exsolution and recrystallization. Obviously K-bearing rocks which can resist as many as possible of these factors are optimal for dating by the K–Ar method and indeed are fairly common. Among igneous rocks, volcanics are especially valuable and whole-rock analyses of basalts yield good results since their minerals include plagioclase feldspar and pyroxene which retain Ar satisfactorily. Volcanics containing devitrified glass are less useful and are best avoided, as is the case for specimens with secondary minerals such as zeolites and xenocrysts. Curiously, basaltic glass seems to retain Ar quite well, but hydrated glass, particularly when devitrified, does not. The presence of such devitrified glass lowers K–Ar ages whereas, if xenocrysts occur which contain excess radiogenic ^{40}Ar, increased K–Ar ages will be obtained. Where basalts have been extruded on the ocean floor under high hydrostatic pressure, excess radiogenic ^{40}Ar may have concentrated in the vitreous crusts of individual pillows (perhaps because Ar in solution in the lava became trapped through rapid quenching). The amount of such excess radiogenic Ar decreases inwards because the slower cooling rate allows the gas to migrate outwards. Holocrystalline samples from the centres of pillows are optimal.

Such excess radiogenic Ar has been recorded in some K bearing minerals and affects K–Ar ages, notably in young minerals or in those having a low K content. However, while beryl, cordierite and tourmaline often contain excess radiogenic ^{40}Ar, many other minerals do not. Such excess radiogenic ^{40}Ar may be present in minerals which have been subjected to a high partial pressure of the gas during an episode of regional metamorphism. It may derive through the outgassing of older K-bearing minerals in the Earth's crust and mantle or from older xenoliths or xenocrysts. Its existence increases K–Ar ages and so provides overestimated ages for such minerals.

In practice, feldspars, biotite and hornblende in volcanics are among the most useful minerals for the K–Ar method. Nevertheless, plutonics and metamorphics can be sampled as well because they too contain these minerals. Among sedimentary rocks, the most appropriate for K–Ar dating are those containing the mineral glauconite, a hydrated Mg–Al–K ferrous and ferric iron aluminosilicate.

8.2.2 Argon Loss

In K–Ar dating, it is assumed that the sample contained no Ar when formed and that radiogenic argon afterwards produced in it was quantitatively retained. However, Ar is easily lost by diffusion, even at temperatures below the melting point, and so it is impossible to be sure that minerals containing the gas record equilibria during cooling when effective closing of the system took place.

In 1964 S.R. Hart described the loss of Ar due to increase in temperature in a contact metamorphic zone resulting from the intrusion 55 Ma ago of the Eldora stock into the Idaho Springs Formation of the Front Range, Colorado, USA. Hornblende, biotite and K feldspar specimens were dated from this Precambrian lithostratigraphic unit. They were collected at increasing distances from the intrusive contact of the quartz monzonite stock which comprises a quartz–feldspar–biotite gneiss with some biotite–quartz–feldspar schist and amphibolite. Hart reviewed age determinations suggesting that the rocks were regionally metamorphosed about 1.35 to 1.4 Ga ago. The much later intrusion of the stock produced only small-scale mineralogical metamorphic effects. Nonetheless, the K–Ar dates were affected, the fraction of Ar lost by each mineral specimen decreasing as a function of distance from the contact and reflecting the different retentivities of the relevant minerals for radiogenic Ar. Apparently the coarse biotite lost most of the gas out to a distance of approximately 30 m from the contact, but the effects of the Ar loss were traced to distances more than 2135 m from it. Thereafter, the loss decreased rapidly. At about 30 m the K–Ar dates reached a plateau value of 1.2 Ga which rose to 1.375 Ga some 4270 m from the contact. It is uncertain whether the latter value indicates the response of hornblende to the episode of contact metamorphism 55 Ma ago or relates to the earlier regional metamorphism of the Idaho Springs Formation 1.35 to 1.4 Ga ago. Figure 8.1 illustrates Hart's results.

It is clear that contact metamorphism causes systematic loss of radio-Ar from minerals and it has been shown than hornblende has a greater

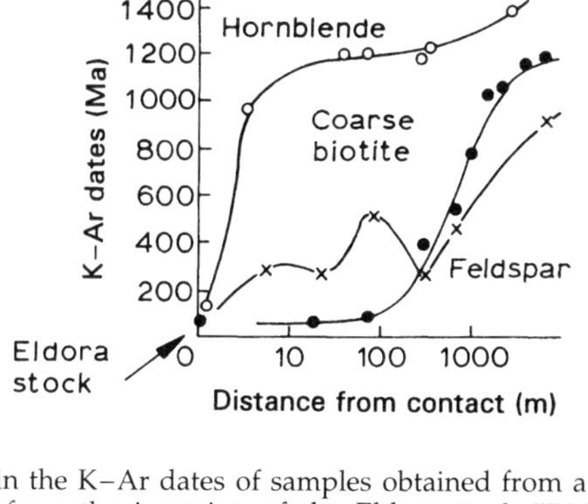

Fig. 8.1 Variations in the K–Ar dates of samples obtained from a zone of contact metamorphism resulting from the intrusion of the Eldora stock 55 Ma ago into the Idaho Springs Formation of the Front Range, Colorado, USA. The latter underwent a much earlier episode of regional metamorphism between 1.35 and 1.4 Ga ago.

retentivity for the gas than biotite under the same physicochemical conditions. Because K feldspar loses radiogenic Ar easily at quite low temperatures, it is not an optimal mineral for K–Ar dating. This may not necessarily indicate a low coefficient of retentivity, however, since F.W. McDowell showed that it is more difficult to extract radiogenic $^{40}Ar^{+}$ from alkali feldspar even at 1670°C, in part because of the lack of structural volatiles such as water which would facilitate flushing the gas out of the melt.

During his work on the Idaho Springs Formation, Hart dated biotite by the Rb–Sr method and obtained a similar pattern of variation in dates which were always greater than the K–Ar ones. The inference is that biotite retains $^{87}Sr^{+}$ better than $^{40}Ar^{+}$. Also, since the K–Ar dates in a contact metamorphic aureole of the type which he investigated relate mainly to variable radiogenic argon losses, they are not too useful in determining the age of the rocks concerned. However, data derived very near the intrusive body and at a considerable distance from it are more valuable. Indeed, the K–Ar dates from biotite samples stabilize at 4270 m from the contact.

To be useful in K–Ar dating, minerals and rocks must be capable of retaining radiogenic Ar quantitatively, be resistant to chemical alteration, contain K and be common enough to provide meaningful results. Dealing first with volcanics, the most valuable minerals are feldspars (sanidine, anorthoclase and plagioclase), the mica biotite, the amphibole hornblende and pyroxenes, but the feldspathoids leucite and nepheline are sometimes useful as well. Among plutonic rocks, biotite and the other micas, muscovite and, to a lesser extent, lepidolite are of service, as are hornblende and pyroxenes. Apropos metamorphic rocks, the most useful minerals are the micas biotite, phlogopite and muscovite together with hornblende and pyroxenes. Finally, among sedimentary rocks, only glauconite, possibly a mica, is of use.

Dating of volcanics by means of whole-rock analyses is a reliable technique for investigating basalts which contain both plagioclase and pyroxene. Of course, altered specimens or those containing devitrified glass, secondary minerals or xenoliths must be avoided. This is partly because, although anhydrous volcanic glass retains radiogenic Ar well, hydration seems to have an adverse effect and would produce lower dates if analysed. In additon, xenoliths may contribute extra radiogenic Ar and thus produce elevated dates, The actual amount of excess Ar found in submarine basalts increases as a function of depth of water, which probably shows that it is controlled by hydrostatic pressure. Excess radiogenic Ar has been recorded in beryl, pyroxene and tourmaline, etc., while hornblende, feldspar and biotite do not usually have such an excess.

Interestingly, excess radiogenic argon has been observed in diamonds, which usually occur in ultrabasic igneous rocks called kimberlites which have been subjected to high partial pressure of the gas during regional metamorphism. The Ar either diffused into or was occluded by them and probably came from the outgassing of K-bearing minerals in the terrestrial crust and mantle.

8.2.3 Isochrons

The quantity of $^{40}Ar^{+}$ in a unit weight of a K-bearing mineral or rock of age t can be expressed as follows:

$$^{40}\mathrm{Ar} = {}^{40}\mathrm{Ar}_i + \left(\frac{\lambda_e}{\lambda}\right) {}^{40}\mathrm{K}\,(e^{\lambda t\,-\,1}) \qquad (8.1)$$

Where $^{40}Ar_i$ represents the number of atoms of ^{40}Ar present per unit weight of sample not formed by the decay of ^{40}K since closure of the mineral or rock sample, so consituting a non-radiogenic component arising from a number of sources. For instance, it may have been released from solution in a magma deriving from the Earth's mantle or it may have been outgassed from old K-bearing minerals in the Earth's crust. Alternatively, it may have been produced in the course of later thermal metamorphism of the rocks and diffused into minerals as this process went on. Lastly, it is also possible that atmospheric argon became adsorbed on grain boundaries and in microfractures during exposure of the rock or mineral specimen after collection., It may even become bound within the crystal lattice.

The usual procedure is to assume that all initial ^{40}Ar derives from the atmosphere and subtract this from the total gas occuring in a specimen on the basis of the relevant $^{40}Ar/^{36}Ar$ ratio of 295.5. Where rocks or minerals contain Ar from magmatic and/or metamorphic sources, they will yield calculated K–Ar dates which are older than their actual ages. The difficulty can be evaded in favourable circumstances by the isochron approach to dating.

Equation (8.1) can be modified by dividing all the terms by ^{36}Ar, the number of ^{36}Ar atoms present in a unit weight of sample, so that

$$\frac{^{40}\mathrm{Ar}}{^{36}\mathrm{Ar}} = \left(\frac{^{40}\mathrm{Ar}}{^{36}\mathrm{Ar}}\right)_i + \frac{\lambda_e}{\lambda}\frac{^{40}\mathrm{K}}{^{36}\mathrm{Ar}}\,(e^{\lambda t-1})\,.$$

This expresses the measured $^{40}Ar/^{36}Ar$ ratio as a sum of two terms, the first relating to Ar donated by igneous, metamorphic and atmospheric sources and the second to the radiogenic component accumulated since closure of the relevant mineral system. In a general way, the magnitude of this ratio depends upon the isotopic composition of the different components of Ar and hence will differ from that of atmospheric Ar. But K-bearing minerals coexisting in a rock sample will have extremely similar, if not identical, such ratios and also the same age. Under such conditions, the equation represents a set of straight lines in coordinates of $^{40}Ar/^{36}Ar$ and $^{40}K/^{36}Ar$. If compared with that of a straight line in the slope–intercept form of $y = b + mx$, then the intercept on the y-axis will be thus:

$$b = \left(\frac{^{40}\mathrm{Ar}}{^{36}\mathrm{Ar}}\right)_{\mathrm{i}},$$

and the slope will be as follows:

$$m = \frac{\lambda_{\mathrm{e}}}{\lambda}\,(e^{\lambda t\,-\,1})\,.$$

Thus coexistent minerals with the same initial $^{40}Ar/^{36}Ar$ ratio as well as the same age can be represented by points defining a straight line in coordinates of the measured $^{40}Ar/^{36}Ar$ and $^{40}K/^{36}Ar$ ratios, and this constitutes an isochron. The slope of such an isochron, m, may be used in order to caluculate the relevant age in this manner:

$$t = \frac{1}{\lambda}\ln\left[m \times \frac{\lambda}{\lambda_{\mathrm{e}}} 1\right].$$

K–Ar isochrons have been extensively utilized and they are formed only where all samples involved comprise closed systems with respect to K and Ar and had the same initial $^{40}Ar/^{36}Ar$ ratios, but different $^{40}K/^{36}Ar$ ones. This ensures that the data points obtained define the relevant slopes well. Consequently, the isochron approach is perhaps more advantageous than conventional K–Ar dating because the extent of good fit of such data points to a straight line expresses the degree of closure of the samples being analysed through geological time and also the similarity or concordance of their initial $^{40}Ar/^{36}Ar$ isotope ratios. Such initial ratios which exceed the atmosperic value reflect the derivation of gas not just from the atmosphere but also from the outgassing of crustal rocks or the terrestrial mantle. Ar arising from such sources is highly enriched in ^{40}Ar relative to ^{36}Ar which causes the initial $^{40}Ar/^{36}Ar$ ratios obtained from K–Ar isochrons to exceed the value of the ratio in atmospheric Ar. Such extraneous radiogenic ^{40}Ar produces conventional K–Ar dates of minerals exceeding the actual ages of the host rocks.

Difficulties occur when the different K-bearing minerals from a given rock sample show different inital $^{40}Ar/^{36}Ar$ ratios induced by varying degrees of atmospheric Ar contamination. They can be minimized or even obviated by removing any adsorbed atmospheric Ar either by leaching with HF acid or by preheating under vacuum before Ar is extracted from the sample. It must be added that some are sceptical about this solution.

An important caveat relates to spurious K–Ar isochrons formed by rocks containing minerals with different closure temperatures. The correlation of ^{40}Ar and ^{40}K here is due to mixing and the resultant lines have no geochronological significance.

In sum, the K–Ar isochron method is appropriate where all rocks or minerals to be analysed incorporate Ar of the same isotopic composition regardless of its source and are sufficiently enriched in radiogenic Ar for any differences in their initial $^{40}Ar/^{36}Ar$ ratios to be insignificant.

8.2.4 Sedimentary Rocks and Minerals

Dating sedimentary rocks is basic to constructing a geological timescale, but it is a difficult task in part because abundant rocks are mostly derivative, e.g. sandstone and shale contain minerals which originate from the erosion of pre-existing rocks so that dates obtained from them only establish their provenance in most cases. Thus their depositional ages remain indeterminate. Another adverse factor is that such rocks composed of authigenic minerals just do not contain enough K for dating purposes or do not quantitatively hold radiogenic Ar formed by ^{40}K decay after their deposition. Hence, dating sedimentary rocks is best tackled using volcanics interbedded with them and some clay minerals formed by chemical reactions in the oceans. These considerations eliminate carbonates, phosphorites and ironstones.

Despite these problems, attempts have been made to apply K–Ar dating to sedimentary rocks. In the case of shales, diagenetic effects on K–Ar dates have been examined and it appears that the whole-rock ages exceed the time of deposition, decreasing with initial buried depth, a result of loss of radiogenic ^{40}Ar from detrital K feldspar and mica which became unstable. Their decomposition may result from an increase in the amount of illite in illite-smectite mixed-layer clay with depth, from reactions exemplifying burial metamorphism. During this process, K is conserved, while radiogenic ^{40}Ar is lost and, as a result, the K–Ar dates of whole-rock samples of shale are lowered.

A decrease in K–Ar dates with depth has been demonstrated in the Gulf Coast region of the USA, a phenomenon not attributable to ^{40}Ar loss through geothermal heating. In one well in Harris County, Texas, whole-rock shale samples showed a decrease in K–Ar dates which stopped at about 3700 m depth where the temperature is 95°C. The smallest grain-size fraction ($< 0.1\ \mu m$), i.e. the one theoretically most liable to loss of radiogenic Ar, actually possessed more of this gas below this depth than above it. It was shown that this coincided with an increase in K concentration which decayed after burial metamorphism. The rises in K and ^{40}Ar concentrations were used to calculate the actual age of this metamorphism as about 18 Ma, which agrees well with the Late Oligocene to Early Miocene age of the rocks.

Sedimentary rocks are often interbedded with layers of bentonite which were formed by the alteration of volcanic ash deposited from eruptions at great distances away. The ash, originally including shards and glass fragments, altered into smectite and other clay minerals. Such bentonites are useful in correlating stratigraphic sections because they were rapidly deposited over wide areas downwind from their sources. Also, they contain residual volcanic materials such as sanidine, quartz and zircon. Sanidine provides the most reliable K–Ar dates as it contains much K, resists weathering and has low initial ^{40}Ar content. Another bentonite mineral is biotite and, if unaltered, this yields reliable K–Ar dates.

Glauconite has been alluded to earlier and is an authigenic K-bearing silicate. It has given rather confusing dates using the K–Ar method and these are sometimes regarded as minimal values which under-estimate the depositional ages of the host rocks. The phenomenon may result from increase in K concentration after burial or from loss of radiogenic Ar after low-grade thermal metamorphism or structural deformation.

8.2.5 Argon from the Mantle

Excess radiogenic ^{40}Ar in sea-floor basalts consists of a mixture of atmospheric and mantle gas with the $^{40}Ar/^{36}Ar$ ratio reaching levels of about 25 000, a level varying according to the geochemical history of the relevant magma sources. Thus analyses of basaltic glasses from individual sites in the Atlantic and Pacific Oceans yielded ratios of approximately 10 000. Gases have been analysed from vesicles in so-called 'popping' rocks from the Mid-Atlantic Ridge at 36°49.3′ N latitude and 33°15′ W longitude. The idiomatic term refers to the high gas contents which cause popping when samples are brought up from 2490 m depth, the rocks themselves being tholeiitic glasses. One contained as much as 0.881 ml g^{-1} of mostly CO_2, H_2, CO, SO_2, HCl and H_2S with roughly 0.1% N_2 and Ar plus small concentrations of hydrocarbons. Block glass samples were stepwise-heated so as to release gases for isotope analyses. Ar liberated from vesicles at 750°C gave a $^{40}Ar/^{36}Ar$ ratio of $10\,524 \pm 1047$ but, if melting was carried out at 1500°C, the ratio fell to 930 ± 160 and a powdered sample gave a value of just 385 ± 14. It was concluded that the dissolved Ar in glass was probably contaminated with atmospheric Ar while that in the vesicles derived from the mantle. The interpretation, if correct, implies that mantle Ar is highly enriched in radiogenic ^{40}Ar as against atmospheric Ar.

The terrestrial atmosphere probably originated by outgassing from the planetary interior shortly after the formation of the Earth, most of the ^{40}Ar being produced by the radioactive decay of ^{40}K. Of course, the rate of such outgassing and its possible variability through time remain unknown. It has been argued that the process could not have been uniform because models entailing this require a present $^{40}Ar/^{36}Ar$ ratio in the mantle less than 2000 and a K concentration under 50 ppm which are both unacceptable. If a model is used which embodies an early outgassing event and later continuous outgassing, the matter is simplified. For instance, the current $^{40}Ar/^{36}Ar$ ratio in the mantle of *c.* 10 000 could be derived using a model in which about 95% of the initial Ar was lost by outgassing around 4.45 Ga ago. Some work supported this concept, other work did not. The latter includes analyses of Hawaiian basalts with consistently lower $^{40}Ar/^{36}Ar$ and $^{4}He/^{3}He$ ratios than glassy basalts from mid-oceanic ridges in the Atlantic, Indian and Pacific Oceans. The He data are particularly important because they rule out atmospheric contamination as a source of the anomaly. The ratio between the two stable He isotopes changes through the formation of ^{4}He emitted as α-particles during the decay of U,Th and their daughters. Much of the radiogenic He escapes into the atmosphere, this having a $^{4}He/^{3}He$ ratio of 722 540. Two glassy basalts had an average $^{40}Ar/^{36}Ar$ ratio of *c.* 350 with a $^{4}He/^{3}He$ ratio of 37 000. The low value of the former is attributable to contamination by atmospheric Ar, but the low He ratio cannot be affected by atmospheric He. The basalts probably derived from magmatic souces in the mantle which lost neither ^{36}Ar nor ^{3}He during outgassing early in the history of the Earth.

8.2.6 Blocking Temperature and the Metamorphic Veil

Rapidly cooling volcanics and shallow igneous intrusions containing K-bearing minerals start accumulating radiogenic Ar as soon as they crystallize. However, deep-seated plutonic and metamorphic rocks cool slowly and in them the retention of Ar is deferred until they reach a temperature sufficiently low to eliminate any diffusion of the gas. That temperature is that at which the loss of ^{40}Ar by diffusion from minerals becomes negligible by comparison to its rate of accumulation. It is termed the blocking temperature. It is assumed that the minerals stayed closed both to Ar and K after crystallization and cooling through the said blocking temperature.

The various minerals used in K–Ar dating have different blocking temperature. There is an equation for the blocking temperature, but the mathematical formulation may not be applicable to slowly cooled terranes where significant age gradients may have developed in individual crystals. These could imply closure temperatures showing variation in a crystal such that estimates of cooling rate based on assumed integrated closure temperatures may be fundamentally wrong. It is interesting that a similar magnetic blocking temperature arises where rocks acquire magnetism by cooling through it. It may be defined as that temperature below which magnetic moments are unable to adjust to changes in the orientation and polarity of the terrestrial magnetic field. Igneous and metamorphic rocks which have cooled through it preserve magnetization imposed at higher temperatures.

In deep-seated plutons containing both biotite and hornblende, the blocking temperature of the former with regard to radiogenic Ar is lower than that of hornblende so that the K–Ar 'clock' of the latter commenced prior to that of the biotite. If both minerals comprise closed systems and if there was a smooth decrease in temperature, then, after the respective blocking temperatures have been reached, the hornblende will record an older date than the biotite. In fact, differences in the K–Ar dates of the two minerals can be used to infer the rate of cooling. But discordances of such dates of coexistent minerals have been noted in metamorphic rocks where two different temperature histories took place and either all the minerals were totally degassed during metamorphism and slow cooling followed or not all gas was lost (in this case the fraction of Ar lost from a particular mineral is a function of the retentivity of that mineral for it).

Many biotites from granitic gneisses and schists of Precambrian shields and orogenic belts have been dated using the K–Ar method, but the dates obtained usually did not indicate the real ages of the host rocks, rather reflecting the times elapsed after cooling though the appropriate blocking temperatures. Interestingly, when the dates from such biotites from metamorphic terranes were plotted on a map, they produced a phenomenon termed a 'metamorphic veil' by R. L. Armstrong. This was so called because the K–Ar dates veil the real ages of the host rocks and indicate only the timing of the regional metamorphic event itself and their later cooling histories. The regional homogeneity of such data from shields and orogenic belts implies that large parts of the continental crust underwent such metamorphism and later cooling induced large-scale systematic patterns. Such cooling might have resulted from uplifting of cratonic blocks bordered by fracture zones extending deep below the surface. It is possible that K–Ar dates of individual minerals from rocks in such a block represent the time interval elapsed since those presently exposed were thrust up across the isotherm corresponding to the blocking temperature of those minerals. Contours can be drawn to represent equal times since cooling through these temperatures and they are termed thermochrons or chronochrons. In the Grenville Province, Canada, they run roughly parallel to the Grenville Front, forming a surface which slopes south-eastwards. It has been inferred that rocks now exposed were uplifted through the blocking temperature isotherms between 850 Ma and 1.125 Ga ago, the western edge of the Grenville craton having been uplifted earlier than the eastern one. But it must not be forgotten that the younger dates along the eastern edge could have been caused by partial Ar losses induced thermally by the Appalachian orogeny.

8.2.7 Geomagnetic Polarity Reversals

Over 30 years ago, it became apparent that the polarity of the Earth's magnetic field has altered repeatedly through geological time and isotopic dating of these resulted in a geomagnetic polarity timescale (GPTS) for the Late Cenozoic. Actually, this synthesized palaeomagnetic and isotopic information and was basic in confirming plate tectonic theory. Ongoing refining of isotopic ages has made geomagnetic polarity reversals valuable as marker horizons in numerous types of geological researches. Detailed knowledge of these has increased considerably since pioneering work in the early 1960s suggested timescales for the reversals in question and proposed that the linear magnetic anomalies along mid-oceanic ridges are explicable by sea-floor spreading recording them. Such reversal geochronology refers to alternations from one polarity to its opposite and, through the analysis of deep-sea sediment cores, has been extended back into the Mesozoic Era and even earlier in Earth history. Of course, the further back in time the matter is examined, the more uncertain become both the information and resulting inferences. Hence the magnetostratigraphic intimations of M.W. McElhinny and other researchers as regards Phanerozoic quiet intervals cannot be regarded as other than provisional in nature. He included among these Cretaceous Normal (KN) between 85 and 110 Ma ago, Jurassic Normal (JN) between 145 and 164 Ma ago, Triassic Normal (TN) between 205 and 220 Ma ago, Permo-Carboniferous Reversed (PCR) between 227 and 313Ma ago and, preceding these at unspecified time intervals, Ordovician Reversed (Early to Middle Ordovician), Cambrian Reversed (Middle to Late Cambrian) and Latest Precambrian Reversed, (Precambrian–Cambrian boundary).

Sedimentary rocks have been used to further geomagnetic reversals investigations by the application of very sensitive magnetometers (superconducting quantum interference devices, SQUIDS). Multiple sampling of suitable sedimentary rock layers is necessary to take into account distortions which might have arisen during or after deposition and been influenced by chemical weathering and alteration. In 1980 relevant data were presented by Ness, Levi and Couch and these are listed in Table 8.2. A detailed treatment for the Jaramillo and the Matuyama–Brunhes geomagnetic boundary is given in section 8.6.1.

8.3 APPLICATION: DIAMONDS

In section 8.2.2 the pressure of excess radiogenic Ar in diamonds was mentioned. Additional work on a group of ten cubic diamonds from Zaire produced correlated concentrations of K and ^{40}Ar which,

Table 8.2 A geomagnetic polarity timescale for the last 5 Ma

Epoch	*Event*	*Geomagnetic polarity*	*Time interval* (Ma)	*Anomaly number*
Brunhes		Normal	0–0.72	1
	Laschamp?	Reversed	0.045–0.025	
	Blake?	Reversed		
Matuyama		Reversed	0.72–2.47	
	Jaramillo	Normal	0.91–0.97	1′
	Olduvai	Normal	1.66–1.87	2
	Reunion	Normal	2.01–2.04	
		Normal	2.12–2.14	
Gauss		Normal	2.47–3.40	2.1′, 2.2′, 2.3′
	Kaena	Reversed	2.91–2.98	
	Mammoth	Reversed	3.07–3.17	
Gilbert		Reversed	3.17–5.41	
	Cochita	Normal	3.86–3.98	3.1
	Nunivak	Normal	4.12–4.26	3.2
	Sidufjall	Normal	4.41–4.49	3.3
	Thvera	Normal	4.59–4.79	3.4

interpreted as a whole-rock K-isochron with the usual assumptions, gave an unreasonable age of 6.0 Ga, according to Podosek *et al.* (1988). The same age was also determined by means of $^{40}Ar/^{39}Ar$ analysis of an additional four diamonds from the same group in earlier work by Y. Takigami and others. The explanation for the data which was presented by Podosek *et al.* is that the K in these diamonds is isotopically anomalous, specifically, enrichment by a factor of two in ^{40}K relative to normal *K* being postulated and perhaps resulting from preservation of cosmic ray-induced spallation K produced in Fe-rich planetesimals prior to these being accreted to the Earth. Examination of the isotopic composition of K in three diamonds from the same group showed that, within analytical uncertainties of about 1%, ^{40}K occurs in normal abundance which does not support the hypothesis of isotopically anomalous K. However, an alternative explanation was put forward, namely that the ^{40}Ar in these diamonds represents a trapped or excess constituent. But this would necessitate the unusual condition of a good correlation between ^{40}Ar and K, for instance by the incorporation of a fluid phase, associated with a lack of correlation between ^{40}Ar and other noble gas species, including ^{39}Ar.

^{40}Ar–^{39}Ar DATING

8.4 GEOCHEMISTRY

The method is based on the formation of ^{39}Ar by irradiation of K-bearing samples using both thermal and fast neutrons in a nuclear reactor as follows:

$$^{39}\text{K (n,p)}^{39}\text{Ar}\,, \tag{8.2}$$

^{39}Ar is radioactive, decaying by beta emission with a half-life of 269 a, which makes it stable in terms of the relatively insignificant analytical times involved in research. A counting technique for assessment of ^{39}Ar was superseded in the middle 1960s by mass spectrometric measurement of the ^{40}Ar/^{39}Ar ratio.

During the irradiation process, reactions occur which involve K, Ca and Cl, but that of interest is the one given in equation (8.2). The number of ^{39}Ar atoms resulting from it is given by the expression

$$^{39}\text{Ar} = {}^{39}\text{K}\Delta T \int \phi(\varepsilon)\sigma(\varepsilon)\,\mathrm{d}\varepsilon\,,$$

where ΔT is the time interval of irradiation, $\phi(\varepsilon)$ the neutron flux density at an energy $\varepsilon, \sigma(\varepsilon)$ is the capture cross-section of ^{39}K for neutrons with energy ε and integration is effected over the whole energy spectrum of the neutrons.

The number of radiogenic ^{40}Ar atoms present in a sample due to the radioactive decay of ^{40}K during its geological history is

$$^{40}\text{Ar}^{+} = \frac{\lambda_c}{\lambda}\,{}^{40}\text{K}\,(\mathrm{e}^{\lambda t - 1})\,,$$

where λ_c is the decay constant of ^{40}K for electron capture, λ is the total decay constant of ^{40}K and t is the age.

After the neutron irradiation process, the ^{40}Ar/^{39}Ar ratio is given by

$$\frac{^{40}\text{Ar}^{+}}{^{39}\text{Ar}} = \frac{\lambda_c}{\lambda}\frac{^{40}\text{K}}{^{39}\text{K}} \times \frac{1}{\Delta T}\frac{(\mathrm{e}^{\lambda t - 1})}{\int \phi(\varepsilon)\sigma(\varepsilon)\,d\varepsilon} \tag{8.3}$$

and a new irradiation parameter can be devised. This may be termed J and is given by the following:

$$J = \frac{\lambda}{\lambda_c}\frac{^{39}\text{K}}{^{40}\text{K}}\Delta T \int \phi(\varepsilon)\sigma(\varepsilon)\,d\varepsilon\,.$$

If J is substituted into equation (8.3), then

$$\frac{^{40}\text{Ar}^{+}}{^{39}\text{Ar}} = \frac{(\mathrm{e}^{\lambda t - 1})}{J}\,.$$

It is difficult to establish the value of the neutron flux density and the capture cross-sections because both the energy spectrum of the bombarding neutrons and the relevant neutron capture cross-sections of ^{39}K are not well known. Nevertheless, J can be assessed by irradiation of a sample with a known age (the flux monitor) together with samples of unknown age. Once the $^{40}Ar/^{39}Ar$ ratio of the flux monitor has been determined, J can be calculated from the following:

$$J = \frac{(e^{\lambda t_m - 1})}{^{40}Ar^+/^{39}Ar} \tag{8.4}$$

where t_m is the known age of the flux monitor and $^{40}Ar^+/^{39}Ar$ is the measured value of the ratio in the monitor.

The energy spectrum of the neutron flux applied to a sample during irradiation is dependent on where it is in the sample holder, hence a number of samples of the flux monitor are inserted into the holder at known positions between unknown samples. Samples may be irradiated for several days in the nuclear reactor in order to permit ^{39}Ar to be produced. Subsequently, the Ar in the flux monitors is liberated by fusion in vacuum. The $^{40}Ar^+/^{39}Ar$ ratios of the samples are measured in a mass spectrometer and the J-values calculated using equation (8.4). They are plotted as a function of position in the holder. J-values for the unknown samples are derived by interpolation of the resultant graph from their known positions in it. The $^{40}Ar^+/^{39}Ar$ ratios of the unknown samples can be used to calculate dates thus :

$$t = \frac{1}{\lambda} \ln \left(\frac{^{40}Ar^+}{^{39}Ar} J + 1 \right).$$

A number of mineral concentrates of different compositions have been employed as flux monitors. Clearly their ages must be accurately known. Precise K–Ar dates of four often used flux monitors have been established and these are Hb3gr, MMHb-1, LP-6 and FY12a. A typical value of J is that of MMHb-1 ($J = 0.0133 \pm 0.0001$).

An estimated analytical error in the calculated date has been given as

$$\sigma_t = \frac{J^2 F^2 (\sigma^2_F + \sigma^2_J)^{1/2}}{t^2 \lambda^2 (1 + FJ)^2}.$$

where $F = {^{40}Ar^+}/{^{39}Ar}$, σ_F^2 and σ_J^2 are the variances of F and J respectively (given as percentages) and t is the age of the sample. Dates so obtained are referred to as total Ar release dates, but are subject to the same caveats which apply to conventional K–Ar dates, namely that no radiogenic ^{40}Ar is present. Dates obtained are advantageous in that they obviate problems arising from inhomogeneous distribution of K and Ar in a particular sample and require only measurement of the isotope ratios of Ar.

In the above argument a simplification exists. It is assumed that all ^{40}Ar in the irradiated sample is either radiogenic or atmospheric in origin, that all ^{36}Ar is atmospheric and that ^{39}Ar is produced by the (n,p) reaction. If this were the case, the measured values of the $^{40}Ar/^{39}Ar$ and $^{36}Ar/^{39}Ar$ ratios could be used in calculating the desired ratio of radiogenic ^{40}Ar to ^{39}Ar as follows:

$$\frac{^{40}Ar^+}{^{39}Ar} = \left(\frac{^{40}Ar^+}{^{39}Ar}\right)_m - 295.5\left(\frac{^{36}Ar^+}{^{39}Ar}\right)_m$$

where m denotes measured values and 295.5 is the ratio between ^{40}Ar and ^{36}Ar in atmospheric Ar. Additional interesting data are given in Table 8.3.

It may be inferred that several corrections are necessary, especially as regards young samples ($\sim 10^6$a) and those having K/Ca < 1.0. Particularly important are interfering reactions involving Ca isotopes, e.g. ^{36}Ar is produced by the reaction ^{40}Ca (n, nα)^{36}Ar and removed by the reaction $^{36}Ar(n,\gamma)^{37}Ar$, both interfering with the correction for atmospheric Ar which is based on ^{36}Ar. Calcium is also responsible for the production of ^{37}Ar through the reaction ^{40}Ca $(n,\alpha)^{37}Ar$ and the abundance of the latter in an irradiated sample indicates that calcium interference has taken place. Yields of ^{37}Ar from ^{39}K and ^{36}Ar are smaller. The radioisotope ^{37}Ar decays by electron capture to the stable isotope ^{37}Cl (T_ε = 35.1d). Therefore a correction for its decay since irradiation is required because it abundance is used to estimate the contributions of $^{40}Ca(n,n\alpha)^{36}Ar$, $^{40}Ca(n,\alpha)^{40}Ar$ and $^{44}Ca(n,n\alpha)^{40}Ar$ affecting the total quantity of ^{40}Ar. ^{38}Ar production is insignificant unless an Ar spike is to be utilized for measuring absolute amounts of Ar. ^{39}Ar and ^{40}Ar are produced by reactions with isotopes of Ca and K. Correction factors for Ca and K-derived Ar in the TRIGA reactor of the US Geological Survey were:

$$(^{36}Ar/^{37}Ar)_{Ca} = (2.72 \pm 0.014) \times 10^{-4},$$

$$(^{39}Ar/^{37}Ar)_{Ca} = (6.33 \pm 0.043) \times 10^{-4},$$

and

$$(^{40}Ar/^{39}Ar)_K = (5.9 \pm 0.42) \times 10^{-3}.$$

The production rate of $^{40}Ar/^{37}Ar$ due to Ca has been given as 4.6×10^{-3}.

The most important of the corrections of Ca-rich samples arises from the production of ^{39}Ar by $^{42}Ca(n,\alpha)^{39}Ar$ and of ^{36}Ar by $^{40}Ca(n,n\alpha)^{36}Ar$ reactions. With young samples having high K/Ca ratios, extra corrections are needed to remove ^{40}Ar produced by the reactions $^{40}K(n,p)^{40}Ar$ and $^{41}K(n,d)^{40}Ar$.

There is an equation relating the age of an irradiated sample to its $^{40}Ar^+/^{39}Ar$ ratio and incorporating corrections for interfering reactions.

Table 8.3 Some interfering nuclear reactions produced by the irradiation of mineral samples by neutrons

Isotope produced	*Element irradiated*			
	K	Ca	Cl	Ar
^{36}Ar	–	^{40}Ca(n, n α)	–	–
^{37}Ar	^{39}K(n, nd)	^{40}Ca(n, n α)		^{40}Ar(n, γ)
^{38}Ar	^{39}K(n, d)	^{42}Ca(n, n α)	^{37}Cl(n, γ β$^-$)	^{40}Ar(n, nd β$^-$)
	^{41}K(n, α β$^-$)			
^{39}Ar	^{39}K(n, p)	^{42}Ca(n, α)	–	^{38}Ar(n, γ)
	^{40}K(n, d)	^{43}Ca(n, n α)		^{40}Ar(n, ndβ$^-$)
^{40}Ar	^{40}K(n, p)	^{43}Ca(n, α)	–	–
	^{41}K(n, d)	^{44}Ca(n, n α)		

If this ratio is expressed as F, then

$$F = \frac{A - C_1B = C_1C_2D - C_3}{1 - 1\,C_4D}$$

where A is the measured value of the ^{40}Ar/^{39}Ar ratio, B is the measured value of the ^{36}Ar/^{39}Ar ratio, C_1 is the ^{40}Ar/^{36}Ar ratio in the atmosphere (295.5), C_2 is the ^{36}Ar/^{37}Ar ratio produced by interfering neutron reactions with Ca ($(2.72 \pm 0.014) \times 10^{-4}$), C_3 is the ^{40}Ar/^{39}Ar ratio produced by interfering neutron reactions with K ($(5.9 \pm 0.42) \times 10^{-3}$), C_4 is the ^{39}Ar/^{37}Ar ratio produced by interfering neutron reactions with Ca ($(6.33 \pm 0.043) \times 10^{-4}$) and D is the ^{37}Ar/^{39}Ar ratio in the given sample after correcting for the radioactive decay of ^{37}Ar.

8.5 METHODOLOGY

8.5.1 Thermochronological data through incremental heating

It is possible to calculate a date from the ^{40}Ar$^+$/^{39}Ar ratio of a sample which has been irradiated by neutrons and so a set of such dates can be obtained for the sample if Ar is liberated from it in steps following temperature increases. If the sample was a closed system for both Ar and K since it first cooled, the dates obtained from each step should be constant. However, if some loss of radiogenic Ar occurred and this took place in regard to some but not all crystallographic sites, the ^{40}Ar/^{39}Ar ratios of the gas released at different temperatures will vary and a range of dates will result. From this, the time elapsed since initial cooling can be derived. The concept was incorporated into a model by G. Turner and this was used to forecast the ^{40}Ar$^+$/^{39}Ar spectrum of a K-bearing mineral grain which underwent a partial loss of radiogenic

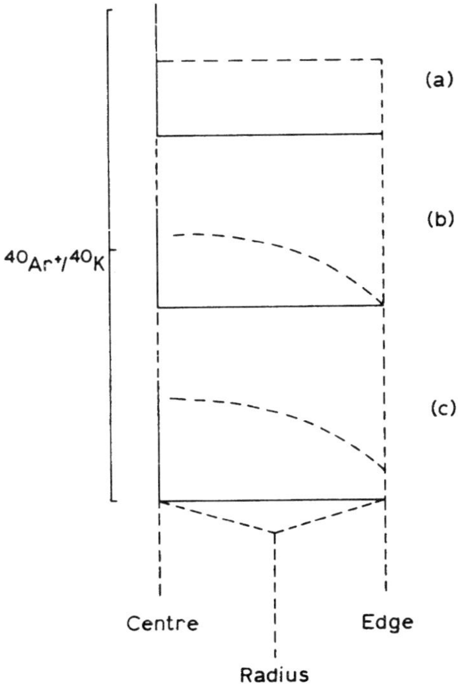

Fig. 8.2 Schematic diagram illustrating Turner's model involving incremental heating, showing (a) the uniform distribution of the $^{40}Ar^{+}/^{40}K$ ratio in a spherical mineral grain comprising a closed system since its formation; (b) partial loss of $^{40}Ar^{+}$ through diffusion induced by heating which reduced the relevant ratio to zero at the surface of the grain while leaving it unaltered centrally; (c) the distribution of the ratio at a time after partial loss of Ar after the system again became closed to $^{40}Ar^{+}$. The value of the ratio on the surface of the grain is a function of that time which elapsed subsequently to partial outgassing and, at the centre of the grain, it corresponds to the age of the mineral sample.

^{40}Ar caused by heating in the course of a metamorphic event. The model is shown in Fig. 8.2.

A mineral grain which has experienced partial loss of ^{40}Ar may become a closed system again when the metamorphic event which caused this loss ends. If uniformly sized grains of a particular K-bearing mineral with such a geological history is incrementally heated, the $^{40}Ar^{+}/^{39}Ar$ ratios of the gas fractions liberated at different temperatures will vary systematically. The first one will emerge at the lowest temperature and come from grain surfaces with a low ratio. This will reflect the time interval elapsed since the accumulation of radiogenic ^{40}Ar restarted after a metamorphic event. Succeeding fractions released through rising temperatures will have progressively higher $^{40}Ar^{+}/^{39}Ar$ ratios representing the removal of the gas from increasingly retentive locations which lost ever smaller fractions of ^{40}Ar during this event. Finally, the $^{40}Ar^{+}/^{39}Ar$

ratios may reach a plateau corresponding to a date approaching the time which has passed after the original cooling of the mineral. So the stepwise heating procedure applied to a mineral sample which underwent partial loss of radiogenic Ar permits a set of dates to be calculated and this ideally includes the time of the metamorphism (low-temperature release) and also the time of the initial cooling (high-temperature release). This is advantageous because a date can be derived which nears, if sometimes underestimating, the original cooling age of K-bearing minerals in rocks which endured partial loss of radiogenic ^{40}Ar during a metamorphic event.

Investigations have been made on stony meteorites including severely shocked chondrites. In addition, the effect of implanting Ar isotopes during a similar shock metamorphism on $^{40}Ar^{+}/^{39}Ar$ dates of both terrestrial and lunar rocks has been studied. Many data have been obtained from samples derived from metamorphic terranes on Earth, for instance Barberton Mountain komatiites.

The validity of the technique was scrutinized by dating minerals and rocks with ages known through alternative isotope methods. A case in point involved age spectra for biotite and microcline from a granite sample taken from crystalline rock in the Marble Mountains of south-eastern California. This had a Rb–Sr date for the biotite of 1.41 Ga $\pm$ 30 Ma and a U–Pb date for a zircon of 1.45 Ga. The biotite gave a K–Ar date of 1.1152 Ga $\pm$ 30 Ma and the microcline was dated at 992 $\pm$ 26 Ma. The K–Ar dates for both biotite and microcline are less than the age of crystallization of the complex, which shows that radiogenic Ar was lost during a later thermal event to an extent of *c.*20% in the biotite and even more so with the microcline (which accords with its lower retentivity). Various age determinations imply that the loss occurred between 160 and 180 Ma ago. The $^{40}Ar^{+}/^{39}Ar$ age spectrum of the biotite agreed with that expected for a mineral undergoing partial Ar loss and the actual dates rose from 223 $\pm$ 30 Ma for the first gas fraction to reach a plateau after about 30% of the gas was freed. The average plateau date is 1.3 Ga, which is about 150 Ma greater than the K–Ar date but lower than the crystallization date by up to 150 Ma. The total Ar release date was 1.251 Ga $\pm$ 61 Ma. The plateau date of the biotite is less than its known age of crystallization by 5 to 10% (predicted by calculations for mineral grains losing 20% Ar). The spectrum from the microcline differed, the first fraction giving a date of 1.063 Ga and the next two fractions having dates around 200 Ma. Thereafter, dates reached a maximum of 1.133 Ga $\pm$ 10 Ma and no stable plateau formed. Hence the maximum date for the highest temperature of release represents a minimum value for the age of crystallization, the high date from the first fraction possible resulting from a small quantity of ^{40}Ar which diffused out of the crystal lattices into non-lattice sites from which it was easily extracted at low temperatures. Such excess ^{40}Ar may also have derived from ambient fluids and stored in the non-lattice sites. The lowest dates for both microcline and biotite are overestimates of the age of the thermal event which led to Ar loss from them. If such excess ^{40}Ar occurs, errors arise in conventional K–Ar dates and cannot be corrected.

Whole-rock samples of diabase containing excess radiogenic ^{40}Ar were analysed in order to discover whether the stepwise heating technique is capable of overcoming the problem. The samples came from dykes in western Libya intrusive in Precambrian crystalline rocks which were Rb–Ar dated at 2.7 to 3.4 Ga old. Conventional K–Ar dates of whole-rock samples and plagioclase from the dykes discord, ranging from 1.2 Ga to 186 Ma. But in other localities in Libya, similar dykes found in Palaeozoic sedimentary rocks yielded more concordant dates of 173 to 193 Ma. The palaeomagnetic pole positions of all these imply that they are post-Triassic and pre-Tertiary in age. Discordant K–Ar dates might be the result of incorporation of varying amounts of ^{40}Ar from country rock into the magma while the dykes were being intruded. One diabase sample gave a conventional K–Ar date of 853 ± 26 Ma, for its total release $^{40}Ar^{+}/^{39}Ar$ date, 838 ± 42 Ma. Both dates exceed the actual age of the rock so that, although the partial release patterns of samples are similar, they are geochronologically meaningless. The high dates recorded are a consequence of the presence of excess ^{40}Ar arising as suggested above. The minimal dates of the central part of the spectrum are above the age of crystallization. While the dates of the high-temperature fractions reach 1.995 Ga $\pm$ 16 Ma, the last 2.8% of that Ar released by complete fusion gave a date of only 79 ± 69 Ma. Clearly, the stepwise heating technique was not useful in solving the problem.

From the above discussion it is apparent that minerals from slowly cooled metamorphic and intrusive igneous rocks normally give $^{40}Ar^{+}/^{39}Ar$ ages younger than their real crystallization ages because radiogenic ^{40}Ar continues to escape from the mineral crystals in appreciable amounts until they cool below some critical temperature range. The ^{40}Ar loss in the course of cooling may be considered to be a volume diffusion process, while the integrated closure temperature of the mineral at that time, which corresponds to the temperature of the mineral at the time conforming with its apparent integrated age, can be regarded as a function of the diffusivity of ^{40}Ar in the mineral, that dimension over which diffusive loss takes place, and the rate of cooling. Previously mentioned and numerous other studies have established that ^{40}Ar diffusivity differs greatly among K-rich minerals, which is why part of the cooling history of a rock can be assessed by determination of the $^{40}Ar^{+}/^{39}Ar$ ages of different minerals with different integrated closure temperatures.

An atmospheric ^{40}Ar correction is relevant both to conventional K–Ar and $^{40}Ar^{+}/^{39}Ar$ dating mehtods. The assumption is made that ^{36}Ar is of atmospheric origin and that the ratio of $^{40}Ar/^{36}Ar$ in the atmosphere is 295.5. However, since ^{40}Ar as well as ^{36}Ar may become incorporated into minerals when they crystallize, the $^{40}Ar/^{36}Ar$ ratio of this Ar may differ significantly from that of atmospheric Ar today. Thus the observed amount of ^{36}Ar in a mineral or rock may not permit an accurate correction to be made for the presence of non-radiogenic ^{40}Ar. Where the $^{40}Ar/^{36}Ar$ ratio of the inherited Ar exceeds 295.5 an apparent excess of radiogenic Ar will result. But if the value of this ratio is below 295.5, an apparent deficiency of radiogenic Ar will be the consequence. This latter might then mistakenly be attributed to a partial loss of ^{40}Ar.

An Ar isotope correlation diagram was proposed in 1966 in an effort to overcome this difficulty. The measured $^{40}Ar/^{36}Ar$ ratio obtained from gas liberated during stepwise heating of an irradiated sample (after correcting for interfacing reactions) may be expressed as follows (m = measured ratio):

$$\left(\frac{^{40}\mathrm{Ar}}{^{36}\mathrm{Ar}}\right)_{\mathrm{m}} = \frac{^{40}\mathrm{Ar}_{\mathrm{c}} + {}^{40}\mathrm{Ar}^{+}}{^{36}\mathrm{Ar}_{\mathrm{c}}},$$

where c refers to contaminant argon (including the atmospheric component as well as argon either occluded during crystallization of the mineral or entering it at a later stage). This can be rewritten thus:

$$\left(\frac{^{40}\mathrm{Ar}}{^{36}\mathrm{Ar}}\right)_{\mathrm{m}} = \left(\frac{^{40}\mathrm{Ar}}{^{36}\mathrm{Ar}}\right)_{\mathrm{c}} + \left(\frac{^{40}\mathrm{Ar}^{+}}{^{39}\mathrm{Ar}}\right)_{\mathrm{k}}\left(\frac{^{39}\mathrm{Ar}}{^{36}\mathrm{Ar}}\right)_{\mathrm{m}},$$

where k indicates Ar produced by K in the sample. This equation represents a group of straight lines in coordinates of $(^{40}Ar/^{36}Ar)_m$ and $(^{39}Ar/^{36}Ar)_m$ of which the slopes are equal to the $^{40}Ar^{+}/^{39}Ar$ ratio. Consequently, the measured $^{40}Ar/^{36}Ar$ and $^{39}Ar/^{36}Ar$ ratios of the incremental gas fractions freed from undisturbed mineral or rock samples will define a set of points fitting a straight line, i.e. an isochron, of which the slope is the $^{40}Ar^{+}/^{39}Ar$ ratio and this relates to the age of the sample t thus:

$$t = \frac{1}{\lambda}\ln\left(\frac{^{40}\mathrm{Ar}^{+}}{^{39}\mathrm{Ar}} \times J + 1\right)$$

The intercept of the isochron will be the $^{40}Ar/^{36}Ar$ ratio of the contaminant, that is to say the ratio of these isotopes in the non-radiogenic gas fraction associated with that sample. It has been shown that the isochron dates of 11 such undisturbed samples are identical within experimental error with the plateau dates, and also that the initial $^{40}Ar/^{36}Ar$ ratio is not usually significantly different from that of the atmophere (295.5). One caveat must be entered–some have argued that Ar isochrons may sometimes be mixing lines so that the slopes will give incorrect dates. It is interesting to note that such isotope correlation diagrams have been applied to orange glass collected in the Taurus–Littrow Valley on the Moon and returned to Earth aboard Apollo 17.

An 85 mg sample analysed by stepwise heating gave an age of 3.71 ± 0.06 Ga which accorded with ages from basalt fragments from the valley floor and ruled out volcanic activity as a source for the said glass. But there are problems with using the $^{40}Ar/^{39}Ar$ approach on the Moon because both isotopes are implanted in rocks there which are exposed to the solar wind. Therefore a correction is necessary to allow for this effect in calculating the $^{40}Ar^{+}/^{39}Ar$ ratios of Ar freed from lunar samples. The work reported was done in the early 1970s and those involved made a least-squares fit of data points on an Ar correlation diagram. This gave an intercept value of 3.3 ± 0.2 and a slope of 154 ± 3. The intercept was utilized to correct each gas increment for the presence of excess ^{40}Ar and the resulting $^{40}Ar^{+}/^{39}Ar$ ratios produced a spectrum of dates. The best estimate of the actual age of the sample was believed to be the mean of dates derived for fractions from 1150°C to 1350°C. The slope of the isotope correlation line corresponded with a date of 3.7 Ga which agreed well with the average $^{40}Ar/^{39}Ar$ plateau age.

^{39}Ar may be lost or relocated in a sample because of recoil of the ^{39}K nucleus arising from the emission of a proton during the (n,p) nuclear reaction. If much is lost from an irradiated sample, the $^{40}Ar/^{39}Ar$ ratio will increase proportionally and so the total release date will exceed the real geological age. The loss of ^{39}Ar from a K-bearing mineral grain is thought to occur from a surface layer about 0.08 μm thick and the effect of such recoil-induced loss is related to the particle size of the mineral, i.e. fractured or fine-grained minerals lose more of the isotope during irradiation than do coarse-grained ones. This implies that the $^{40}Ar/^{39}Ar$ method may be unsuitable if applied to dating of K-rich clay minerals of the glaucony type which normally comprise green pellets made up of aggregates of platy crystals about 1 μm in thickness. Very old total release $^{40}Ar/^{39}Ar$ dates for Early Ordovician glaucony from Vastergotland, Sweden, were recorded and its conventional K–Ar dates were about 8% lower than the known geological age. Anomalous dates were attributed to the loss of ^{39}Ar through recoil during irradiation. Such ^{39}Ar recoil loss was also reported from glauconite samples which were well crystallized.

The recoil loss may result in gas expelled from K-rich grains entering adjacent grains of K-poor minerals. Then, the $^{40}Ar/^{39}$ratio of the former would increase and that of the latter would decrease. The amount of increase would relate to grain sizes, namely the surface-to-volume ratio of the relevant grains. Where this is large, i.e. where the grains are small, there should be a larger increase in the $^{40}Ar/^{39}Ar$ ratio due to the ^{39}Ar recoil loss than in a similar sample with larger grains of K-bearing minerals. Potassium-depleted minerals which would take up expelled ^{39}Ar include pyroxene and they are more retentive as a rule than K-rich minerals. Hence Ar freed at high temperatures may have low $^{40}Ar/^{39}Ar$ ratios because it is derived from pyroxenes containing ^{39}Ar expelled from K-bearing minerals during neutron irradiation of the sample. This may explain the fact that the stepwise release spectra of some lunar rocks showed a decrease of this ratio at the highest gas release temperatures.

Another important topic is the detection of excess ^{40}Ar, that Ar, which becomes incorporated into minerals and rocks by processes other than *in situ* radioactive decay of ^{40}K and which must be distinguished from inherited Ar deriving from the same radioactive decay before the rock-forming event affecting minerals containing ^{40}Ar. Both excess and inherited Ar may be referred to as extraneous Ar. The conventional K–Ar dates of minerals which include excess ^{40}Ar exceed their cooling ages and are not geochronologically valid. The incremental heating technique of the $^{40}Ar/^{39}Ar$ method can sometimes detect the presence of excess ^{40}Ar by a saddle-shaped spectrum of dates. Similar patterns have been recorded for biotite, pyroxene, hornblende and plagioclase from rocks known to contain such excess ^{40}Ar. However, while these reveal the existence of excess ^{40}Ar, they do not provide information enabling the age of the host rocks to be determined. A further adverse factor is the observation that biotites from metamorphic rocks known to contain excess ^{40}Ar have partial release spectra with flat plateaux which exceed the known geological ages of these rocks. Then the incremental heating approach cannot distinguish radiogenic ^{40}Ar formed by radioactive decay of ^{40}K in the rock from excess ^{40}Ar added after crystallization because the latter is more or less homogeneously distributed in the sample with respect to the K in it. So a plateau in the spectrum of partial release dates of a sample is not necessarily indicative that its $^{40}Ar/^{39}Ar$ plateau date is valid geochronologically.

8.5.2 Argon release by lasers

The $^{40}Ar/^{39}Ar$ age spectra of a whole-rock sample and mineral concentrates may be altered by the occurrence of minerals which may lose Ar at a continuous rate even at low temperatures, and perhaps also of minerals younger than the main minerals. Such minerals can be produced by episodes of thermal metamorphism entailing low temperatures. Alternatively, they may exist as inclusions in the main minerals. Either way it is hard to separate them from the main minerals and, since they are characterized by their inability to retain Ar, their presence in irradiated samples contributes to low dates commonly derived from low temperature-derived gas fractions freed by incremental heating.

The matter has been tackled by using a precisely focused laser beam to release gas from individual grains of retentive minerals. In one experiment, a pulsed ruby laser melted less than 1 μg of the neutron-irradiated sample to create a pit 10 to 100 μm wide, the beam being focused through a petrographic microscope onto a polished surface of the sample sealed in a vacuum chamber on the stage. An individual grain can be selected and impinged on often enough to liberate enough Ar for isotope analysis. The gas is collected on charcoal at the temperature of liquid N_2 and later a gas source mass spectrometer may be used to provide the $^{40}Ar/^{39}Ar$ ratio required.

Another method involved applying continuous Ar-ion lasers to melt individual mineral grains previously irradiated by fast neutrons. Dates were obtained from biotite, hornblende and muscovite of Precambrian age and could not be distinguished from total release dates resulting from fusion of the relevant minerals by radio frequency induction. Argon was also recovered from a 3 mm slate chip by stepwise heating at different wattages from 1 to 2.65 W, the resulting spectrum of dates agreeing well with data obtained already. Further laser work includes that recorded by Hames and Hodges (1993). They made a $^{40}Ar/^{39}Ar$ evaluation of slow cooling and episodic loss of ^{40}Ar from a sample of polymetamorphic muscovite.

Using lasers in this way is valuable because individual mineral grains can be dated so that their inclusions and any alteration products can be excluded. In addition, it is feasible to analyse several such grains in a rock sample and compare them in order to deduce the effects of temperature, pressure and alteration on the retention of Ar and K. Lastly, polymictic breccias and similar rocks which contain minerals of different ages can be examined without the necessity to separate individual minerals (a task frequently hard or even impossible to carry out). The laser technique has been used to study the occurrence of He, Ne and Ar in lunar rocks from the Mare Serenitatis, but there were drawbacks. These arose from the fact that the date obtained is a total release one which could underestimate the ages of samples which lost ^{40}Ar after crystallization. Also the laser beam may stimulate the release of Ar from other minerals when penetrating the target mineral. However, it has been stated that less than 1% of the gas freed from pits comes from an area more than 10 μm from their peripheries.

Dating sulphide minerals is complicated because they lack naturally occurring radioactive elements such as U, Th, K and Rb, but sometimes they contain K, e.g. in the mineral rasvumite, KFe_2S_3. Samples of this collected from an alkali-rich igneous rock at Coyote Peak near Arcata in northern California gave a date of 26.5 $\pm$ 0.5 Ma, while associated phlogopite mica yielded a total release date from $^{40}Ar/^{39}Ar$ measurements of 28.3 $\pm$ 0.4 Ma. These results confirmed that K-bearing sulphide minerals can be dated on the basis of the radioactive decay scheme $^{40}K \rightarrow {}^{40}Ar$. Both K–Ar and $^{40}Ar/^{39}Ar$ ages can be used to examine the physical parameters involved in diffusion processes, such ages being usable in determining temperatures that prevailed in rocks when dated minerals crystallized in them if grouped into Ar isotope thermometric categories. However, it must be certain that diffusion is the sole transportation mechanism for the gas, i.e. that it is not removed by melting or water loss. The approach was employed by heating hornblende under hydrothermal conditions at various temperatures and calculating Ar diffusion coefficients from the fractions of radiogenic ^{40}Ar freed as a function of temperature. The Fechtig–Kalbitzer formula may be used. The relationship between D, the diffusion coefficient (in $cm^2\,s^{-1}$), and temperature T (K) is given by the Arrhenius equation:

$$D = D_o e^{(-E/RT)}$$

where D_o is the frequency factor ($cm^2\,s^{-1}$), E the activation energy (kcal mol^{-1}), R the gas constant (8.314 510 $J\,K^{-1}\,mole^{-1}$). The equation can be tranformed into a straight line by taking the natural logarithms of both sides and converting E and R to compatible units, so that

$$\ln D = \ln D_o - 10^3 E/RT$$

whereby the numerical values of the diffusion coefficients obtained from the partial argon release data at stepwise increasing temperatures should generate a set of data points defining a straight line in coordinates of ln D and $10^3/T$. E is determined from the slope m of such a line since

$$E = mR$$

and the frequency factor is given by the intercept with the ordinate. Also, the Arrhenius equation can be expressed in terms of D/a^2, where a is the radius of the grains from which Ar is diffusing. Harrison (1981) gave the procedure for ascertaining D_0 and E by diffusion data from hornblende. If the diffusion coefficients in this mineral for Ar obtained empirically are combined, then an activation energy is derived so that $E = 64.1 \pm 1.7$ kcal mol^{-1} and D_0 is $0.024^{+0.053}_{-0.011}\,cm^2\,s^{-1}$. These results were used to calculate closure temperatures between 578°C and 490°C for hornblende in relation to grains possessing effective diameters of 80 μm and cooling rates ranging from 500°C Ma^{-1} to 5°C Ma^{-1}, using Dodson's formula:

$$T_c = \frac{E/R}{\ln\left(\dfrac{ART_c^2 D_0/a^2}{E\ dT/dt}\right)}$$

where T_c is the closure temperature, A a geometric factor (55 for spherical grains and 27 for cylindrical ones) and dT/dt is the cooling rate at the closure temperature. The solution of this equation may be attempted by substituting an arbitrary value for T_c into the logarithmic term, this then being used to calculate T_c for suitable values of A, R, D_0/a^2, E and dT/dt. In this manner, a value for T_c is derived and then back-substituted into the logarithmic term until the calculated value for the parameter becomes successively more constant. This approach showed that the closure temperatures of the mineral hornblende increase with increasing grain size and rate of cooling, but appear to be independent of the chemical composition of the hornblende.

Frequency factors and activation energies for the diffusion of radiogenic ^{40}Ar in hornblende, biotite, K feldspar and plagioclase in rock samples from the Grenville structural province in Canada were determined in 1981. Also, coefficients of diffusion from relative volumes of ^{40}Ar released at different temperatures were calculated using two criteria, namely that at least five data points form a reliable plateau in the spectrum of $^{40}Ar/^{39}Ar$ dates and that they must constitute a straight line in the Arrhenius plot in coordinates of D_0 and $10^3/T$. The results are shown in Table 8.4.

They show a sequential decrease in closure temperatures for radiogenic ^{40}Ar in minerals in the range hornblende to plagioclase. The numerical values for the closure temperatures were calculated on the assumption that the gas was lost in a diffusion process and they are dependent upon considerations such as grain sizes, geometry and radii as well as upon the cooling rate at the actual closure temperature. Of 18 mineral concentrates studied, 11 gave 'reliable' results from which a cooling curve was constructed for the Haliburton Highlands of the Grenville Province in Ontario. Rates of cooling implied from this decrease geochronologically from *c.* 20°C Ma^{-1} at 975 Ma to *c.* 2°C Ma^{-1} at 925 Ma with a rate of *c.* 0.3°C Ma^{-1} at 700 Ma. Lithostatic pressures were estimated from the observed metamorphic grade of the rocks in the appropriate site and an estimate made that the burial depth *c.* 1 Ga ago was *c.* 21 km with an associated temperature of *c.* 700°C. This suggests a high geothermal gradient, perhaps of the order of 33°C km^{-1} compared to the existing one of between 15°C and 20°C km^{-1}. A cooling curve was selected showing that rocks reached 140°C or so at 550 Ma. From all these data it was calculated that the rocks were probably uplifted from a depth of 21 km 1 Ga ago to *c.* 4.3 km depth at 550 Ma, the rate of uplift

Table 8.4 Closure temperatures calculated from the rate parameters for the diffusion of Ar in various minerals at rates of cooling of 5°C Ma^{-1}

Mineral	*Closure temperature* (°C)	*Grain size* (μm)
Hornblende	490	80
Hornblende	685 ± 53	210–840
Biotite	373 ± 21	500–1410
K feldspar	230 ± 18	125–840
Plagioclase	176 ± 54	125–250
Microcline	132 ± 13	125–250

being 37 m Ma^{-1}. Later the rate of uplift may have declined to *c.* 8 m Ma^{-1} during the Phanerozoic Eon. This scenario was though to be consistent with Grenville Province geology and, if this is the case, it would demonstrate the reliability of the Ar isotope thermochronometer approach.

Closure temperature in cooling geochronological and petrological systems, kinetic processes and the thermal history of slowly cooling solids and the theory of cooling ages are topics dicussed by Dodson (1973, 1976, 1979). Cooling histories from $^{40}Ar/^{39}Ar$ age spectra and their implications for Precambrian plate tectonics have been examined by York (1984). Harrison *et al.* (1985) considered the diffusion of ^{40}Ar in biotite and temperature, pressure and compositional effects.

Hodges *et al.* (1994) carried out a study of $^{40}Ar/^{39}Ar$ age gradients in micas from a high-temperature and low-pressure metamorphic terrain as part of a larger attempt to evaluate two rival hypotheses concerning the tectonothermal evolution of a 1300 km wide Proterozoic orogenic belt in the south-western USA which comprises mainly juvenile terrains accreted to the southern periphery of the Wyoming craton between 1.8 Ga and 1.6 Ga. The first hypothesis is based on conventional $^{40}Ar/^{39}Ar$, K–Ar and Rb–Sr data from New Mexico and asserts that tectonic assembly took place in the course of a number of thermally separate deformational events ranging in age from ~ 1.7 Ga to < 1.2 Ga. The second hypothesis was derived pricipally from U–Pb data from Arizona and New Mexico and suggests early tectonic assembly at 1.7 Ga to 1.6 Ga and cooling at < 1 K Ma^{-1} over several hundred milion years. Hodges *et al.* set out to assess $^{40}Ar/^{39}Ar$ systematics in relation to these two hypotheses and studied mucovite and biotite from the Crazy Basin monzogranite of central Arizona. This is a ~ 1.7 Ga granitoid pluton which was emplaced at shallow crustal levels of 10 to 12 km into an assemblage of 1.74 to 1.72 Ga old volcanic and sedimentary rocks which were at ambient temperatures of 675 to 775 K. When this pluton intruded, it caused the temperature of the country rocks to rise enough to produce a well-defined metamorphic field gradient from sillimanite grade at the pluton margins to garnet grade within some kilometres. The monzogranite is believed to have cooled down relatively rapidly to ambient temperatures. However, previous geochronological data were regarded as consistent with either later slow cooling for at least 300 Ma or a reheating event at 1.4 Ga to 1.45 Ga (Karlstrom and Williams, 1994).

Mica samples came from two different samples of rock collected from adjacent outcrops of the Crazy Basin pluton and originally crushed for U–Pb geochronology. Hand picking of relatively unbroken, $\sim 700\,\mu m$ diameter phenocrysts of muscovite and biotite was carried out and major element compositions determined using the Jeol 733 electron microprobe at MIT. Samples were irradiated in the McMaster University Reactor along with the flux monitor MMhb-1, irradiation parameter

$J = 0.0133 \pm 0.0001$ (Samson and Alexander, 1987), together with synthetic salts to allow corrections for interfering nuclear reactions. Incremental heating involved bathing individual crystals with the defocused beam of a 10W Ar-ion laser for 2 mins at successively higher power levels ranging from 0.07 to 0.95 W until all Ar was released. For spot fusion analyses, these researchers focused the beam at its minimum diameter ($\sim 25\,\mu$m) on an $\{001\}$ cleavage plane. Melt pits were made by firing the laser for 0.3 to 2.0 s at 1 W. The result was approximately hemispherical melt pits with diameters of 50 to 100 μm. Isotopic analyses were carried out on a MAP 215-50 mass spectrometer with an electron multiplier. System blanks were measured prior to and after each sample and were typically two orders of magnitudes smaller than the signals. All data were corrected for blanks, interfering reactions and mass fractionation effects and apparent ages for each increment calculated assuming that the initial ^{40}Ar/^{39}Ar was that of the modern atmosphere (295.5). However, the non-radiogenic component of ^{40}Ar proved to be so low ($< 2\%$) that the calculated age is insensitive to the choice of initial ratio. Precision limits represent propagated measurement and J-value uncertainties and were reported by Hodges *et al.* (1994) at the 2σ level.

Interesting incremental heating results were reported, for instance that the Crazy Basin muscovite sample experienced several textural alterations during the process. A frosted appearance at 0.12 W, after 5.9% of its total ^{39}Ar was released, was attributed to dehydration on a large scale. At 0.19 W (cumulative total ^{39}Ar = 42.4%) glazing of the crystal occurred and showed that melting had begun. This only changed after an output power of 0.36 W was reached (cumulative total ^{39}Ar = 90.8%) at which point collapse into a glass bead took place. The integrated or total gas age for this sample was assessed as 1.420 Ga $\pm$ 10 Ma.

The Crazy Basin biotite sample showed similar textural changes in the course of heating, but these correlated differently with gas emission. An age of 1.410 Ga $\pm$ 10 Ma was obtained with an initial ^{40}Ar/^{39}Ar ratio of 16 ± 9. This completely non-atmospheric ratio was taken to reflect the fact that practically all ^{40}Ar released during analysis was radiogenic.

Hodges *et al.* (1994) suggested that a traditional interpretation of their data could be that the micas passed through their integrated closure temperatures (nominally $\sim$ 630 K for the muscovite and $\sim$ 580 K for the biotite) at 1.412 Ga $\pm$ 5 Ma and 1.410 Ga $\pm$ 10 Ma, respectively. This would suggest a rather rapid average cooling rate, which would be 25 K m^{-1} using the actual ages (though uncertainties enter into the matter because of combined age uncertainties). In view of the fact that the U–Pb zircon age of the monzogranite is almost 300 Ma older, the ^{40}Ar/^{39}Ar ages obtained through incremental heating appear to reinforce one of the hypotheses put forward by Karlstrom and Williams (1994), namely that the pluton suffered a short-lived reheating event just before 1.410 Ga which was followed by relatively rapid cooling.

Hodges *et al.* (1994) continued by recording a surprising consequence of spot fusion analyses of the same micas, i.e. the data derived from them imply a very different thermal evolution. Apparent ages from the muscovite range from 1.650 Ga $\pm$ 10 Ma to 1.270 Ga $\pm$ 10 Ma in roughly concentric manner within the $\{001\}$ cleavage plane, although the oldest ages are off-centre with respect to the crystal fragment so that ages up to 1.600 Ga $\pm$ 10 Ma occur at the rim. But the old rim ages are absent from pristine crystal faces, which implies that the off-centre zonation may be an artefact related to breakage in the course of sample preparation. These researchers regarded the observed zoning pattern as suggestive of Ar loss by a volume diffusion process with an effective diffusion dimension of at least 500 μm.

Two (simplest) interpretations of the muscovite spot fusion data were given. One is that a short thermal pulse took place at $\leqslant$ 1.270 Ga and caused episodic loss of radiogenic argon. However, this is untenable because no suitable heat source was identifiable in the local geological framework. The other is that extremely slow cooling took place during the time interval between 1.650 Ga (the oldest preserved age in the crystal core) and the final closure at 1.270 Ga. If this happened, the integrated cooling rate can be estimated by following the procedure given by Dodson (1986). Hodges *et al.* (1994) assumed an effective diffusion dimension of 500 μm, a cylindrical diffusion geometry and diffusion coefficients from the experimental work of Robbins (1992), from which they calculated a closure temperature for the core of 600 K and for the rim of 480 K with an integrated cooling rate of 0.30 K Ma^{-1} over the time interval of the closure. The integrated closure temperature of this sample is given as only 570 K, much lower than the nominal closure temperatures frequently quoted (McDougall and Harrison, 1988), in part because the apparent cooling rate is so low.

Hodges *et al.* (1994) regarded the contrast between cooling rates estimated from conventional incremental heating and age zoning studies of their Crazy Basin samples, i.e. ~ 25 against ~ 0.3 K Ma^{-1}, as having significant implications for the general interpretation of $^{40}Ar/^{39}Ar$ data. A lot of thermochronological work is based on incremental heating analyses of multigrain samples effected using a resistance furnace. These samples are usually crushed into fragments far smaller than the physical grain size. Their results support findings of previous researchers such as Scaillet *et al.* (1990) that age zoning at the scale of the physical grain size is possible in micas. The apparent integrated ages derived from multi-grain furnace analyses of such samples can be different from the true integrated ages of the crystals as a consequence of the sampling bias perhaps induced by the processes of crushing, sieving and picking. However, the laser incremental heating of individual crystals reduces the problem. The integrated ages of the Crazy Basin crystals mapped by spot fusion ($\sim$1.460 Ga for muscovite and $\sim$1.380 Ga for biotite) estimated from the areas between contours are similar to the integrated ages derived from laser incremental heating (1.420 Ga and 1.400 Ga, respectively), even though these micas were broken during preparation. But incremental heating did not show the remarkable 300 Ma to 400 Ma age gradients preserved in these samples in any systematic manner. Hodges *et al.* (1994) retrospectively realized that the very old and very young ages derived from some of the heating steps could represent subtle indications of the degree of age zonation. It is unfortunate that these increments were obtained near the commencement and the end of their experiments because they make up only small fractions of the total ^{39}Ar released, and conventional wisdom would imply that these steps should be disregarded in the course of interpreting the data. Taking into account earlier published data showing that amphiboles strongly zoned in radiogenic ^{40}Ar can give simple release spectra with well-defined plateaus, it may be inferred that using bulk incremental heating data for hydrous phases to quantify the thermal evolution of very slowly cooling ($< 1\ K\ Ma^{-1}$) in terrains is not advisable. They stressed that this does not constitute a blanket condemnation of the step-heating approach in $^{40}Ar/^{39}Ar$ thermochronology; unless unsupported radiogenic Ar occurs, significant age gradients will not be discovered in minerals cooling at rates exceeding $\sim 10\ K\ Ma^{-1}$ which do not undergo episodic reheating.

Hodges *et al.* (1994) considered the tectonic implications and took thermobarometric evidence for high temperatures at shallow depths in the crust as suggesting the presence of anomalously steep geotherms in high-temperature and low-pressure terrains. The causes for metamorphism associated with these entail either large-scale magmatic advection of heat at shallow crustal levels or extensional deformation causing almost isothermal decompression of previously deep crustal rocks (Brown, 1993). Both processes induce transient and short-lived (perhaps $<$100 Ma) thermal anomalies. The $^{40}Ar/^{39}Ar$ muscovite data suggest that the Crazy Basin pluton had cooled to 600 K below intrusion ambient conditions by 1.650 Ga within 50 Ma of the crystallization age. Of course, the micas possessed no clear record of the pluton's thermal history prior to 1.650 Ga and so, according to Hodges *et al.* (1994), it is possible that cooling to ambient conditions occurred within 1 Ma of the intrusion, as was proposed by Karlstrom and Williams (1994). But, Hodges *et al.* (1994) pointed out that both their muscovite and biotite spot fusion data showed that the pluton's rate of cooling dropped by at least an order of magnitude after the initial phase of fast cooling and stayed very low for hundreds of millions of years. They added that, accepting the apparent core-closure temperature of the muscovite (600 K) as a pointer to the ambient temperature 1.650 Ga ago and assuming that there was no up-roofing between the time of intrusion at 11 km depth (based on thermobarometric data reviewed in Karlstrom and Williams, 1994) and 1.650 Ga, the nominal geothermal gradient would have been $\sim 27\,K\,km^{-1}$ at the time of muscovite core closure. This would be a minimum estimate since probably at least some uproofing took place over the time interval 1.700 to 1.650 Ga. However, even this is twice as great as the nominal conductive geotherm for continental crust. Other calculations relating to the rim of the muscovite and both the core and the rim of the biotite gave minimum estimated geothermal gradients of $24\,K\,km^{-1}$ at 1.480 Ga, $16\,K\,km^{-1}$ at 1.270 Ga and $15\,K\,km^{-1}$ at 1.150 Ga. If these are correct, then the thermal structure of the Crazy Basin region needed some 400 Ma after Early Proterozoic high-temperature, low-pressure metamorphism in order to decay to a state nearing a nominal conductive geotherm. Hodges *et al.* (1994) regarded their preliminary results as implying that the thermal stabilization of newly formed lithosphere requires hundreds of millions of years. In addition, their presentation of large, quantifiable age gradients resulting from slow cooling intimates that age mapping of micas could be a significant new way of investigating the post-crystallization thermal history of igneous and metamorphic terrains and it should later be possible to use such data to construct time–temperature curves.

8.6 APPLICATIONS

8.6.1 The Jaramillo Normal Subchron and The Matuyama–Brunhes Geomagnetic Boundary

Izett and Obradovich (1994) presented $^{40}Ar/^{39}Ar$ data and ages for some key volcanic units which support changes in the isotopic age geochronology of the geomagnetic polarity timescale (GPTS) for the Late Cenozoic. The volcanic rocks were erupted near the time of the geologically youngest two geomagnetic polarity reversals.

Technically, the introduction of ultralow background, ultrasensitive inert gas mass spectrometers coincident with the development of the $^{40}Ar/^{39}Ar$ method offered new resolving potential for geochronological studies. Later use of a continuous Ar-ion laser fusion $^{40}Ar/^{39}Ar$ system permitted minute quantities (~ 1 mg) of K-rich geological materials to be dated with a precision approaching 0.2%. Izett and Obradovich pointed out that the $^{40}Ar/^{39}Ar$ approach has a number of advantages over the K–Ar method. Possibly the greatest of these is that only ratios of argon isotopes have to be measured in order to calculate an age rather than absolute quantities. As result, it is not necessary to extract all radiogenic Ar from a mineral to derive an accurate age. They added that this mitigates a problem inherent in the K–Ar dating of sanidine, namely that K–Ar ages of this mineral tend to be anomalously young since the melt usually forms a viscous mass which retains some radiogenic Ar under ultrahigh vacuum. Even at maximum attainable temperatures (1800°C) in standard RF-induction heaters and resistance furnaces, 5–10% of the gas can stay trapped in such viscous sanidine melts. The problem was obviated by employing a $^{40}Ar/^{39}Ar$ analytical system (Dalrymple, 1989).

A lithostratic framework for a large, complex volcanic centre of Pleistocene age was founded on geological mapping in the Jemez Mountains in northern Mexico. This showed that after eruption of tuff, rhyolite dome flow complexes were emplaced sequentially in and along the ring fracture zone of the Valles caldera. It was integrated with K–Ar age study and palaeomagnetic investigations to form the basis for recognizing the Jaramillo Normal Subchron (JNS) and the work also yielded calibration points near the Matuyama–Brunhes Boundary (MBB). Sanidine and obsidian were used. Later the isotopic ages of volcanic rocks significant for both the JNS and the MBB were reported following K–Ar and $^{40}Ar/^{39}Ar$ ages analyses of rhyolite domes of the Valles caldera which formed just before the MBB. Baksi *et al.* (1992) contributed new information on its exact age by measuring $^{40}Ar/^{39}Ar$ ages of basalt flows extruded during the Matuyama–Brunhes field reversal. The age of the MBB was further constrained by the dating by Izett and Obradovich (1994) of sanidine from the reversely magnetized Oldest Toba Tuff of Sumatra. Deino and Potts (1990) gave a $^{40}Ar/^{39}Ar$ age of 0.974 ± 0.007 Ma (1σ) for anorthoclase from pumice from Kenya. The host unit has reverse magnetic polarity and was presumably deposited just after the end of the Jaramillo.

In their researches, Izett and Obradovich used sanidine and a rhyollite sample, some of the former coming from pumice lumps which floated in water. Such lumps were either rolled in a ballmill or trimmed with a diamond saw to remove their rinds, then ultrasonically scrubbed in dilute HF (5%) to remove possible near-surface contamination. Sanidine concentrates were obtained by means of heavy liquid separation techniques. Glass adhering to crystals was removed by etching in 24% HF accompanied by ultrasonic scrubbing for up to 3 min. A typical irradiation packet comprised a milligram-sized aliquot of sanidine loaded into a 10 mm diameter Al foil cup and covered with a 10 mm Al foil cap. Such flattened packets were placed between similar packets of sanidine neutron fluence monitors, arranged in a vertical stack in a 10 mm diameter quartz tube and the positions of the packets in this were measured. The distance between adjacent packet centres was *c.* 0.5 mm and the lengths of the irradiation packages did not exceed 3 cm. The sealed quartz vials were wrapped in Cd foil to ensure that the samples and the fluence-monitor minerals were irradiated only with fast neutrons. Quartz vials were irradiated in the core of the US Geological Survey's TRIGA reactor and the received doses of fast neutrons were either 2.0×10^{17} total integrated neutron flux or 3.0×10^{17} total integrated neutron flux (nvt). A few milligrams of each of the irradiated samples were loaded into wells in a copper disc and placed in the sample chamber of the extraction line. In ultrahigh vacuum, small clusters of sanidine crystals or single chips of obsidian were heated with a 5 W Ar-ion laser at the maximum attainable temperature of 1500°C. The liberated gas was cleaned with Zr–Al and Zr–V–Fe getters and the isotopic composition of the released Ar determined with an ultrasensitive inert gas, computer-controlled mass spectrometer (Mass Analyzer Products 216 and Bauer-Signer ion source). The absolute quantities of radiogenic ^{40}Ar measured varied from *c.* 2.0×10^{-15} to 2.7×10^{-13} mol.

Izett and Obradovich entered a caveat about the advantages of the ^{40}Ar/^{39}Ar method over the K–Ar one. This is that the former is a relative method, ages of unknown samples being relative to the assigned ages of fluence-monitor minerals. One such mineral which is widely used is a hornblende (MMhb – 1) from syenite of the McClure Mountain Complex of the Wet Mountains in Colorado. Its published weighted-mean age based on K and Ar measurements in 18 worldwide laboratories is 520.4 ± 1.7 Ma (Alexander, Michelson and Lanphere, 1978). But the age calibrated in the laboratory where measurements were made of the Izett and Obravodich material is 513.9 Ma, *c.* 1.26% younger. For this reason, they used sanidine from the Fish Canyon Tuff (FCT) of Colorado and the Taylor Creek Rhyolite (TCR) of New Mexico as neutron fluence-monitor minerals because they are chemically uniform and have isotopic ages within an acceptable range of the suspected ages of the volcanic rocks selected for dating. Sanidine from the FCT is widely used by geochronology laboratories whereas that from the TCR is mainly an intralaboratory standard (Dalrymple and Duffield, 1988). Ages reported by Izett and Obradovich were calculated using fluence-monitor mineral ages as follows: FCT sanidine, 27.55 Ma; TCR sanidine, 27.92 Ma; both relative to an age of 513.9 Ma for MMhb-1. Comparison of their ^{40}Ar/^{39}Ar ages with those of other researchers entailed use of the following equation, which gives an exact mathematical expression for converting from one monitor age to another:

$$t_2 = \frac{1}{\lambda}\log_e\left[\frac{e^{\lambda t_{m2}} - 1}{e^{\lambda t_{m1}} - 1}(e^{\lambda t_1} - 1) + 1\right],$$

where t_1 is the original age, t_2 is the converted age, t_{ml} is the monitor age used for t_1, t_{m2} is the monitor age used for t_2 and λ (the decay constant) = $5.543 \times 10^{-1}\,a^{-1}$.

Errors connected with individual ages in tables of results are stated to be estimates of the analytical precision at the 1σ level and include these authors' evaluation of the precision error (0.5%) of the fluence-calibration parameter *J*. On average, they made five determinations for each fluence-monitor package in the irradiation stack and they estimated precision for each monitor as the error of the mean at the 95% level, which they propagated into the unknowns. Summary ages were given and comprised weighted means, weighted by the inverse of the variance. The error calculated for a group of ages was the error of the mean at the 95% confidence level. All K–Ar and $^{40}Ar/^{39}Ar$ ages were either calculated or recalculated using decay constants recommended by the Subcommission on Geochronology of the International Union of Geological Sciences.

As regards the JNS, it was concluded by Doell and Dalrymple (1966) that it began at 0.99 Ma and ended at 0.90 Ma. However, on the basis of $^{40}Ar/^{39}Ar$ ages of Ivory Coast tektites plus sanidine and obsidian ages from rhyolite dome flow complexes in the Valles caldera, Izett and Obradovich placed the inception of the JNS at *c.* 1.11 Ma and its ending before 0.92 Ma, probably near 0.97 Ma. This geochronology is older than most of the other frequently quoted geomagnetic polarity timescales and almost compatible with that suggested by Shackleton, Berger and Peltier (1990). This latter was based on a phase relationship match between the astronomical timescale and oxygen isotope calibrated climatic records in deep sea cores; it extends from 1.07 Ma to 0.99 Ma.

Apropos the onset of the JNS, K–Ar ages constraining this were first put forward by Doell and Dalrymple (1966) and documented by Doell *et al.* (1968). Obsidian sampled from two of three reversely magnetized rhyolite flows of the Cerro del Medio I and Cerro del Medio II domes gave ages of 1.06 ± 0.05 Ma and 1.17 ± 0.03 Ma, respectively, but these do not agree with the field stratigraphic relationships of the rhyolite flows which indicate that Cerro del Medio I is older than Cerro del Medio II, according to Izett and Obradovich. Their ages may by discussed. The weighted mean for three obsidian ages from the reversely magnetized rhyolite dome of Cerro del Medio II is 1.161 ± 0.010 Ma, which is in close accord with the Doell *et al.* (1968) value. The onset of the JNS is more widely constrained by a single age of 1.207 ± 0.017 Ma from a small sanidine concentrate from three reversely magnetized palaeomagnetic cores derived from Cerro del Medio I rhyolite.

The beginning of the JNS is constrained more widely by ages of sanidine from pumice lumps from the basal air fall units of the two units of the tuff mentioned above as having been erupted and succeeded by the rhyolite dome flow complexes around the Valles caldera. This

Bandelier Tuff has reversely magnetized upper and lower units called the Tsankawi and Guaje Pumice Beds. The weighted mean ages for these came out as 1.223 ± 0.018 Ma and 1.613 ± 0.011 Ma, respectively, Interestingly, these are some 10% older than $^{40}Ar/^{39}Ar$ ages measured by Spell, Harrison and Wolff (1990) and Spell and McDougall (1992). Spell, Harrison and Wolff (1990, p. 179) used an inhomogeneous fluence-monitor mineral (Bern4M muscovite) and an irradiation package geometry resulting, according to Izett and Obradovich, in a large uncertainty for *J*. They quoted Baksi (1987) as claiming that the age of the BernM muscovite standard is uncertain by 3.5%.

The JNS onset may be calibrated through the isotopic ages of Ivory Coast tektites. Submillimetre equivalents (microtektites) have been recorded near the base of the JNS in deep sea cores in the equatorial Atlantic Ocean. Fission track and K–Ar ages of 1.02 ± 0.1 Ma and 1.1 ± 0.1 Ma were measured on Ivory Coast tektites by Gentner, Kleinmann and Wagner (1967). Later data were recorded by Koerbel *et al.* (1989), and fission track and $^{40}Ar/^{39}Ar$ ages of 1.05 ± 0.11 Ma and 1.10 ± 0.10 Ma, respectively, were given. However according to Izett and Obradovich these have a high analytical uncertainty. From three Ivory Coast tektites (14 analyses), the $^{40}Ar/^{39}Ar$ method gave a weighted mean age of 1.15 ± 0.05 Ma (95% confidence limit for the error of the mean). A plot of the analytical data gave an intercept on the *x*-axis of 1633, and the inverse of this used in the age equation resulted in an age of 1.10 ± 0.08 Ma. This is a maximum age since the glass may contain minute quantities of inherited Ar. In sum, the onset of the JNS may be given as 1.11 Ma, 'questionably' according to Izett and Obradovich.

Turning to the end of the Jaramillo, Doell *et al.* (1968, Table 3) placed this at *c.* 0.90 Ma, based on a K–Ar sanidine age of 0.907 ± 0.028 Ma for the reversely magnetized rhyolite dome of Cerra Santa Rosa I, assuming that this was extruded near the end of the JNS because of its intermediate magnetic polarity. Izett and Obradovich data for six splits of the sanidine concentrate dated by Doell *et al.* (1968, Table 3) show a weighted mean age of 0.916 ± 0.017 Ma, not significantly older than their K–Ar age of 0.907 ± 0.028 Ma. In fact, the two statistically concordant ages offer firm calibration points in the Matuyama Chron for delineation of the end of the JNS. Some of the sanidine ages from the reversely magnetized rhyolite domes of the Valles caldera are older than K–Ar sanidine ages given by Doell *et al.* (1968, Table 3) and compatible with some ages of Spell and McDougall (1992). By contrast, the $^{40}Ar/^{39}Ar$ age of sanidine from Cerro Santa Rosa I and obsidian from Cerro del Medio II were statistically concordant with the K–Ar ages of Doell *et al.* (1968). Mexican, Sumatran and Ivory Coast data indicated that the JNS started *c.* 1.11 ± 0.08 Ma and ended before 0.92 Ma, perhaps as early as 0.97 Ma (based on the age and magnetic polarity of the Lewis Canyon rhyolite).

As regards the Matuyama–Brunhes boundary, Izett and Obradovich based their isotopic age calibration of the geomagnetic polarity timescale in part on $^{40}Ar/^{39}Ar$ ages of sanidine from three reversely magnetized rhyolite domes of the Valles caldera, namely Cerro San Luis, Cerro Seco and Cerro Santa Rosa II. Nine sanidine analyses from the first gave a weighted mean age of 0.812 ± 0.023 Ma. This is *c.* 9% older than the analytically best K–Ar age of 0.710 ± 0.015 Ma of Doell *et al.* (1968, Table 3) who gave in addition another K–Ar age of 0.845 ± 0.074 Ma which had a high uncertainty. From the other two rhyolite domes, results gave 0.794 ± 0.007 Ma, which is older than the K–Ar ages of 0.745 ± 0.015 Ma of Doell *et al.* (1968). Another calibration point was obtained from sanidine from the reversely magnetized Oldest Toba Tuff on Sumatra and the weighted mean age derived was 0.789 ± 0.006 Ma.

Establishing a calibration point within the earliest Brunhes time entailed analyses of 23 sanidine concentrates of Bishop Tuff in eastern California, which has normal magnetic polarity and was emplaced just after the Mauyama–Brunhes Boundary. From 11 concentrates an age of 0.764 ± 0.005 Ma was obtained and the other 12 gave an age of 0.757 ± 0.009 Ma. The former and the latter are generally older than K–Ar ages reported by Izett and Wilcox (1982), which were 0.738 ± 0.003 Ma and 0.736 ± 0.005 Ma, respectively. Both $^{40}Ar/^{39}Ar$ ages are also older than $^{40}Ar/^{39}Ar$ plateau and total fusion ages of 0.73 Ma and 0.74 Ma given by Hurford and Hammerschmidt (1985).They used Bern4M muscovite and Bern 4B biotite fluence monitors and the first has an age uncertainty of 3.5%.

Izett and Obradovich noted that the isotopic ages of reversely and nomally magnetized volcanic rocks on each side of the MBB bracket but do not date the geomagnetic polarity change. Nevertheless, an estimate for its age is possible by utilizing its stratigraphic position in continental sedimentary deposits relative to a dated marker horizon, e.g. the Bishop ash bed (distal Bishop Tuff tephra) which is about 0.65 m above the MBB in sediments of the ancient Lake Tecopa, California. It is 1.83 m above the boundary in sediments of the ancient Lake Bonneville, Utah. From the average rate of sedimentation of 1 m (6.5 K a^{-1}) between two isotopically dated volcanic ash beds in cores from the latter and the stratigraphic separation of 1.83 m between the Bishop ash (0.76 Ma) and the underlying MBB in a core, it may be inferred that the MBB took place *c.* 12 ka before the deposition of the Bishop ash bed, which gives it an age of 0.77 Ma. Izett and Obradovich noted that in the Pacific and Indian Oceans Australasian microtektites are closely associated with the MBB, and 54 analyses of these gave a weighted mean $^{40}Ar/^{39}Ar$ age of 0.76 9 ± 0.021 Ma for them with a 95% confidence limit for the errors of the mean (Izett and Obradovich, 1992). The inferred age of 0.77 Ma for the MBB, statistically accords with $^{40}Ar/^{39}Ar$ ages of basalt lava flows (0.783 ± 0.011 Ma) erupted in the course of the Matuyama-Brunhes transition on Maui, according to them, although one of the Maui samples contained excess Ar and others lost some Ar after crystallization. So, while the isochron age reported for Maui basalts by Baksi *et al.* (1992) is analytically sound, other work suggests that incremental $^{40}Ar/^{39}Ar$ ages of whole-rock basalts can give anomalous results (e.g. McDougall and Harrison, 1988).

8.6.2 A Burdigalian (Miocene I) Age for Extensional Ductile Tectonics in the Edough Massif, Kabylies, Algeria, from $^{40}Ar/^{39}Ar$ data

Monié, Montigny and Maluski (1992) noted that the Edough massif, an Alpine chain feature in north-eastern Algeria, was subjected to a powerful synmetamorphic ductile deformation which they interpreted in terms of normal faulting that seemingly controlled both the exhumation and cooling of the metamorphic rocks. Mica dates of 16 Ma to 17 Ma were cited as suggesting a late Burdigalian age for extensional tectonics. The termination of the underthrusting process in the massif and the commencement of exhumation probably occurred in associated with $^{40}Ar/^{39}Ar$ amphibole dating of 24.1 ± 1.4 Ma and U–Pb monazite dating of 18 Ma $\pm$ 5 Ma. Monié, Mortigny and Maluski compared these ages with the ages of post-nappe marine sediments and magmatism which led to cooling rate estimates between 200°C and 300°C Ma^{-1} with rates of uplift of about 6.8 km Ma^{-1}. These researchers proposed that, within the geodynamical framework of the western Mediterranean, the new data might be associated with the opening of the South Balearic basins and crustal extension on their margins. Actually, the Edough massif is a window beneath a Numidian flysch nappe emplaced at the end of the Burdigalian.

8.6.3 Preliminary $^{40}Ar/^{39}Ar$ Age Spectrum and Laser Probe Dating of the MI Core of the Manson Impact Structure–a K–T Boundary Crater Candidate

Kunk *et al.* (1993) took results of these investigation as indicative of a reasonable possibility that the Manson Impact Structure (MIS) is a Cretaceous–Tertiary (K–T) boundary impact event. Analyses were made of several types of samples from a melt-matrix breccia, a unit of apparent crater fill intersected by the Manson Impact (MI) core. The $^{40}Ar/^{39}Ar$ results from these samples showed a maximum age for the MIS of *c.* 65.4 ± 0.4 (2σ) Ma and petrographic analyses gave a high probability that all of the dated samples from the breccia contain relict grains not wholly melted or degassed at the time of impact. This suggested that the actual age of the MIS might be somewhat younger than the above-mentioned preliminary data implied. The results were taken to be consistent with a previously published age estimate for shocked microcline from the MIS central uplift of 65.78 ± 1.0 Ma presented by the same authors.

In the western interior of the USA, the boundary layer comprises a two-part body perhaps recording two or more bolide impacts and, if this is so, multiple impacts may have taken place at the K–T boundary. The lower unit is the thicker of the two and is a claystone made mostly of kaolinite, but also including spherules which may be the alteration product of impact-produced tektites. The lower layer might be ejecta from the Chicxulub crater, Mexico, of which the age is practically the same as that of the K–T boundary. The upper unit is called the 'magic layer' and includes both the iridium anomaly and shocked mineral grains characterizing the K–T boundary worldwide. If such multiple impacts really took place at the K–T boundary, the MIS (located in north-central Iowa) becomes a K–T boundary crater candidate because of its earlier published age of 65.7 ± 1.0 Ma and its proximity to K–T boundary sites that have the highest concentrations and largest sizes of shocked mineral grains.

The samples chosen for the preliminary $^{40}Ar/^{39}Ar$ dating included feldspar grains from clasts in the melt-matrix breccia which were melted and recrystallized to varying degrees together with the matrix itself. The fluence monitors used were sanidine from the Fish Canyon Tuff (27.79 Ma; cf. Izett and Obradovich, 1994, value of 27.55 Ma) and MMhb-I (519.4 Ma, cf. Izett and Obradovich age of 513.9 Ma, taken from Lanphere *et al.*, 1990). Tektite glass from the K–T boundary layer in Haiti was irradiated at the same time and dated at 65.0 ± 0.4 Ma, a result indistinguishable from earlier published data.

The $^{40}Ar/^{39}Ar$ age spectrum results derived from feldspar from variable melted and recrystallized clasts from the melt-matrix breccia did not develop age plateaux. Their overall pattern implied the presence of incompletely degassed protolith feldspar or the presence of some extraneous Ar in the samples. Ages of individual steps increase with temperature of extraction and range from *c.* 68 Ma to 123 Ma and the relevant age spectra were interpreted by Kunk *et al.* (1993) as representing a maximum age for the MIS of *c.* 68 Ma. $^{40}Ar/^{39}Ar$ laser probe results from a bulk K feldspar sample from the 420.5 ft level (128.2 m) level of the MI core were consistent with conventional age spectrum total gas results from the same sample at ~ 74 Ma. Similar results from multiple one- and two-grain aliquots from the same sample produced an age range from 65.4 ± 0.4 Ma to 97.5 ± 2.6 Ma. The 65.4 ± 0.4 Ma result is in accord with previous data, but may nonetheless represent an overestimate since part of the individual grains may be protolith material not totally degassed. Kunk *et al.* regarded the older ages of some analyses as almost centainly being due to included and incompletely degassed material. The $^{40}Ar/^{39}Ar$ age spectra of melt-matrix samples were more complex than those of the feldspars and were interpreted as representing the effects of some combination of the presence of excess ^{40}Ar in the samples, ^{39}Ar recoil during irradiation and the inclusion of incompletely degassed protolith material. They imply a time of impact less than $\sim$73 Ma ago. The work of Kunk *et al.* suggested a maximum age for the MIS of 65.4 $\pm$ 0.4 Ma, from which it may be inferred that it was formed during K–T boundary times. Special interest has centred on this boundary ever since it was suggested that it marked a giant impact leading to the extinction of over half the species on Earth. As a result, a search has been on for the source crater left by this and as noted earlier, Kunk *et al.* (1993) thought that the MIS might be just that.

However, another possible candidate is the Chicxulub crater in Mexico and the case for this was put by Blum *et al.* (1993). The structure has a diameter of more than 200 km, hence it is larger than the Manson structure which is about 35 km across, but it has a similar $^{40}Ar/^{39}Ar$ age of 65 Ma. This is indistinguishable from that of impact glass spherules found in K–T boundary sediments, strongly suggesting that it is a source crater for the K–T catastrophe. By contrast, the Manson melt rocks showed different isotopic composition when Sr, Nd and O isotopes were investigated. The identification of impact glass spherules and fragments at the K–T boundary in Haiti gave the first unambiguous sampling of the elemental and isotopic composition of at least part of the target material. But oxygen isotope study of these showed formation by impact-induced mixing of a siliceous target rock end member with a mixed carbonate–evaporite rock end member and both candidate craters have this rock combination in their target stratigraphy. Hence this did not enable a choice to be made between them. High-precision $^{40}Ar/^{39}Ar$ dating of Haitian K–T impact glasses and of melt rock from drill holes into both the Manson and Chicxulub structures also proved to be incapable of distinguishing the source crater because all three are of identical age within the analytical uncertainties, according to Blum *et al.* Therefore they measured Sr, Nd and O isotopic compositions of whole-rock samples of melt rock, which appeared to be unaltered taken from the C-l drill hole into the central peak of the Mexican crater and the M-l drill hole into the central peak of the US one. Strontium and Nd isotopic compositions were assessed by thermal ionization mass spectrometry after clean laboratory digestion in HF–$HClO_4$ and chemical separation in quartz cation exchange columns. Data were presented as measured and corrected for 65 Ma of radioactive decay. Haitian glass has a very variable composition, falling within two general compositional ranges, namely the abundant high-silica black glass and less abundant ($\sim$2%) high-CaO yellow glass. Major element and S concentrations and the O isotopic compositions of the glasses had been examined previously and the conclusion reached that they formed through impact-induced mixing of $\sim 58\%$ silicate rock (with the average composition of the black glass) and $\sim 42\%$ platform carbonate containing $< 10\%$ sulphate-rich evaporite. From the Sr and O isotopic composition of the average black and yellow glass and their mixing model, these researchers calculated a Sr isotopic composition of a hypothetical platform carbonate end member which, when mixed in the above proportion with average black glass, will give a Sr and O isotopic composition identical with that of the yellow glass. The $^{87}Sr/^{86}Sr$ ratio of the carbonate end member calculated above is 0.70776, which coincides with the range of values for Late Cretaceous marine carbonates. Such a mixture of silicate (low Sr concentration) with carbonate (high Sr concentration) will give a mixing curve when $^{87}Sr/^{86}Sr$ is plotted agains $\delta^{18}O$. Also, end member like these which formed the Haitian glasses might have produced impact melts from either of the two crater candidates and, if so, should plot on the mixing curve. Actually the melt rock samples from Chicxulub do, whereas the two melt rock samples from Manson have very high $^{87}Sr/^{86}Sr$ ratios plotting far above the curve, which implies that they are not related to the Haitian glass. This would seem to establish a genetic link betweeen the latter and the Chicxulub melt rock while excluding the MIS.

8.6.4 $^{40}Ar/^{39}Ar$ Fusion ages from the Polish Sudetes

Investigations by Oliver and Kelley (1993) established a reconnaissance study of $^{40}Ar/^{39}Ar$ mineral ages from the Sudeten Mountains of southwestern Poland in order to constrain the latest tectonothermal events in the various terranes now known to have undergone a Caledonian history. Regarding the geology, U–Pb zircon dating and considerations of faunal provincialism enabled the region to be interpreted as a possible Lower Palaeozoic suture zone of the Cambro-Ordovician Tornquist Sea. Several suspect terranes were defined and accreted when Gondwana and Baltica collided initially in the Upper Ordovician with partial closure and later in the Lower Devonian with total closure. A fundamental fault zone occurs in the region. Sampling covered a batholith, some gneisses and a terrane. The dating technique used involved irradiating samples at the Scottish Universities Reactor, East Kilbride. The resulting J-value using the Berne biotite standard was $0.000\,25 \pm 0.000\,002$. Samples of two or three grains were placed in small glass tubes open at one end and secured with glass wool. Then irradiation took place, the samples and tubes later being loaded into a laser port. A pulsed Nd–YAG laser was used so as to fuse grains, releasing the Ar for analysis on a MAP 215 high-sensivity gas spectrometer applying techniques given by Kelley and Bluck (1989). The blank levels for the extraction system were stated to have been typically about 10^{-12} cc S–P ^{40}Ar. The approach facilitated analysis of large numbers of samples in a relatively short time so that a regional picture of the ^{40}Ar/^{39}Ar total fusion ages could be obtained. Data were corrected for blank, discrimination, ^{37}Ar decay and reactor-induced interference. Ages were calculated using decay constants recommended by Steiger and Jäger (1977). The low flux levels produced high ^{40}Ar/^{39}Ar ratios and consequently errors in the final ages were higher than usually anticipated for this type of analysis, but the level of precision, it was claimed, permitted the regional picture of the Variscan overprint in the Polish Sudetes to be observed.

The above-mentioned work coupled with published U–Pb, Rb–Sr and K–Ar ages and ensuing analysis of the new ^{40}Ar/^{39}Ar fusion dates, together with an examination of field relationships and petrography, produced interesting results. The Ar data, etc. showed that the Sowie Gory terrane underwent firstly Ordovician and then Middle Devonian regional metamorphic episodes followed by Upper Devonian to Lower Carboniferous uplift, cooling and erosion. The Ordovician Sudeten Batholith terrane and the Devonian Moravian–Silesian terrane both went through Carboniferous uplift and cooling immediately after a major episode of Lower Carboniferous granite intrusion. The uplift in these terrane synchronized with mylonitization along the Intra-Sudetic Fault. Some biotite ^{40}Ar/^{39}Ar fusion ages turned out to be 'older' than coexisting muscovites from the mylonites, which implies that ^{40}Ar gas infiltrated the fault zone.

8.6.5 Using the ^{40}Ar/^{39}Ar method to date Mylonitic Deformation

West and Lux (1993) tested the effects of mylonitization on argon isotope systems in muscovite using strongly mylonitic rocks associated with the regionally extensive Norumbega fault zone in south-central Maine. Argon age spectra from samples obtained outside the zone of mylonitization are relatively undisturbed and possess well-defined Early Carboniferous plateau ages, according to these researchers. However, all age spectra for muscovite from the mylonites are highly discordant and characterized by young ages at low extraction temperatures, these systematically increasing to ages equalling the plateau ages for muscovite collected outside the mylonite zone. It was inferred that such systematic discordances relate to a mixing of Ar components from older, relict, muscovite prophyroclasts and fine-grained white mica aggregates which recrystallized in the course of mylonitic deformation.

The total gas ages of five different grain size fractions separated from one mylonite sample became progressively younger with decreasing grain size, which was taken to indicate a larger component of the recrystallized grains in the finer grain size fractions. In spite of the fact that the three finest grain size fractions gave different total gas ages and do not show overlap in age for most of their release spectra, their initial increments coincide, at about 290 Ma. West and Lux took this to indicate a minimal older age contribution from the relict prophyroclasts in the inital increments and suggested that the 290 Ma age offers a good estimate for the time of mineral growth connected with mylonitic deformation. These data, in association with kinematic analysis, showed, according to West and Lux, that the segment of the Norumbega fault zone examined, that is to say the Sandhill Corner fault, is a Lae Carboniferous–Early Permian dextral strike-slip fault. Because of lack of any meaningful offset in regional Early Carboniferous mineral age patterns across the fault, they inferred that displacement was probably less than 30 km.

The results of West and Lux (1993) may show that $^{40}Ar/^{39}Ar$ dating methods are applicable to the dating of deformational events effectively if certain criteria are met. Perhaps most importantly, the samples must be well characterized before analysis. Dynamic recrystallization must have taken place at or below the closure temperature of the mineral to be analysed. In addition, regional cooling patterns must be established by means of detailed thermochronology in order that mineral ages and age spectra from the deformed rocks can be compared with regional cooling ages of the same mineral. Lastly, the effects of excess Ar must be negligible.

8.6.6 Meteorites

8.6.6(a) $^{39}Ar/^{40}Ar$ Ages of Acapulcoites and Lodranites

Acapulcoites and lodranites are samples of one parent body which is though to have undergone low degrees of partial melting. They have roughly chondritic compositions with achondritic textures. Acapulcoites have been heated to just above the Fe, Ni–FeS eutectic ($\sim 980°C$) and lodranites to $\sim 1000°C$ to $1050°C$; they underwent silicate partial melting. The temperatures mentioned are broadly consistent both with oxygen isotopic and two-pyroxene geothermometry. Depending on the methods applied, various cooling rates have been reported, e.g. for Acapulco, values of $\sim 100°C\ Ma^{-1}$ from 900 to 1000°C, $20000 - 70000\ °C\ Ma^{-1}$ from 500 to 750°C and $\sim 1.7°C\ Ma^{-1}$ at 300–100°C have been estimated. These suggest very different times from <1 Ma to >400 Ma to cool to low temperatures and maybe also a complex thermal history, according to Bogard *et al.* (1993a).

Bogard *et al.* (1993a) pointed out that, although isotopic chronology can offer assistance in elucidating both the formation and the thermal history of these meteorites, earlier researches gave uncertain results.

A 4.60 ± 0.03 Ga Sm–Nd isochron age for Acapulco could well be an indication that this might be the oldest precisely dated object, exceeding, as it does, the usually accepted 4.55–4.56 Ga age of the solar system. U–Pb studies merely implied an early re-equilibration. $^{39}Ar/^{40}Ar$ profiles for the acapulcoites Y-74063 and ALH 78230 suggested 'plateau' ages of 4.556 ± 0.053 Ga and 4.531- $\pm$ 0.021 Ga, respectively, but individual temperature extractions possess wider age ranges. No age dating of lodranites has yet been reported.

Results presented by Bogard *et al.* (1993a) gave $^{39}Ar/^{40}Ar$ ages and K/Ca ratios as a function of cumulative release of ^{39}Ar for stepwise temperature extractions of Acapulco, Monument Draw and Gibson. From six extractions of Acapulco (releasing 22.97% of the ^{39}Ar), a 'plateau' age of 4.50 ± 0.01 Ga (error is 1σ) was defined. Thirteen extractions of Mountain Draw, releasing 0.2–99% of the ^{39}Ar, defined an age of 4.52 ± 0.02 Ga. Seven extractions of Gibson, freeing 8–99% of the ^{39}Ar, defined a 'plateau' age of 4.48 ± 0.02 Ga. Additionally, all extractions of Monument Draw gave a highly linear ($r^2 = 0.99995$) isochron of $^{40}Ar/^{36}Ar$ versus $^{39}Ar/^{36}Ar$ which passed through $^{40}Ar/^{36}Ar = 0$ and had a slope which corresponded to an age of 4.523 ± 0.005 Ga, i.e. was identical to the 'plateau' age. Bogard *et al.* indicated that the errors quoted for the ages included all analytical uncertainties, but did not include an extra 1% uncertainty in the age of the hornblende monitor. The ages of Monument Draw and Gibson, which were irradiated together, differed at the limits of their analytical uncertainties, which Bogard *et al.* took as implying that Gibson is 20 to 40 Ma younger.

Both the rate of Ar release and the K/Ca ratios showed that each meteorite contained two K-bearing phases, high K/Ca plagioclase and low K/Ca pyroxene. A diffusivity parameter D/a^2 was calculated, using temperature release data for Ar from each phase. If a cooling rate of $10^\circ C\ Ma^{-1}$ is assumed and diffusivity data from an Arrhenius plot for Monument Draw are used, closure temperatures of $\sim 275^\circ C$ for plagioclase and $\sim 470^\circ C$ for pyroxene would result from the calculation. In view of the much higher heating temperatures, total Ar degassing from these meteorites would be anticipated, even at moderately rapid cooling rates. Since the two phases have different closure temperatures but no observable age difference, i.e. < 30 Ma, the parent body must have cooled down from $\sim 470^\circ C$ to $275^\circ C$ at $> 7^\circ C\ Ma^{-1}$. Bogard *et al.* inferred that if cooling from the peak temperature of $980^\circ C$ of Monument Draw to the $275^\circ C$ closure temperature in $\sim$ 30 Ma, that is to say 4.55 Ga minus 4.52 Ga, occurred, then the average cooling rate of the parent body may have been $\sim 25^\circ C/Ma^{-1}$. From this they opined that neither very rapid nor very slow cooling rates predominated in the course of cooling of these meteorites. They also suggested that cooling may have taken place at tens of $^\circ C/Ma^{-1}$ above $500^\circ C$, decreasing fast to a few

$^{\circ}C/Ma^{-1}$ below 300°C. Gibson seems to have been completely degassed of planetary ^{36}Ar, but Acapulco and Monument Draw released ^{36}Ar at the highest extraction temperatures in concentrations comparable with type 5–6 ordinary chondrites.

In a discussion, Bogard *et al.* noted that the $^{39}Ar/^{40}Ar$ data given by them are consistent with the petrological and oxygen isotopic information about these meteorites. They continued by presenting a reconstruction of the history of the acapulcoite–lodranite parent body in a chronological setting. Three stages were envisaged. The first was accretion. The parent body was of chondritic composition and accreted 4.55 to 4.56 Ga ago. Unless it is substantiated by additional data, the 4.60 ± 0.03 Ga Sm–Nd age for Acapulco is assumed not to represent the accretion of the parent body. However, that there was a common single parent body for these meteorites is believed to be consistent with similarities in oxygen isotope compositions, mineralogy and mineral compositon and cosmic ray exposure ages indicative of excavation in a single impact event.

The second stage involved heating by non-collisional sources. After accretion, the parent body was heated to the point of partial melting by such sources, e.g. ^{26}Al and electromagnetic induction. There is no evidence for shock in the silicates and this accords with well-defined $^{39}Ar/^{40}Ar$ release patterns. High two-pyroxene geothermometric temperatures ($\sim$950°C to 1000°C) imply that partial melting did not take place from local excursions in shock pressures. Also, evidence for 'early re-equilibration' observed in the U–Pb system probably reflects such early heating. A higher temperature estimated for Gibson may have produced total loss of ^{36}Ar and lower K concentrations. The third stage of Bogard *et al.* is that of cooling. They asserted that the parent body cooled below the Ar closure temperature about 4.51 Ga ago for acapulcoites and about 4.48 Ga ago for the Gibson lodranite, cooling decreasing slowly from tens of $^{\circ}C/Ma^{-1}$ at higher temperatures to a few $^{\circ}C/Ma^{-1}$ at low ones. As noted, neither very rapid nor very slow cooling predominated. Lodranites were heated about 100°C higher than acapulcoites and hence needed longer cooling times, which explains the younger age. No evidence was found that the meteorites underwent any significant thermal event after the closure of the Ar chronometer. In addition, the $^{39}Ar/^{40}Ar$ ages of 4.48 Ga to 4.52 Ga were stated to be identical to the upper range of $^{39}Ar/^{40}Ar$ ages of unshocked ordinary chondrites. The acapulcoites and lodranites seemingly formed in a manner and on a timescale analogous to metamorphism on the parent bodies of ordinary chondrites.

8.6.6 (b) $^{39}Ar/^{40}Ar$ Dating of Shock–Melted Phases of the Chico Chondrite

Dating impact events both on Earth and on the Moon often involves using $^{39}Ar/^{40}Ar$ determinations of shock-melted materials believed to be most likely to have been degassed of ^{40}Ar. While some terrestrial craters have the remains of massive melt sheets exposed and lunar features such as Fra Mauro may have undergone significant resetting of a number of isotope chronometers when ejected by large basin impacts such as Imbrium, meteorite samples connected with impacts on meteorite parent bodies frequently lack such melt material or have only small melt veins. However, the 105 kg Chico L6 chondrite is unique in being composed of 50% to 75% impact melt, which has been interpreted as having been impact-injected into host chondrite material beneath a large crater on the L6 chondrite parent body. Garrison and Bogard (1993) effected $^{39}Ar/^{40}Ar$ determinations using a variety of melt and chondrite samples of apparently differing shock temperatures and thermal histories in order to elucidate the resetting of the K–Ar system in such melts.

Cutting the Chico chondrite in half produced one side (called south) which is chondritic while the opposite (north) side consists of melt invading highly fractured chondrite. Analyses were made on samples of melt and chondrite from both, as well as from two interior samples (IS) located about 1 cm apart on either side of a distinct melt–chondrite boundary. In addition, analysis was carried out of a melt sample (IN) from the middle of the melt vein and chondritic sample from the bottom. Both exterior samples were weathered, but the subsurface ones were not.

Apropos $^{39}Ar/^{40}Ar$ ages and K/Ca ratios, profiles of the former against fractional release of K-produced ^{39}Ar were made for IS and S (chondrite) and for all four chondrite samples changes in the K/Ca ratios during sample degassing took place sharply between phases and show a high K/Ca phase of 0.3 to 0.4 at lower extraction temperatures and a low K/Ca phase of 0.02 to 0.04 at higher extraction temperatures. Four melt samples gave similar K/Ca results, but they had overall about twice as great K concentrations as the chondrites, and the K/Ca ratios in the low temperature extractions have higher values exceeding 1. It seems that impact melting increased (K) in the melt over L6 chondrite material. The significant difference in temperature necessary to degas the two phases together with other considerations suggest that they may have undergone different degrees of loss of radiogenic ^{40}Ar. Garrison and Bogard regarded this as obviously true for all chondritic samples and most melt samples. In order to evaluate separately the degree of resetting of the $^{39}Ar/^{40}Ar$ system in the high-K/Ca and low-K/Ca phases, they divided the $^{39}Ar/^{40}Ar$ release spectrum for each sample into these two phases, using as criteria the sudden change in K/Ca and the distinct peaks in the release curve of ^{39}Ar against temperature. For each phase, the lowest $^{39}Ar/^{40}Ar$ age measured and the average age were determined. Both ages, together with the assessed (K) and (Ca), led to interesting data, including plateau ages summarized by Garrison and Bogard (1993), which are shown in Table 8.5. They gave an estimated low-temperature 'plateau' age stated to be a common technique for obtaining an event 'age' from such spectra. Data were reported too for Point of Rocks, an L6 discovery thought to be paired with Chico (2), for which a Rb–Sr isochron age of 0.46 Ga has been recorded.

The 'plateau ages' vary in 'quality' between samples and do not imply a common value, most suggesting an event age of 0.5 to 0.6 Ga. The average of the low-temperature phase of the four chondrite samples was 0.61 Ga, which was close to that derived for the melt ones (0.63 Ga). But these 'ages' were much higher than the inferred event age of 0.46 Ga, and only chondritic B and S samples and the N exterior melt sample (which is more weathered and hence may have lost ^{40}Ar recently) gave ^{39}Ar/^{40}Ar ages of individual temperature extractions as low as the Point of Rocks Rb–Sr age. The average age of the high-temperature phase of the chondrite samples (1.25 Ga) exceeds that of the melt samples (0.80 Ga). Higher ^{39}Ar/^{40}Ar ages for the high-temperature phase compared with the low-temperature one could be anticipated in view of the greater resistance to Ar diffusion from the former. The ages for the low-temperature phase of chondrite and melt samples also exceed the event age; this was not anticipated because of the high temperature necessary to produce the melt, and nor was the likelihood that it might have been superheated in the course of injection into chondrite material. If the impact event really took place 0.46 Ga ago and involved L6 material not degassed earlier, the apparent 0.62 Ga K–Ar ages of the low-temperature phases show that ~1% of the radiogenic ^{40}Ar present at that time was not degassed and this would make up ~25% of that present today. As an exact Rb–Sr isochron for Chico is not available, it is feasible that the event age is older, perhaps within the 0.50 to 0.55 Ga range implied by several 'plateau ages'. However, as Garrison and

Table 8.5 Age determinations and geochemical data of low and high temperature samples from Chico

Chico sample	*Low temperature*			*High temperature*			
	Plateau age (Ga)	*Low age*	*Average age*	*Low age (Ga)*	*Average age* (Ga)	(K) (ppm)	(Ca) (%)
Chond B	−0.5	0.45	0.56	0.62	1.02	685	1.20
Chond IS	0.77	0.71	0.78	0.70	0.87	641	0.86
Chond S	0.50	0.47	0.54	0.65	1.27	630	1.02
Chond N$^+$	0.53	0.53	0.56	0.93	1.86	448	0.66
Melt IS	0.66	0.66	0.67	0.77	0.82	1534	1.58
Melt IN	0.55	0.53	0.60	0.57	0.77	1040	1.06
Melt N	−0.6	0.55	0.71	0.92	1.01	1022	1.06
Melt N$^+$	0.57	0.44	0.56	0.56	0.61	1624	0.93
Pt Rocks	0.54	0.52	0.55	0.56	0.72	891	0.54

Bogard went on to point out, even this age necessitates that some melt samples such as IS retained ^{40}Ar.

Consideration of cooling rates against Ar degassing led to the opinion that even sizeable quantities of impact melt (e.g. that in Chico) may have cooled too fast totally to degas radiogenic argon. Thus $^{39}Ar/^{40}Ar$ ages would be obtained which are older than the impact age. That an impact melt sample exists is insufficient to produce a completely reset K–Ar age which could be used to date the event. Chico melt and chondrite samples demonstrated that non-melted material which was not heated as hot as the melt may be totally reset. Garrison and Bogard regarded the cooling rate of shocked or ejected material after thermal equilibrium was reached as more important in impact age resetting.

8.6.7 $^{40}Ar/^{39}Ar$ Isotope Study of Lunar Meteorite Asuka 881757

Asuka 881757 (A881757) was collected in Antarctica and it is a basaltic meteorite believed to come from lunar mare regions. The importance of it and other such material is that their petrology and chemistry differ from those of the Apollo and Luna samples and so provide information on the chemistry of lunar areas which were not sampled during those missions. A881757 is a new type of lunar meteorite, an unbrecciated, coarse-grained mare basalt comprising pyroxene (59%), plagioclase (30%) and ilmenite (6%) plus minor quantities of troilite, olivine, etc.

Misawa *et al.* (1993) effected U–Th–Pb, Sm–Nd, Rb–Sr and $^{40}Ar/^{39}Ar$ isotopic studies on this meteorite which has a chemical composition similar to low-Ti and very low-Ti mare basalts. The $^{40}Ar/^{39}Ar$ age spectra for a glass fragment and a maskelynitized plagioclase were relatively flat and gave a weighted mean plateau age of 3.798 Ga $\pm$ 12 Ma. This was interpreted as indicating that the basalt crystallized from a melt 3.78 Ga ago (this is the Sm–Nd age) and an impact event disturbed the Rb–Sr system and totally reset the K–Ar one at 3.80 Ga. A slightly higher Pb–Pb age compared with the Sm–Nd age might be the result of secondary Pb (from terrestrial and/or lunar surface Pb contamination) which stayed in the residues after acid leaching. Another scenario is possible, namely that the meteorite crystallized at 3.94 Ga (the Pb–Pb age) and the Sm–Nd, Rb–Sr and K–Ar isotopic systems were disturbed by an impact event at 3.80 Ga. The crystallization age derived here was older than those reported for low–Ti basalts (3.2 to 3.5 Ga) and for very low–Ti basalts (3.4 Ga), but resembled ages of some mare basalts. This may show that the basalt formed from a magma related to a basin–forming event which Misawa *et al.* believed might have been Imbrium. The age span for the very low–Ti basalts from various sampling sites implies that they were erupted over a wide area during a time interval of at least $\sim$500 Ma. The impact event which thermally reset the

K–Ar system of A881757 must have been post–Imbrium and possibly Orientale in age, according to these researchers.

The U–Th–Pb, Sm–Nd and Rb–Sr data on this meteorite suggest that the basalt came from a low-U/Pb, low–Rb/Sr and high-Sm/Nd source region primarily comprising olivine and orthopyroxene with small amounts of plagioclase or clinopyroxene and with sulphides enriched in volatile chalcophile elements. The basalt source might be deep in origin and different chemically from those previously estimated from examination of Apollo and Luna mare basalts. This, if true, would imply heterogeneous sources for mare basalts.

8.6.8 Diamonds

The work of Podosek *et al.* (1988) was discussed in section 8.3. Dodson (1989) noted that some diamonds are known to be 2 to 3 Ga old, adding that forming these samples of the Earth's mantle may entail incorporating large quantities of 'excess' radiogenic Ar in association with K. Hence dating projects may turn into studies of ancient mantle fluids.

The origin of anomalous $^{40}Ar/^{39}Ar$ ages of Zaire cubic diamonds and excess ^{40}Ar in pristine mantle fluids were examined by Ozima *et al.* (1989). These diamonds showed excellent correlations between K content and the $^{40}Ar/^{39}Ar$ ratio as well as between $^{40}Ar/^{36}Ar$ and $^{39}Ar/^{36}Ar$. These data were intrepreted so as to give an isochron age of about 6 Ga. The discovery of a correlation between Cl content and ^{40}Ar together with recent mineralogical and geochemical work regarding the origin of cubic diamonds imply that the ^{40}Ar is an excess component without any age significance, also that the ^{40}Ar and its associated K are contained in submicrometre inclusions of mantle-derived fluid.

The technique used entailed irradiating diamond samples by means of a total neutron flux of $\sim 3 \times 10^{18}$ neutrons cm^{-2} (irradiation parameter $J = 1.709 \times 10^{-3}$ for sample 87N01 and $J = 1.91 - 1.83 \times 10^{-3}$ for sample 88N01–03) with standard samples MMHb–1(6), K_2SO_4 and CaF_2. Prior to the stepped heating analysis, all samples were heated to 200°C for 6 h in an effort to eliminate adsorbed atmospheric Ar. Two temperature fractions were taken from sample 87N01 and four such fractions from the others. The samples were maintained at each temperature step for 30 min for noble gas extraction. More than 90% of the Ar from the diamonds was released in a single temperature step around 1900°C, by which the diamonds are graphitized. Hence the data were not presented in this form, but Ozima *et al.* noted that in most cases the individual ages agree, within error, with previously reported K–Ar isochron ages of $\sim$6 Ga. They used instead isotope–correlation plots of which one was a conventional isochron plot of $^{40}Ar/^{36}Ar$ against $K/^{36}Ar$. Molar ratios were employed as the most logical manner of comparing elemental ratios with isotope ratios (invariably so expressed). In this instance the linear correlation implied that ^{40}Ar derived from sites occupied by K, which makes it an age indicator. From the slope of the line of best fit, an age of $\sim$5.7 Ga was obtained and regarded as consistent with earlier K–Ar measurements. Thermal release patterns for samples were plotted and degassing patterns for ^{40}Ar, ^{39}Ar and ^{38}Ar were almost identical, but different from those for other noble gases, including ^{36}Ar. Ozima *et al.* concluded that this showed that ^{40}Ar, ^{39}Ar and ^{38}Ar were degassed from similar trapping sites. They pointed out that, because ^{39}Ar is produced by the reaction ^{39}K (n,p) ^{36}Ar and ^{38}Ar by (n,γ,e) reactions of ^{37}Cl, the identical thermal release patterns for ^{38}Ar, ^{39}Ar, and ^{40}Ar show that ^{40}Ar was trapped at the same sites as K and Cl.

Ozima *et al.* (1989) explained the good correlation between ^{40}Ar, K and Cl by assuming that these elements existed in the fluid phase prior to becoming trapped in diamonds and that there they could have been rather easily homogenized to give a uniform $^{40}Ar:K:Cl$ ratio. As a consequence of its incompatible mature, ^{40}Ar would be more easily leached from the country rocks into the fluid phase than K and this would account for the excess ^{40}Ar. If the diamonds trapped such a fluid component to varying extents, then identical thermal release patterns for ^{40}Ar, K, and Cl would result and so would a linear correlation in ^{40}Ar–K and ^{40}Ar–Cl diagrams since they would be trapped as fluid inclusions at the same sites.

As to the difference between the thermal release patterns of ^{40}Ar, ^{38}Ar and other noble gases (Ne, ^{36}Ar, Kr and Xe), Ozima *et al.* thought this explicable in terms of a many–stage growth model. Carbon and N stable isotope studies suggest that cubic diamonds underwent two-stage growth, initially a transparent core with octahedral morphology growing to common diamonds and rimmed by a cubic-habit coat that grew later in contact with the protokimberlite host. These researchers argued that their cores trapped noble gases from the mantle and that micro–inclusions later became incorporated during their fast growth in the presence of fluids. In this way, the diamonds contain two generations of noble gases of which one (trapped in the cores) comprises mostly normal mantle components and the second (contained within micro-inclusions in the coats) consists of the excess and *in situ* radiogenic ^{40}Ar. If it is assumed that the cores crystallized in the deeper parts of the mantle and were later transported to the upper mantle where their coats developed, these two generations of noble gases would represent the respective mantle regions. This concept, according to Ozima *et al.* (1989), fits suggestions on the state of noble gases in the mantle made by many authors; whereas the degassed mantle is characterized by a high $^{40}Ar/^{36}Ar$ ratio and hence a high abundance of ^{40}Ar

relative to other noble gases, with the less degassed deeper mantle retaining the primordial noble gases. That Ar which correlates with K and Cl has a high $^{40}Ar/^{36}Ar$ ratio like the values recorded from mid–ocean ridge basalts and contrasts with a relatively low $^{40}Ar/^{36}Ar$ ratio in uncorrelated Ar. This latter ratio may correspond to the lower mantle component. A solar-type Ne recorded in the core of Zairean cubic diamonds hence would also represent the deeper mantle.

9

Carbon–14 dating

9.1 BACKGROUND

A large number of radionuclides are produced in the terrestrial atmosphere as a result of nuclear reactions involving cosmic rays. These comprise very energetic radiation falling upon the Earth from outer space and consisting mainly of charged particles of which the majority are probably protons, although alpha particles and electrons are also present. There is some evidence that approximately 2% of the primary radiation consists of heavy atomic nuclei. Cosmic rays originate from the Sun as well as from galactic sources. Their flux depends upon the kinetic energies of their components which may be as great as 10^2 GeV (1 GeV = 10^9 eV); it appears that about half of the flux has a value between 0.5 and 5 GeV.

The primary particles present in cosmic radiation initiate spallation effects when incident on atoms of O, N and Ar, high-energy protons and secondary neutrons bombarding their nuclei to emit one or more nucleons and form product nuclei of lower atomic number. Proton–neutron collisions in the stratosphere give rise to mesons and their decay produces high-energy electrons which are also the result of further knock-on collisions of mesons with electrons. The high-energy electrons in question produce cosmic ray showers which, in turn, create more electrons, photons and positrons. In fact, energies as high as 10^8 GeV have been recorded from some cosmic-ray particles. Interestingly, the noted intensity of cosmic rays coming fron the west is greater than that coming from the east at any given latitude on Earth; this asymmetry is ascribed to the deflection of the primary charged cosmic-ray particles by the terrestrial magnetic field. This indicates a preponderance of positively charged particles in the incoming radiation.

Cosmogenic radionuclides originating by spallation reactions either reside for long periods of time in the atmosphere or undergo rapid removal by precipitation. Some of them have half-lives which are long enough to be of use in studying various geological processes: among these are ^{3}H, ^{10}Be, ^{14}C, ^{26}Al, ^{32}Si, ^{36}Cl, ^{39}Ar and ^{81}Kr. In this chapter, ^{14}C (radiocarbon) will be discussed and it is interesting to begin with an account of its discovery.

9.2 DISCOVERY

It was observed in 1934 by Grosse that the mineral eudialyte, a basic Ca, Na, Ce, Zn and Fe^{2+} chlorosilicate, is more radioactive than would be expected from its content of U, Th and their daughters; he explained this by suggesting that radioisotopes produced by the interaction of cosmic rays with elements such as O, Si, Zr, etc., were present and contributed this excess radioactivity. Later, with Libby, he established that cosmic ray-produced ^{14}C occurs in CH_4 arising from sewage gas. The actual existence of ${}^{14}_{6}C$ in nature was demonstrated in 1934 by Kurie, who exposed N_2 to fast neutrons and noted that some N_2 nuclei emitted particles with long thin tracks in the cloud chamber, attributing these to an (n, p) reaction, thus

$${}^{14}_{6}N(n, p){}^{14}_{6}C ,$$

this also taking place with slow neutrons. In 1940 a measurable amount of ^{14}C was made by Kamen using a (d, p) reaction, a tentative value for the half-life of 4000 a being given. Subsequently, it became clear that the (n, p) reaction is far more efficient in producing ^{14}C than the (d, p) one,

$${}^{13}_{6}C(d, p){}^{14}_{6}C .$$

The first to suggest that ^{14}C might be produced in the atmosphere by the interaction of cosmic ray neutrons with ^{14}N were C.G. and D. D. Montgomery in 1939 following the discovery of cosmic ray neutrons by Locker (1933) and his further work with Rumbaugh (1936). By 1946 it had been estimated that the average global production rate of ^{14}C was sufficient for detection in natural samples, and 1 year later Anderson *et al.* (1947) demonstrated the presence of ^{14}C in CH_4 extracted from sewage gas mentioned above. Its radioactivity was measured using a screen wall counter and, in 1949, the feasibility of radiocarbon dating was shown through the analysis of a suite of archaeological samples of known age. In 1951 Anderson and Libby reported that the specific activity of ^{14}C in the biosphere is constant within rather narrow limits and also independent of latitude, their average value for the disintegration rate of the radioisotope in samples of modern plant tissue being

15.3 ± 0.5 dpm g^{-1} of C. Additionally, they analysed a sample of seal oil from Antarctica and this gave a specific activity of 15.69 ± 0.3 dpm g^{-1} (which is within the range of activities of plant tissue samples).

9.3 PRODUCTION OF ^{14}C AND ITS USES IN DATING

The most significant process of production of radiocarbon in the atmosphere is the reaction between slow cosmic ray neutrons and the nuclei of stable ^{14}N:

$$^{14}_{7}N(n, p)^{14}_{6}C ,$$

which may be expanded to

$$^{14}_{7}N + ^{1}_{0}n \rightarrow ^{1}_{1}H + ^{14}_{6}C .$$

The ^{14}C produced in this manner is subsequently incorporated into molecules of CO_2 by reactions with O_2 or by exchange reactions with stable C isotopes in molecules of CO or CO_2. Molecules of $^{14}CO_2$ are swiftly mixed throughout the atmosphere and the hydrosphere to reach a constant level of concentration which represents a steady-state equilibrium. This is maintained as a result of the production of ^{14}C in the atmosphere and its ongoing decay. The molecules concerned will enter into plant tissue through photosynthesis and also by absorption through roots. Of course, the concentration of ^{14}C in living green plants is kept at a constant level as a consequence of its continuous absorption from the atmospere and its continuous decay. In fact, those animals which feed on such plants or absorb C-bearing molecules or ions from the atmosphere or from the hydrosphere also obtain a constant level of radioactivity due to ^{14}C. When the plants and animals in question die, they are no longer absorbing any of the radioisotope from the atmosphere directly or indirectly, and therefore their activities diminish because of radioactive decay. If the activity of ^{14}C in living materials is known, then its activity in dead plant tissues may be employed in order to calculate the time which has elapsed since they died. This is the ^{14}C date of the sample.

Decay of ^{14}C takes place through the emission of a negatron and produces stable ^{14}N, thus:

$$^{14}_{6}C \rightarrow ^{14}_{7}N + \beta^{-} + \bar{\nu} + Q ,$$

the end-point energy Q being 0.156 MeV. Decay occurs directly to the ground state of ^{14}N without gamma emission. The radioactivity of a C sample taken from a plant or animal tissue which died t years ago is given by

$$A = A_0 e^{-\lambda t} ,$$

in which A is the measured activity due to ^{14}C in units of dpm g^{-1}C and A_0 is the activity of this radioisotope in the same specimen when the plant or animal died.

Karlen *et al.* (1966) estimated the specific activity of ^{14}C in equilibrium with the atmosphere, A_0, as $13.56 \pm 0.07\,dpm\,g^{-1}$.

Godwin (1962) gave a half-life for the radioisotope of $5730 \pm 40\,a$: this was adopted at the Fifth Radiocarbon Dating Conference held in that year in Cambridge, UK, and indeed restated in 1982 at the Radiocarbon Conference in Seattle, Washington, USA. A previous and less accurate value of $5568 \pm 30\,a$ is still in use today, however, and dates calculated employing it appear in the journal *Radiocarbon*.

A survey of ^{14}C in hydrogeological studies was given by Mook (1980). A sample containing C which is no longer in equilibrium with the ^{14}C of the atmosphere or hydrosphere can be dated by solving for t:

$$\ln\left(\frac{A}{A_0}\right) = -\lambda t,$$

$$t = \frac{1}{\lambda}\ln\left(\frac{A_0}{A}\right).$$

Converting to logarithms to base 10 and substituting $\lambda = 0.693/5730 = 1.209 \times 10^{-4}a^{-1}$, the following expression is obtained:

$$t = 19.035 \times 10^3 \log\left(\frac{A_0}{A}\right) a.$$

Figure 9.1 shows the relationship between the ^{14}C activity (A_0) of a sample and the time which has elapsed since exchange stopped with the reservoir (t).

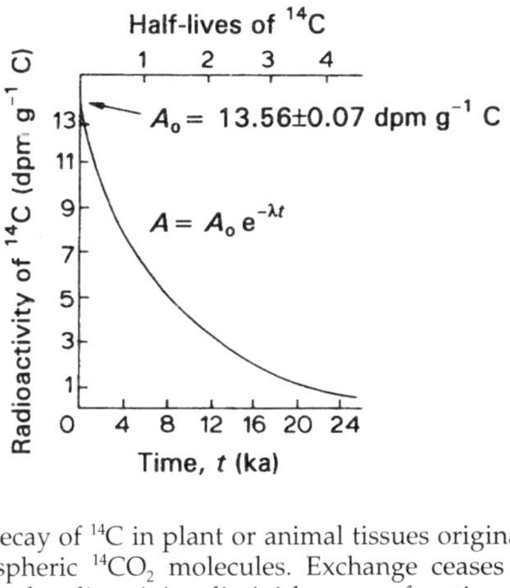

Fig. 9.1 Radioactive decay of ^{14}C in plant or animal tissues originally in equilibrium with atmospheric or hydrospheric $^{14}CO_2$ molecules. Exchange ceases after the death of the organisms concerned and radioactivity diminishes as a function of time (the half-life of radiocarbon is 5730 a). If the residual radiocarbon activity is measured, the age (t) of the relevant specimen is obtainable from the graph.

A number of necessary assumptions have to be considered with reference to the accuracy of the method. It is assumed that the initial activity of ^{14}C in plant and animal tissues (A_0) is a known constant which has been independent of time during the past 70 ka and that its value is also independent of geographical location and the species of plant or animal of which the dead tissues are under examination. It is also assumed that the sample which is to be dated has not been contaminated by modern ^{14}C and that the observed activity is unaffected by any radioactive impurities in it.

Researchers have demonstrated that the initial ^{14}C activity of modern plant and animal tissues is, in fact, constant to a first approximation, and also that the activity of the radioisotope in archaeological materials of known age is measurable with enough precision to give ^{14}C dates agreeing quite well with their known historical ages. However, it is apparent that there have been systematic variations in the ^{14}C content of the atmosphere in the past and these cause inaccuracies in application of the method. It is important to describe their discovery. As noted earlier, the rate of formation of ^{14}C by the (n,p) reaction in the upper atmosphere is dependent mainly on the density of the cosmic ray-produced neutron flux which increases with altitude to attain a maximum between 12 000 and 15 000 m above the terrestrial surface, also being greater by a factor of four in polar regions than at the equator. From this, it is clear that the production rate of radiocarbon is significantly greater at the poles than at the equator and this should be reflected in a greater ^{14}C of plants and animals in the former than in the latter areas. This effect was investigated by analysing plant and animal tissues collected at different geomagnetic latitudes. The data obtained demonstrated that, in fact, the specific radiocarbon activity seems to be independent of this locational factor. The geomagnetic latitude was utilized because the neutron flux density depends upon the proton flux, which is of course modified by the magnetic field of the Earth. The apparent uniformity of the ^{14}C activity is said to be due to the rapid mixing that takes place in the atmosphere, the process taking under 2 years.

The intensity of the above-mentioned proton flux is related to the numbers of protons emitted by the Sun and, as it varies, so will the neutron flux; therefore, the ^{14}C content of the terrestrial atmosphere no doubt varied with time as a function of the alternating activity of the Sun. Also, the cosmic ray proton flux is modified by changes in the planetary magnetic field. These factors must have had a considerable effect on the radiocarbon content of the Earth's atmosphere during the past 40 ka or so. However, Arnold and Libby (1949) carried out researches which implied that the initial ^{14}C activity in a set of archaeologically dated samples was quite compatible with the present-day value.

Later, de Vries (1958), demonstrated that the radiocarbon content of the atmosphere certainly varied in a systematic manner in the past, and also that the ^{14}C activity around AD 1700 and 1500 was as much as 2% more than in the 19th-century. This phenomenon is termed the de Vries effect. Another such effect was noted by Suess (1955). He found that the activity of 20th-century wood is about 2% lower than that of wood from the 19th-century and attributed this to the introduction of 'dead' CO_2 into the atmosphere through the combustion of fossil fuel after the commencement of the Industrial Revolution. This Suess effect has been exceeded since 1945 because of the explosion of thermonuclear devices in the atmosphere and the operating of nuclear reactors and particle accelerators since that date, these distributing artificially produced radiocarbon everywhere. Interestingly, the Suess effect is not always apparent, as Druffel and Griffin (1993) indicated. Details of the data which they obtained from corals of the southeren Great Barrier Reef, Australia, and their resulting high-precision $\Delta^{14}C$ record are given below. Bomb-produced radiocarbon is detectable in the oceans and also in the biosphere (Stenhouse and Baxter, 1977; Linick, 1980). It is valuable in examination of the interaction between atmosphere, surface and deep ocean waters.

Analysis of old trees has facilitated investigation of the secular variation of the radiocarbon content of the atmosphere. Such trees can be dated since they grow by annual accretions of woody tissue to the circumferences of their trunks, and layers deposited earlier can no longer absorb ^{14}C from the atmosphere; their radiocarbon content decreases by radioactive decay. The variations in the content of this radioisotope in the atmosphere over the past several millenia can be measured, therefore, by analysing samples of wood of which the age is known by dendrochronology (counting of tree rings). The bristlecone pine, *Pinus aristato*, and the sequoia, *Sequoia gigantea*, were suitable for this purpose and in 1970, Michael and Ralph utilized the observed age-corrected deviations of the radiocarbon contents of such trees in order to calculate corrections applicable to conventional radiocarbon dates. These are shown in Fig. 9.2.

Such earlier work was done when it was believed that dendrochronology could extend back only *c.* 8 ka. Subsequent researches showed that it could be used up to 11.4 ka dendro BP This designation refers to absolute dates based on tree rings and radiocarbon dates may be termed ^{14}C a BP. Interesting data were presented by Becker and Kromer (1993) who constructed two such long chronologies from 5 000 trees, the time interval from now to 9950 dendro BP being covered by the German oak chronology and the earlier part going back to 11.4 ka dendro BP consisting of a 1768 a long pine chronology (*Pinus sylvestris*). Apart from a short interval *c.* 500 BC bridged by Irish tree-ring series and another at 6200 BC, replication usually exceeded ten cross-sections and reached several hundred samples in the AD period. Confirmation of a high degree of probability was obtained through ^{14}C analyses of overlapping sections and this enabled these researchers to prolong the oak chronology back

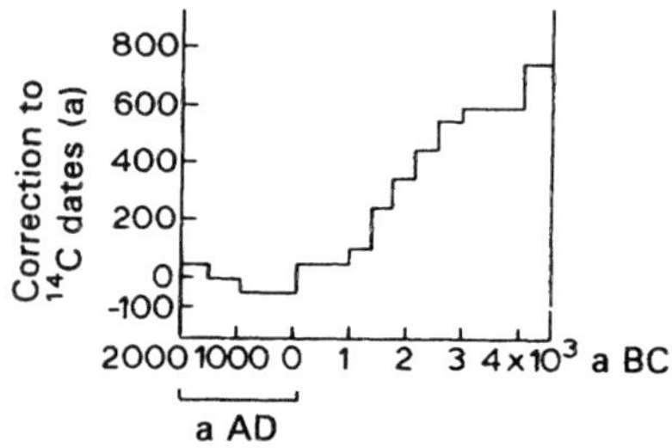

Fig. 9.2 Corrections which have to be added to radiocarbon dates so as to bring them into agreement with tree-ring dates based on 143 analyses of the wood of bristlecone and sequoia dated by dendrochronology. Systematic deviations in radiocarbon dates are caused by variations in the atmospheric content of ^{14}C in the atmosphere over the past 6 Ka.

to 8021 BC when the ^{14}C age must, according to the youngest pine sections, exceed the ^{14}C age plateau of 8800 ^{14}C a BP which had obstructed linking of the oak and pine chronology by ^{14}C. Actually, the earliest oak sections before 8000 BC demonstrated the rapid rise of ^{14}C ages consistent with the corresponding pine sections. As regards dendroclimatological evidence, Becker and Kromer (1993) pointed out that it might be suspected that deposition of trees by river floods took place randomly over the past ten millenia, as was implied by inspection of the distribution of the end-years of the trees in the series. High counts in the AD age range were due perhaps to human-induced increased fluvial activity, e.g. by forest clearance during Roman and early Mediaeval times. Since more than one factor is involved in alluvial plains, trees there do not record long-term changes in climatic conditions. They adapt rapidly to environmental changes such as increased flood loam along these features. This enabled these authors to locate strong indicators of a continental type summer climate during the Preboreal (10 to 9 ^{14}C ka BP) which entailed sudden growth depressions, often with rings lacking, frost rings and frost splitting of stems. Such anomalies are not apparent today in southern German valley plains, but occur only in northern latitudes. They reflect severe winter or early spring conditions in south-central Europe during Preboreal times.

The German oak chronology and its widening through pine chronology proved basic to calibrating a radiocarbon timescale and the stable isotope content of tree-ring cellulose has been analysed so as to permit inferences about past temperature and humidity to be made. From absolutely dated oak and pine chronologies, atmospheric ^{14}C can be estimated and data given by Stuiver, Long and Kara (1993) showed variations in the ^{14}C level lasting for up to 500 a superimposed on a long-term trend due to changes in the magnetic dipole of the Earth. The most recent variation has been found to correlate well with low solar activity during the Maunder minimum *c*. AD 1700. Indeed, most of the secular-scale variation is explicable by harmonic oscillation with fundamental frequencies of *c*. 400 and 200 a. Extending the absolutely dated record by the earliest oak and by the pine chronology made it clear that the strong coherence noted for the post-9000 dendro BP section of the oscillations is not kept up in the Early Holocene. The century-scale variability of the early ^{14}C record may well have been the result of climatic (oceanic) forcing.

As regards the radiocarbon timescale, century-scale ^{14}C variations show up most strongly in the 10 to 8 ^{14}C ka BP time interval (comparable with the ^{14}C anomaly at 500 BC; Kromer and Becker, 1993). They cause a considerable distortion of the radiocarbon timescale which has to be taken into account when either absolute ages or the duration of events are inferred from ^{14}C dates. Becker and Kromer (1993) gave a calibration curve of ^{14}C age against absolute age. On this, two 'plateaux' of seemingly constant ^{14}C age with respect to true age were found at 10 ^{14}C ka BP and at 9.6 ^{14}C ka BP respectively, and each lasted more than 400 a. At these two time intervals, exact ^{14}C dating was regarded as impossible. Also, the rapid change from 9900 to 9700 ^{14}C a BP could induce incorrect assumptions regarding a possible 'hiatus' in ^{14}C-dated sequences, because ^{14}C ages over this interval in fact span a true interval of under 100 a. Contrariwise, two samples which were dated as 10000 ^{14}C a BP and 9500 ^{14}C a BP respectively might be between 150 and 1000 true years apart if an age error of ± 50 a is assumed. Apropos the stable isotopes, the ^{13}C and D of the local atmosphere and groundwater are recorded, but trees may alter the original signal through physiological processes which mask it through extra, frequently stronger, isotopic fractionation. This matter is discussed below.

The strongest climatic signal was anticipated from the pine series, centred in the Preboreal. From this, the ^{13}C and D of 5 a samples were measured by Becker, Kromer and Trimborn (1991). The samples in question were cellulose nitrate of total wood and both isotope series obtained gave low values of -95 to -100 and -24 to -24.5% respectively for more than 200 a before 11100 dendro BP (D) and 11000 dendro BP(^{13}C). Then both rose coincidentally in overlapping trees until *c*. 10.9 Ka dendro BP. Becker and Kromer (1993) took the strong isotopic signal at 11 ka dendro BP as indicating a climatic amelioration during the transition from the Younger Dryas to the Preboreal. If true, the alteration in the ^{13}C signal from the pine trees would suggest a greater humidity stress by lower relative humidity (= higher temperature) in the Preboreal summers (when the tree rings formed) as well as a potential global signal from changes in the temperature of the sea surface and also in carbonate chemistry. With regard to absolutely dating the above-mentioned transition, a key factor is the isotopic signal in the pine chronology from the Younger Dryas to the Holocene. This transition is recognizable on a global scale from a number of sources such as pollen sequences, lake and marine sediments, ice cores from Antarctica and Greenland, etc. Apart from the ice cores, all other sequences can be dated by ^{14}C. However, the ^{14}C plateau at 10000 ^{14}C a BP is a difficulty since it does not allow accurate dating of this significant transition by using ^{14}C. Here the stable isotope results are important because they permit the assigment of the absolute date of the commencement of the climatic transition near to the end of the ^{14}C plateau *c.* 11100 to 11000 dendro BP. Such a timing coincides with available observations from the beginning of the deposition of the pine trees. Becker and Kromer (1993) noted that the earliest trees of the chronology started growing *c.* 11400 dendro BP but were deposited 200a later *c.* 11 150 dendro BP. This marked the start of the Holocene type of hydrological regime of the River Danube, that is to say the change from Late Glacial braided river systems with perhaps shallow channels frozen at the ground level where tree trunks were not buried to the Holocene river meander systems. Interestingly, Becker and Kromer (1993) gave a determination of the absolute age of the end of the Younger Dryas hundreds of years younger than that from the marked ^{18}O signal observed in high-resolution ice cores from Greenland dated to *c.* 11550 cal BP by counting annual dust layers (Tayor *et al.* 1993). Dates determined by layer counting in ice cores or varves are denoted 'cal BP'. But the shape of the stable isotope signal and data from the deposition frequency of the pine trees suggest a real lag of climatic change between Central Europe and Greenland. Varve chronologies of sediments deposited in glacial lakes may be used to examine variations in radiocarbon activity and results obtained can be expressed as 'cal BP'.

The ^{14}C timescale over the past 30 ka has been examined by Bard *et al.* (1990) using mass spectrometric U–Th ages from Barbados corals. A suite of coral cores raised off this island yielded over 30 *Acropora palmata* samples later dated by conventional ^{14}C techniques and these showed that the last deglaciation started *c.* 15 000 ^{14}C a BP and proceeded in two stages. This agreed with ^{18}O records from deep-sea cores dated by accelerator mass spectrometry of ^{14}C. Bard *et al.* (1990) regarded the offshore corals as appropriate for investigating the timing of the last glacial cycle and the calibration of the ^{14}C timescale back to 30000 a BP. They noted that radiocarbon dating cannot be fully accurate since the atmospheric $^{14}C/C$ reference level has altered with time, evidence of differences between the true age and the ^{14}C age ($\Delta^{14}C$) coming from the dating of tree rings. Through the past 9 ka, the $\Delta^{14}C$ curve shows a decreasing trend having a maximum difference of *c.* 100% around 7 ka BP. This has been correlated with alterations of the magnetic dipole which shields production of ^{14}C by cosmic rays. Overlaying the trend are oscillations believed to arise through short-term variations in the magnetic properties of the solar wind. So far, the radiocarbon timescale is not well calibrated prior to 9 ka BP through insufficient data from fossil trees suitable for dendrochronological dating. In an attempt to remedy this situation, varved sediments, U–Th and ice cores have been analysed, but with rather unsatisfactory results which often are incompatible. However, before 9 ka BP the ^{14}C ages appear to be consistently younger than the true ages, which implies higher $^{14}C/C$ ratios than those which exist now (a positive $\Delta^{14}C$). Bard *et al.* (1990) measured both Th and ^{14}C ages in the same coral samples in order to be able to compare two geochronometers and concentrated on the U–Th–^{14}C comparison for the past 30 ka. As a first check on the precision of the dating technique employed, these researchers dated samples from Barbadian and Haitian uplifted reefs corresponding to the last interglacial and the relevant measurements accorded with earlier α-counting ones and were taken as confirming that Th ages measured in samples from the last interglacial highstand are limited to the time interval between 120 and 130 ka BP. The precision of the U–Th dating was given as typically better than 100 a (at 2σ) for ages between 6 and 20 ka, noticeably better than that reached in most ^{14}C dating. Also, replicate analyses of the same hand samples showed that the method is reproducible within the 2σ errors. Initial $^{234}U/^{238}U$ ratios calculated utilizing the Th ages fell into the range of modern sea water, i.e. 1.14 to 1.15 for all of the deglaciation samples and this demonstrated that their U–Th systematics were not strongly altered subsequently to deposition. A plot of the coral record of the last deglaciation dated by U–Th and ^{14}C showed no stratigraphic inversions in the U–Th data and indicates that erosion and redeposition of the Barbados core collection coral fragments constitute minor problems only. Measurements of radiocarbon in corals cannot be used directly for calibration against absolute geochronometers because of the fact that the oceanic surface is not in equilibrium with the atmosphere, a steady state balance existing between the ^{14}C input from the atmosphere and its removal through advection and radioactive decay in the water column. Consequently, the ^{14}C ages from corals have to be corrected for the age difference between the atmosphere and the surface ocean, a parameter dependent mainly on upwelling and mixing with old subsurface water masses. In the low latitude range, the reservoir age is given as $\sim$ 400 a (50%) and regarded as unlikely to have altered by more than 100 to 200 a during the late Pleistocene, according to Bard *et al.* (1990). These researchers altered raw ^{14}C ages by 400 a. Interestingly, the Th ages were always older than the ^{14}C ages with a maximum difference of $\sim$3.5 ka at $\sim$20 Th ka BP. The three youngest radiocarbon-dated coral samples from Barbados fell within the range of exact tree-ring calibration and the U–Th ages of these samples accorded with the dendrochronological ages, the calibrated

^{14}C dates being 7440–7140, 8970–8310 and 9420–8970 a, respectively. The agreement between the two independent calibrations was regarded as confirming the accuracy of the U–Th chronometer in corals, at least for the Holocene and probably earlier as well. Using varve counts and ^{14}C dating in Lake of the Clouds, the calibration curve was extended back to 11 ka BP and there was good accord between the varve data and the U–Th–^{14}C comparison.

At *c.* 10 ^{14}C ka BP (Preboreal–Younger Dryas boundary), the U–Th calibration (~11 500 Th a BP) accords well with dendrochronological estimates (~11 300 dendro a BP), but both disagree with the correlation between climatic spectra dated by the annual counting of seasonal variations in the ^{18}O content, dust and acidity curves from the Dye III and Camp Century ice cores (~10 700 a BP). Consideration of the U–Th–^{14}C calibration showed that the real length of the Younger Dryas climate event (10 000 to 11 000 ^{14}C a BP) could have been as much as 1700 a and would not be inconsistent with a plateau in ^{14}C ages at ~9500 ^{14}C a BP observed in the dendrochronological data. Turing to the 12 000 to 25 000 ^{14}C a BP time interval, sizeable discrepancies have been recorded between ^{14}C and K–Ar dating, also between ^{14}C and thermoluminescence dating. Bard *et al.* (1990) took these as according with their results from Barbados.

A possible reason for an increasing lag between radiocarbon and Th ages might be contamination of samples by recent C either *in situ* or in the course of radiocarbon measuring. In order to correct the radio-carbon ages back to the Th ages, a contamination of *c.* 4% modern C would be necessary. Bard *et al.* (1990) thought this unlikely for several reasons, e.g. 2% contamination is the maximum permissible so as to allow the three recent ages to remain in agreement with tree-ring ages. Another factor which could be significant in producing the discrepancy between the two geochronometers is a gradual loss of U of which the dissolution can be modelled with first-order kinetics. Bard *et al.* (1990) stated that a U loss rate equivalent to $k = 7\lambda_{234}$, where λ_{234} is the decay constant of ^{234}U, could reconcile the ^{14}C ages with the U–Th ones. They added that U loss is unlikely since a strong drop in U concentration with time ought to show in the record, noting that, at ~30 ka BP the decrease should be 50%. But their data demonstrated that the U concentration in the corals is almost constant (3 to 3.5ppm) and this agrees with other published estimates for *A. palmata*. Of course, the radiocarbon age calculation assumes that the atmospheric ^{14}C/C ratio was constant through time and may be used to normalize the measured ratio in fossil C. However, the atmospheric content is under 1 tonne of radiocarbon, most of this radio-isotope being in the ocean (> 40 tonnes). Change in the rates of exchange between reservoirs as well as alterations to their sizes through time could well have influenced the ^{14}C content of the atmosphere and CO_2 measurements in old polar ice indicated that the atmospheric concentration was lower during the last glacial than during the Holocene. The steady-state $\Delta^{14}C$ might increase by between 25% and 75%. Additionally, the atmospheric $\Delta^{14}C$ could rise if the rate of penetration of $^{14}CO_2$ in the deep sea fell. But changes in the C cycle can be responsible only for a ~100 to 150% rise in the $\Delta^{14}C$ of the atmosphere as compared with the 400 to 500% which Bard *et al.* (1990) inferred from their Barbados data.

In sum, U–Th mass spectrometry was shown to be capable of achieving an age precision and reproducibility perhaps better than that obtained by the radiocarbon geochronological approach and U–Th ages measured in corals drilled off the Barbadian shore demonstrated a significant compression of the ^{14}C timescale during the late glacial time interval. The data of Bard *et al.* (1990) could be explained primarily as a result of changes in the cosmogenic nuclide production connected with the gradual fall of the Earth's magnetic field. Alterations in the C cycle may account for ~10 to 20% of the age discrepancy recorded between the two geochronometers. It must be emphasized that the accuracy of the U–Th ages has been established only up to 10 ka BP. However, these researchers claimed that all available geophysical information such as dendrochronology, varve dating and palaeomagnetism imply that their U–Th results are reasonable and that they may be employable as a first-order tool to calibrate the radiocarbon timescale beyond the range of the high-resolution dendrochronological comparison.

As regards the manner in which C is cycled in the interior of the oceans, there are important implications arising from the relatively low quantity of atom bomb-derived ^{14}C in deep-ocean particulate organic C (POC) and Druffel and Williams (1990) concluded that the low $\Delta^{14}C$ which occurs in POC suspended in the deep ocean may result from the effects of dilution due to incorporation of ^{14}C-depleted organic C (DOC) into deep POC. In addition to this DOC→POC transformation, dissolved inorganic carbon (DIC) may well constitute another large source of C for such dilution and the conversion rate of DIC to POC in the deep sea is both quantitatively and isotopically important, as Rau (1991) indicated. He warned that DIC uptake should not be neglected when interpreting the paucity of ^{14}C in some deep-sea organisms and POC pools. He regarded it as premature to ascribe such depletion in ^{14}C to the incorporation of DOC alone.

Bomb ^{14}C studies became significant because of nuclear weapons testing in the atmosphere during the time interval between the mid-1950s and the mid-1960s, after which the radiocarbon concentration almost doubled there. Obviously the slow transfer of the bomb-produced ^{14}C to the oceans made it into a very useful tracer of atmosphere–ocean exchange processes and also of oceanic circulation. Measurements of DIC $\Delta^{14}C$ content in surface sea water from the Atlantic Ocean during GEOSECS (Geochemical Ocean Sections Study) in the early 1970s showed the bomb 'pulse' when compared with pre-1958, i.e. pre-bomb, sea water measurements. The low equatorial and high latitudinal $\Delta^{14}C$ values of surface waters may be due to a deep water supply since this has such values caused by dilution (see above) and/or radioactive decay. Of course, variation in such deep water supply has not been at all adequately examined as the relevant measurements have been made in only a relatively small number of locations, e.g. in the North Atlantic, they were restricted mainly to two post-bomb basin-wide ones, namely GEOSECS (1972–1973) and Transient Tracer in the Ocean (TTO, 1981–1982). Otherwise, some information was obtained from a pre-bomb set of measurements ranging only to 39°N. However, continuous $\Delta^{14}C$ data have been derived from banded corals and found useful indicators of variations in oceanic ventilation on 10 a, annual and seasonal timescales. Unfortunately, this application is restricted to those low latitudes where shallow-living corals grow.

Utilizing molluscs was an approach first effected by Weidman and Jones (1993) and they produced the first time history of bomb ^{14}C from the high-latitude North Atlantic Ocean from a 54 a old mollusc specimen of the (Bivalvia) *Arctica islandica* collected alive from the Georges Bank (41°N) in 1990. Annual growth bands of the shell were analysed for $\Delta^{14}C$ using accelerator mass spectrometry and yielded a $\Delta^{14}C$ time history from 1939 to 1990. The Georges Bank ^{14}C signal was depleted relative to two coral-derived North Atlantic $\Delta^{14}C$ time histories and this was taken as implying a significant deep-water source for the waters on Georges Bank. The actual time history showed a strong phase similarity to two other North Atlantic $\Delta^{14}C$ time histories obtained from banded corals from Florida and Bermuda and all three demonstrated minimum $\Delta^{14}C$ in the early 1950s with a maximum in the mid-1970s, the first large increase in $\Delta^{14}C$ having taken place in the late 1950s. In their work Weidman and Jones (1993) age-corrected sample $\Delta^{14}C$ values to 1950 and also corrected for fractionation using a constant $\delta^{13}C = +2.3\%$, which is the mean of 214 analyses carried out on a nearby contemporary *A. islandica* shell of similar age. It has been suggested that a correction for terrigenous-source C might well be applied as well because at least some estuarine molluscs incorporate up to ~50% metabolic C into their shells. Such terrigenous-source C is usually enriched in $\Delta^{14}C$ and these researchers considered it unnecessary to allow for it since the Georges Bank site is >250 km off shore and hence unlikely to be affected by it. The Georges Bank $\Delta^{14}C$ signal may be approximated as linear functions either of the Florida or of the Bermuda $\Delta^{14}C$ time histories thus:

$$^{14}C_{\text{Georges Bank}} = 0.66(^{14}C_{\text{Elorida}}) - 26\% \ (r = 0.993)$$

$$^{14}C_{\text{Georges Bank}} = 0.68(^{14}C_{\text{Bermuda}}) - 28\% \ (r = 0.967)$$

Buttressed by earlier work which linked the origin of waters on Georges Bank to the Labrador Sea, the $\Delta^{14}C$ budget on Georges Bank was modelled as Labrador sea water which becomes mostly confined to the shelf and in part equilibrates with the atmosphere in the course of a 1 year transit time from the sea to the bank. This model was utilized so as to obtain a time history of bomb ^{14}C for the Labrador Sea and when this was compared with pre-bomb records, the inference was that a larger inventory of bomb ^{14}C accumulated in it than had been reported previously. But the estimated pre-bomb average $\Delta^{14}C$ (−70.6%) for this period was almost the same as the −70% earlier calculated for the pre-bomb source of North Atlantic Deep Water and accords with TTO subsurface tritium data from the central Labrador Sea. Variations in the ventilation and/or $^{14}CO_2$ rates of uptake in this sea correspond to observed surface salinity changes there and this implies a reduction in deep water formation during the late 1960s and 1970s.

Many other investigations on the oceans have been made, but only a few more can be discussed here. However, they may give an idea of exciting advances. Shackleton *et al.* (1988) discussed the radiocarbon age of last glacial Pacific Deep Water, noting that tandem accelerator mass spectrometric determinations of radiocarbon made such endeavours feasible. At the start, difficulties arose because of bioturbation in a core, V28-238, used by previous researchers. This had an accumulation rate of $\sim 1.5\,\mathrm{cm\,ka^{-1}}$, hence selecting a core with a higher rate became necessary. The glacial section of core TR163-31B was accumulated at a rate of $10\ \mathrm{cm\ ka^{-1}}$ and Shackleton *et al.* (1988) showed from this that the ventilation age of Pacific deep water was *c.* 500 a greater in last glacial times than it is now. Druffel and Griffin (1993) examined large variations of surface ocean radiocarbon and discussed evidence for circulation changes in the south-western Pacific. They presented radiocarbon $\Delta^{14}C$ and stable isotope ($\delta^{18}O$ and $\delta^{13}C$) records obtained from a 323 a banded coral series from the southern Great Barrier Reef, Australia. A high-precision $\Delta^{14}C$ record contained variations on an interannual timescale and they are especially great between AD 1680 and 1730. They compared these with $\Delta^{14}C$ records from tree rings and found that probably such shifts are not the result of alterations in the $\Delta^{14}C$ of atmospheric CO_2. More likely causes are changes in vertical mixing and large-scale advective changes involving source waters to the western Coral Sea region. Most of the low values of $\Delta^{14}C$ for the time interval AD 1635 to 1875 coincide with El Niño/Southern Oscillation events in the eastern tropical Pacific. But these do not account for all variations, particularly those of AD 1875 to 1920 when $\Delta^{14}C$ values stayed high. Cross-spectral analysis of the early half of the $\Delta^{14}C$ and $\delta^{18}O$ records (AD 1635 to 1795) showed that the 6-year period is coherent, a coherency absent from the latter half (AD 1797 to 1957) of the isotopic records. The data imply 100 a timescale changes in the nature of El Niño/Southern Oscillation in the south-western Pacific.

Curiously, the coral record of Druffel and Griffin (1993) did not demonstrate the Suess effect, i.e the diminution in $\Delta^{14}C$ from the late 1800s to 1955 caused primarily by CO_2 input through burning of fossil fuel, and they noted this as coinciding with the change observed in the nature of the El Niño/Southern Oscillation as well as offering extra evidence that some long-term alteration in the mixing of upper waters took place in the south western Pacific.

Rock paintings have been dated using radiocarbon. Until recently, determination of the age of such pictographs entailed using indirect evidence such as dating deposits covering them *in situ* or analysing the migration of ions between the bulk rock and the natural coatings which form on a freshly exposed surface. Radiocarbon was not used principlally because of the difficulty of separating inorganic C from the organic material in the pigments. However, Russ *et al.* (1990) reported a new technique allowing this to be done by employing a low-temperature, low-pressure oxygen plasma to oxidize selectively the organic component which can then be analyzed utilizing standard ^{14}C methods. The approach was applied to a part of a pictograph from the Lower Pecos region, south-western Texas, and the date derived was 3685 ± 100 a which accorded with that anticipated on the basis of archaeological inferences. These researchers pointed out that, as organic C is a universal constituent of pictograph paints, it should be possible to apply their technique throughout the world. Direct radiocarbon dates were obtained for prehistoric paintings at the Altamira, El Castillo and Niaux caves in Spain and the French Pyrenees by Valladas *et al.* (1992). They pointed out that indirect methods being unsatisfactory, dating the relevant pigments is necessary, adding that these are most commonly red Fe_2O_3 and black MnO_2 or charcoal. Dating charcoal using β-activity measurement is inimical to the art work, but accelerator mass spectrometers can be employed since they can separate and count individual C isotopes. Additionally, they require far less material. Valladas *et al.* (1992) reported the first radiocarbon dates for charcoal used to draw stylistically similar bisons in the selected caves. These were 14000 ± 400 a BP for the Spanish caves of Altamira, 12990 ± 200 a BP at El Castillo and 12890 ± 160 a BP for a differently styled bison in the French Pyrenean cave of Niaux. The results showed the imprecise nature of stylistic dating, and also that those dates of paintings derived indirectly from the relics of human activities must be received cautiously.

The radiocarbon dating of the Shroud of Turin reported by Damon *et al.* (1989) is highly controversial, too much so to be discussed here. These researchers claimed that small samples dated by accelerator mass spectrometry in laboratories at Arizona, Oxford and Zurich and monitored using three samples, of which the ages had been derived independently, provided conclusive evidence that the linen in the shroud is mediaeval. Hence it could not have been used to wrap Christ's body and indeed first entered history when displayed at Lirey, France, in the 1350s. An idea of the resultant academic turmoil may be given by Busson's comments in 1991 that one of the relevant researchers, Woelfli, asserted that the size and weight of the shroud samples alluded to by Damon *et al.* (1989) were erroneous. He admitted that none of the participants spotted this and there was no challenge to that date made either by Damon or by any other of the 19 authors concerned. Other researchers have drawn attention to serious anomalies in the procedure involved in the radiocarbon dating of the shroud.

Radiocarbon dating has been used to date the Santorini eruption, although more analyses might be valuable. However, Hammer *et al.* (1988) noted that they have shown that ice core dating and a tree-ring (frost-ring) date agree with radiocarbon dates and indicate a 17th century BP age for this event.

The ^{14}C terrestrial ages of two lunar meteorites, ALHA 81005 and EET 87521 were investigated by Jull and Donahue (1993). Ten such meteorites have been identified and are interesting because their parent body is known and therefore the systematics of their transfer to the surface of the Earth are important. Their terrestrial ages together with other radionuclide information permit constraints to be imposted on the origin, transit time (T_{space}) and terrestrial residence time (T_{terr}) of these objects. All of them may have been ejected during two events and nine meteorites of lunar composition have been obtained from Antarctica. The data obtained were as follows. ALHA 81005 was found to have a finite ^{14}C age of 18 ka if T_{space} is sufficiently long to saturate ^{14}C in a 4π irradiation. If ejected recently with a brief T_{space}, the terrestrial age would be calculated from the quantity of ^{14}C from the lunar irradiation. A previously estimated ejection depth gave a lunar saturated ^{14}C activity of 21 dpm kg^{-1} and a ^{14}C age of *c.* 9 ka. Intermediate conditions with varying values of T_{space} or T_{terr} would be between these limits and, including errors it is possible to state that for ALHA 81005: $7700 < T_{terr} < 19\,600$ a. The terrestrial age range is similar to that for other meteorites from the Allan Hills Middle Western Icefield. The ^{14}C age buttresses shorter estimates of the terrestrial age rather than some earlier values ranging up to 170 ka. The irradiation time in space is not constrained solely by ^{14}C and ^{41}Ca data show that this radionuclide did not decay appreciably from the lunar value. Hence the time for both transit and terrestrial residence cannot exceed *c.* 50 ± 30 ka (and TL and track data suggest a similar transit time). Jull and Donahue (1993) set limits to T_{space} by subtracting measured T_{terr}, assuming that the production of ^{14}C took place either by irradiation on the Moon (2π) or in transit through space (4π). Assuming

2π irradiation, T_{space} would be 41 ± 30 ka. The length of 4π irradiation might produce a significant quantity of ^{14}C. Assuming a 4π irradiation in space, using T_{terr} of 18.3 ± 1.3 ka, $T_{space} < 62\,000$ a.

As regards EET 87521, radionuclide data on ^{10}Be, ^{26}Al and ^{36}Cl demonstrate that this meteorite was irradiated at great depth in the Moon. Jull *et al.* (1993) suggested a terrestrial age of $> 44\,000$ a with the assumption that T_{space} was long enough (c. 20 ka) to saturate ^{14}C. Other radionuclide data gave low values for the ^{10}Be, ^{26}Al and ^{36}Cl consistent with irradiation at depths of *c.* 560–590 g cm^{-2}. At this depth range, the ^{14}C production rates were probably *c.* 2.1 dpm kg^{-1}. Perhaps the negligible ^{14}C observed is the result of decay of ^{14}C produced on the Moon.

Finally, mention must be made of an interesting new discovery which followed the initial suggestion that appreciable ^{14}C might be produced *in situ* in polar ice. This arose from studies of ^{14}C in CO_2 extracted from both accumulation and ablation samples. In the former, ^{14}C activities were observed to be between four and six times greater than would have been anticipated due to trapped atmospheric CO_2. Lal *et al.* (1990) later reported detecting an unambiguous signal of *in situ* cosmogenic ^{14}C in ice samples obtained from two Antarctic ablation sites. The ^{14}C was ascribed mostly to nuclear spallations of O in ice. The concentration of ^{14}C in ablation ice samples was noted as $1–3 \times 10^3$ atoms g^{-1} ice, i.e. three orders of magnitude greater than could be expected from trapped atmospheric CO_2 in this ice. The *in situ* ^{14}C showed a unique signature, *c.* 60% existing as ^{14}CO and the rest as $^{14}CO_2$. Lal *et al.* (1990) regarded this as being consistent with that anticipated from the examination of artificially produced ^{11}C in solid targets and the ^{14}C concentration decreased with depth as surmised for *in situ* production. Calculated model ablation rates were 5.8 ± 0.7 and 7.6 ± 0.8 cm a^{-1} at two sites in the Allan Hills Main Ice Field and these accorded with rates assessed using the stake method. The inference of these researches is that the ^{14}C age of accumulation ice based on atmospheric-originated, trapped CO_2 would be an underestimate of the true age if a correction is not made for *in situ* produced $^{14}CO_2$. Lal *et al.* (1990) asserted that this can be made easily since the ^{14}C activities of both the CO and CO_2 phases, as well as the trapped CO_2 concentration, can be measured.

The *in situ* rate of production of ^{14}C in ice resulting from O spallation has been estimated to be 84 atom g^{-1}a^{-1} at geomagnetic latitudes $> 60°$ and 2 000 m altitude. Using a mean altitude–pressure relationship, revised values at 0, 1, 3, 4 and 5 km altitude and the same geomagnetic latitudes are 15, 38, 84, 164 and 480 ^{14}C atoms g^{-1}a^{-1}, respectively. A small contribution (under 2%) from thermal neutron capture in ^{14}N present in the ice as a constituent of the trapped 'air' has to be added. The total *in situ* rate of production of ^{14}C in ice at Allan Hills (altitude 2 km and geomagnetic latitude $\sim 60°$) is estimated to be 85 atoms (g ice)$^{-1}$ a^{-1}. Without the spallation of O, the ice, over 100 ka old according to U-series dating, the ^{14}C concentration from atmospheric trapped CO_2 would be *c.* 2 atoms $^{14}C\,g^{-1}$ ice if this be assumed to be 50 ka old. The residual *in situ* ^{14}C produced in the ice during accumulation could be expected to be <2 atoms $^{14}C\,g^{-1}$ ice. So the signals noted in the CO_2 and CO fractions are almost completely due to ^{14}C produced *in situ* in the ice during its outcropping stage.

An interesting application of radiocarbon was recorded by Nelson *et al.* (1995). In the Cascadia subduction zone, a region of converging tectonic plates along the North American Pacific coast, very large-scale plate boundary earthquakes occurred through Earth history, but none has taken place since Euro-American settlement some 150 a ago. Geophysical estimates of the moment magnitudes of the biggest earthquakes range from 8 to 9.5. It is possible to date earthquake-killed vegetation using ^{14}C, a procedure which can set upper bounds on earthquake size by constraining the length of plate boundary which ruptured in individual earthquakes. High-precision radiocarbon ages were measured in Seattle by proportional counting of gas obtained from blocks of tree rings from the outer and inner parts of tree stumps killed by submergence at four sites. Other high-precision ages from a northerly site originated in a stick in tsunami-deposited sand above the youngest soil and from undergound stems and attached leaf bases of a tidal-march herb. Accelerator mass spectrometry was employed at the Rafter Radiocarbon Laboratory, Lower Hutt, New Zealand. Radiocarbon dating had already shown that the most recent rupture (or rupture series) extended at least 55 km along the Washington coast within a few decades *c.* 300 a ago. Nelson *et al.* (1995) reported 85 new radiocarbon ages which suggest that this most recent rupture or rupture series extended at least 900 km between southern British Columbia and northern California. It was concluded that one magnitude 9 earthquake or several smaller ones ruptured most of the length of the Cascadia subduction zone between the late 17th-century and early 19th century (probably the early 18th century).

Additionally, ^{14}C constitutes an important tool for the dating of groundwater. This application was first proposed by Münnich in 1957 on the basis that soil-zone CO_2 is of biogenic origin, resulting from the respiration of plant roots as well as from plant decay and hence containing the radioisotope in question, this having been derived from the atmosphere. The radiocarbon content decreases by radioactive decay: that fraction of the original content which remains is a measure of the time which has elapsed since its removal from the soil zone, i.e. the time since the associated water infiltrated, *t*, which is given by

$$t = 8270 \ln (C_0/C),$$

where t is in years, 8270 is the mean life of ^{14}C in years, C_0 is the initial concentration of the radioisotope and C is the ^{14}C concentration in the sample. Radiocarbon is measured relative to the total C content of the sample, so the origins of both this and the stable C present must be examined. Not all the stable C is of the same origin as the ^{14}C because infiltrating water containing CO_2 dissolved out of the soil zone dissolves carbonate minerals in the soil:

$$H_2CO_2 \rightleftharpoons HCO_3^- + H^+ ,$$

$$CaCO_3 + H^+ \rightleftharpoons HCO_3^- + Ca^{2+} .$$

However, water deriving from limestone does not contain radiocarbon and therefore water reaching the water table contains dissolved C in the chemical forms H_2CO_3, HCO_3^- and CO_3^{2-} with a ^{14}C content lower than that occurring in the soil (biogenic) CO_2. The evaluation of the dilution of soil CO_2 originally containing 100% of modern ^{14}C with ^{14}C-free (dead) carbonate in order to estimate the initial radiocarbon concentration in recharge water reaching the water table is one of the most difficult problems of ^{14}C age determination of water.

9.4 CARBONATE SAMPLES

These may arise from biogenic or chemical processes; their ^{14}C contents are much more dependent upon local conditions than that of plant material which has equilibrated with the atmosphere. In fact, the initial ^{14}C content of $CaCO_3$ deposited in water reflects the abundance of this radioisotope among the ionic and molecular species of carbonate in the water. That the specific activity of radiocarbon in carbonates exceeds that occurring in plants by a factor of approximately 1.05 was first supposed to be the result of enrichment by isotope fractionation, but this explanation is too simplistic. The matter may be investigated with reference to the two types of carbonate-depositional environment, namely freshwater and marine.

Modern freshwater molluscs have ^{14}C contents which relate mainly to the source of the HCO_3^- in solution. Where regions are underlain by rocks composed mostly of silicate minerals, the HCO_3^- ion derives mostly from atmospheric CO_2. For instance, aqueous CO_2 and water react with K feldspar thus:

$$\underset{\text{orthoclase}}{2KAlSi_3O_8} + 2CO_2 + 11H_2O \rightarrow 2K^+ + \underset{\text{kaolinite}}{H_4Al_2Si_2O_9} + 4H_4SiO_4 + 2HCO_3 .$$

All HCO_3^- ions so derived from weathering come from atmospheric CO_2 and therefore the specific activity of radiocarbon in such mollusc shells, when corrected for isotope fractionation to $\delta^{13}C = -25\%$, is near that of plant material in equilibrium with the atmosphere. However, in freshwater bodies which contact ancient carbonate rocks, as much as half to the HCO_3^- ion content may comprise 'dead' carbon from these rocks, following the reaction

$$CaCO_3 + CO_2 + H_2O \rightarrow Ca^{2+} + 2HCO_3^- .$$

Clearly, one of two HCO_3^- ions in such a reaction will derive from the $CaCO_3$, of which any original radiocarbon will have disappeared by radioactive decay in the time interval through which it has passed since it was deposited. As a result, the radiocarbon content of modern mollusc shells or of inorganically precipitated $CaCO_3$ in hard-water areas having lakes and rivers is normally lower than that predicted from consideration of equilibrium with atmospheric CO_2. Consequently, carbonate samples of this origin yield fictitious ^{14}C dates which are too old by thousands of years, the size of the effect varying greatly with local circumstances. It is not surprising, therefore, that ^{14}C dates from freshwater molluscs often do not accord with those obtained from terrestrial plants or animals. If the relevant radiocarbon contents are compared with similar material of known age, the initial activity can be calculated, but even then the activity of freshwater mollusc shells is liable to the secular variations of the ^{14}C content of the atmosphere.

There are many records of fictitious dates: one of the most interesting is that of Keith and Anderson in 1963. Analyses of freshwater and marine molluscs gave an average age of 1733 a for the former and only 155 a for the latter. It was suggested that the results reflected depletion in ^{14}C of the freshwater shells as a consequence of the oxidation of humus (up to 3000 a old). Broecker (1964) argued against this proposed link, but nevertheless it may well be valid.

With marine mollusc shells and seawater, the radiocarbon activity is lower than might be anticipated, probably because of water circulation in the ocean basins whereby cold water sinks in the polar regions as a result of its higher salinity and flows along the bottom towards lower latitudes. The flow rate involved is very low and hence causes a significant diminution in the ^{14}C content of the cold bottom water. The ^{14}C dates of this water range from some hundreds of years to a millenium. Sometimes, for example in the west coastal areas of both North and South America, this bottom water ascends because surface water is displaced by prevailing winds. This mixing process depletes the surface water with respect to the radioisotope. It has been shown that the prebomb radiocarbon content of marine shells collected in the coastal waters of California and southwards to Peru is up to 8.5% lower than that of biospheric C, the corresponding fictitious radiocarbon dates of modern marine mollusc shells from the same regions being as much as 300 a.

Obviously, a number of caveats may be given in regard to dating mollusc shells which have been collected in such areas. Their exact localities of origin must be known to ensure that they have not been transported. Then, it is necessary to study the ^{14}C contents of pre-bomb shells from the same localities. Shells coming from the open ocean in which the ^{14}C has not undergone depletion by mixing with old bottom water or with carbonate-rich river water are usually reliable for dating. There is a possibility that the ^{14}C content of a mollusc shell has been altered by the deposition of secondary calcite or aragonite from percolating groundwater. Despite the problems, good agreement has been obtained between results of wood dating and shell dating using this method.

9.5 ISOTOPE FRACTIONATION

The fractionation of the isotopes of C occurs in nature and is instigated by various physicochemical reactions. The result is that the abundances of the stable isotopes of the element, i.e. $^{12}_{6}C$ and $^{13}_{6}C$, which become incorporated into plants during photosynthesis, are different from those in the atmosphere. Plants usually become enriched in ^{12}C with a corresponding depletion in ^{13}C and ^{14}C. In a parallel manner, the activity of ^{14}C in plant tissue is different from that of chemically or biologically precipitated $CaCO_3$. Consequently, the fractionation of the C isotopes in nature entails small systematic errors in radiocarbon dates which can be eliminated where the isotopic composition of stable C to be dated is assessed using a mass spectrometer. The isotopic composition is expressed by means of the $\delta^{13}C$ value:

$$\delta^{13}C = \left[\frac{\left(\frac{^{13}C}{^{12}C}\right)_{\text{sample}} - \left(\frac{^{13}C}{^{12}C}\right)_{\text{standard}}}{\left(\frac{^{13}C}{^{12}C}\right)_{\text{standard}}}\right] \times 10^3\ \%\ ,$$

which is the parts-per-thousand difference between the $^{13}C/^{12}C$ ratio of a sample and that of a suitable standard; enrichment of the sample in ^{12}C relative to the standard produces a negative value for $\delta^{13}C$, and vice versa. The standard used is the equivalent of CO_2 made from *Belemnitella americana* collected from the Peedee Formation of the Cretaceous of South Carolina, USA, the PDB standard. Correction of radiocarbon activities for isotope fractionation involves the oxalic acid ^{14}C standard.

The National Bureau of Standards has prepared an amount of this containing a known quantity of ^{14}C, i.e. NBS No. 4990. It has been demonstrated that the age-corrected activity of wood grown between AD 1840 and 1860 is equal to 95% of the activity of the oxalic acid standard, so that $A_0 = 0.95A_{ox}$ dpm g^{-1}. It is a pity, but the C isotopes are fractionated in the course of the combustion of the oxalic acid and to different extents depending upon the experimental conditions.

Craig (1954, 1961) found that samples of CO_2 gas prepared from oxalic acid have $\delta^{13}C$ values of $-19.3‰$ (dry combustion) and $-19.6‰$ (wet combustion) relative to PDB. A value of $-190‰$ was informally accepted generally later, and the activity of every batch of CO_2 gas prepared from oxalic acid may be corrected on the basis of the measured $\delta^{13}C$ value for the particular batch. The correction assumes that the fractionation depends upon mass differences, also that radiocarbon is enriched or depleted twice as much as ^{13}C relative to ^{12}C. The appropriate expression is

$$A_{ox} = A'_{ox}\left[1 - \frac{2(19 + \delta^{13}C')}{1000}\right] \text{dpm g}^{-1} \tag{9.2}$$

in which the first term refers to the activity of the oxalic acid corrected for fractionation to $\delta^{13}C = -19‰$ relative to PDB, the second to the recorded activity of a particular batch of CO_2 gas made from the oxalic acid standard, and $\delta^{13}C$ is the measured value of this parameter of the CO_2 gas relative to PDB. Using these corrections, the initial activity of 19th century wood corrected for decay is

$$A_o = 0.95A'_{ox}\left[1 - \frac{2(19 + \delta^{13}C')}{1000}\right] \text{dpm g}^{-1}.$$

If samples of wood or shells with C isotope compositions differing from that of average wood ($\delta^{13}C = -25‰$) have to be measured, bearing in mind that wood on average is depleted in ^{13}C relative to oxalic acid by 6‰, an additional correction must be used. The activity demonstrated by such samples, which have a $\delta^{13}C$ value different from $-25‰$, is corrected using the following expression:

$$A_{\text{corrected}} = A_{\text{measured}}\left[1 - \frac{2(25 + \delta^{13}C_{\text{PDB}})}{1000}\right] \text{dpm g}^{-1} \tag{9.3}$$

In order to date a C-bearing sample by the ^{14}C method, the specific activity of the oxalic acid standard must be known, corrected for any isotope fractionation which took place during its combustion to CO_2 gas [see equation (9.2)]. Then, the specific activity of the sample to be dated must be measured and corrected for possible fractionation of C isotopes, using equation (9.3) above. The radiocarbon date is finally calculated from the fractionation-corrected activity of the sample and an initial activity equalling $0.95A_{ox}$. The ultimate step is to make any necessary corrections to this calculated date for secular variations in the radiocarbon content of the atmosphere.

9.6 ANALYTICAL METHODS

The major problem connected with the analyses is the difficult task of detecting and accurately measuring the rather low radioactivity level that occurs in natural samples. The original approach was to combust the sample to be dated so as to produce CO_2 gas, which was then reduced over hot Mg to solid C, the radioactivity of which was determined by using a shielded Geiger counter. The solid C was believed to give sufficient of the element to obviate having to effect isotopic enrichment of the sample in ^{14}C. The drawback was that the efficiency to be expected was only 5% because the very low-energy β particles emitted by ^{14}C (average 50 KeV) are adsorbed by the layer of solid C and thus lower the efficiency of the counter.

In addition, it soon became obvious that the method produced erroneously elevated counting rates because contamination from bomb fallout products took place. Consequently, by 1956, gas counting techniques had been substituted and liquid scintillation techniques were being investigated.

Liquid scintillation entails counting chemical synthesis of an organic compound which can be used as a solvent for a suitable phosphor or itself functions as a phosphor. The optimum approach seems to be to convert CO_2 samples to benzene, which both has a high content of C and is a good solvent. Currently, after purification, C is liberated as CO_2 by combustion of organic samples with O_2 or treatment of carbonates with H_3PO_4 or HCl. The resultant gas is again purified to remove such compounds as oxides of N_2, O_2 and halogens. Attempts are made to remove Rn as well. Subsequently, CO_2 can be converted to other gases such as CH_4 or C_2H_2, although this procedure is dangerous because of the risk of explosion and it also has the disadvantage that the compounds concerned do not always form with 100% yields. When the yields are lower, isotope fractionation may take place, enriching or depleting the product gas in ^{14}C. Nevertheless, both CH_4 and C_2H_2 are better counting gases since they are less influenced by impurities and need lower voltages in the proportional region of the gas counting tubes. Purified CO_2 or any other appropriate gaseous compound is inserted into a counting tube, which, in order to increase the rate of counting, is usually large in volume (up to 8 l); the gas in question is compressed to 1 – 3 atm so as to increase the number of ^{14}C atoms in it.

Naturally, background radioactivity must be shielded off. This may arise from a number of sources: perhaps the most important of these is natural or artificial radioisotopes in the environment, although gamma rays are also present as a component of cosmic rays. Gamma radiation may be reduced by shielding the counter with steel plates up to 30 cm thick, but this will not eliminate cosmic ray mesons, which require a ring of Geiger counters around the counter, connected to it by means of an anti-coincidence circuit. It is possible to reduce the background even more by utilizing a neutron trap made up of a mixture of boric acid and paraffin, which prevents any locally produced neutrons from entering the counter. In 1971 Ralph gave rates of counting for equipment operated at the Radiocarbon Laboratory of University of Pennsylvania in Philadelphia, USA, as listed in Table 9.1.

Usually, samples are counted for a sufficient time to derive favourable counting statistics. A standard deviation of 1% is attained by counting 10000 events, which normally needs 12–24 h. The measurement is repeated, and a statistical test effected in order to confirm that the two measured rates, corrected for background, are the same. This also ensures that no contamination with Rn has occurred.

Table 9.1 Background radiation in liquid scintillla-tion counter equipment

Equipment	*Counting rate (cpm)*
Unshielded counter filled with CO_2 made from anthracite	1500
Shielded by Fe and Hg	400
Shielded by Fe and with anti-coincidence circuit functioning	8

The recent development of ultrasensitive high-voltage tandem mass spectrometers which can count ions has had a tremendous influence on radiocarbon dating. The direct counting of ^{14}C ions represents a major advance in the technology of the method and the sole drawback is the economic factor. In addition, the conventional approach is perfectly adequate for samples less than 30 ka old from which enough material is obtainable.

10

Tritium dating

10.1 BACKGROUND

Tritium, designated T, is a short-lived radioisotope of hydrogen, ${}^3_1H$, produced in the atmosphere by the interaction of cosmic ray-produced neutrons with N:

$$ {}^{14}_{7}N + n \rightarrow {}^{3}_{1}H + {}^{12}_{6}C \,. $$

It also results from solar emission and the explosion of thermonuclear devices, as well as from the operation of nuclear reactors and particle accelerators on Earth. It decays by beta emission, thus:

$$ {}^{3}_{1}H \xrightarrow{\beta^-} {}^{3}_{2}He \,. $$

The rate of natural production of T is *c.* 0.25 atom cm^{-2} s^{-1}, according to Lal and Suess (1968). Once formed, T is quickly incorporated into H_2O molecules and resides in the lower stratosphere for 1 – 10 years. Later, these molecules enter the lower troposphere and rainout of the tritiated water takes place in a time interval of between 5 and 20 d. In this way, HTO is introduced into the hydrologic cycle, where it acts as a valuable label for water which has been in the atmosphere for the past 30 years or so. The T content of natural waters is expressed in T units, TU, 1 TU being equal to an abundance of one atom of tritium per 10^{18} atoms of H (equivalent to 7.1 dpm of T per litre of water). Such a great dilution means that measurement of low tritium contents in water samples entails preceding isotope enrichment. In fact, 1 TU is about the lowest limit of feasible detection.

Cosmic ray-produced T established a pre-bomb level of about 10 TU in temperate-zone meteoric waters. This was assessed by examination of the T contents of wine samples, since these do not depend only on water intake but also on the atmospheric water content exchanging with the grapes. In fact, wines laid down before the nuclear-bomb era represent almost the only undisturbed and well-dated samples of atmospheric water from those times, because part of their water content arose from tropospheric moisture, thus obviating the delay and dispersion due to soil action. However, after 1953 the T concentration in precipitation increased through thermonuclear testing and values as high as 10 000 TU were recorded in the northern hemisphere by 1963 when they reached their maxima. Thereafter, the values fell steadily until 1968 as a result of the moratorium on atmospheric bomb testing. Afterwards they remained almost stable, but current levels are still slightly elevated.

Most of the bomb-produced T was deposited in the stratosphere and about half of the stratospheric inventory is transferred to the troposphere in the spring and summer months of the year, so that maxima in the stratosphere occur in winters and minima in summer. The T content of precipitation varies considerably with latitude, lower values being recorded at oceanic and coastal sites compared with continental ones. This is because the oceans serve as sinks for HTO through isotope exchange between atmospheric water vapour and ocean water, which has a low T content. Low concentrations are found in equatorial and southern hemispheric areas; this is because most T is released in the northern hemisphere by transfer from the stratosphere to the troposphere preferentially at high latitudes. In addition, the higher proportion of ocean surface in the southern hemisphere and the high vapour pressure in equatorial regions provide more of a washout and dilution effect. The T content in the southern hemisphere is in fact only about one-tenth of that in the northern hemisphere.

The concentration of the radioisotope in meteoric waters is dependent upon several factors, such as the natural stratospheric production rate and the decay of T to the stable isotope ^{3}He, the half-life $T_{1/2}$ being 12.3 a. Also important are the seasonal injections of T from stratosphere to troposphere mentioned above, and the pulses introduced into the atmosphere from the nuclear facilities scattered around the world.

10.2 TRITIUM DATING

Bomb-produced T constitutes a natural labelling tracer for studying atmospheric, hydrological and oceanographical processes, but numerous complicating factors have to be taken into account. These include the ongoing decay of the radioisotope, its very uneven global distribution and the localized pulses into the environment of human-generated T.

However, the T content of precipitation has been used for estimating the input of the radionuclide to groundwater systems, allowance being made for the seasonal factor and also for the fact that, in vegetated areas, most precipitation (and T) returns to the atmosphere by evapotranspiration. In semi-arid locations, light precipitation is evaporated from the soil before it is able to infiltrate. In analysis of the T content of groundwater, a number of situations may arise. Absence of T indicates that the recent recharge water has not reached the point of sampling and that more than 20 years are necessary for water to reach that aquifer. The presence of an appreciable quantity of T, which varies with time, demonstrates that water younger than 20 years of age is present, the variations implying a short circulation period of only a few years. Another possibility is that water from two or more different sources occurs, perhaps involving (for example) a mixture of old, T-dead water and young water containing T. The variations in the respective contributions of two such components through time determine the changes in the overall content of the radioisotope. On the other hand, there may be a T content which is constant with time: this is taken to indicate good mixing between the young and old waters with the size of the aquifer masking the fluctuation of the recharge. Consequently, in groundwater studies, T measurements provide data on the time of recharge to the system. Of course, the above argument depends to some extent on a discrete segregation of groundwater of different ages during transit, except in the cases mentioned of incomplete or complete mixing with older, T-dead waters. The occurrence of such cases reflects the dispersive effect in some aquifers of granular materials, which smooth out T valleys and peaks. The well-mixed reservoir shows exponential discharge and a mean residence time of water within the system can be estimated (Dinçer and Davis, 1967).

Tritium has been utilized to study the downward progress of water in the unsaturated zone, the common means of recharging a groundwater system. Tritiated water, enriched not only in T but also in D, was employed in labelling a specific horizon of soil moisture; its rate of downward movements as well as the dispersion of the radiotracer were monitored in core samples subsequently taken from the test plot (Zimmerman, Münnich and Roether, 1967).

Smith *et al.* (1970) showed that the downward progress of environmental T represents infiltration over years in a completely undisturbed system. Figure 10.1b shows the T profile in the Upper Chalk in the Lambourne catchment, UK. There is a sharp maximum at *c.* 4 m depth with a maximum T concentration of 600 TU. A second, broader peak occurs between 7 and 9 m. At greater depths, there are variable quantities of T recorded (up to 40 TU) down to the water table at 27m depth. The two maximal peaks stem from the very high T concentrations in rainfall in 1963–1964 and the high values of 1958–1959. The existence of these peaks confirms that downward water transport is predominantly through intergranular seepage. From the position of the peaks in question, it was calculated that the mean seepage velocity is 2.8×10^{-8} m s^{-1} (0.88 m a^{-1}).

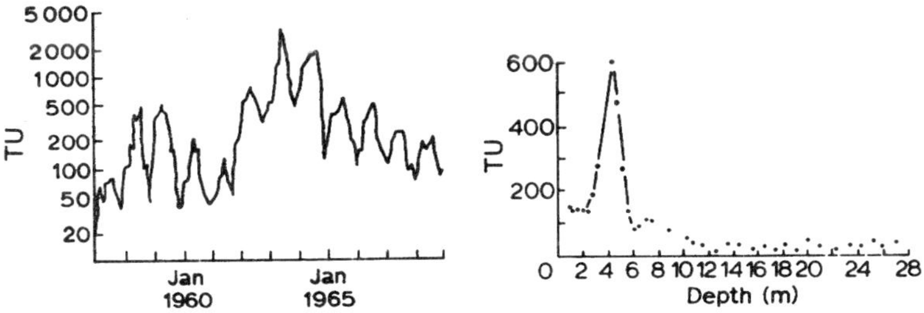

(a) **(b)**

Fig. 10.1 (a) Observed or computed values of T concentration in rainfall in the United Kingdom from 1957 to 1969 (uncorrected for radioactive decay); (b) a Tr profile in a borehole into the Upper Chalk in the Lambourne catchment in Berkshire, England.

Such percolation of water through overlying soil material is an important component of recharge to unconfined aquifers in humid and semi-humid regions and Allison and Hughes (1976) made significant contributions to its study. Their area of investigation was the Gambier Plain in Australia, which is underlain by an unconfined aquifer of great economic importance that receives nearly all its recharge by such percolation through the soil. The amount of recharge varies considerably and 11 cores were taken from three different sites, all within 30 km of each other. They were 'undisturbed', with 100 mm diameters, involving depths of up to 8 m. Water was extracted from each core section by vacuum distillation at 200°C. The T concentrations were obtained by 50- to 80-fold electrolytic enrichment of the soil water samples in constant-feed cells, followed by conversion to ethane, and gas proportional counting. The T concentration of rainfall in southern Australia generally and at Mt Gambier particularly has been well-known since 1965, one of the relevant monitoring station being at the mountain and within 20 km of the sampling stations. For earlier years, T concentrations were estimated by analysing wine samples dating back to 1950. There is a good correlation between wine and rainfall concentrations and the mean annual rainfall at the three selected sampling sites is ~ 750 mm. A comparison of observed and model T profiles showed that piston-type flow of soil water dominates under the root zone (~ 1.5 m) but, above this, direct input to beneath the soil surface becomes significant. Estimation of the recharge to groundwater from the T profiles obtained, using both their shapes and the total quantities of T held in them, demonstrated that the mean annual recharge veries between 40 and 140 mm a^{-1} for the sites examined. Multiple coring at two of the sites established that a recharge estimate made from a single core has a coefficient of variation of 15–20%.

11

Other methods

11.1 $^{137}Cs/^{135}Cs$ AS A CHRONOMETER TRACER

11.1.1 Geochemistry

Caesium is an element with relative atomic mass of 132.905 and atomic number 55. It is a very reactive silvery-white metal which resembles sodium and has a relative density of 1.87. Its melting point is 28.4°C and its boiling point is 678°C. The natural isotope ^{133}Cs (abundance 100%) is stable and there are 34 radioisotopes as well. Compounds are rare and the element is used in photoelectric cells and as a catalyst. The interesting isotopes are ^{135}Cs and ^{137}Cs. Both are radioactive with half-lives of 2.3 Ma and 30 a, respectively. They are produced with equal yields of *c.* 6 to 7% in binary fission of actinide nuclides. Caesium nuclei enter the environment through atmospheric testing of nuclear weapons and release from nuclear industry. ^{137}Cs is a well-known tracer and time marker whereas, according to Lee *et al.* (1993). ^{135}Cs has been practically disregarded because its long half-life precludes detection by counting, and also that in conventional mass spectrometers the tail from the stable ^{133}Cs swamps its signal (since its abundance is typically 10^{-9} smaller).

11.1.2 Application

Lee *et al.* (1993) reported the first measurement of the Cs isotopes 133 (stable), fission 135 and fission 137 using two coastal sediment samples and a thermal ionization mass spectrometer equipped with a retarding potential quadropole lens filter. The $^{135}Cs/^{133}Cs$ ratio was *c.* 1×10^{-9} and the $^{137}Cs/^{135}Cs$ ratio was about 5 due to the decay occurring through the past 30 a from their production ratio of *c.* 1. These authors noted that this appeared to be the first detection of fallout ^{135}Cs in nature, adding that, being an isotope ratio, ^{137}Cs/^{135}Cs is much more powerful as a chronometer-tracer than ^{137}Cs on its own. They presented two models illustrating how the ratio may be employed in order quantitively to estimate recent sedimentation and rates of erosion.

11.2 I–Xe DATING

Iodine is an element with relative atomic mass of 126.9045 and atomic number 53. A black to grey crystalline solid, its relative density is 4.953, its melting point is 113.6°C and its boiling point is 184°C. Highly volatile, it is essential for the functioning of the thyroid gland and its lack in the diet produces goitre. Hence it is used in medicine, and also in chemical analysis and photography. The radioisotope ^{131}I ($T_{1/2}$ 8.6 d) is used specifically to diagnose and treat disorders of the thyroid gland. In total, I has 33 isotopes of which 16 are radioactive. The one of interest is ^{129}I ($T_{1/2} = 15.7 \times 10^6$ a). Common I is ^{127}I (relative abundance 100%).

Xenon is an element with relative atomic mass of 131.29 and atomic number 54. An inert (noble gas), it is present in the atmosphere in minute amounts (*c.* 0.006 ppm) and has a melting point of −111.9°C and a boiling point of −108.1°C. It is used to fill some types of fluorescent tubes and light bulbs. A number of its compounds are known, e.g. XeF_2 and XeO_3. Xenon has 36 isotopes of which ^{129}Xe is the one of interest. It has an abundance in the naturally occurring element of 26.4 (atom %).

The I–Xe system is based on the radioactive decay of 15.7 Ma ^{129}I and could be a potentially precise geochronometer, especially of early events in the solar system. Lack of knowledge as to how it responded to thermal events during the post-formational histories of ordinary chondritic meteorites restricts its applicability.

11.3 La–Ce DATING

11.3.1 Geochemistry

Both of these elements are lanthanides (rare earths), and hence belong to that group of rare metals with atomic numbers ranging from 57 to 71 inclusive. All members of the group have similar properties and these resemble those of Al. The elements occur in monazite and comprise transition elements (d block elements). Cerium belongs to the inner transition elements.

Cerium has a relative atomic mass of 140.12 and atomic number 58. Its relative density is 6.7 and it has a melting point of 798°C. Its boiling point is 3433°C. Used in pyrophoric alloys such as misch metal for lighter flints, its compounds are employed in manufacturing some glasses.

Lanthanum has a relative atomic mass of 138.9055 and atomic number 57. Its relative density is 6.146 and it has a melting point of 920°C. Its boiling point is 3464°C. It too is used in pyrophoric alloys and also as a catalyst in oil cracking.

11.3.2 Theory

Tanaka and Masuda (1982) described the La–Ce geochronometer as a new dating method, applying it to gabbro samples from the Bushfeld Complex in South Africa. The theoretical basis of the approach is that ^{138}La, natural abundance 0.09% compared with 99.91% for the other naturally occurring stable isotope ^{139}La, undergoes branched radioactive decay by negatron emission to produce stable ^{138}Ce and by electron capture to yield the stable isotope ^{138}Ba. Initial values for partial half-lives were given as $(2.69 \pm 0.24) \times 10^{11}$ a (β^--decay) and $(1.51 \pm 0.10) \times 10^{11}$ a (electron capture), respectively. It quickly became apparent that both of the decay constants had to be refined before wider exploitation of the technique could be achieved.

11.3.3 Constraints on the ^{138}La β-decay half-life

Dickin (1987) recognized that the La–Ce isotope scheme can be a powerful adjunct to the Sm–Nd method for petrogenetic studies, but gave a best determination by counting of the β-decay half-life of La of $3.02 \times 10^{11}\,a^{-1}$ which did not accord with geological half-life determinations from La–Ce and Sm–Nd mineral isochrons. In examining the matter, he determined a six-point La–Ce isochron on whole-rock samples of Lewisian granulite facies Archaean gneiss from north-west Scotland. Based on the above-mentioned half-life, he obtained a La–Ce age of 2.99 ± 0.4 Ga (2σ) which came within the error of a whole-rock Sm–Nd date on Lewisian granulites of 2.895 ± 0.1 Ga (2σ). This half-life, which is equivalent to a La β-decay constant of $2.30 \times 10^{-12}\,a^{-1}$, can be recommended for geological applications of the Ce isotope method, according to Dickin (1987).

11.4 Lu–Hf DATING

As with Re–Os and K–Ca, Lu–Hf dating depends upon the β-decay of a naturally occurring radioactive isotope; all three techniques are meritorious in that they can be applied to rocks and minerals which cannot be dated by other conventional methods. In addition, studying their radiogenic daughters provides information regarding the terrestial mantle as well as the accreting of continental crust.

For a long time the Lu–Hf approach was difficult to implement, in part because of the problem of determining the half-life of ^{178}Lu by dating minerals containing this element, their age being already known. However, since the work of Boudin and Deutsch (1970), Owen and Faure (1974) and Owen we have been able to utilize the method for dating common minerals and rocks; nevertheless, they found it hard to separate Hf from other elements in pure form and this obstacle was overcome by Patchett and Tatsumoto (1980a). As a result, great progress was made in the geochronometry of both terrestrial and extraterrestrial rocks. In fact, employing the theoretical basis developed for the interpretion of the isotope compositions of Sr and Nd, it is now feasible to utilize the isotope composition of Hf in igneous rocks in order to examine their origin.

11.4.1 Geochemistry

Lutetium (cassiopeium) is a lanthanide (REE) with an atomic weight of 174.97 and an atomic number of 71. Its relative density is 9.842 and its melting point 1652°C. It is a d-block element with electronic configuration [Xe] $4f^{14}3d^{12}6s^2$. It is widely dispersed in all types of rock and has 29 isotopes of which two are found naturally, namely ^{175}Lu with an abundance of 97.4% and ^{176}Lu (abundance 2.6%). The ratio ^{175}Lu/^{176}Lu is 37.46, although, from analyses of very pure Lu reagents and natural samples, Patchett and Tatsumoto (1980) obtained a value of 37.701 ± 0.07, and used this in Lu determinations by means of isotope dilution. Curiously, this contradicts an earlier value of 37.36 ± 0.07 given by McCulloch, de Laeter and Rosman (1976) and there is as yet no explanation for the disparity. Subsequent researchers have utilized the value given by Patchett and Tatsumoto.

^{176}Lu is radioactive and subject to branched decay by β-emission. ^{176}Lu decays to ^{176}Yb, the frequency of the process being about $3 \pm 1\%$. This decay to ^{176}Yb can be ignored because of the long half-life of ^{176}Lu. The decay scheme of interest is

$$^{176}_{71}\text{Lu} \rightarrow ^{176}_{72}\text{Hf} + \beta^- + \bar{\upsilon} + Q$$

and this leads to an excited state of ^{176}Hf which is succeeded by γ-emission. Two approaches were used in determining the half-life of ^{176}Lu, namely the analysis of Lu-bearing minerals with known ages, and direct counting. There have been variable results which range from 2×10^{10} to 7×10^{10} a. Patchett and Tatsumoto (1980) first gave a value of $T_{1/2} = 3.53\,(\pm 0.14) \times 10^{10}$ a calculated from the slope of a Lu–Hf isochron formed by ten achondrites (eucrites) with an age of 4.55×10^9 a. The corresponding decay constant for the decay of ^{176}Lu to ^{176}Hf was stated as $(1.96 \pm 0.08) \times 10^{-11} a^{-1}$ but subsequently this was revised after analyses of three extra eucrites with low Lu/Hf ratios and a figure of $(1.94 \pm 0.07) \times 10^{-11} a^{-1}$ was substituted (Patchett *et al.*, 1981).

Lutetium and Hf occur in most rocks and it is interesting that the average Lu concentrations in igneous rocks do not increase much with increasing differentiation from basalt to granite. The important Lu-bearing minerals are zircon, biotite and apatite together with some REE ones such as the phosphates monazite and xenotime, the silicates allanite and gadolinite and the oxides samarskite and euxinite. Lutetium concentrations in ppm in rocks have been given as 0.447 for MORBS, 0.55 for basalt, 0.68 for granite, 23.65 for zircon, 0.0376 for chondrites and 0.251 for achondrites, while Hf concentrations in ppm in rocks have been given as 2.408 for MORBS, 2.44 for basalt, 6.93 for granite, 11 270 for zircon, 0.167 for chondrites and 0.251 for achondrites. Relevant Lu/Hf ratios for these rocks are 0.185 for MORBS, 0.22 for basalt, 0.098 for granite, 0.00209 for zircon, 1.255 for chondrites and 0.200 for achondrites. From lunar basalt Lu was recorded as 1.745 ppm, and Hf as 11.75 (Lu/Hf 0.148) and from lunar rocks (Tranquillity) Lu was recorded as 1.85 ppm, and Hf as 12.7 (Lu/Hf 0.15). These data imply that the Lu/Hf ratio of igneous rocks is *c.* 0.2 and less, basalts having higher values than acidic igneous rocks. It has been inferred that these values make dating using whole-rock samples impracticable.

Hafnium (celtium) has an atomic weight of 178.49 and an atomic number of 72. It is very rare and has a relative density of 13.3 with a melting point of 2227°C. It has been utilized as a neutron absorber in nuclear reactors as well as in the manufacture of W filaments. Hf has 31 isotopes of which six occur naturally and have the following abundances (%), according to Tuli (1985): ^{174}Hf, 0.162 (radioactive, $T_{1/2} = 2 \times 10^{15}$ a); ^{176}Hf, 5.206; ^{177}Hf, 18.606 (with metastable states); ^{178}Hf, 27.297 (with metastable states); ^{179}Hf, 13.629 (with metastable states); ^{180}Hf, 35.100 (with metastable state).

Hafnium is a dispersed element found in many minerals where it replaces Zr and, to a lesser extent, Ti. It occurs in Group IVB of the Periodic Table with a valency of +4 and both chemical properties and ionic radius paralleling those of Z with which, indeed, it is closely associated in nature. It is a d-block element with electronic configuration [Xe] $4f^{14}5d^{2}6s^{2}$. Erlank *et al.* (1978) have given details concerning these two elements.

Zircon is the most abundant mineral containing Hf, and generally alkali-rich igneous rocks are rich in it. For instance, average concentrations of 33 and 128 ppm Hf have been recorded from amphiboles and pyroxenes obtained from the Kangerdlugssuaq alkaline intrusion of East Greenland, while a sphene sample from the same region gave a value of 12.4 ppm Hf. An ilmenite sample from the famous Skaergaard Complex in Greenland gave a value of 12.5 ppm Hf. Geochemically, both Lu and

Hf resemble Sm and Nd; Hf concentrates relative to Lu in silicate liquids produced by partial melting in the mantle. Consequently, basalt magmas deriving from the mantle usually possess lower Lu/Hf ratios than their source rocks. Of course, residual solids left behind after magma extraction will be correspondingly depleted in Hf, thus obtaining higher Lu/Hf ratios than the original magma before melting occurred. It is highly possible that garnet may have a significant role in all these effects because of its strong affinity for Lu: this may act to prevent, or at least hinder, Lu from entering the melt and so increase the Lu/Hf ratios of the residual solids to a greater degree than those of other minerals in the mantle. As the partition coefficients for Hf are not yet accurately known, it is not feasible to model the partitioning of Lu and Hf between silicate minerals and melts. By comparison with Sm–Nd, some progress in this direction has been made and it was inferred that the ratio of the bulk distribution coefficients for magma formation can be expressed for the mantle thus:

$$D_{\mathrm{Hf}}/D_{\mathrm{Lu}} \simeq 2.3$$

in which D is the liquid–solid partition coefficient and the remaining solid is taken as comprising 60% olivine, 30% orthopyroxene and 10% clinopyroxene with the amount of partial melting falling in the range 5–50%. Therefore, the ratio of the concentrations is given by

$$\frac{C_{\mathrm{l}}^{\mathrm{Hf}}}{C_{\mathrm{s}}^{\mathrm{Hf}}} \simeq 2.3\frac{C_{\mathrm{l}}^{\mathrm{Lu}}}{C_{\mathrm{s}}^{\mathrm{Lu}}}$$

and it may be seen from this that Hf is concentrated more strongly into the liquid phase than Lu by a factor of 3.

11.4.2 Assessment of Ages

Because of the radioactive decay of ^{176}Lu, the abundance of ^{176}Hf in Lu-bearing rocks increases as a function of time and the actual age of these may be obtained by solving an equation of similar type to that for Sr, namely

$$\frac{^{176}\mathrm{Hf}}{^{177}\mathrm{Hf}} = \left(\frac{^{176}\mathrm{Hf}}{^{177}\mathrm{Hf}}\right)_{\mathrm{i}} + \frac{^{176}\mathrm{Lu}}{^{177}\mathrm{Hf}}(\mathrm{e}^{\lambda t} - 1) \qquad (11.1)$$

in which the first term refers to the relevant ratio in present-day samples, the second is the initial ratio which obtained at the time the system formed, and the third (the ratio of ^{176}Lu to ^{177}Hf) refers to the sample now; λ is the decay constant of ^{176}Lu ($(1.94 \pm 0.07) \times 10^{-11}\ \mathrm{a}^{-1}$) and t is the time which has elapsed since the formation of the mineral or rock.

In order to determine the age of a mineral or rock sample, it is necessary to measure the concentrations of Lu and Hf and also the ^{176}Hf/^{177}Hf ratio. The preceeding equation can then be solved for t (if the initial $^{176}Hf/^{177}Hf$ ratio is accurately known, the decay constant is accurately known and the system under examination had remained closed to both Lu and Hf during its geological history). If a suite of samples has a common age, but differing Lu/Hf ratios with the same initial $^{176}Hf/^{177}Hf$ ratio, then all will lie on a straight line isochron in coordinates of $^{176}Hf/^{177}Hf$ and $^{176}Lu/^{177}Hf$, its slope being proportional to their age (its intercept is the initial $^{176}Hf/^{177}Hf$ ratio).

The analysis involves dissolving rock powders in a mixture of HF and HNO_3 at 160°C over 4 days in a sealed teflon-coated bomb, a process which makes sure that Hf-rich minerals (especially zircon) decompose. Hafnium and Lu are separated and purified by ion exchange techniques employing HF and other acids as eluants. Isotope dilution is utilized in order to determine the concentrations of Lu and Hf and the measured $^{176}Hf/^{177}Hf$ ratios are corrected for isotope fractionation to $^{179}Hf/^{177}Hf = 0.7325$. A very pure HfO_2 reagent (Johnson Matthey Co. 475) was recommended as an interlaboratory Hf standard and has a $^{176}Hf/^{177}Hf$ ratio of 0.282 195. The technique was used to date a suite of total-rock samples and separated zircons from the Amitsoq gneiss of West Greenland by Pettingill *et al.* (1981). The relevant isochron gave a date of 3.59 ± 0.22 Ga with an initial $^{176}Hf/^{177}Hf$ ratio of 0.280482 ± 0.000033. This agrees well with dates obtained by application of the Rb–Sr method and fairly well with U, Th–Pb dating as well. It seems to be a reasonable estimate of the time which has passed since the sources of the Amitsoq gneiss were transferred up from the mantle. There is not good reproducibility in the present case because of the heterogeneous distribution of Lu and Hf in granitic rocks. They occur in accessory minerals–Lu mostly in sphene and apatite, Hf in zircons–which have an uneven distribution in the relevant rocks.

Zircon crystals may contain detrital cores with ages preceding the crystallization of igneous rocks containing them and, in addition, they can acquire overgrowths during the metamorphism of their host. In such zircons, the Hf may well have a wide range of isotope compositions which produces a scattering of data points about the isochron and introduces systematic errors into the derived ages. However, zircons have low Lu/Hf ratios and hence can be utilized in order to extend the range of Lu/Hf ratios of data points on isochrons, as well as to calculate the initial $^{176}Hf/^{177}Hf$ ratios of igneous rocks by using their ages after these have been determined by alternative methods.

11.4.3 Hafnium in Earth History

Stony meteorites have been used in describing the variation of the terrestial $^{176}Hf/^{177}Hf$ ratios as a function of time. The primordial ratio was stated to be 0.279 78 ± 0.000 09 from an isochron of 13 achondrites (eucrites) by Patchett and Tatsumoto (1980) and Tatsumoto, Unruh and Patchett (1981). The Lu/Hf ratio of chondritic meteorities is 0.240, based upon duplicate analyses of the chondrite Murchison; its corresponding $^{176}Lu/^{177}$ Hf ratio is 0.0334. Assuming that the age of a chondritic reservoir is 4.55 Ga, then the present value of the $^{176}Hf/^{177}Hf$ ratio in it is 0.282 86, these data defining a Hf growth curve which serves as a reference for the isotope composition of Hf in terrestrial minerals and rocks. The relationship can be expressed in terms of an epsilon parameter thus:

$$\epsilon^{0}(\mathrm{Hf}) = \left[\frac{(^{176}\mathrm{Hf}/^{177}\mathrm{H})^{0}_{\mathrm{sa}}}{(^{176}\mathrm{Hf}/^{177}\mathrm{Hf})^{0}_{\mathrm{ch}}} - 1 \right] \times 10^{4}$$

This notation of ϵ^0(Hf) shows that present Hf isotope ratios are being compared, while the parameter ϵ^t(Hf) relates to a comparison in the past which is based upon initial $^{176}Hf/^{177}Hf$ ratios:

$$\epsilon^{t}(\mathrm{Hf}) = \left[\frac{(^{176}\mathrm{Hf}/^{177}\mathrm{Hf})^{\mathrm{i}}_{\mathrm{sa}}}{(^{1776}\mathrm{Hf}/^{177}\mathrm{Hf})^{t}_{\mathrm{ch}}} - 1 \right] \times 10^{4}$$

in which $(^{176}Hf/^{177}Hr)^{t}_{sa}$ is the initial value for this ratio in a sample corrected for decay after it formed, and $(^{176}Hf/^{177}Hf)^{t}_{ch}$ is the value of this ratio in the chondritic reservoir *t* years ago. This value may be calculated from the equation

$$\left(\frac{^{176}\mathrm{Hf}}{^{177}\mathrm{Hf}}\right)^{t}_{\mathrm{ch}} = 0.282\,86 - 0.0334\,(e^{\lambda t} - 1)$$

in which *t* is the time which has elapsed since the formation of the relevant sample. Positive epsilon values mean that the sample is enriched in radiogenic ^{176}Hf compared with the chondritic reservoir and hence it probably orginated from a source with a higher Lu/Hf ratio than the chondrites. Likewise, negative epsilon values result from a deficiency of ^{176}Hf and from this may be inferred to have derived from a source possessing a lower Lu/Hf ratio than this reservoir.

In basalts from mid-oceanic ridges (MORB) and oceanic islands (OIB) the $^{176}Hf/^{177}Hf$ ratios range from 0.2828 to 0.2835, usually exceeding the present ratio in the chondritic reservoir. The inference is that such basalts derived from mantle sources depleted in Hf relative to Lu during earlier episodes of partial melting and hence they have higher Lu/Hf ratios than chondrites. Interestingly, the $^{176}Hf/^{177}Hf$ ratios of oceanic basalt correlate negatively with $^{87}Sr/^{86}Sr$ ratios and positively with $^{143}Nd/^{144}Nd$ ratios. Consequent Hf–Sr and Hf–Nd isotope mantle arrays can be seen in Fig. 11.1. Such correlation diagrams are divisible into

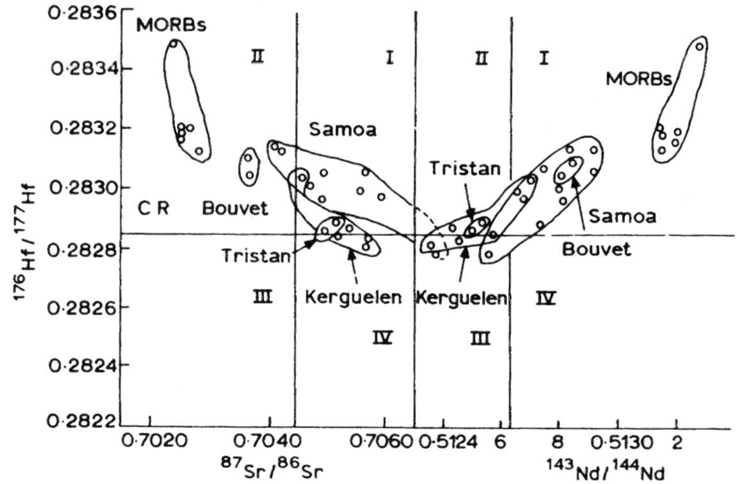

Fig. 11.1 Hf–Sr and Hf–Nd isotope arrays from basalts from mid-oceanic ridges and oceanic islands (MORBs). The quadrants contain the isotope ratios of basalts originating from sources either depleted or enriched in the radioactive parents compared with the chondritic reservoir. In the upper quadrants (from left to right), the fractionation factor f may be stated as follows: II, $f_{Lu} > 0$, $f_{Rb} < 0$; I, $f_{Lu} > 0$, $f_{Rb} > 0$; II, $f_{Lu} > 0$, $f_{Sm} < 0$; I, $f_{Lu} > 0$, $f_{Sm} > 0$. In the lower quadrants, the situation is as follows: III, $f_{Lu} < 0$, $f_{Rb} < 0$; IV, $f_{Lu} < 0$, $f_{Rb} > 0$; III, $f_{Lu} < 0$, $f_{Sm} < 0$; IV, $f_{Lu} < 0$, $f_{Sm} > 0$. The ^{87}Sr/^{86}Sr ratio of the chondritic reservoir is 0.7045 and the conclusion is that the basalts originated from sources having different Lu/Hf, Sm/Nd and Rb/Sr ratios in the past.

quadrants, using the isotope ratios of the chondritic reservoir which are:

$$^{176}\text{Hf}/^{177}\text{Hf} = 0.282\,86;$$

$$^{87}\text{Sr}/^{86}\text{Sr} = 0.7045;$$

$$^{143}\text{Nd}/^{144}\text{Nd} = 0.512\,638.$$

The geochemical histories of the magma sources of the igneous rocks found in these quadrants are expressible in terms of a fractionation factor f:

$$f_{Lu} = \left[\frac{^{176}\text{Lu}/^{177}\text{Hf}}{(^{176}\text{Lu}/^{177}\text{Hf})^0_{\text{ch}}} - 1 \right]$$

in which ^{176}Lu/^{177}Hf is the value of this ratio in a rock or magma source now and the chondritic reservoir has $(^{176}\text{Lu}/^{177}\text{Hf})^0_{\text{ch}} = 0.0334$. Obviously, source regions enriched in Lu and correspondingly depleted in Hf will have $f_{\text{Lu}} > 0$, whereas sources depleted in Lu and enriched in Hf compared with chondrites will have $f_{\text{Lu}} < 0$.

The range of $^{176}Hf/^{177}Hf$ ratios in MORBs extends from 0.283 112 to 0.283 476, placing them in quadrants characterized by $f_{Lu} > 0$. The mean Lu/Hf ratio possessed by a source region of the mantle with $^{176}Hf/^{177}Hf = 0.0400$ is calculable from equation (11.1) assuming the following values for the relevant parameters: $(^{176}Hf/^{177}Hf)^{i}_{ch} = 0.279\,78$; $\lambda = 1.94 \times 10^{-11}\ a^{-1}$; $t = 4.55\,Ga$.

$$\frac{^{176}Lu}{^{177}Hf} = \frac{(^{176}Hf/^{177}Hf)_{MORB} - (^{176}Hf/^{177}Hf)^{i}_{ch}}{(e^{\lambda t} - 1)}$$

This ratio emerges as 0.0400, thus:

$$\frac{Lu}{Hf} = 0.0400 \times \frac{174.97 \times 18.5}{2.60 \times 178.49} = 0.279$$

Actually, the range of $^{176}Hf/^{177}Hf$ ratios for MORBs suggests that the mean time-integrated Lu/Hf ratios of the relevant sources vary from 0.251 to 0.279. The chondrite reservoir has a Lu/Hf ratio of 0.240; hence these sources must have been enriched in Lu before MORB magmas formed. This probably occurred after the Earth formed. Because MORBs contain 11.8 times more Lu and 14.4 times more Hf than chondrites, it may be inferred that, even though both of these elements are strongly partitioned into the silicate liquid, Hf is concentrated to a greater extent than Lu.

On the Hf–Sr and Hf–Nd diagram, oceanic island basalts (OIBs) are strongly correlated with $^{176}Hf/^{177}Hf$ ratios lower than those of MORBs. This probably means that they came from sources which were less enriched in Lu. Interestingly, each island group seems to be quite distinctive as regards both the range and magnitude of the isotope ratios of Hf, Sr and Nd. Since the slope of the Hf–Sr array given by rocks from Kerguelen, Samoa, Tristan da Cunha and Bouvet is lower than that of MORBs and the same rock samples form a better collinear array on the Hf–Nd diagram, it may be that, geochemically, the Lu–Hf couple resembles Sm–Nd more than it does Rb–Sr, or perhaps the $^{87}Sr/^{86}Sr$ ratios of oceanic basalts could have increased through interaction with sea water, the isotope ratios of both hafnium and neodymium being unaffected by this factor as a result of their extremely low concentrations. It has been stated that the concentration of Hf in sea water is less than 8×10^{-6} ppm.

It is not as yet possible to explain the observed Hf–Sr and Hf–Nd arrays. They may be attributable to contamination of magma with Hf, Sr, and Nd of various isotope compositions coming from crustal and mantle sources or perhaps they arose from the origination of each individual flow from its own characteristic sources, i.e. lava flows derived from heterogeneous sources. This latter explanation is rather plausible because some of these lava flows on the Samoa–Society Islands and also on the Kerguelen Islands have low $^{176}Hf/^{177}Hf$ and $^{143}Nd/^{144}Nd$ ratios, but high $^{87}Sr/^{86}Sr$ ones, possibly also incorporating sedimentary rocks earlier subducted into the mantle. This phenomenon relates to the assimilation of wall rock discussed by White and Patchett (1984) with reference to incompatible elemental abundances in island arcs. Of course, petrogenetic models developed using isotope data have to be monitored by chemical information if any one of them is to be generally acceptable. However, it is clear that the Lu–Hf decay may well be optimally applicable to the study of mantle evolution and the genesis of magma.

Turning now to the subject of lunar basalts, their $^{176}Hf/^{177}Hf$ ratios have been measured and found to be much more variable than those recorded from MORBs, varying from 0.2816 to 0.2834 and having corresponding magma-source Lu/Hf ratios which range from 0.230 to 0.544 (again exceeding the range believed appropriate in the case of the terrestrial MORBs). It seems apparent that basalts comprising the lunar mare areas came from interior lunar sources which are considerably more heterogeneous than those of the terrestrial mantle. However, the lunar basalts have low and uniform Lu/Hf ratios which range from 0.0985 to 0.1700. All these data facilitate the reconstruction of the geochemical history of Earth's satellite if it is assumed that the Lu/Hf ratio of the Moon as a whole is the same as that of chondrites. In addition, it must be taken as read that the basalts in question stayed closed to both Lu and Hf, their isotope composition changing solely as a consequence of the decay of ^{176}Lu to ^{176}Hf. Finally, it is presumed that the isotope compositions of both elements in the basalt magma were the same as those of the solids from which they formed. Taking all this into account, it is probable that the relevant chain of events started with a geochemical differentiation of the lunar interior relatively soon after the satellite formed about 4.55 Ga ago, this altering the Lu/Hf ratio from the chondritic value (0.240) to the range 0.230–0.544. Afterwards, the $^{176}Hf/^{177}Hf$ ratios altered at different rates which depended upon the Lu/Hf ratios involved. Between 3.2 and 3.9 Ga, magma oceans arose on the lunar surface (now constituting the mare areas), the basalt being enriched in Hf compared with its source and having a mean Lu/Hf ratio about 2.5 times less than the Lu/Hf ratios of the source region prior to melting. Remaining interior solids were Lu enriched and may now possess higher Lu/Hf ratios than those obtaining before mare formation. Although no data are available, it is likely the the existing $^{176}Hf/^{177}Hf$ ratios of the Hf depleted and Lu–enriched interior lunar regions are high. Fujimaki, Tatsumoto and Unruh (1983) discussed this matter in detail.

From the terrestrial crust, igneous and metamorphic rocks have been examined in order to derive appropriate data bearing on the origin of their Hf; i.e. did it arise from sources in which the Lu/Hf ratio altered by earlier magma–creating episodes, or did it come from the unfractionated, pristine chondritic reservoir? Most such rocks have a low Lu/Hf ratio and this, together with the slow decay rate of ^{176}Lu, makes detection of any differences between the initial ^{176}Hf/^{177}Hf ratio of a sample of the rock from that of the chondritic reservoir complicated. It becomes more feasible in cases where the alteration of the Lu/Hf ratio in the magma source took place a minimum of 150 Ma before the crystallization of the rock. Concomitantly, it may be inferred that the initial ^{176}Hf/^{177}Hf ratio of a particular igneous rock derived from the melting of volcanics and volcaniclastic sediments derived from mantle sources less than 150 Ma before remelting is indistinguishable from that of the mantle reservoir. The same constraints operate for Nd and Sr in granitic rocks.

It is possible to determine the initial isotope composition of Hf in igneous rocks by analysing Hf-rich minerals such as zircon or baddeleyite, the former being the commonest of such minerals and extensively employed for dating by the decay of U and Th isotopes to Pb. It is not surprising, therefore, that pure zircons of known age are readily available and they have Lu/Hf ratios which are so low (0.0029) that no corrections for the decay of ^{176}Lu since crystallization are required (except for extremely old zircons formed more than 3 Ga ago). However, it is necessary to remember that zircons have the disadvantage that they may have cores which are older than the host rock. In addition, some possess metamorphic overgrowths which are younger. The former will contain Hf with a ^{176}Hf/^{177}Hf ratio lower than that of initial Hf in the rock and the young overgrowths have ^{176}Hf/^{177}Hf ratios which are higher. It is advisible to avoid such zircons.

Initial ^{176}Hf/^{177}Hf ratios for many granitic rocks from Precambrian to Phanerozoic in age were determined and found to increase with time, plotting above the chondritic growth curve. Very old rocks (*c.* 3 Ga) only deviate slightly and may have originated from unfractionated mantle sources. The younger the rock, the greater the deviation; they seem to have come from fractionated sources, i.e. those with Hf depletion or Lu enrichment.

An extremely old rock which probably derived from a fractionated source is a metal tholeiite dyke of the Suomussalmi–Kuhmo greenstone belt in Finland. The dyke is 2.27 $\pm$ 0.03 Ga old, with an initial ^{176}Hf/^{177}Hf ratio of 0.281 421 (corresponding to ϵ^{t}(Hf) = +14.3). From this, it is inferred that geochemical differentiation in the mantle started early in Earth history and promoted increases in its Lu/Hf ratio.

The situation with Hf is not paralleled by Nd, the initial ^{143}Nd/^{144}Nd ratios of both igneous and metamorphic rocks implying that the recycling of old crustal rocks played an important part in the petrogenesis of granitic-type rocks (Hf isotopes do not imply this at all). Why this difference exists is not at all clear, although it must relate in some manner to the different geochemical properties of the two elements involved. For instance, it was shown that Hf concentrates more than Nd in silicate liquids formed by partial melting, (Patchett, 1983). As a consequence of this, the mantle could be more depleted in Hf than in Nd, and therefore ϵ^t(Hf) values should exceed ϵ^t(Nd) ones. In fact, the ratio ϵ(Hf)/ϵ(Nd) varies between 1.3 and 2.7, and $\epsilon^t(\mathrm{Hf}) = 2.0\,\epsilon^t(\mathrm{Nd})$. The conclusion may be drawn that during the same geochemical differentiaion processess in the mantle the Lu/Hf ratios of magma sources have been increased more than Sm/Nd ones; hence the initial Hf isotope ratios of mantle-derived rocks exceed those of Nd when compared with the ratios of both elements in the chondritic reservoir.

The involvement of Hf and Nd in petrogenesis is dependent upon the way in which they occur in crustal rocks. Neodymium is found in all the main rock-forming silicate minerals, but Hf is concentrated in zircon in which it replaces Zr^{4+} in the lattice. This reflects the greater mobility of Nd which might explain the fact that Nd isotope studies provide definite evidence of crustal assimilation by magma which Hf studies do not. Crustal Hf with low ^{176}Hf/^{177}Hf ratios is not as easily extracted as Nd during petrogenesis so that Hf isotopes do not evidence crustal contamination of magma at all satisfactorily. However, they are able to assist in detecting mantle process which cause an increase in the Lu/Hf ratio by the preferential removal of the latter element. Isotope studies of Nd and in particular Sr may in the future aid in discernment of crustal contributions to the process of magma formation.

It is of interest to add that Hf isotope studies of the Earth were few until recently because of the problem of assessing Hf isotope ratios, a difficult matter due to low ion beam intensities and insufficient analysis times (due to the refractory nature of Hf and its low ionization efficiency). In the early 1990s, however, microgram-sized Hf samples were analysed in the Radiogenic Isotope Laboratory, University of Wisconsin–Madison, using a VG Sector 54 thermal ionization mass spectrometer. Both the usual Re triple filament assemblage and a novel single Re filament method were evaluated. Standard filament posts were modified with a special jig so as to produce about twice the standard Re filament length, which extends the filament life both of the triple filament and single filament assemblies. Hf was loaded onto triple filaments using dilute Hf and H_3PO_4. For single filaments, Hf was loaded in a slurry of Ir and C. As regards double filament analysis, this was examined, but Hf ion beam intensities for the single filament method were usually 1×10^{-12} A ^{177}Hf compared with higher intensities for a similar signal life of $2–3 \times 10^{-12}$ A ^{177}Hf obtainable using the triple filament method. Rhenium ion beam intensities for the single filament method were often excessive towards the end of filament life and poor quality data were obtained when ^{187}Re exceeded *c.* 10^{-11} A. By contrast, the modified filament design for triple filament assemblies stops excessive Re ion beam growth and hence the analyses are restricted solely by the quantity of Hf on the actual filament.

11.5 ^{53}Mn DATING

The element manganese (relative atomic mass 54.938, atomic number 25) has a naturally occurring stable isotope ^{55}Mn and 12 radioactive isotopes of which one, ^{53}Mn with a half-life of 3.7×10^6 a, decays by K-capture into the stable ^{53}Cr and is of interest geochronologically. Possible applications are to the terrestrial ages of meteorites and extraterrestrial dust abundances in ice and sediments.

11.6 ^{59}Ni DATING

Nickel (relative atomic mass 58.69, atomic number 28) has five stable isotopes and eleven unstable ones, of which one, ^{59}Ni is of interest as regards dating. Its half-life is 7.5×10^4 a and it decays by electron capture. It is the longest lived of the radioactive nuclides of the element. Its applications include assessment of the terrestrial ages of meteorites and extraterrestrial dust abundances in ice and sediments.

11.7 Po–Pb DATING

11.7.1 Geochemistry

The geochemistry of lead is discussed in Chapter 4.

Polonium is a radioactive element with 27 isotopes of which the longest-lived has a mass number of 209, atomic number 84 and a half-life of 103 a. It is a rare metal with a relative density of 9.4 and a melting point of 254°C. The isotope of interest in this new geochronometer, first presented by Rubin, Macdougall and Perfit (1994), is ^{210}Po, which occurs in some U ores to an extent of about 1 part in 10^{10} parts (mass of Po to mass of U). Its half-life is 138.4 d and it is a member of the ^{238}U decay series. It has the same activity as all other nuclides in this series in an undisturbed system, i.e. one at secular or radioactive equilibrium. Po is moderately volatile and hence gaseous at 400°C in volcanic emissions. It is possible that Po volatilizes as a halogen-containing species and it appears to be completely degassed in freshly erupted subaerial lavas at various volcanic sites such as Kilauea and Etna. In these, its activity shifts towards equilibrium with the parents in the decay series following the half-life, the immediate one being ^{210}Bi ($T_{1/2} = 5.013$ d). After a fractionation event, this latter rapidly equilibrates with its parent which is

^{210}Pb. The sequence is illustrated in Fig. 4.1. Therefore, the rate of change of ^{210}Po in such newly erupted lavas is effectively controlled by ^{210}Pb decay within half an hour or so of the events taking place. Unlike ^{210}Po, almost no Pb degasses during them so that the degassing of ^{210}Po sets the geochronometer for volcanic rocks at such periods and the times since occurrence can be assessed using the degree of ingrowth to equilibrium with ^{210}Pb.

11.7.2 Methodology

Rubin, Macdougall and Perfit (1994) used a simplified form of the Bateman equation for closed-system decay to describe the ^{210}Po content of degassed lavas as a function of time thus:

$$(^{210}\mathrm{Po})_t = (^{210}\mathrm{Po})_{t=0} \times [\mathrm{e}^{-\lambda t}] + (^{210}\mathrm{Pb}) \times [1 - \mathrm{e}^{-\lambda t}]$$

where t is the time elapsed since the eruption and λ is the decay constant. Parentheses are utilized to show activities. It is clear that to calculate t from a single measurement of (^{210}Po) it would be necessary to know both (^{210}Pb) and $(^{210}\mathrm{Po})_{t=0}$. While, for a totally degassed lava, the latter term would be zero, it is probable that only partial degassing may occur in the case of submarine lavas which erupted under high hydrostatic pressures. However, it is feasible to calculate a unique maximum value for t by assuming that total degassing took place and fitting an exponential function to a set of measurements made on a sample as ^{210}Po grows into equilibrium with ^{210}Pb. The maximum age corresponds with (^{210}Po) = 0 on the best-fit ingrowth curve. Rubin, Macdougall and Perfit (1994) applied the approach to East Pacific Rise (EPR) lavas and made repeat measurements of (^{210}Po) over several years on aliquots of each sample together with ingrowth curves having the functional form of the above equation, fitting these to the data using the time interval between sequential measurements in order to constrain t. The regression results gave a function for (^{210}Po) and the known analysis times linked the results to calendar time.

If Po degassing was not total at eruption, the eruption time cannot be precisely ascertained, but is bracketed in every instance by the window between collection of the sample and the calculated maximum age. Such windows (eruption dates) are most accurately determinable for samples either having the lowest (^{210}Po) when collected or for which the initial (^{210}Po) can be estimated independently.

Samples came from mid-ocean ridge basalts (MORBs) and these have similar chemical compositions and were erupted at the same water depth, both of these factors perhaps affecting the efficiency of degassing. Thus it can be anticipated that all will have had more or less the same initial ^{210}Po. But the degassing content may not be as high as the subaerial value of 100%. Two of the samples analysed by Rubin, Macdougall and Perfit (1994) had ^{210}Pb contents at the time of collection which were only *c.* 25% of the secular equilibrium value (25.1% and 25.6%, respectively), which implies that all samples were minimally 75% degassed in the course of eruption.

Best-estimate eruption windows are *c.* 2 months long and permit investigation of MORB eruption chronology on a far finer scale than hitherto possible. If Po degassing turns out usually to be larger than 'conservative' minimum estimate of 75% of Rubin, Macdougall and Perfit, then the resolution of the technique may improve, although this would necessitate analyses of ^{210}Po in a number of very young MORBs having eruption dates which are well known so that the normal degree of degassing could be established. These authors proposed that making remote observations of potential eruption indicators falling within an eruption window, e.g. seismicity or megaplumes, could be additionally valuable in constraining eruption dates and the range of initial ^{210}Po in MORBs. Some of their eruption windows were uncertain because they were not able to make more than two ^{210}Po analyses on each of the relevant lavas. Nevertheless, the results demonstrated that they are younger than the eruptions of early 1991 and must have erupted between the 1991 and 1992 expeditions. Consequently the Rubin, Macdougall and Perfit (1994) ^{210}Po/^{210} Pb dating work explicity showed that a small part of the EPR underwent eruptions during the first few months of 1991 and again during late 1991/early 1992. Clearly, such information is invaluable for evaluating the frequency and volume of eruptions within an eruptive cycle as well as short-term variations in MORB chemistry in a small area. Also, it can shed light on such matters as whether changes in lava chemistry relate to hydrothermal vent fluid chemistry and how rapidly biological communities respond to eruption-induced environmental changes.

All of the lavas with ^{210}Po eruption windows in the early months of 1991 are mafic (> 8.5 wt% MgO) MORB with almost the same chemical composition. Rubin, Macdougall and Perfit (1994) regarded the Po data as consistent with them being separate samples of one single flow only if the degree of degassing was very variable when eruption occurred, actually from total to 75%. In addition, although all samples were collected close to each other, at least one came from a physically different flow unit. From this they inferred that the relevant samples originated from a number of separate flows arising from a homogeneous magma reservoir through about 4 months, but comprising one volcanic episode. Two younger lavas were distinguishable from the above set and had distinct eruption windows, which was taken as evidence of origin as separate events either stemming from a different magma reservoir or having an intervening alteration in the lava composition. One of these younger lava samples was collected from the same area as the 1991 flows and was practically chemically identical to them. The implication was that the magma reservoir contributing lavas to what Rubin, Macdougall and Perfit (1994) termed the 'tubeworm barbecue' site retained an essentially constant chemical composition over a time interval of 1 year or more. However, one sample was obtained approximately 5 km to the north so that it remained unclear whether its different chemical composition is temporal or spatial in nature. The younger age of the flow accords with other observations suggesting that volcanicity shifted northwards into a previsously inactive region. That Po degasses from erupting mid-oceanic ridge lavas intimates the operation of similar mechanisms involving other volatile and fairly volatile metals and, if true, estimates of their fluxes into sea water based solely upon information from mid-oceanic ridge hydrothermal systems would be lower limits. Of course, the degassing process may well affect just the surfaces of flows since Po in the one microcrystalline sample analysed by Rubin, Macdougall and Perfit (1994) was $< 20\%$ degassed on eruption. It is apparent that further work aimed at evaluating fine-scale variations in Po and other volatiles, e.g. CO_2 and volatile metals, with depth in submarine flows is mandatory.

Rubin, Macdougall and Perfit (1994) regarded the significance of their work on $^{210}Po/^{210}Pb$ dating as twofold, namely that it confirmed suspicions of mid-oceanic ridge eruptions shortly prior to the April 1991 visit to 9° 50′ N on the EPR of the submersible *Alvin* and also produced the only information regarding both the number and timing of these phenomena. Revisiting the area after a 9 months lapse revealed that a lot of the physical evidence for recent volcanic activity had disappeared, although other adjacent regions seemed to be active. It is clear that the $^{210}Po/^{210}Pb$ chronometer potentially can provide data apropos the temporal variability of such submarine eruptions with excellent time resolution, especially when used together with detailed direct and remote observations. Rubin, Macdougall and Perfit (1994) concluded that this chronometer, in conjunction with other radioactive tracers of MORB eruptions, may well improve understanding of the temporal variability of submarine eruptions and their effects on both the biology and geology of the ridge ecosystem.

11.8 K Ca DATING

The K–Ar geochronometer is based upon the radioactive decay scheme of ^{40}K to ^{40}Ar. However, ^{40}K also decays to ^{40}Ca, although this has not yet been widely utilized as a dating technique because Ca isotopes are fractionated by natural and industrial processes as well as in the mass spectrometer.

Calcium is a soft white metallic element with an atomic weight of 40.08 and an atomic number of 20. It rapidly tarnishes in air, has a relative density of 1.55 and a melting point of 845°C. Its compounds are very abundant and extremely widely distributed; in addition, they are essential to life. Its commonest minerals are calcium carbonates such as chalk and limestone, and calcium sulphates like anhydrite and gypsum; Ca is an essential constituent of bones and teeth. Its compounds are industrially important.

It has 17 isotopes of which six are stable, with the following abundances (%): ^{40}Ca, 96.9821; ^{42}Ca, 0.6421; ^{43}Ca, 0.1334; ^{44}Ca, 2.0567; ^{46}Ca, 0.0031; ^{48}Ca, 0.1824. The isotope composition of calcium in minerals and rocks varies because of the formation of ^{40}Ca by β-decay of naturally occurring ^{40}K and also because of the fractionation of Ca isotopes during certain physicochemical processes. Of course, the chance of Ca isotopes fractionating is improved by the considerable differences in their masses (approximately 20% for ^{48}Ca relative to ^{40}Ca).

11.8.1 Calcium Isotope Fractionation

The existence of the large mass difference between the isotopes of Ca made it obvious that fractionation of Ca in nature would occur, but the various attempts to detect these in natural samples at first gave rather confusing results (e.g. Heumann and Luecke, 1973). This natural fractionation of Ca is significant in terms of conveying information about processes which have affected a particular sample. However, it is also important because of its interference with the K–Ca dating technique. In addition, the differences in the isotope composition of Ca that have been observed in meteorites may have been caused by incomplete mixing of the element injected into the solar nebula from exploding stars near it, either before or during the first formation of planetary bodies.

Various analytical techniques have been employed in investigation of the matter, among them mass spectrometry, neutron activation and deuterium activation (Meshceryakov, R.P. and Stolbov, 1967; Corless, 1968). The first is probably optimal since it is both direct and extremely sensitive, but it does produce isotopic fractionation in the element during analysis, thus masking any pre-existing natural isotope fractionation. However, it is possible to eliminate this problem using the double-spike isotope dilution approach first applied by Hirt and Epstein (1964) and more recently by Russell, Papanastassiou and Tombrello (1978) and by Lee and Papanastassiou (1978). Russell pointed out that evaporation of CaF_2 in a vacuum at 1200°C produces changes in the $^{40}Ca/^{44}Ca$ ratio ranging from +4.5 to −7.0‰, where the parameter δ^{40}Ca is defined as follows:

$$^{40}\mathrm{Ca} = \left[\frac{(^{40}\mathrm{Ca}/^{44}\mathrm{Ca})_{\mathrm{spi}}}{(^{40}\mathrm{Ca}/^{44}\mathrm{Ca})_{\mathrm{std}}} - 1 \right] \times 10^3 .$$

The standard is CaF_2 reagent. ^{40}Ca was found to be lost preferentially during vacuum evaporation so that the first Ca distilled is enriched in ^{40}Ca and the later fractions are depleted in this isotope. In addition, two different batches of spectrographically pure $CaCO_3$ demonstrated isotope fractionation. One was also enriched in ^{40}Ca ($\delta^{40}\mathrm{Ca} = +12.8‰$) and the other showed a more normal isotope composition ($\delta^{40}\mathrm{Ca} = +1.0‰$). The former was prepared from calcium purified by a vacuum evaporation–distillation process and the latter was derived from the English Derbyshire chalk and had not been purified by vacuum distillation. Since another sample of Ca metal from a different manufacturing source proved to be enriched in ^{40}Ca, with $\delta^{40}\mathrm{Ca} = +6.8‰$, it was inferred that vacuum distillation definitely causes Ca isotopes to fractionate. Of course, this is very important because previous attempts to find fractionation of the Ca isotopes in nature involved comparison with Ca reagents; thus observed differences in isotope ratios may have been compared with natural samples, reflecting a so-called fractionation mirage, simply an illusion produced by isotopic fractionation occurring during the industrial purification of Ca. Russell and Papanastassiou (1978) showed too that Ca isotopes are fractionated by ion-exchange columns which are employed in separating the element from others prior to mass spectrometric analysis, the effect sometimes being as great as 3‰ per mass unit. It is ascribed to the preferential retention of lighter isotopes on the column so that the Ca initially eluted is depleted in ^{40}Ca and enriched in ^{44}Ca, resulting in ^{40}Ca values which are less than they ought to be. It is possible to circumvent this difficulty by recovering all the Ca located on the ion-exchange column or, alternatively, by adding the double spike prior to the passage of the Ca-bearing solutions through the column. In mass spectrometers it is essential to eliminate the effects of instrumental mass fractionation: an exponential fractionation law has been formulated in which the correction per mass unit decreases with increasing mass difference. This empirical relationship fits experimental data better than a linear mass fractionation equation or one based upon Rayleigh distillation.

Such technical improvements facilitated the comparison of the isotope compositions of Ca from a broad spectrum of both terrestrial and extraterrestrial samples, resultant data indicating that the δ^{40}Ca parameter varies significantly but within the narrow limits of *c.* 2.5‰. Ca which has been biologically processed shows no greater isotope fractionation than that of inorganic samples. It seems, therefore, that the isotope fractionation of the Ca isotopes in nature is much less than was at one time anticipated.

However, despite the fact that Ca in bulk meteorites, lunar samples and terrestrial rocks is not highly fractionated, the Ca in some Ca–Al-rich inclusions in the carbonaceous chondrite Allende demonstrates a wide range of isotope compositions. In 1984 Niederer and Papanastassiou believed that these variations may have been the result of isotope fractionation produced by kinetic effects in the course of a complicated history of condensation, evaporation and recondensation of Ca in parts of the solar nebula. In fact, some of these inclusions contain the FUN isotope anomalies mentioned in Chapter 7.

11.8.2 K–Ca Dating Approach

This method depends upon the branched decay of ^{40}K to form ^{40}Ar by electron capture and ^{40}Ca by beta (negatron) decay, the total decay constant of ^{40}K being the sum of the decay constants for these two process, which are 0.581×10^{-10} and $4.962 \times 10^{-10}\,a^{-1}$, respectively. Therefore,

$$\lambda = \lambda_{ec} + \lambda_{\beta-}\,,$$

$$\lambda = 5.543\,X\,10^{-1}\,a^{-1}\,.$$

The following expression can be written for the $^{40}Ca/^{44}Ca$ ratio of a K-bearing system with an age of t:

$$\frac{^{40}Ca}{^{42}Ca} = \left(\frac{^{40}Ca}{^{42}Ca}\right)_i + \left(\frac{\lambda_{\beta}-}{\lambda_{ec} + \lambda_{\beta}-}\right)\left(\frac{^{40}K}{^{42}Ca}\right)(e^{\lambda t} - 1)\,,$$

where the first term is the ratio of isotopes in question now, the second their initial ratio when the relevant system formed or was last isotopically homogenized, the λ values give the appropriate decay constants, and $^{40}K/^{42}Ca$ being the ratio of the respective isotopes in the existing sample.

Cogenetic rocks or minerals which possess the same initial $^{40}Ca/^{42}Ca$ ratio and are of the same age lie on an isochron in coordinates of $^{40}Ca/^{42}Ca$ and $^{40}K/^{42}Ca$, this having a slope given by

$$m = \left(\frac{\lambda_{\beta}-}{\lambda}\right)(e^{\lambda t} - 1)\,,$$

with an intercept equal to the original initial $^{40}Ca/^{42}Ca$ ratio.

Many attempts to utilize the K–Ca geochromometer have been made (Polevaya *et al.*, 1958; Coleman, 1971; Heumann *et al.*, 1979). However, the most interesting is that of Marshall and DePaolo (1982) who employed the analytical techniques of Russell, Papanastassiou and Tombrello (1978). They obtained high-precision values of the ratio $^{40}Ca/^{42}Ca$ relative to $^{42}Ca/^{44}Ca = 0.31221$, actually a reference for making fractionation corrections. These researchers dated the Pikes Peak batholith of Colorado, USA, by analysing minerals and a whole-rock sample of granite from it. The K–Ca isochron gave a date of 1041 ± 62 Ma, which agrees closely with the results from other isotope geochronometers. The relevant initial $^{40}Ca/^{42}Ca$ ratio is 151.024 ± 0.0116, which is practically indistinguishable from that of two stony meteorites, namely the achondite Norton Country and the chondrite Murchison as well as from three mantle-derived rocks and minerals. In all of these samples, the mean value of the initial $^{40}Ca/^{42}Ca$ ratio is 151.016 ± 0.011. This means that the ratio is so low in the terrestrial mantle that the primordial isotope composition of Ca has been preserved. It was also preserved in the batholith, and hence it may be inferred that the isotope composition of Ca in the magma was unaltered and, consequently, that assimilation of crustal rocks, if it took place, had no effect. That such a process actually did occur is evidenced by a study of Sr isotopic compositions and trace element concentrations carried out by Barker *et al.* (1976).

The differences in the isotopic composition of Ca which result from the decay of ^{40}K or by isotopic fractionation can be expressed using an epsilon parameter which is defined thus:

$$\epsilon_{\mathrm{Ca}} = \left[\frac{(^{40}\mathrm{Ca}/^{42}\mathrm{Ca})_{\mathrm{spi}}}{(^{40}\mathrm{Ca}/^{42}\mathrm{Ca})_{\mathrm{mantle}}} - 1 \right] \times 10^4 ,$$

in which the $^{40}Ca/^{42}Ca$ ratio of the mantle is 151.016 ± 0.011. In the case of the Pikes Peak batholith, the ϵ_{Ca} values of the minerals obtained from it range from $+61.3 \pm 2.9$ for biotite to $+1.6 \pm 2.8$ for plagioclase. Usually, micas possess a high K/Ca ratio which can exceed 100, and hence acquire extremely positive ϵ_{Ca} values during their existence. On a graph with coordinates ϵ_{Ca} and time in the past (in Ga), development lines for these minerals and also for the whole rock are curved, not straight as is the case with Sr and Nd. The reason for this is that ^{40}K has a much shorter half-life with respect to ^{40}Ca than have ^{87}Rb and ^{147}Sm with respect to their stable daughters. Consequently, the approximate equivalence of $(e^{\lambda t} - 1)$ to t mentioned in sections 11.9.2 and 11.9.3 (on Re–Os dating) is invalid in the present case.

11.8.3 K–Ca Chronology of Lunar Granites

Russell, Papanastassiou and Tombrello (1978) introduced high-precision measurements of Ca isotopic compositions in geological samples, which were the basis for applying the $^{40}K/^{40}Ca$ method to dating granitic rocks, e.g. by Marshall and DePaolo (1982) as noted above. The advantage of this approach over the similar $^{87}Rb/^{87}Sr$ method is that the total decay constant of ^{40}K has been assessed more accurately than that of ^{87}Rb and is $0.5543 \pm 0.009 \times 10^{-11}\,a^{-1}$. However, the accuracy of the K–Ca age is affected by its poor precision due to the limited fractionation of K–Ca ratios in most igneous rocks, but not granite. Shih *et al.* (1993, 1994) and Bogard *et al.* (1993) dated an impact-affected lunar granite using the K–Ca, Rb–Sr and K–Ar methods and their researches showed that the K–Ca system is more resistant to impact reheating than either the Rb–Sr or K–Ar ones. This means that the K–Ca age is more likely to date the original formation event. Shih, Nyquist and Wiesmann (1993) effected K–Ca age determinations for three lunar granite samples (14321, 1062, 14303, 206 and 12013, 141). They also used the technique to analyse bulk samples of pristine lunar igneous rocks which included four mare basalts (10017; 15555; 14321, 1394 and 14305, 304, 371), a norite (k5445, 17), an anorthosite (60025), a KREEP basalt (72275, 543) and seven meteorites (Juvinas, Pasamonte, Moore County, Stannern,Y75011, 84B, Zagami and Shergotty). The K–Ca mineral isochron for lunar granitic clast 14321, 1062 gave an age of 4.06 ± 0.007 Ga for (^{40}K) = 0.5543 Ga^{-1} and an initial $^{40}Ca/^{44}Ca$ ratio of 47.141 ± 0.010 (normalized to $^{42}Ca/^{44}Ca = 0.31221$). Shih *et al.* claimed that the K–Ca age agreed well with the Rb–Sr and Sm–Nd ages and is slightly older than the average $^{39}Ar/^{40}Ar$ and U–Pb zircon ages. Low-precision K–Ca ages of 4.04 ± 0.064 Ga and 3.76 ± 0.72 Ga were derived from lunar granites 14303, 206 and 12013, 141, respectively, and ascribed to their lower K–Ca ratios. The initial values of $^{40}Ca/^{44}Ca$ for seven lunar and seven meteoritic bulk samples are similar and gave an average initial $^{40}Ca/^{44}Ca$ value of $47.136 \pm 0.002\,(2\sigma_m)$ and $\pm 0.006\,(2\sigma_p)$. If this value is taken as the lunar initial $^{40}Ca/^{44}Ca$, a $\times$ 10 to $\times$ 20-fold fractionation for K/Ca in the course of formation of the granite 14321 is calculable for a two-stage model. The large K/Ca-enrichment is consistent with a model of granite genesis entailing a silicate–liquid immiscibility process. Shih, Nyquist and Wiesmann (1993) referred to the range of lunar granite ages in the literature as being wide, extending from 3.9 to 4.4 Ga. Hence they suggested that lunar granites may well represent samples from about eight different intrusive bodies. The positive K and K/Ca correlations of lunar rocks was taken strongly to support an early global lunar differentiation, although most of the granites were unlikely to have been produced directly from this event.

11.9 Re–Os DATING

The β-decay scheme of ^{187}Re to ^{187}Os is applicable to the study of sulphide minerals of Mo and Cu as well as of Os-rich minerals such as iridosime, a mixture of Ir and Os and iron meteorites. In addition, it may be utilized in conjunction with Rb–Sr, Sm–Nd, Lu–Hf and U–Pb dating methods in order to examine mantle differentiation and the accretion of continental crust. The main problem is that of the half-life of ^{187}Re, but another, lesser, one relates to the fact that Os occurs in silicate minerals in very low concentrations.

11.9.1 Geochemistry

Rhenium is an element with an atomic weight of 186.20 and an atomic number of 75. It is a hard, heavy, grey metal with a relative density of 21.023 and a melting point of 3180°C and is used in thermocouples and as a catalyst. It was discovered in 1925. In the Periodic Table it belongs to Group VIIB together with Mn and Te, but geochemically it is more closely related to Mo than to Tc. It is a dispersed element and therefore is found almost exclusively in minerals of other elements, being most abundant in molybdenite, MoS_2, which is often found in pegmatites, aplites and granites. The Re concentrations in molybdenite vary from merely a few parts per million to as much as 1.88%. It has been found that molybdenites associated with Cu sulphide minerals possess unusually high concentrations of Re. There is only a single mineral in which the metal can be a major constituent, namely dzhezkasganite, $CuReS_4$, first recorded in cupriferous sandstone of the Dzhezkasgan deposits in Kazakhstan, former USSR. Re also occurs in the rare earth mineral gadolinite, Be_2Fe^{2+}, $Y_2Si_2O_{10}$, in tantalite and of course in the Cu sulphides. In igneous rocks, its abundance is approximately 0.5 ppb.

Osmium is an element with an atomic weight of 190.2 and an atomic number of 76. It is actually the most dense substance known, with a relative density of 22.58 and a melting point of 3030°C. It occurs together with Pt and is used in alloys with that element and Ir. Actually, it is a member of the Pt metals and belongs to Group VIII of the Periodic Table. It is siderophilic, occurring mainly in the mineral iridosmine. The Pt metals are associated with ultramafic rocks and gabbros which contain Cu and Ni sulphides and in addition they are found in tellurides, selenides, arsenides and antimonides. The terrestrial abundance of Os in the planetary crust is approximately 0.4 ppb. Table 11.1 provides relevant data.

In the solar system, the abundances of Re and Os are 5.07×10^{-2} and 7.17×10^{-1} atoms per 10^6 Si atoms, respectively, and this yields a Re/Os ratio of approximately 0.072. Interestingly, rocks comprising silicate minerals on the Earth and on the Moon show marked depletion in both elements as compared with iron meteorites; this reflects their siderophilic character. The high Re/Os ratio of crustal rocks on Earth is the result of their very low Os concentrations (*c.* 0.05 ppb), rather than of any increase in rhenium content. Extrapolating, it is assumed that the mantle has a much lower Re/Os ratio. Consequently, crustal Os ought to be significantly enriched in radiogenic ^{187}Os compared with Os in the mantle.

11.9.2 Assessment of Ages

Rhenium has 29 isotopes, but only two occur naturally, the stable ^{185}Re (with a relative abudance of $37.398 \pm 0.016\%$) and the radioactive ^{187}Re (with a relative abundance of $62.602 \pm 0.016\%$). The decay scheme mentioned above is as follows:

$$^{187}_{75}Re \rightarrow {}^{187}_{76}Os + \beta^- + \bar{\upsilon} + Q\,,$$

and the end-point energy of the beta spectrum is only *c.* 2.5 keV. This is why it is difficult to determine the half-life of ^{187}Re by direct counting. Various values have been produced; among the most recent is that reported by Luck, Birck and Allègre (1980) who obtained $T_{1/2} = (4.27 \pm 0.20) \times 10^{10}$ a and $\lambda = (1.62 \pm 0.08) \times 10^{-11}\,a^{-1}$ from the slope of a Re–Os isochron for several iron meteorites. Subsequently Luck and Allègre (1983) revised this to $(4.56 \pm 0.11) \times 10^{10}$ a and $\lambda = (1.52 \pm 0.04) \times 10^{-11}\,a^{-1}$ and Yin *et al.* (1993) gave $\lambda = (1.64 \pm 0.05) \times 10^{11}\,a^{-1}$.

Os has 34 isotopes of which seven occur naturally and all are stable. Luck and Allègre (1983) gave isotope ratios of Os in $(NH_4)_2OsCl_6$ (Merck) from which the following abundances were obtained (%): ^{184}Os, 0.023; ^{186}Os, 1.600; ^{187}Os, 1.510; ^{188}Os, 13.286; ^{189}Os, 16.251; ^{190}Os, 26.369; ^{192}Os, 40.957.

The growth of Os in a Re-bearing system may be expressed thus:

$$\frac{^{187}Os}{^{186}Os} = \left(\frac{^{187}Os}{^{186}Os}\right)_i + \frac{^{187}Re}{^{186}Os}(e^{\lambda t} - 1) \tag{11.2}$$

in which the first term is the ratio of the isotopes now, the second is that obtaining initially (when the system closed to both elements), and the third includes the present ratio of ^{187}Re to ^{186}Os; λ is the decay constant of ^{187}Re and t is the time which has elapsed after the system closed to both elements.

This parallels a similar expression for Rb–Sr dating; of course, the initial $^{187}Os/^{186}Os$ ratio must be assumed. An alternative is to apply equation (11.2) to a suite of samples with a common age and the same initial ^{187}Os ratio. Such samples form an isochron in coordinates of $^{187}Os/^{186}Os$ and $^{187}Re/^{186}Os$ of which the relevant slope is proportional to their age. The zero intercept of this isochron will give the initial $^{187}Os/^{186}Os$ ratio. The usual caveat applies, namely that the relevant system has been closed to Re and Os and, of course, the decay constant of ^{187}Re must be accurately known.

Rhenium and Os concentrations were determined in iron meteorites by means of isotope dilution using an ion-probe mass spectrometer and reported by Luck, Birck and Allègre (1980) who bettered previous analytical precision not only for these meteorites but also for the metallic phases in stony meteorites. Further refinement of the spike calibrations

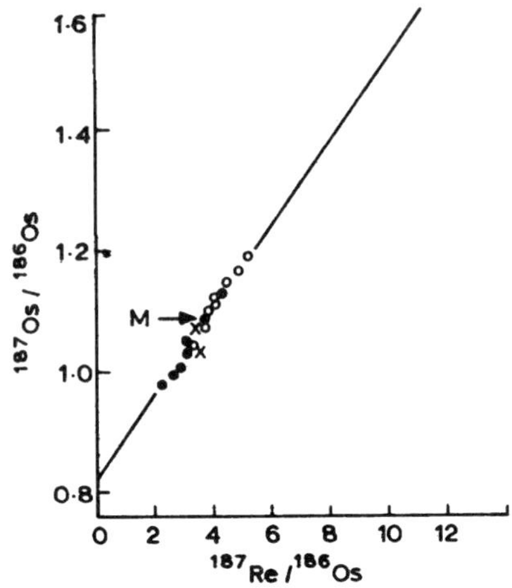

Fig. 11.2 The Re–Os isochron for iron meteorites and for the metallic phase of chondrites. ● = Iron meteorites; O = chondrites; × = stony irons. $(^{187}Os/^{186}Os)_i = 0.807 \pm 0.006$.

later allowed Luck and Allègre (1983) to plot the isochron shown in a rather short time interval, *c.* 90 Ma, from the solar nebular (which contained isotopically homogeneous Os). The primordial $^{187}Os/^{186}Os$ ratio is 0.807 ± 0.006. The isotope compositions of both Re and Os in meteorites are quite compatible with terrestrial samples.

Several measurements have been made of the Re contents of molybdenites and Hirt, Herr and Hoffmeister (1963) recorded concentrations of 2.0–230 ppm, with an average of *c.* 42 ppm for 19 samples. However, the Os content is under 0.2 ppm and due mainly to the presence of radiogenic ^{187}Os formed by *in situ* decay of ^{187}Re. As a result of its extremely high Re/Os ratio, molybdenite is especially well suited for dating by the Re–Os technique. The researchers in question obtained ages ranging from 105 to 2540 Ma, but the validity of these is uncertain because of the difficulty of determining the ages of hydrothermal vein deposits. In fact, the Re–Os technique is particularly interesting as it may afford a means of dating such molybdenite-bearing ore deposits. The major problem is that the low Os concentrations are difficult to determine accurately.

11.9.3 Osmium through Earth History

Osmium-rich minerals mentioned earlier together with others such as the sulphide laurite are found in ultramafic igneous rocks or their derived deposits. While these have high Os contents, they contain almost no Re and therefore the Os isotope composition has remained practically unaltered since the relevant minerals formed. Notwithstanding this, Hirt, Herr and Hoffmeister (1963) showed that iridosmine samples from variously aged deposits possess different $^{187}Os/^{186}Os$ ratios which significantly exceed the primordial value determined from meteorites.

Osmium-bearing minerals may record the isotope evolution of this element in the mantle. This is implied by the genetic association of iridosmine with ultramafics: to test this hypothesis, Allègre and Luck (1980) measured $^{187}Os/^{186}Os$ ratios from samples of this mineral having known ages. Their data plotted in coordinates of $^{187}Os/^{186}Os$ and time fit a straight line, its slope being $1.040 \pm 0.050\,786t$. The collinearity of all the data points shows that the mantle sources of these Os samples possessed a homogeneous Re/Os ratio which altered solely through the decay of ^{187}Re. It is possible to calculate this ratio from the slope of the Os development line by taking $(e^{\lambda t} - 1)$ in equation (11.2) as approximately λt. Then $^{187}Re/^{186}Os = \text{slope}/\lambda$, and, if $\lambda = 1.52 \times 10^{-11}\,a^{-1}$, then this ratio in the mantle is 3.34 ± 0.14 and the corresponding Re/Os ratio turns out to be *c.* 0.085. The intercept of the Os development lines gives a value for the $^{187}Re/^{186}Os$ ratio in the mantle at present, i.e. at $t = 0$, of 1.040 ± 0.010. Both present day ratios in the mantle now are the coordinates of a point which lies on the Re–Os isochron for meteorites shown in Fig. 11.2 The agreement of all these data is strongly confirmatory that the Earth and the parental bodies of the meteorites formed at approximately the same time from a primordial source which contained Re and Os, which both had a homogeneous isotope composition at that time.

Since a unique Os development line exists, this leads to the inference that, despite internal magmatic activity which formed the crust, the mantle is homogeneous as regards its Re/Os ratio. Information in Table 11.1 shows clearly that granites as well as ultramafics and basalt possess similar Re concentrations, suggesting that this element is not much partitioned during partial melting and fractional crystallization. Hence the geochemical differentiation of the mantle probably did not change its Re concentration much. On the other hand, Os is depleted in

Table 11.1 Average concentrations of Re and Os in terrestrial and extraterrestrial materials (from Allègre and Luck, 1980)

Material	*Re (ppb)*	*Os (ppb)*	*Re/Os*
Ordinary chondrites	57 (30–80)	660 (300–900)	0.086
Iron meteorites	2500 (5–5000)	30000 (7–50000)	0.083
The Moon	0.002–2	0.02–0.9	
Peridotite	0.43	5.3	0.081
Tholeiitic basalt	0.84	0.03	28
Granite	0.6	0.06	10

granites and basalts compared with its concentration in ultramafics and, therefore, it is believed that only a small fraction of mantle Os has been transferred into the crust. As a result of all this, the Re/Os ratio seems to have been stable throughout geological time; this is in contrast with the situation in the crust, which is marked by higher, variable Re/Os ratios. Consequently, it is likely that variability may constitute a valuable indicator of crustal contamination of magma. Allègre and Luck (1980) examined the $^{187}Os/^{186}Os$ ratio of a laurite sample from the Merensky Reef in the Bushveld Complex, South Africa, and found that it occurs significantly above the Os evolution line of the mantle. It may reflect contamination of the complex magma by Os which arose from the intruded country rocks of the crust, this element later concentrating in the reef zone and depositing, together with Ru and Ir, in laurite. This would accord with Sr and Pb isotope studies made in the Bushveld Complex. Allègre and Luck (1980) doubted this, however, because the Os concentration in ultramafics exceeds that of crustal rocks by a factor of 100; hence the former are practically unaffected by crustal contamination of the type proposed. Against this objection may be brought the composition of the Bushveld magmas, which were basaltic in nature. Thus, assimilation of a mere 5% of granite, with its higher $^{187}Os/^{186}Ps$ ratio of 10, would suffice to account for the enrichment of laurite in radiogenic ^{187}Os. Of course, it has also been argued that the Bushveld magma came from anomalous mantle sources having a Re/Os ratio which had been increased by earlier magmatic activity or, alternatively, that this magma chamber underwent an increase in the $^{187}Os/^{186}Os$ ratio through decay of ^{187}Re during a 25–30 Ma time interval in basaltic magma with a Re/Os ratio of *c.* 28.

11.9.4 Common Os Dating

The equation of the Os development line on a plot with coordinates $^{187}Os/^{186}Os$ and time is as follows:

$$\frac{^{187}\mathrm{Os}}{^{186}\mathrm{Os}} = 1.040 - 0.050\,768\,t$$

where the $^{187}Os/^{186}Os$ term is the isotope ratio of Os in an Os-rich, Re-poor mineral, t being the time which has elapsed since the Os was removed from its mantle source (expressed in Ga).

This equation constrains the isotopic composition of Os in such Os-rich, Re-poor minerals and is employable as a geochronometer for dating common Os minerals if the elements came from a mantle source having the same Re/Os ratio as the sources of the iridosmine samples on which the equation is based, and if the isotope composition of the Os remained unaltered after its removal from the mantle. In addition, the Os mineral to be determined should have formed quite soon after the removal of the Os from its mantle sources. If all of these requirements are satisfied, then Os minerals can be dated by their $^{187}Os/^{186}Os$ ratios only, obviating the necessity of measuring the concentrations of Re or Os. The relevant equation for t (in Ga) is

$$t = \frac{1.040 - \dfrac{^{187}\mathrm{Os}}{^{186}\mathrm{Os}}}{0.050768}.$$

11.9.5 Osmium Isotope Analyses using a Finnigan MAT 262

New methods supplement less satisfactory ones (such as secondary ion mass spectrometry (SIMS) and resonance ionization mass spectrometry (RIMS)) and a detailed discussion of these follows. The assessment of Os isotope ratios using negative thermal ionization (NTI–MS) was discussed by Tuttas (1992) who quoted unpublished researches by Wachsmann and Heumann (1989) which first demonstrated the feasibility of this approach. The technique was later improved e.g. by introducing O_2 or freon into the ion source during heating up and analysis. The advantage of it lies in its high ion yield of several percent. In association with static Faraday multicollector techniques, it is possible to analyse Os samples of < 3 ng with a precision of $< 0.02\%$ (1σ, external) for the $^{187}Os/^{188}$ Os ratio. The advantages of dynamic ion counting (single collector) or multiple ion counting (MIC, static) permit analysing sample quantities < 1 pg Os with adequate precision. Tuttas' analytical work was effected on a standard MAT 262 which had seven Faraday collectors for static ion collection as well as the optionally available device for negative ionization. It is a simple matter to switch from positive to negative ionization mode. The Os isotopes can be measured as O_5O_3 (masses 234–242). The loading procedure involved using Pt double filaments previously cleaned and degassed under vacuum at 3 A for 30 min. The Os solution used was Osmium DTM standard from MPI Mainz with a concentration of 0.92 ng μ 1^{-1}. 3μ 1 of the solution ($= 2.76$ ng Os) were loaded onto the evaporation filament together with *c.* 15 g of Ba (as hydroxide) and the ionization filament was also coated with a solution of 15 g Ba. This Ba coating reduces the work function of actual filament material and hence is vital to the formation and stability of negative ions. During sample heat-up and data acquisition, O_2 was introduced through an inlet system and this resulted in a source vacuum of *c.* 5×10^{-7} m bar. All the analyses were carried out in simultaneous ion collection mode. The Faraday cups were moved automatically to the requisite positions. The data acquisition time was *c.* 26 min per sample (plus a sample heating time of *c.* 30 min) for 100 individual data sets of 8 s each. Faraday baselines were measured between the data blocks. Stability was so good that amplifier gain factor calibration was effected only once prior to the first sample analysis. A total of nine samples (bends) were measured. After data collection, the isotopes were corrected for mass fractionation using $r(240/236) = 3.092202$. After the fractionation correction, data were corrected for O using $^{18}O/^{16}O = 0.00\,2045$ (Nier, 1940). That for $^{192}OsO_3/^{188}OsO_3$ is 3.092 203. Final results obtained for the following Os ratios were as follows: 184/188 = 0.001 383 1, 185/188 = 0.120 001 0, 187/188 = 0.173 9280, 189/188 = 1.219 6661, 190/188 = 1.983 7885.

Data from runs 1 and 3 were significantly offset (ascribed to varying interferences from WO_3^- ions. Measurement of a sample of only 1 pg of Os was done using a single collector peak jumping and ion counting. The following results were obtained: $187/188 = 0.17370 \pm 19$, $190/188 = 1.98296 \pm 76$ (2 × standard error for both). The same standard error was obtained in the preceding measurements. Plots of data showed that variance was improving during the run. It is possible that this reflects the decaying of interferences in the course of a run and Tuttas (1992) showed this with regard to ^{185}Re.

The conclusions were that N-TIMS of Os gave excellent precision, reproducibility and ion yield employing quantities of sample of only 2.76 ng Os, static multicollection results in external precisions of 12 ppm for the ratio $^{190}Os/^{188}Os$ or 160 ppm for the ratio $^{186}Os/^{187}Os$ and, for 1 pg samples. High-precision results ($\pm 0.1\%$) are obtainable in ion counting mode. Other Os isotopic analysis, in this case using a VG Isotech Sector 54, was effected, the Os solution employed being UMCP standard B. A sample containing 9 ng Os was loaded on a single Pt filament, reduced at a temperature of 600–700°C by heating the filament with a current of 1.9 A and then 10 g of $Ba(NO_3)_2$ was added. An O_2 bleed was used and the opening source pressure was 3×10^{-7} m bar. A stable 1 V (1×10^{-11}) $^{192}OsO_3$ beam was obtained. Employing static multicollectors, five blocks of 20 ratios were obtained; the main peak was at 1×10^{-11} A and the baselines were measured once per block at −0.5 amu with respect to each mass. Each sample took 1 h to run and, using the multidynamic spreadsheet, the data were corrected on-line for Re interference by $^{185}ReO_3$ and corrected for ^{17}O and ^{18}O. The data were normalized to $^{192}OsO_3/^{188}OsO_3 = 3.082\,710$. This work showed that the precision, sensivity and ionization efficiency of N-TIMS are better than alternative techniques available for the isotopic analyses of Os. The VG Sector 54 in negative ion mode yields large stable Os ion beams and can analyse them both in a static mode and a peak-jumping single collector approach on the channeltron. In the former, the external reproducibility on the $^{190}OsO_3/^{188}OsO^3$ ratio is 18 ppm and $^{187}Os/^{188}Os$ is 130 ppm. Such a level of precision suggests that N-TIMS will prove useful in $^{187}Re/^{187}Os$ geochronometry as well as for other negative ion applications.

11.10 'Os–Os' ($^{187}Os/^{186}Os$ AND $^{187}Os/^{188}Os$) METHODS OF DATING

This is a useful adjunct to the beta decay scheme of ^{187}Re to ^{187}Os which is discussed in section 11.9 and can be utilized for the assessment of ages of minerals containing it and ^{187}Os. Creaser, Papanastassiou and Wasserburg (1991) described the negative thermal ion mass spectrometry of Os, Re, and Ir while Walczyk, Hebeda and Heumann (1991) noted improvement in precision and enhancement in emission by introducing $\hat{O}_2$ or freons into the ion source. Since the development of negative thermal ion mass spectrometry (N-TIMS), relatively few publications demonstrating its wide range of applicability have appeared, which Yin *et al.* (1993) attributed mostly to what they termed the 'embarrassingly' high Re blank from the current generation of clean Pt filaments (from 30 m V down to 10^3 to 10^4 cps at mass 251) and also to problems in attaining isotopic exchange and equilibrium between sample and spike for Os. Another difficult is the inhomogeneity of samples for Re–Os systems. This may cause Re and Os concentrations to vary by 10% to 40% in the same sample. Avoiding or eliminating these complications was partly overcome by Yin *et al.* (1993). Irradiating a Re-containing sample with thermal neutrons in a nuclear reactor produced the following reactions with high cross-section: $^{185}Re(n,\gamma)$ ^{186}Re and $^{187}Re(n,\gamma)$ ^{188}Re. ^{186}Re ($T_{1/2} = 90.6$ h) and ^{188}Re ($T_{1/2} = 16.7$ h) decay to the stable isotopes ^{186}Os and ^{188}Os, these enriching in proportion to the neutron fluence and the Re/Os ratio of the sample being irradiated. If only the isotopic composition of the sample is measured, two ages can be derived and should be consistent. In an irradiated molybdenite sample from the Kingsgate molybdenum mine in New South Wales, Australia, these researchers obtained values for $^{186}Os/^{192}Os$ and $^{188}Os/^{192}Os$ of 0.3284 and 0.6299, respectively. In an unirradiated sample, these values were assumed to be 0.0390 and 0.3248. The Mainz TRIGA reactor was used for irradiation and the relevant ratios were measurable with N-TIMS to a precision of 1‰. Derived ages for the sample in question were $t_{186} = 220.5 \pm 8.7$ Ma and $t_{188} = 214.9 \pm 8.9$ Ma, respectively. These closely agreed with data from Evernden and Richards (1962), Kleeman (1988) and Gilligan and Barnes (1990) presenting 221 Ma to 238 Ma K-Ar ages for the leucogranite emplacement which caused the Mo mineralization at Kingsgate. Yin *et al.* (1993) observed that the main limiting factor on the precision of this age determination is the continuing uncertainty regarding the half-life of ^{187}Re. The errors deriving from the irradiation parameters are considerably reduced by selection of a suitable neutron flux monitor. The advantage of their approach is that the Re/Os ratio is determined from the same sample and only a single measurement of the isotopic composition of Os is necessary, this surmounting a number of experimental problems in conventional Re–Os analysis. Among these is the problem of sample inhomogeneity and the requirement to measure the absolute concentrations of Re and Os. The approach also makes it feasible to obtain an age spectrum for the disturbed Re–Os system through successive leaching steps or through *in situ* analysis.

Yin *et al.* (1993) made numerous interesting observations in connection with their new high-precision dating technique called the $^{187}Os/^{186}Os$ and $^{187}Os/^{188}Os$ or 'Os–Os' method, the application of which to a molybdenite sample implied that such dating can be readily achieved so that the system is usable for geochronology. They noted that it elimates experimental difficulties involved in conventional Re–Os analyses because both the Re/Os ratio and the Os isotopic composition are derived from one measurement alone, as was mentioned above. They discussed the theoretical basis of their method in detail, providing suitable equation for the quantities of ^{186}Os and ^{188}Os induced by thermal neutrons and defining neutron fluence terms by analogy with the parameter *J* in the $^{40}Ar/^{39}Ar$ method. They recorded no change due to neutron irradiation in the isotopic composition of the MPI-Os standard. In order to produce significant quantities of neutron-induced ^{186}Os and ^{188}Os, samples with high Re/Os ratios are necessary and this makes the method applicable to most crustal materials. Molybdenite was claimed to be an ideal sample for dating using it. Since most ore-forming minerals have low concentrations of lithophile elements and so the Sm–Nd, Rb–Sr, K–Ar and U–Pb methods are often inappropriate, the Os–Os method is even more valuable and may indeed come to be widely used in economic geology. It could date molybdenites and yield important informtion about associated mineralization episodes. Even more Re-enriched minerals include calcite, K–Al silicate phases and powellite, and their occurrences relate to phenomena such as alteration and metamorphic events in hydrothermal systems. With these, direct Os–Os dating of individual grains of unaltered and secondary minerals is feasible. Lin, Goresy and Fang (1991) reported high Re/Os nuggets in Ningqian (CV3) chondrite and Os–Os dating of these would be possible and probably show whether they are 'fremdlinge' (formed under presolar conditions) or were formed within the solar nebula. Numerous molybdenite grains have been found in meteorites, e.g. in Ca–Al-rich inclusions in the Allende meteorite. Theoretical considerations demonstrate that isochron construction from a series of cogenetic samples is possible and Ravizza (1990) applied the Re–Os system to black shale geochronometry. However, Yin *et al.* (1993) opined that a dramatic improvement in age and derived initial value is expected with Os–Os isochron dating on such shales.

11.11 Sm–Nd DATING

These elements belong to the lanthanides or rare earths (REEs), a group of metallic elements with atomic numbers from 57 to 71 inclusive which all have similar properties resembling those of Al. Samarium has an atomic weight of 150.36 and an atomic number of 62 with a relative density of 7.536 and a melting point of 1050 °C. It has 26 isotopes of which seven are found in nature with the following abundances: ^{144}Sm, 3.1%; ^{147}Sm, 15%; ^{148}Sm, 11.3%; ^{149}Sm, 13.8%; ^{150}Sm, 7.4%; ^{152}Sm, 26.7%; ^{154}Sm, 22.7%. The one of interest is $^{147}_{62}$Sm, which is radioactive, decaying by α-emission to a stable isotope of neodymium, $^{143}_{60}$Nd. Neodymium has an atomic weight of 144.24 and an atomic number of 60 with a relative density of 6.9 and a melting point of 1024 °C. It also has 26 isotopes of which again seven occur naturally; they have the following abundances: ^{142}Nd, 27.13%; ^{143}Nd, 12.18%; ^{144}Nd, 23.80%; ^{145}Nd, 8.30%; ^{146}Nd, 17.19%; ^{148}Nd, 5.76%; ^{150}Nd, 5.64%. The half-life, $T_{1/2}$, of ^{147}Sm is long, 1.06×10^{11} a, and $\lambda = 6.54 \times 10^{-12}\,a^{-1}$. Despite this, the decay scheme is applicable to the dating of terrestrial rocks as well as stony meteorites and lunar materials. In addition, the growth of both radiogenic ^{143}Nd and ^{87}Sr together affords a valuable perspective on the geochemical evolution of planetary objects. In addition, the Sm–Nd decay scheme provides a reliable basis for the determination of crustal residence ages since, once these elements have been incorporated into continental crust, they are not lost either by sediment subduction or by weathering and leaching (Sm and Nd being practically insoluble in water).

11.11.1 Geochemistry

As noted above, both elements belong to the REEs of group IIIB of the Periodic Table and these usually form ions with a charge of 3+ with radii which decrease with increasing atomic number from 0.115nm in lanthanum (Z = 57) to 0.093 nm in lutetium (Z = 71). REEs are found in high concentrations in a number of economically valuable minerals such as monazite ($CePO_4$), but they are also present as trace elements in common rock-forming minerals such as biotite and plagioclase feldspar. These two, with apatite, tend to concentrate the light REEs of the Ce group. However, the pyroxenes, amphiboles and garnet usually integrate the heavy REEs of the Gd group.

Both Sm and Nd belong to the light REEs and have very similar ionic radii, namely 0.104 nm for Sm^{3+} and 0.108 nm for Nd^{3+}, respectively.

According to Anders and Ebihara (1982) the relative abundances (atoms per 10^6 Si atoms) in the solar system are 2.61×10^{-1} Sm and 8.36×10^{-1} Nd, the lower abundance of Sm being consistent with the general decrease of the cosmic abundances of the elements with increasing atomic number. The atomic Sm/Nd ratio in the solar system is 0.31, and the ratio of Sm to Nd in terrestrial rocks and minerals is only sightly variable, from *c.* 0.1 to 0.5, as a result of their chemical similarity which precludes a greater separation through geological processes.

Bowen's reaction series indicates the order of crystallization of common rock-forming silicate minerals from magma: the concentrations of both Sm and Nd in these increase in the order in which they crystallize. Increases are demonstrable from olivine through augite, amphibole, biotite to K feldspar. The phosphates apatite and monazite have high concentrations of both elements, but nevertheless their Sm/Nd ratios are not appreciably different from those found in other rock-forming minerals. The Sm/Nd ratio of most minerals ranges from *c.* 0.37 in pyroxene to *c.* 0.15 in K feldspar. The concentrations of the elements in igneous rocks increase with their increasing differentiation, but the Sm/Nd ratios decrease; for example, the value for mid-oceanic ridge basalt (MORB) is 0.32 compared with a value of 0.19 for granites. The alkalic igneous rocks are enriched in Sm and Nd compared with acid ones, and the Sm/Nd ratios range from 0.10 to 0.20. Some sedimentary rocks also have low ratios, but in meteorites the ratios are higher – around 0.32 (Ca-enriched achondrites having higher concentrations of Sm and Nd than the chondrites). It may be inferred that Nd is concentrated relative to Sm during fractional crystallization of magma, and also that crustal rocks are normally possess lower Sm/Nd ratios than rocks originating from the upper mantle such as oceanic tholeiites. In an analogous manner, it is thought that during the formation of silicate liquids by partial melting of rocks in the terrestrial crust or mantle the liquid phase becomes enriched in Nd relative to Sm. This is probably because Nd^{3+} has a larger ionic radius than Sm^{3+}, this giving it a lower ionic potential (e/r) than the Sm. As a result, Nd^{3+} forms weaker ionic bonds which are more easily ruptured than those of Sm^{3+}.

It is also possible to describe the geochemical characteristics of the two elements in terms of their distribution coefficients D_1^m, where

$$D_1^m = \frac{C_m}{C_1}$$

in which C is the concentration of a given element in a mineral, m crystallizing in equilibrium with a silicate liquid, l. Partition coefficients of lithophile trace elements are functions of temperature and pressure as well as O_2 fugacity and the chemical composition of the liquid.

Consequently, there is sometimes diagreement between empirical partition coefficients and those derived by analysing phenocrysts and the matrices of porphyritic volcanics. It was demonstrated by Nakamura, Unruh and Tatsumoto (1982) that such partition coefficients of the REEs for clinopyroxene satisfy the following expression:

$$\log_{10} D_1^{\mathrm{m}} = \mathrm{a} \log_{10}\left(\frac{\mathrm{Fe}}{\mathrm{Fe-Mg}}\right) + \mathrm{b}$$

where Fe/(Fe–Mg) is the atomic ratio in the clinopyroxene and a and b are constants (for Sm, 1.438 and 0.670, respectively and for Nd, 1.348 and 0.438, respectively). Consequently, the partition coefficients of a clinopyroxene in which Fe/(Fe + Mg) = 0.2 are D(Sm) = 0.46 and D(Nd) = 0.31.

11.11.2 Assessment of Ages

The two elements Sm and Nd are interrelated by parent–daughter links through the alpha decay process through which ^{143}Nd derives from ^{147}Sm, and by the decay from ^{146}Sm, itself produced by the alpha decay of ^{150}Gd, to the stable isotope ^{142}Nd. The following expressions are appropriate.

$$^{147}_{62}\mathrm{Sm} \rightarrow {}^{143}_{60}\mathrm{Nd} + \alpha + Q \quad T_{1/2} = 1.06 \times 10^{10}\ \mathrm{a}$$

$$^{150}_{64}\mathrm{Gd} \rightarrow {}^{146}_{62}\mathrm{Sm} + \alpha + Q \quad T_{1/2} = 1.8 \times 10^{6}\ \mathrm{a}$$

$$^{146}_{62}\mathrm{Sm} \rightarrow {}^{142}_{60}\mathrm{Nd} + \alpha + Q \quad T_{1/2} = 1.0 \times 10^{8}\ \mathrm{a}$$

Q is the decay energy and $T_{1/2}$ the half-life.

Interestingly, another REE geochronometer based upon the decay of $^{138}_{57}$La to $^{138}_{58}$Ce was employed by Tanaka and Masuda (1982) to date gabbro samples from the Bushveld Complex in South Africa. ^{138}La undergoes decay by negatron emission to produce stable ^{138}Ce and by electron capture to the stable isotope $^{138}_{56}$Ba. The partial half-lives are respectively $(2.69 \pm 0.24) \times 10^{11}$ a (β^- decay) and $(1.51 \pm 0.10) \times 10^{11}$ a (electron capture). Both of the decay constants need to be refined prior to wider exploitation of the La–Ce geochronometer.

The decay of ^{147}Sm and the growth of radiogenic ^{143}Nd may be expressed as follows:

$$\frac{^{143}\mathrm{Nd}}{^{144}\mathrm{Nd}} = \left(\frac{^{143}\mathrm{Nd}}{^{144}\mathrm{Nd}}\right)_0 \frac{^{147}\mathrm{Nd}}{^{144}\mathrm{Nd}} (\mathrm{e}^{\lambda t} - 1)\ ,$$

in which ^{144}Nd acts as the reference since the number of its atoms in a unit quantity of any given sample remains constant so long as the host system stays closed to this element. The agreed value for the decay constant is $6.54 \times 10^{-12} a^{-1}$, determined by measurements discussed by Lugmair and Marti (1978). The normal approach to determining ages by the Sm–Nd method is to analyse either individual minerals or cogenetic rock suites in which the ratios between the two isotopes are sufficiently diverse to define the slope of an isochron in coordinates of ^{143}Nd/^{144}Nd and ^{146}Sm/^{144}Nd. The technique is appropriate for mafic and ultramafic igneous rocks, as opposed to the Rb–Sr one which is best suited for acidic and intermediate igneous rocks enriched in Rb and depleted in Sr. In addition, the REEs possess a lower mobility than the alkali metals and alkaline earths, so that phenomena such as regional metamorphism have less effect. This means that suitable rocks can be dated by the Sm–Nd method although they may have gained or lost Rb and Sr. This makes it a useful complement to the Rb–Sr technique.

Hamilton *et al.* (1979a) applied the Sm–Nd technique to volcanic rocks from the Onverwacht Group of Swaziland and South Africa, this comprising the basal part of the Early Precambrian Barberton greenstone belt of Swaziland and the Transvaal. The relevant rocks are ultramafics with overlaying mafic-to-felsic materials (the so-called 'mafic-to-felsic unit'). Included volcanics were altered by low-grade regional metamorphism and hence they were difficult to date because of its influence on the U–Pb and Rb–Sr systems. However, the Sm–Nd approach enabled Hamilton and his colleagues to calculate an age of 3.540 Ga $\pm$ 30 Ma with an initial ^{143}Nd/^{144}Nd ratio of 0.50890 $\pm$ 0.000 04. This was confirmed by Jahn, Gruau and Glickson (1982). Interestingly, the Sm/Nd ratio of komatities in the lower ultramafics exceed those of the felsic volcanics, as do their ^{143}Nd/^{144}Nd ratios. The initial ^{143}Nd/^{144}Nd ratio of the volcanics of the Onverwacht Group is similar to that of other Early Precambrian rock suites and can be employed in order to make an estimate of the Sm/Nd ratio of the upper mantle of the Earth, the probable source of the lavas. A time-integrated Sm/Nd ratio of 0.297 $\pm$ 0.005 is obtained by assuming that the primordial ^{143}Nd/^{144}Nd ratio of the planet equals that of the achondrite Angra dos Reis, and that the age of the Earth is about 4.55 Ga. As the value obtained is close to that for the Sm/Nd ratios of stony meteorites, it may be inferred that the Sm/Nd ratio of the Earth when it formed resembled that of the parental bodies which gave rise to the meteorites.

Hamilton *et al.* (1978, 1979) also worked on the Isua supracrustals in West Greenland and the Lewisian gneisses of Scotland. In addition, Nyquist *et al.* (1979) worked on the Sr and Nd isotopic record of Apollo 12 basalts from the Moon.

11.11.3 Neodymium through Earth History

The abundance of ^{143}Nd as well as the $^{143}Nd/^{144}Nd$ ratio of the planet increased with time as a result of the radioactive decay of ^{147}Sm to radiogenic ^{143}Nd. This phenomenon can be represented by a model which derives from the age and Sm/Nd ratio of the Earth together with its primordial $^{143}Nd/^{144}Nd$ ratio. Of course, both of these are known from studies of stony meteorites, many of which have been dated by the Sm–Nd method by analysing minerals fractions which have been separated from individual specimens and considering the results on Sm–Nd isochron diagrams. For instance, Hamet *et al.* (1978) produced a mineral isochron for an achondrite (eucrite) Moama showing an age of 4.58 ± 0.05 Ga and a primordial $^{143}Nd/^{144}Nd$ ratio of $0.506\,84 \pm 0.000\,08$. This is closely similar to data recorded from Angra dos Reis in 1981 by Jacobsen and Wasserburg (1981) who obtained an age of 4.562 ± 0.031 Ga and a primordial $^{143}Nd/^{144}Nd$ ratio of 0.506 64. The Sm–Nd approach has been utilized to examine several other achondrites, among them Shergotty, Allan Hills 77005 and Nakhla. These have peculiar chemical and mineralogical compositions and they have given curiously low ages by the Rb–Sr method. Many uncertainties exist regarding the shergottites; some have been resolved by application of the Sm–Nd technique. Nakamura, Unruh and Tatsumoto (1982) obtained an age of 1.26 ± 0.07 Ga for Nakhla from analysing mineral separates and a whole-rock sample. Shih *et al.* (1982) obtained a similar age of 1.34 ± 0.06 Ga from a whole-rock isochron of Shergotty and Allan Hills 77005; this value also resulted from an analysis of another achondrite, namely Zagami. Actually, their Sm–Nd data confirmed Rb–Sr dates which showed that the samples had experienced an episode of shock metamorphism some 180 Ma ago. This is part of the evidence tending to the conclusion that shergottites may have derived from a large, geologically dynamic planet such as Mars.

A model is available for the isotope evolution of Nd on Earth based on the assumption that the Sm/Nd ratio of the planet equals that of chondritic meteorites. Numerous analyses yielded an average $^{147}Sm/^{144}Nd$ ratio of 0.1967 for five chondrites and an achondrite, and this may be used to calculate the $^{143}Nd/^{144}Nd$ ratio of a chondritic sink at the present, assuming the terrestrial age to be about 4.6 Ga. Results embodied in Table 11.2 are based on Wasserburg *et al.* (1981).

The correction for isotopic mentioned in the footnote to Table 11.2 is an important consideration. Naturally, there is uncertainty regarding the values of the primordial $^{143}Nd/^{144}Nd$ ratio and the present $^{143}Nd/^{144}Nd$ ratio of the chondrite reservoir. This is due to variations in analytical procedures and the data reductions of various laboratories. The question involves the correct calibration of spike solutions which are employed in

Table 11.2 Compatible sets of present-day isotope ratios for Nd in a reservoir of chondritic composition

Row[a]	$\frac{^{142}Nd}{^{144}Nd}$	$\frac{^{143}Nd}{^{144}Nd}$	$\frac{^{145}Nd}{^{144}Nd}$	$\frac{^{146}Nd}{^{144}Nd}$	$\frac{^{148}Nd}{^{144}Nd}$	$\frac{^{150}Nd}{^{144}Nd}$
1	1.138 305	0.511 847	0.348 956	0.724 134	0.243 075	0.238 619
2	1.141 827	0.512 638	0.348 417	0.721 9	0.241 578	0.236 418

[a] Row 1 comprises isotope ratios of Nd in the chondritic uniform reservoir now, after correcting NdO^+ mass scans for O possessing $^{18}O/^{16}O = 0.002\,11$ and $^{17}O/^{16}O = 0.000\,387$, and after correction for isotopic fractionation to $^{146}Nd/^{142}Nd = 0.63\,151$. Row 2 includes the same isotope ratios, but normalized to $^{146}Nd/^{144}Nd = 0.7219$.

order to determine concentrations of Sm and Nd by isotope dilution as well as corrections for interfering isobaric ions in the mass spectra of these elements. Also of significance are the form of the isotope fractionation law utilized to correct for mass-dependent fractionation in the mass spectrometer and the numerical value of the isotope ratio selected for this purpose. At the California Institute of Technology, workers initially normalized measured $^{143}Nd/^{144}Nd$ ratios to $^{150}Nd/^{142}Nd = 0.2096$, but subsequently decided to use $^{146}Nd/^{142}Nd = 0.636\,151$. Other laboratories have employed $^{148}Nd/^{144}Nd = 0.241\,578$ or $^{146}Nd/^{144}Nd = 0.7219$. However, not all of these isotope ratios are compatible. Table 11.2 contains two sets of ratios.

A model called CHUR is proposed for describing the isotopic evolution of neodymium in the Earth, this acronym standing for CHondritic Uniform Reservoir ; it assumes that terrestrial neodymium has evolved in a uniform reservoir in which the Sm/Nd ratio equals that of chondritic meteorites. The present value of the $^{143}Nd/^{144}Nd$ ratio of this reservoir is 0.512 638, relative to a $^{146}Nd/^{144}Nd$ ratio of 0.7219. The present value of the $^{147}Sm/^{144}Nd$ ratio of CHUR is 0.1967, allowing calculation of the $^{143}Nd/^{144}Nd$ ratio of CHUR in any past time interval from the following equations:

$$I^t_{\mathrm{CHUR}} = I^o_{\mathrm{CHUR}} - \left(\frac{^{147}\mathrm{Sm}}{^{144}\mathrm{Nd}}\right)^0_{\mathrm{CHUR}} (e^{\lambda t} - 1) \tag{11.3}$$

in which I^t_{CHUR} is the $^{143}Nd/^{144}Nd$ ratio of CHUR at a past time t, I^0_{CHUR} the present ratio of CHUR (0.512 638 normalized to $^{146}Nd/^{144}Nd = 0.7219$) and $(^{147}Sm/^{144}Nd)^0_{\mathrm{CHUR}}$ is the present value in CHUR based upon analyses of stony meteorites, this being 0.1967. It is believed that the Sm/Nd ratio of CHUR is lower than that of the solar nebula. Taking the age of the Earth to be 4.6 Ga, the following argument can be presented. The partial melting of CHUR produces magmas with lower Sm/Nd ratios so that igneous rocks derived from them also have lower $^{143}Nd/^{144}Nd$ ratios now than CHUR. A corresponding effect ensures that residual solids left behind after withdrawal of magma possess higher Sm/Nd ratios than the chondritic reservoir, and of course such depleted zones will have higher $^{143}Nd/^{144}Nd$ ratios than CHUR now. All this is quite consistent with the known geochemical properties of Sm and Nd.

Some portions of the chondritic reservoir will not be affected by such magma-forming episodes and they will contain Nd of which the isotope composition has evolved right up until the present without any interference; this can be represented linearly on a graph with coordinates $^{143}Nd/^{144}Nd$ and age (Ga). On such a graph, the residual solids after partial melting will lie along a steeper line, in contrast with the partial melts which lie along a line with slope less than either. It may be added that the age selected for the Earth, i.e. 4.6 Ga, is greater than the normal 4.5 Ga: this is in order that consistency can be maintained with the numerous papers dealing with Nd evolution. Attempts have been made to compare the initial $^{143}Nd/^{144}Nd$ ratios of both igneous and metamorphic rocks with the corresponding ratios of CHUR at the times when these crystallized in the terrestrial crust. Of course it would be interesting to determine whether these initial ratios in rocks were higher or lower than those of CHUR at the relevant times. The observed differences in the isotope ratios are very small and, for this reason, DePaolo and Wasserburg (1976) introduced an epsilon parameter which they defined in the following manner:

$$\epsilon^{t}_{\mathrm{CHUR}} = \left[\frac{(^{143}\mathrm{Nd}/^{144}\mathrm{Nd})_i}{I^{t}_{\mathrm{CHUR}}} - 1 \right] \times 10^4 \tag{11.4}$$

$$\epsilon^{0}_{\mathrm{CHUR}} = \left[\frac{^{143}\mathrm{Nd}/^{144}\mathrm{Nd})_{\mathrm{measured}}}{I^{0}_{\mathrm{CHUR}}} - 1 \right] \times 10^4 \tag{11.5}$$

The parameter $\epsilon^{0}_{\mathrm{CHUR}}$ is calculated from (11.3) and (11.4) by substituting the initial $^{143}Nd/^{144}Nd$ ratio of the rock obtained from a whole-rock isochron and the same ratio of CHUR, i.e. I^{t}_{CHUR}, caculated for the time t from equation (11.3) derived from the whole-rock isochron. In an analogous fashion, $\epsilon^{0}_{\mathrm{CHUR}}$ is calculated from the measured $^{143}Nd/^{144}Nd$ ratio of the rock sample and the present value of this ratio in CHUR, i.e. $I^{0}_{\mathrm{CHUR}} = 0.512\ 638$ (cf. Table 11.2, row 2). Therefore, I^{t}_{CHUR} expresses the difference between the initial $^{143}Nd/^{144}Nd$ ratio of a rock suite and the corresponding value of this ratio in CHUR when this crystallized, and $\epsilon^{0}_{\mathrm{CHUR}}$ provides a similar comparison valid today. Consequently, a positive value for epsilon shows that the rock samples came from residual solids in the reservoir subsequent to earlier magma withdrawal, such portions of the reservoir being depleted in the large ion lithophile (LIL) elements which are partitioned preferentially into the liquid phase in the course of partial melting. On the other hand, a negative value for ϵ^{t}_{CHUR} demonstrates that the rock samples concerned originated in sources with a lower Sm/Nd ratio than the chondritic reservoir and either stemmed from or incorporated old crust of which the ratio had been lowered on separation from CHUR. One final case is possible: a zero value for the parameter. Here, the isotope composition of the Nd in the rock samples cannot be distinguished from that obtaining in the chondritic reservoir–hence, they probably originated directly from that reservoir.

Clearly, ϵ^{0}_{CHUR} has several advantages. First, it facilitates comparison of the isotope compositions of Nd in rocks with CHUR, thus aiding their interpretation. Then, its value is independent of the convention utilized in order to normalize the isotope ratios of Nd. These conventions must be used carefully; isotope ratios consistent with row 2 of Table 11.2 are comparable solely with $I^{0}_{CHUR} = 0.512\,638$ in calculations of the epsilon values, and those normalized following row 1 are applicable only to $I^{0}_{CHUR} = 0.511\,847$.

CHUR may be employed in order to calculate the time at which the Nd in a crustal rock probably separated from the chondritic reservoir. The various times constitute model dates, which are significant when the Sm/Nd ratio of the samples did not alter after the separation event took place. It must be remarked that this requirement appears to be satisfied better by this Sm/Nd ratio than by Rb/Sr. No doubt this reflects the similarity in geochemical properties between the REEs, which does not exist with the alkali metals and the alkaline earths. Usually, the Sm/Nd ratio in the rocks in question does not change by metamorphism or erosion and redeposition. In consequence, model dates represent estimates of the residence time of Nd in a rock sample in the crust.

The trends of Sr and Nd isotopes through time near the Japan Sea in north-eastern Japan were examined by Nohda and Wasserburg (1986). They mentioned that since preliminary work of DePaolo and Wasserburg (1977), the Nd and Sr isotope compositions of practically all major island arcs have been reported in the hope that the relevant data would assist identification of the magma source materials in such environments. A previous exercise in 1981 by the same researchers on the Japanese islands revealed two definite trends, namely a regular decrease in ϵ_{Nd} from the Pacific side of Myoko in the central section (Hakone–Asama–Eboshi–Myoko) with a corresponding increase in the ϵ_{Sr} values, and a reverse trend in the traverse across north-eastern Japan (Nachimantai–Moriyoshi–Kampu). Three volcanic rock samples from the volcanic front to the back-arc basin along 40°N showed a regular increase in ϵ_{Nd} from 2.0 to 8.3 and a decrease in ϵ_{Sr} from −6.1 to −25.6. The most westerly site sampled was the Kampuzan volcano on the Japan Sea, a location of recent volcanism. Both the Nd and Sr isotopic signatures demonstrate that it has oceanic mantle characteristics. Interestingly, Notsu (1983) confirmed the regular decrease from the volcanic front to the Japan Sea through analysis of Sr isotope compositions from appropriate other Recent volcanics. The regular isotopic variations across the Japanese island arc might correspond to lateral compositional variations in the magma source region.

Nohda and Wasserburg (1986) noted that Nd and Sr isotopic analyses were effected on selected volcanics of known stratigraphic age from a restricted area in the so-called 'Green Tuff' region of north-eastern Japan as well as from adjacent areas. The actual ages of these volcanics range from 27 Ma to 6 Ma and they vary in ϵ_{Nd} from -0.9 to 5.7 and in ϵ_{Sr} from 14.1 to -16.3. The isotopic data plot generally along the region of the 'mantle array' and, when plotted against time, are found to be a regular function of geological age. The results indicate that the volcanics in question comprise part of a 'related' sequence or evolutionary pattern covering a time interval of 27 Ma, and also that this proceeds from an initially enriched source to that of a very depleted one. The shift in isotope composition seems to reflect the ongoing lining of magma conduits or the decreasing extraction of lower crustal contributions with increasing infiltration of oceanic magma types in an environment of crustal thinning. Two different stages were proposed, based upon the geotectonic development of the area due to the back-arc spreading of the Japan Sea. The first is a pre-opening stage (>15 Ma) and the second a post-opening one (<15 Ma). The volcanic rocks of the former are characterized by an enriched isotopic signature, while those of the latter possess a depleted signature. The host rocks of the famous Kuroko ore deposit lie within this evolutionary trend and contain a substantial quantity of continentally derived materials processed through an oceanic environment. Generally, these authors inferred that the degree of continental crust involvement is great during the pre-opening stages of a back-arc basin when it is associated with a rifted continental mass.

Palmer and Elderfield (1985) made important observations on the Nd composition of foraminifera from Atlantic Ocean sediments. They stated that the mean oceanic residence time of dissolved Nd is less than the mixing time of the planetary oceans and that this is shown by the non-uniform $^{143}Nd/^{144}Nd$ ratios of sea water and hydrogenous sediments, although individual oceans possess distinct Nd isotope signatures. However, within each basin, the isotopes are relatively well mixed. The $^{143}Nd/^{144}Nd$ ratio in the Pacific is higher than in the Atlantic, and the Indian Ocean has intermediate values. These differences were interpreted in terms of the relative proportions of Nd supplied to individual ocean basins from mantle (with high $^{143}Nd/^{144}Nd$ ratio, ϵ^{0}_{Nd} MORB $= +10$) and crustal (low $^{143}Nd/^{144}Nd$ ratio, ϵ^{0}_{Nd} average crust $= -12.4$) sources, the Nd in the Atlantic deriving practically entirely from the continents in contrast to the situation in the Pacific which contains a significant quantity of Nd coming from more radiogenic basalt sources.

The authors intended to evaluate the variability of Nd isotopes within an ocean basin. This was considered to be important in order to determine the limits of the oceanic residence times of Nd and other REEs as well as to discover whether distinct isotope provinces exist for Nd in hydrogenous deposits, as has been found for Pb. These could have implications in deduction of the sizes of ocean basins in the past, and in consideration of recycling of subducted sediments into the mantle. Hydrogenous sediments, because they contain higher concentrations of Nd than the oceans, were sampled, and foraminifera were selected since carbonate-rich sediments represent the most widely distributed sediment type in the Atlantic. This permits the Nd isotope ratios to be mapped in greater detail than is feasible using more areally restricted ferromanganese nodules and hydrothermal crusts. To a first approximation, calcareous sediments are usually concentrated in different areas of the oceans from ferromanganese deposits. Also, foraminifera were expected to yield a more easily interpreted Nd isotope signal than nodules and crusts. A detailed study of REEs in foraminifera, however, clearly showed that it is not justifiable to assume that foraminiferal $^{143}Nd/^{144}Nd$ ratios merely reflect the Nd isotope compositions of the sea waters in which these organisms acquired their calcitic tests. After removal of adherent clay, it was noted that most of the REEs in foraminifera and 90% of Nd do not occur in the calcite lattice at all, but are situated in the Fe, Mn-rich authigenic phase which coats the tests. The origin of REEs in this coating phase is not yet sure. Probably Nd derives mostly from pore and bottom waters, as is suggested by the mobility of the REEs during diagenesis. Samples were obtained from the top 1 cm of carbonate-rich sediments and the raw material was sieved ($> 170\,\mu m$ portion retained), the foraminifera being separated by hand picking. Tests were crushed and then ultrasonically cleaned alternately in quartz-distilled (QD) H_2O and QD methanol until the supernatant remained clear. The samples were further examined under the microscope in order to make sure that all visible adherent debris was removed before dissolution overnight in 25 ml of QD 0.1 M HCl. Typical sample sizes were 100–250 mg of $CaCO_3$ (200–500 ng of Nd). Where enough sample sizes were available, REE, Ca, Fe, Mn, Cu, PO_4 and Al concentrations were determined. The Nd isotope ratios were assessed using standard mass spectrometric techniques following the chemical separation of Sm and Nd. All ratios were normalized to a $^{146}Nd/^{144}Nd$ ratio of 0.7219. About 600 $^{143}Nd/^{144}Nd$ ratios were obtained for each sample. The La Jolla standard gave a $^{143}Nd/^{144}Nd$ ratio of $0.511\,895 \pm 14$ ($\epsilon^0_{Nd} = -14.5$) and sample blank ratios generally exceeded 2×10^4. Some of the results are quoted in Table 11.3. No significant relationships were noted

Table 11.3 Neodymium isotopic ratios[a]

Core	*Location (lat., long.)*	*Water depth (m)*	$^{144}Nd/^{144}Nd$	$\epsilon^{0}_{\mathrm{Nd}}$	$^{148}Nd/^{144}Nd$
9940K	34.2°S, 10.0°E	4 980	0.512 248 ± 16	−7.7 ± 0.3	0.241 563 ± 11
KNR 54-6-100 27BC	63.0°N, 14.2°W	1 526	0.512 355 ± 18	−5.6 ± 0.3	0.241 580 ± 22
CHN 82-6-26 6PG	44.2°N, 31.6°W	3 132	0.511 968 ± 22	−13.1 ± 0.4	0.241 569 ± 17
CHN 99-3-15 12PC	4.6°S, 19.1°W	4 353	0.512 146 ± 16	−9.6 ± 0.3	0.241 578 ± 15

[a] ^{148}Nd/ ^{144}Nd ratio (no mass fractionation) = 241 578 ± 18.

between the Nd isotope composition and the concentrations of any of the elements mentioned above. This implies that the Nd isotope compositions reflect the $^{143}Nd/^{144}Nd$ ratios near the sediment–water interface in various parts of the ocean; that is to say these ratios are not due to sample-specific effects such as the degree of residual detrital contamination.

The results obtained by these researchers would appear to support the conclusion that continental input is the dominant factor in determining the isostopic composition of Nd in the Atlantic Ocean. The average ϵ^0_{Nd} value obtained (-10.1 ± 2.3, $n = 25$) does not differ significantly from that of average Atlantic atmospheric dusts and river particulates (-10.4 ± 2.6, $n = 14$), Atlantic ferromanganese deposits (-11.0 ± 1.3, $n = 11$) and Atlantic sea waters (-11.8 ± 1.5, $n = 21$). The Atlantic data demonstrated that ϵ^0_{Nd} is higher in surface waters than in deep ones. Of course, the results from sea water come from relatively few locations and therefore it is not possible to compare them with the much more widespread carbonate data. However, the implication is inescapable that the foraminifera and the ferromanganese deposits both contain Nd derived from the same part of the ocean system, perhaps reflecting mainly a benthic signal. Although most of the analysed foraminifera were planktonic, their Nd isotope compositions were practically wholly determined by their coatings. The greater range in the ϵ^0_{Nd} values (-14.5 to -4.5) compared with Atlantic ferromanganese deposits (-14.0 to -8.5) in part reflects both the larger data base and the greater geographic sample coverage. It also reflects differences in the relative importance of different sources of Nd recorded by the sample types. The foraminiferal coatings are in close contact with sediment pore waters so that the Nd isotope ratios recorded probably reflect point sources of Nd. In sharp contrast, the Nd associated with hydrothermal crusts is derived almost entirely from overlying sea water. Ferromanganese nodules may be described by two end members representing an hydrogenous and a diagenetic source; the $^{143}Nd/^{144}Nd$ ratios of those from the Atlantic and Indian Oceans suggest a dominantly diagenetic source, those from the western Pacific nodules implying a significant hydrogenous one. Not enough is known of Nd isotope compositions near the sediment–water interface to decide whether sea water has a narrower range of $^{143}Nd/^{144}Nd$ ratios than pore waters in a particular ocean basin.

Additional research on the geochemical balance of the REEs and Nd isotopes in the oceans was recorded by Bertram and Elderfield (1993). Noting that the isotopic composition of Nd in sea water is different in the Atlantic and Pacific oceans and also shows small variability within individual oceans, they observed also that the relative abundance of LREEs to HREEs too are different in these two oceans. Fractionation of REEs from one another is frequently shown by shale-normalized REE abundance patterns which, for sea water, are marked by a monotonic increase in HREEs concentration relative to LREEs plus the depletion of Ce relative to its REE neighbours. As the pattern is different from those found in river waters, aeolian dust or hydrothermal fluids, this is taken to indicate significant fractionation of the REEs between their source input signal and the oceans. The range of Nd isotopic compositions measured for sea waters implies that the dominant source is continental. Profiles of REE concentrations and Nd isotope measurements in the western Indian Ocean demonstrate a strong hydrographic control with the influence of intermediate water masses significant. The Nd isotopic compositions for this ocean was assessed as $\epsilon_{Nd(0)} = -8.3$, i.e. intermediate between the Atlantic and Pacific Oceans. Sm/Nd ratios are almost constant for all oceans, whereas Er/Nd ratios vary between all three. A seven-box model demonstrated that the interoceanic variations in Nd isotopic compositions and Nd concentrations are irreconcible unless particle–water exchange is used. The amount of exchange required is consistent with particle residence times of $\sim 1-10$ a. The concentration of particulate REEs is less than 5% of dissolved concentration, except for Ce where up to $\sim 20\%$ of total REEs occur as $> 0.4\,\mu m$ particles.

The Sm–Nd technique has been used in an effort to resolve the matter of the evolution of the continental lithosphere, for which two models have been proposed. One is a steady-state model entailing differentiation into a constant-volume crust early in Earth history associated with a depleted mantle and mostly recycling from then on. The other suggests that continents have grown either steadily or episodically through geological time. Clearly orogenic belts represent possible regions of continental accretion and growth. They usually consist of thick sedimentary sequences later deformed by orogenic events comprising metamorphism and magmatism. To constrain crustal growth models, it would be necessary to decide whether the evolution of such orogenic belts entails fresh additions from the mantle or just record crustal reworking. Turner *et al.* (1993) examined samples from suitable rocks from south-eastern Australia and the Adelaide Fold Belt with the objective of constraining source rock provenance regions and so assessing lithospheric evolution in this orogen. In South Australia, Late Proterozoic and Cambrian sediments accumulated in basins which formed in Archaean to Early Mid-Proterozoic basement and were later deformed and uplifted by the Cambro-Ordovician Delamerian orogeny to create the Adelaide Fold Belt. These researchers noted that, in order to utilize the fold belt in regard to models for lithospheric evolution, the mechanisms of basin formation must be understood and also whether the relevant sequences represent new crustal additions or the reworking of existing cratonic material. To assist in this, Turner *et al.* (1983) offered geochemical and isotopic data on these sediments and basement rocks so as to complement already existing information on the Delamerian igneous rocks.

These showed that the Archaean–Mid-Proterozoic basement rocks of the Gawler Craton have Nd-depleted mantle model ages averaging *c.* 2.6 Ga and Nd values at 800 Ma within the range −9 to −24 or at 500 Ma from −11 to −28. The Late Proterozoic Adelaidean sediments gave an average model age of 1.9 Ga and initial Nd values within the range −5 to −11. In fact, these authors divided the sequence into lowermost rift-phase sediments stemming from the erosion of local Gawler Craton material and later, widespread, thermal sag-phase sedimentation of which the isotopic signature does not accord with that of the basement. It may be inferred that a more juvenile (primitive) source was involved through a broadening of the provenance region. The Cambrian sediments have somewhat older model ages *c.* 2.1 Ga and lower Nd values of −9 to −13. They can be regarded as mixtures of basement and reworked Adelaidean detritus.

The Lachlan Fold Belt to the east is composed of Early Palaeozoic sedimentary sequences with isotopic compositions similar to those of the South Australian Late Proterozoic and Cambrian sequences. Their average model age is 1.85 Ga and their K_2O/Na_2O contents are considerably higher. Turner *et al.* (1993) took this to mean that the widespread flysch of eastern Australia was caused by recycling of Late Proterozoic and Cambrian sediments through erosion of the Adelaide Fold Belt (with modification due to the mixture of the Cambro-Ordovician mafic and felsic magmatic rocks in the region). They claimed that their data imply regional intracontinental crustal evolution with new additions from the mantle in the course of orogenesis as well as episodes of extension and sedimentation.

As noted above, mantle contributions are made to the crust, but does a reverse process occur? When formed, continental crust gains a much lower Sm/Nd ratio than that of its mantle source so that the rate of change of $^{143}Nd/^{144}Nd$ ratios is significantly less in the crust than in the mantle. Also, separation of the crust from the mantle effectively commences the isotope clock, crust–mantle differentiation apparently being the major process during which fractionation of Sm/Nd ratios takes place. Any subsequent reworking events affecting the crust, whether sedimentary, igneous or metamorphic, seem to have almost no effect on these ratios and therefore the memory of the initial crustal formation age becomes retainable. It is interesting that Rb–Sr measurements yield apparent crustal residence ages usually much younger than Sm–Nd model ages, particularly in the case of clastic sedimentary rocks. This mirrors the fact that Rb and Sr can be strongly fractionated by intra-crustal processes, espicially in the sedimentary cycle (weathering, erosion, transportation, deposition and diagenesis). Also, weathering processes can remove Sr in solution and concentrate Rb in clays (cf. the insolubility in water of both Sm and Nd). Clearly this leads to a marked fractionation of Rb and Sr and shifts radiogenic Sr from the continental crust to the ocean basins. Some of this can then be loaded onto the subduction conveyer belt through the hydrothermal interaction of sea water with oceanic basalts, or by low temperature reactions on the oceanic floor or by the deposition of pelagic carbonates. However, the available evidence suggests that the return of this Sr to the mantle may not take place to any great extent, for instance systematic loss of radiogenic Sr from the continental crust can account for the near-stagnation of $^{87}Sr/^{86}Sr$ ratios in sea water through most of the later history of the Earth and for the lowering of initial $^{87}Sr/^{86}Sr$ ratios of numerous rock units formed by crustal reworking. Because of this latter, the Rb–Sr technique results in underestimation of continental crustal residence ages which has important consequences for some continental growth models.

Continental crust (sial) is buoyant which will inhibit bulk return of it to the mantle, although the descending plate at subduction zones might convey some downwards. However, Taylor (1988) suggested that, for elements conveyed in sediments on the oceanic conveyer belt, i.e. Pb, Ce, Lu and Hf together with Sm and Nd, it is improbable that they are deeply recycled into the convecting mantle, the incorporation of subducted sediments into magmas for fast return to the crust through island-arc volcanism perhaps acting as a filter inhibiting such deep recycling.

Clearly, coupled Sm–Nd systematics constitute an important means for investigation of the geodynamic history of the Earth; in 1992 evidence suggested a 33 ppm difference of an Isua sample relative to a terrestrial standard. Jacobsen and Harper (1993) interpreted this to give a formation age range for the depleted mantle source reservoir of 4.45 Ga to 4.55 Ga, regarding this as consistent with a time of LREE-enriched melt extraction from the mantle (proto-crust formation) shortly following magma ocean freeze-up succeeding a putative Moon-forming giant impact. They proposed evaluating the rate of crustal recycling inferred from $^{143}Nd/^{144}Nd$ evolution in the depleted mantle through Earth history using a two-reservoir recycling model and the following equation:

$$\frac{\epsilon_{^{143}\mathrm{Nd}}}{\mathrm{d}t} = -\left(\frac{\varphi}{Y_{\mathrm{Nd}}}\right)\epsilon_{^{143}\mathrm{Nd}}/Q_{^{143}\mathrm{Nd}}\, f^{\,\mathrm{Sm/Nd}},$$

where φ is the fractional rate of recycling, Y_{Nd} is the mass fraction of Nd in the depleted mantle, $Q_{^{143}\mathrm{Nd}} = 25.13\ \mathrm{Ga}^{-1}$ and $f^{\,\mathrm{Sm/Nd}}$ is the Sm/Nd fractionation factor in the depleted mantle. To use this equation necessitates detailed knowledge of the $f^{\,\mathrm{Sm/Nd}}$ evolution of the depleted mantle, but their mantle source (of Isua 715–28) may be interpreted by means of a two-stage model having an initial time differentiation of 4.45 Ga to 4.54 Ga. The work of these researchers demostrated that the coupled geochronometer ($^{146,147}Sm/^{142,143}Nd$) is potentially valuable as a tracer of geodynamic mixing. Jacobsen and Harper (1993) noted that new results on Acasta samples do not support ^{142}Nd enrichments as a widespread effect in the early crustal rocks coming from inferred high ^{143}Nd sources. Clearly, the bulk Earth value for $^{142}Nd/^{144}Nd$ has to be established and their researches represent only the start of utilization of the coupled Sm–Nd systematics. However, they emphasized that all measurement problems can be solved through the development of better techniques and mass spectrometers. Hence they focused their efforts on producing extremely high-precision Nd mass spectrometry and regard the method as very promising for tackling a wide range of fundamental issues in global geodynamics.

Neodymium crustal residence ages were calculated for Upper Cretaceous Pierre Shale (Colorado) samples analysed to examine the possibility that this material (a main component of the uppermost strata at the Manson impact site (*c.* 65 Ma ago) was the source for dark brown to black Haitian tektites. This research was carried out by Premo (1993). He found out that the shale was not the source. Initial ϵ_{Sr} (64.5 Ma) values for four out of five samples of it lay between +200 and +300 (the exception giving a value of +80). These isotopic values are not similar to the initial Nd and Sr isotopic values are not similar to the initial Nd and Sr isotopic signatures (64.5 Ma) of the Haitian tektites (–2.5 to –2.9 and +52 to +62, respectively). Neodymium crustal residence ages calculated for the shale range from 1.23 Ga to 1.37 Ga possibly show significant contributions from the 1.3 Ga to 1.8 Ga old Rocky Mountain Front Range (this component is probably absent from the Pierre Shale in Iowa). Premo (1993) asserted that, if the Silurian Nd model age (400 Ma) is a maximum for the deposition for the target material at Manson, *c.* 6.22 km^3 of Palaeozoic and Mesozoic strata may have been vaporized during the event.

12

Radiation damage dating methods

12.1 ELECTRON SPIN RESONANCE

When dielectric solids are exposed to ionizing radiation, metastable electrons occur and can be detected by electron spin resonance (ESR). Measurements are possible by employing commercial ESR spectrometers which can sense the absorption of microwave radiation by the trapped electrons when they are subjected to a strong magnetic field, which promotes splitting of the energies of unpaired electrons. How this phenomenon takes place is dependent upon whether their magnetic moments are aligned parallel to or opposite the applied field. ESR dating is based on such measurements of unpaired electrons trapped in crystal defects. Such quasi-'free' electrons are produced through geological time by natural radiation arising from the decay of radioactive elements and also cosmic radiation.

The approach was first utilized by Ikeya (1975), who applied it to a stalactite. Later, he analysed the Petrolna stalagmite (also using fission track detection) and, in a third paper which appeared in 1978, summarized the approach as a dating method. Its advantages over conventional thermoluminescence dating are threefold. First, it is less sensitive to surface effects; hence the manner of grinding samples and their resulting grain sizes is not important. Secondly, since the concentration of traps containing the electrons in question is measurable without draining them, duplication of analyses is feasible and leads to greater accuracy. Finally, the lower sensitivity of samples simplifies their preparation, as noted above, and the measurements can be effected at room temperatures. The principal disadvantage of the method is that the relevant concentration of defects which contain unpaired electrons must be high and thus young calcite crystals of great purity which received only a small dose of radiation (< 10 Gy) are very difficult to date. The radiation exposure age of a powdered sample dated by the method is calculated from the relationship:

$$t = \text{total radiation dose/annual dose}$$

in which the annual dose is obtained from the measured concentrations of U, Th and K. Alternatively, the annual dose may be derived by the 'addition method', which utilizes a plot of ESR intensities induced by γ-ray irradiation from a ^{60}Co source.

In addition, Ikeya and Miki (1980) investigated human and animal bones; in succeeding years, their research embraced fossil shells, quartz grains in fault gouge and corals. Also, Shimokawa, Imai and Hirota (1994) dated early Pleistocene volcanic rocks by the use of quartz.

ESR dating without the determination of annual dose was suggested as a possibility by Debuyst *et al.* (1984). This depends upon the occurrence of two ESR centres of different thermal stabilities; the annual doses can be correlated with the thermal stability of an ESR centre and the equivalent dose (ED) which can be obtained from this centre. The approach was applied to mollusc shells by Grün (1985). He used well-dated aragonitic mollusc shells and ESR centres selected with g-values of 2.0018 and 2.0014. These shells normally displayed five ESR signals in the region of $g = 2.00$, g being given in coordinates of log I (where I is the ESR intensity) against annealing time.

The two values mentioned are sensitive to gamma irradiation. It was shown that the ESR signal with $g = 2.0018$ is superimposed on a broader peak with $g \simeq 2.0014$, with a satellite peak at $g = 2.0007$. Annealing experiments gave the following results:

1. the retrapping process of electrons from the centre with $g = 2.0018$ is of second-order kinetics;
2. the ESR signal $g = 2.0007$ decreases with first-order kinetics;
3. the total peak with $g \simeq 2.0014$ decreases at almost the same rate as the signal with $g = 2.0007$.

The signal with $g = 2.0018$ decreases at the commencement of annealing and does so very rapidly until the sharp peak is no longer definable. The rest of the signal in this region decreases at the same rate as $g = 2.0007$.

An unstable trap accumulates electrons according to the equation

$$N = N_0 \pm kD_0(1 - e^{-t/\tau}),$$

in which N is the number of trapped electrons, N_0 the number of these at time $t = O$, D_0 is the annual dose, k is the radiolytic sensitivity, τ is the mean life and t is the age. If N is proportional to the measured ESR intensity, *I*, rewriting gives

$$I = I_0 + ED_0\tau(1 - e^{-t/\tau}),$$

E being the ESR sensitivity. If $I_0 = 0$ (i.e. no ESR signal at the time of formation, the fundamental assumption for ESR dating), then

$$(ED)_t \propto E(I)_t .$$

A simplification for measuring the equivalent dose is

$$(ED) = D_0\tau(1 - e^{-t/\tau}) .$$

This model may be expanded to the general case so that an age can be determined directly when measuring the lifetimes and EDs of the traps, giving

$$(ED)_t = f(\tau)_t .$$

A mathematical treatment was provided by Debuyst *et al.* (1984). To date a mollusc shell through the unstable trap with $g = 2.0018$, the following assumptions are required.

1. The centre correlated with the ESR signal $g = 2.0014$ is stable over the whole dating range and retrapping can be neglected.
2. The alpha efficiency of both signals is equal. This is permissible within given uncertainties.
3. The steady-state density of spins at the centre with $g = 2.0018$ is well below the saturation level for this centre.

Normally, the age is calculated through the stable trap ($g = 2.0014$):

$$\text{Age} = ED_{14}/D_0$$

and

$$ED_{18} = D_0\tau_{18}(1 - e^{-t/\tau_{18}}) .$$

For $t \geqslant 5\tau$, ED_{18} is less than 1% from a steady-state value and the following expression is obtained:

$$D_0 \simeq ED_{18}/\tau_{18} .$$

The age then results from

$$\text{Age}(A) = ED_{14}\tau_{18}/ED_{18} .$$

This was utilized to give an age of 84 400a for the aragonitic mollusc shells under examination. The expected age was *c.* 125 ka, so this is obviously too low. The variation is explained as due to the fact that the signal height of $g = 2.0018$ does not display thermal equilibrium since, during the last five mean lives (*c.* 60 ka), the average temperature was lower than it is now as a consequence of the last glaciation. Thus, a temperature model has to be established in order to obtain an appropriate correction; this was achieved by Grün (1985b). It is based on various palaeoclimatic indicators such as the $\delta^{18}O$ record of the deep sea core V28–238, speleothem growth rates, average temperature data for July and specific details of the Mediterranean. Before dealing with this temperature correction, it must be explained that the 'mean lives' mentioned above are based upon the correspondence of the ESR signals $g = 2.0018$ of the sample in question to that of a sample of age 12 100a according to the recent annual average temperature. Now, during an interval of time t_x with the average temperature T_x, a given ED_0 from, a centre with τ_x decreases to

$$ED_d = ED_0(e^{-t_x/\tau_x}) ,$$

and, if K is the ratio *ED/ED* (at 16.2°C, the recent annual average temperature, in thermal equilibrium), then

$$K_d = K_0(e^{-t/\tau_x}) .$$

In the same interval, electrons are trapped, producing a dose $(ED)_n$. Normalized on ED (16.2°C in equilibrium):

$$K_n = \tau_x / \tau_{16.2}(1 - e^{-t/\tau_x}) ,$$

and therefore the temperature correction factor K_T at the end of the time interval is

$$K_T = K_d + K_n .$$

From calculations of K_T for each temperature interval (assuming $K_0 = 1$), it was observed that the measured ED_{18} is 1.97 times greater than it would be if ED_{18} showed the steady-state dose corresponding to the modern annual temperature. This gives for the age:

$$A = (ED_{14}/ED_{18})\tau_{18}K_T = 166\,300a .$$

This work of Grün (1985a) implies that mollusc shells can be dated by ESR without determination of the annual dose. The approach has the disadvantage that it does not take into account the temporal variation of the annual dose, e.g. if caused by a disequilibrium in the ^{238}U decay.

The uncertainty in the calculation based upon the climatic model is *c.* 10% (in this case, for the calculation, only the last five intervals are needed). The uncertainty in the estimation of τ is *c.* 18% C^{-1} and the error in determining ED_{18} and ED_{14} is *c.* 5% each. The total error due to all of these is between 20 and 30%, but this could still permit satisfactory results. However, there remains a more significant problem. It has been found that the determination of τ can lead to values which are too large or too small by a factor of 3. Of course the inference is that, so long as it is impossible to assess τ with smaller uncertainty, the method can yield only very imprecise ages. Nevertheless a reverse usage may be feasible, i.e. the method could be applicable to determining the temperature history at a site where, for instance, precise radiocarbon ages are available. Finally, the lifetime of an electron captured by a trap can be calculated according to the Arrhenius equation:

$$1/\tau = V_0(e^{-E_a/kT}),$$

in which V_0 is the pre-exponential factor, E_a is the activation energy, k is the Boltzmann constant and T is the temperature (K). The preceding caveat will obviously affect this calculation, therefore, and has to be borne in mind.

ESR dating has been applied to fault movement, again using various defect centres in quartz by Fukuchi, Imai and Shimokawa (1986). These researchers investigated the dates of such movements of the Itoigawa–Shizuoka Tectonic Line (I–STL) and Minobu Fault in the western South Fossa Magna, Japan. The reliability of the method was tested in two ways, namely by determining the ESR age of quartz veins thought to have intruded at the fault movement and also by determining the ESR dates of quartz grains included in the crushed material; various lattice defects in quartz grains contained in the same fault clay and quartz veins were also utilized in order to evaluate the degree of resetting. In addition to the most commonly used E′ centre, ESR signals of Ge, Al and OHC centres were employed. The Ge centre is a lattice defect in which a Ge ion replaces a Si site and traps an electron. The Al centre is a defect in which an Al ion replaces a Si site and traps a hole. The OHC (non-bonding oxygen hole centre) is a defect which has a hole on an oxygen that is presumed non-bridging. The fault in question constitutes a feature adjacent to the I–STL which itself is a boundary fault between the Neogene and Palaeogene, considered to have been active during the Quaternary, mostly during the late Pleistocene. All ESR dates obtained from these two faults at two localities fell into the range 0.30–0.55 Ma, suggesting that the centres were annihilated completely and that the age derived can be admitted as the actual date of the latest fault movement. Meanwhile, ESR dates of quartz vein and fault clay in other localities are divided into two ranges, namely that given above and a second, 1.5–2.0 Ma. This implies incomplete resetting. The dates of 0.30–0.55 Ma and 1.5–2.0 Ma might correspond with clino-unconformities observed in the Plio-Pleistocene sequence of the region studied.

Grün (1989) noted that restricting ESR dating to the Quaternary, i.e. the past 1.6 Ma or so, may be unnecessary since other researchers showed that two defects in quartz lasted for at least 1.5 Ga. As quartz is the most abundant mineral in the crust of the Earth, this work is the basis for a new geochronometer applicable to dating many types of geological feature through a substantial part of the planet's history. Until the recognition of the two ESR centres in the crystal lattice of quartz, it was believed that the upper dating limit of the technique was constrained by the thermal stability of the paramagnetic centres which persisted for time intervals of up to 10^7a (consequently confining it to the Quaternary). A different view is now permissible if certain queries can be answered satisfactorily. The defects alluded to are Schottky–Frenkel defects and estimating their number precisely is difficult because the characteristic ESR lines are frequently small and near the background signal. Also, the exact quantities of α-particle emitting isotopes have to be determined. The relevant isotopes occur in the radioactive decay chains of U and Th, but their concentrations in quartz are usually in the ppb range, again difficult to measure. Finally, the number of centres per unit of radioactive elements can vary according to the variety of quartz being investigated so that this dependency and saturation effects both should be examined. One further point – the defects appear to be stable for 10^9a, but this needs confirmation. Grün believed that these problems can be overcome and that soon ESR dating can be extended from the Quaternary to the entire geological timescale.

Rink *et al.* (1995) used ESR together with U-series dating in order to assess the age of over 850 bones from a number of hominids (Neanderthals) recovered from the Krapina cave near Zagreb. Many were broken, scratched and mixed with animal remains. A particularly interesting skull, that of a child of between 6 and 8 years, showed characteristics approximating conditions of early modern humans and another lacked any Neanderthal features and could be transitional to modern hominds. The dating results assigned some to *c.* 130 ka BP and were compatible with data from enamel and dentine of the teeth. Hence it is probable that some of the remains were deposited in the cave in the course of the last interglacial time interval. The significance of these findings is that the hominids in question are contemporaneous with South African discoveries claimed to represent the 'African Eve' hypothesis for the origins of *Homo sapiens sapiens*, which presents a scenario regarding Neanderthals as an evolutionary dead-end, modern humans emerging in sub-Saharan Africa and thereafter moving northwards into Europe comparatively late in the last ice Age. Of course, the unquestionable Neanderthals from Krapina had no regional continuity with modern hominids. But a multiregional model for modern human origins envisaging derivation from earlier people in Africa, Asia and Europe seems to be a distinct possibility in view of the ESR/U-series dating results from Krapina.

12.2 FISSION TRACK DATING

Energy is transferred from charged particles passing through a solid to its atoms, and a damage trail is left behind. Tracks of this type were noted by Silk and Barnes (1959) during the examination of irradiated solids using an electron microscope with high magnification (× 50 000). Later it was found that the tracks could be both enlarged and made visible under the optical microscope by etching with an appropriate chemical solution; this reflects the preferential solubility of the damaged areas. The tracks have dimensions of about 10 μm and are formed by fission fragments of U which not only have a large mass individually, but also release a great quantity of energy amounting to *c.* 200 MeV during the process.

Interestingly, such tracks were found in natural samples of mica; in these and other natural minerals they are a consequence of the spontaneous fission of ^{238}U. It was proposed that their densities might be utilized for the dating of various types of mica such as biotite and muscovite which contain track densities up to 5000 cm^{-2}. Fleischer and Price (1964b) applied the approach to dating tektites and Libyan Desert Silica Glass, obtaining results consistent with K–Ar dates for the samples. A year later these researchers (with others) reported tracks in meteorites that were also attributed to the spontaneous fission of ^{238}U and, in addition, extinct ^{244}Pu (Fleischer *et al.* 1965; Fleischer, Price and Walker, 1965). However, their origin is a cosmic ray-induced fission coupled with a cosmic ray-induced spallation reaction, the components of such tracks comprising the nuclei of heavy elements. Subsequently such tracks have been found also in lunar samples and they can act as a means for the recording of heavy cosmic rays (e.g. Rajan and Tammane, 1982).

The fission track method is extensively utilized in dating a great variety of minerals as well as both natural and synthetic glasses. It is particularly suitable for fairly young samples which have not been reheated since formation; this is of interest in archaeology as well as geology. Other applications include the determination of uranium in crystals and in natural waters as well as in deep sea sediments. The method involves preparatory work on samples of glass or mineral: this comprises initial grinding followed by polishing and etching with a suitable solvent, with appropriate selection of temperatures and duration of treatment (Table 12.1). Small grains of sample minerals can be mounted in epoxy resin on a glass slide and ground and polished. After etching, the polished surface can be examined with a petrographic microscope (magnification up to × 1800) with a flat-field eyepiece with graticule so that counting of tracks in a specific area is feasible. They can be spotted by a characteristic tubular shape which etch pits deriving from imperfections do not possess, and their density can be determined by counting a statistically significant number. Difficulty may arise if this parameter is under 10 tracks cm^{-2}, but usually there are many more than this. Of course, the track density actually recorded reflects the time interval during which these features accumulated and also the U concentration in the particular sample. This concentration can be assessed through counting of

Table 12.1 Etching procedures for volcanic glass and various minerals

Mineral	*Etching solution*	*T* (°C)	*Duration*	*Reference*[a]
Apatite	Conc. HNO_3–5% HNO_3	25	10–30 s	1
Epidote	6g NaOH + 4 ml H_2O (37.5 M NaOH)	159	150 min	2
Muscovite	HF (48%)	20	20 min	3
Sphene	Conc. HCl	90	30–90 min	4
	1HF : 2HCl : $3HNO_3$: $6H_2O$	20	6 min	5
Volcanic glass	HF (24%)	25	1 min	6
Zircon	100 M NaOH	270	1.25 h	7

[a] 1. Naeser (1967); 2. Wagner and Reimer (1972); 3. Gleadow and Lovering (1974); 4. Bar, Kolodny and Bentor; 5. Lakatos and Miller; 6. Lal *et al.*, (1968); 7. Burchart, Dakowski and Galazka (1975).

fission tracks produced by the induced fission of ^{235}U as a result of irradiation of the sample with thermal neutrons in a nuclear reactor in a procedure described by Hurford and Green (1982). This is the so-called 'zeta' method of calibrating fission track dating against other radiometric dating techniques, and is discussed (with others) below. Here it is necessary only to mention that alternatives include destroying the spontaneous fission tracks by heating the sample to promote annealing, thereafter exposing it to thermal neutrons so as to produce new tracks through the induced fission of ^{235}U. Later, a new surface can be polished and etched. Following this, the density of the induced tracks can be measured. If the effective thermal neutron flux and the length of time involved in the irradiation are known, it is possible to determine the U concentration of the sample from the noted density of the induced fission tracks.

Another approach is to register such induced fission tracks in an external detector pressed against the same surface in which the spontaneous tracks are counted. After neutron irradiation, this detector, which comprises a sheet of low-U muscovite or even a suitable plastic, is etched and induced fission tracks counted. Of course, this approach has the advantage that it eliminates difficulties arising from the irregular distribution of U and also from the alteration in etch behaviour stemming from the annealing of the mineral prior to the neutron irradiation. Naturally, the induced tracks in the external detector are produced solely from U atoms within the sample (2π geometry) in contrast to the spontaneous tracks in the polished surface formed by U atoms situated on both sides of that surface (4π geometry). Hence a correction factor must be introduced in order either to reduce the recorded density of the spontaneous tracks to about half, or to increase the U concentration by a factor of two. The size of this factor has been discussed by Reimer *et al.*, (1970), who queried the utilization of external detectors, and by Green and Durrani (1978), who favoured them. In practice, such external detectors are widely employed.

The neutron dose (flux density × length of irradiation) is assessed using various flux monitors. One technique is to include foil pieces of Au or Cu of known weight with the mineral sample undergoing irradiation. Afterwards the neutron dose can be deduced from induced gamma-ray activity of these monitors if the neutron capture cross-section and the efficiency of the counting system are known. It is also possible to irradiate glass with known U concentration (NBS series SRM 610 to 617 or Corning glasses U1 to U7) and count the tracks resulting from induced ^{235}U fission in an external detector. Nevertheless, the relationship between the neutron dose and the consequent track density in such U-bearing glass must be determined by independent calibrations and there has been found to be considerable scatter in it over a period of 6 years. It was concluded that measurement of the neutron dose is, in fact, extremely difficult and may have an uncertainty as much as $\pm 10\%$ of the absolute value. The greatest source of systematic errors in fission track dating of natural materials is undoubtedly the fading of tracks through the annealing of samples at high temperatures, so such dates must be regarded as 'cooling ages' which do not necessarily reflect the actual times of crystallization of the minerals involved in igneous or metamorphic rocks.

In order to apply the method, the selected mineral or glass must contain enough U to give a track density greater than 10 tracks cm^{-2} during the time interval from cooling of the sample until now. These tracks must be stable at ordinary temperatures for periods of time comparable with the age being measured. Also, the selected sample must be fairly free of defects and lattice ruptures, i.e. clear enough for the fission tracks revealed by etching to be counted. A final caveat – the distribution of U in the sample must be uniform, thus allowing the concentration of ^{235}U to be determined form the density of induced fission tracks in another part of the sample in question.

An adverse factor affecting the tracks is chemical weathering: its effects on retention of tracks in apatite, sphene and zircon have been investigated. Interestingly, while it reduces the date of apatite, it seems to have almost no effect on the other two minerals. The reduced date mentioned was in part caused by difficulty in identifying tracks in corroded crystals and in part the result of partial fading of tracks as a consequence of groundwater action. The apparent loss of tracks was to some extent compensated for by a lowering of the U concentration.

In deriving the age equation, it is necessary to bear in mind that a number of naturally found radioisotopes of high atomic number decay by spontaneous fission and hence are capable of producing fission tracks in materials such as minerals and glass. Nevertheless, it has been shown conclusively that such tracks arise mostly from the spontaneous fission of ^{238}U, so that the effects of ^{235}U and ^{232}Th may be neglected. Additionally, the induced ^{235}U fission caused by absorption of thermal neutrons produced by spontaneous fission of ^{238}U is not significant unless the mineral undergoing dating was actually embedded in U ore. The decay constant for spontaneous fission of ^{238}U, λ_f, is not as well known as could be desired, but the various alternative values are listed in Table 12.2.

If a mineral grain be considered in which ^{238}U atoms are evenly distributed throughout the volume, then the total number of decays of this radioisotope in a time interval t is given by

$$D = {}^{238}\mathrm{U}(\mathrm{e}^{-\lambda_a t} - 1)\,,$$

in which D is this number of decays events, ^{238}U is the number of atoms of this radioisotope occurring per cm^3 of the sample now, and λ_a is the decay constant of ^{238}U for alpha emission ($1.55125 \times 10^{-10}\,\mathrm{a}^{-1}$). The fraction of ^{238}U that decays due to spontaneous fission, leaving fission tracks, is

$$F_s = (\lambda_f/\lambda_a)\,{}^{238}\mathrm{U}(\mathrm{e}^{\lambda_a t} - 1)\,,$$

and a certain proportion of such tracks will pass over the polished surface and be counted after etching. Therefore, the area density of spontaneous fission tracks in the surface in question is given by

$$\rho_s = F_s q = (\lambda_f/\lambda_a)\,{}^{238}\mathrm{U}(\mathrm{e}^{\lambda_a t} - 1)q\,, \qquad (12.1)$$

q being the proportion of tracks crossing after etching.

The number of ^{235}U fission events induced by thermal irradiation per cm^3 of samples is

$$F_i = {}^{235}\mathrm{U}\phi\sigma\,,$$

where ^{235}U is the number of atoms of this radioisotope per cm^3 of sample now, ϕ is thermal neutron dose (neutrons cm^{-2}) and σ is the cross-section for induced fission of ^{235}U by thermal neutrons (580.2×10^{-24}

Table 12.2 Values of λ_f

$\lambda_f(a^{-1})$	*Reference*
8.4×10^{-17}	Gentner, Storzer and Wagner(1969)
$8.46 \pm 0.06 \times 10^{-17}$	Galliker, Hugentobler and Hahn (1970)
$8.46 \pm 0.06 \times 10^{-17}$	Storzer (1970)

cm^2). The fraction of induced tracks crossing an interior surface and being counted after etching is also equal to q, if the U atoms are indeed evenly distributed throughout the sample and the same etching procedure is applied. The density of the induced tracks is then

$$\rho_i = F_i q = {}^{235}\text{U}\phi\sigma q\,, \tag{12.2}$$

and, combining equations (12.1) and (12.2), the following expression is obtained:

$$\rho_s/\rho_i = \lambda_f/\lambda_a(e^{\lambda_a t} - 1)\Phi\sigma I\,, \tag{12.3}$$

in which I is the atomic ratio of $^{235}\text{U}/^{238}\text{U}$. Hence an age equation can be derived, namely

$$t = \frac{1}{\lambda_a}\ln\left[1 + \left(\frac{\rho_s}{\rho_i}\right)\left(\frac{\lambda_a}{\lambda_f}\right)\phi\sigma I\right].$$

By substituting $\lambda_a = 1.55125 \times 10^{-10}\ \text{a}^{-1}$, $\lambda_f = 8.46 \times 10^{-17}\ \text{a}^{-1}$, $\sigma = 580.2 \times 10^{-24}\ \text{cm}^2$, $I = 1/137.88$, then

$$t = 6.446 \times 10^9 \ln\left[1 + 7.715 \times 10^{-18}\frac{\rho_s}{\rho_i}\phi\right].$$

This is soluble for t after measurement of ρ_s/ρ_i and the neutron dose ϕ from the monitor.

In cases where $t < 500$ Ma, the equation can be simplified by setting $e^{\lambda t} - 1 \simeq \lambda t$ when $\rho_s/\rho_i = \lambda_f t/\phi\sigma I$. Then

$$t = \frac{\rho_s}{\rho_i}\frac{\phi\sigma I}{\lambda_f} = 4.974 \times 10^{-8}\phi\frac{\rho_s}{\rho_i}\,. \tag{12.4}$$

Dates calculated from this equation are about 4% too low for $t = 500$ Ma and about 0.7% too low for $t = 100$ Ma. The analytical uncertainty, on the other hand, is usually 5% or more. Equation (12.4) is a means of adjusting the neutron dose so that the density of induced tracks roughly equals that of the spontaneous fission tracks and, if $\rho_s/\rho_i = 1$, then

$$\phi = 2 \times 10^7 t$$

where t is in years. In order to minimize errors in measuring the ρ_s/ρ_i parameter, it is necessary to equalize the densities of both the induced and the spontaneous tracks.

Turning again to zeta calibration, the need for this procedure arose from certain systematic errors stemming from the uncertainty of the decay constant for spontaneous fission of ^{238}U, i.e. λ_f, which affects fission track dates. It also relates to the problem of measuring ϕ, the neutron dose, which must be known in order to determine λ_f by counting fission tracks in U-bearing glasses of known age of manufacture.

To circumvent the problem, Hurford and Green (1982) proposed calibrating the fission track dating method against minerals of known age which had been dated using alternative methods. Thus, in equation (12.3), the parameters of dubious magnitude are combined into a single calibration factor, zeta **Z**:

$$\mathbf{Z} = \frac{\phi \sigma I}{\lambda_f},$$

and the age equation for a sample of unknown age becomes

$$t_{\text{unknown}} = \frac{1}{\lambda_a} \ln \left[1 + \left(\frac{\rho_s}{\rho_i} \right)_{\text{unknown}} \right] \mathbf{Z} \lambda_a . \tag{12.5}$$

The value of the factor **Z** is established by irradiation of a mineral standard of known age with every batch of unknown samples. Later, **Z** is calculated from the measured ratio of the spontaneous and induced track densities of the standard. From the combined equation:

$$\mathbf{Z} = \frac{e^{\lambda_a t_{std}} - 1}{\lambda_a (\rho_s / \rho)_{\text{std}}},$$

The value of **Z** derived is used in calculation of dates for unknown samples which were irradiated with the standard; the consequent fission track dates depending upon accurate knowledge of the age of the standard, the determination of the fission track densities of both the sample and the standard by the same approach, and irradiation of both sample and standard simultaneously.

If the age of the standard is considerably less than the half-life of ^{238}U, then it is permissible to simplify matters by equating $(e^{\lambda_a t_{std}} - 1)$ to $\lambda_a t_{std}$. Then

$$\mathbf{Z} = t_{\text{std}} / (\rho_s / \rho_i)_{\text{std}},$$

and the fission track date of unknown samples becomes

$$t_{\text{unknown}} = \frac{1}{\lambda_a} \ln \left[1 + \frac{(\rho_s / \rho_i)_{\text{unknown}}}{(\rho_s / \rho_i)_{\text{std}}} \lambda_a t_{\text{std}} \right].$$

It is axiomatic that minerals which are to be employed as calibration standards in this technique must have cooled quickly and give concordant dates by other suitable isotope methods. Of course, they must be easily obtainable, in adequate quantities and with sufficient purity. The best geological materials available appear to be rapidly cooled volcanic rocks which have not been changed after crystallization.

Hurford and Green (1983) used zircons from such rocks and from kimberlite pipes in South Africa. The suggested standards are as follows.

1. Zircon from Fish Canyon tuff, San Juan Mountains, Colorado, USA. Age = 27.9 ± 0.7 Ma.
2. Zircon from Tardree rhyolite, Antrim Lava Group, N. Ireland. Age = 58.7 ± 1.1 Ma.
3. Zircon from Bishop tuff, USA. Age = 0.74 ± 0.05 Ma.
4. Zircon from kimberlite pipes, South Africa (including the DeBeers, Bultfontein and Dutoitspan pipes). Age = 82 ± 3 Ma.

Alternatively, standard minerals can be employed in order to calibrate U-bearing dosimeter glasses, the neutron dose being related to the density of induced fission track in them (ρ_d) thus:

$$\phi = B\rho_d$$

in which B is a calibration constant and ρ_d is assessed using an external mica track detector. If this is substituted into equation (12.5) then the following expression may be derived:

$$t_{\text{unknown}} = \frac{1}{\lambda_a}\ln\left[1+\left(\frac{\rho_s}{\rho_i}\right)_{\text{unknown}}\left(\frac{\lambda_a}{\lambda_f}\right)B\rho_d\sigma I\right].$$

By redefining zeta using the symbol ζ:

$$\zeta = \frac{B\sigma I}{\lambda_f},$$

the age equation now becomes

$$t_{\text{unknown}} = \frac{1}{\lambda_a}\ln\left[1+\left(\frac{\rho_s}{\rho_i}\right)_{\text{unknown}}\right]\lambda_a\rho_d\zeta,$$

the new zeta being determined by irradiation of dosimeter glasses with mineral standards which have known ages. After this has been effected, the induced track density in the dosimeter, ρ_d, and in the standard mineral selected, ρ_i, are determined by simply counting tracks in the external detectors which are used with both. The value of zeta for any given glass dosimeter is calculable from equation (12.3) by replacement of ϕ by $B\rho_d$, thus

$$(\rho_s/\rho_i)_{\text{std}} = \lambda_f/\lambda_a\,(e^{\lambda_a t_{\text{std}}})-1)/B\rho_d\sigma I$$

and, inserting ζ as defined above, this becomes

$$\left(\frac{\rho_s}{\rho_i}\right)_{\text{std}} = \frac{(e^{\lambda_a t_{\text{std}}}-1)}{\lambda_a\zeta\rho_d},$$

so that finally

$$\zeta = \frac{(e^{\lambda_a t_{std}}) - 1}{\lambda_a (\rho_s/\rho_i)_{std}\,\rho_d}.$$

From the age equation given above, $t_{unknown}$ may be obtained by using the modified zeta parameter for a glass dosimeter as long as the latter is irradiated with the unknown mineral and ρ_d is determined by counting the resulting tracks in an external detector. Consequently, the zeta parameter facilitates the determination of a neutron dose directly from the density of such induced tracks in the dosimeter glass. Some relevant results obtained recently are included in Table 12.3, which shows that each of the dosimeters has a particular value for zeta, but the values obtained from South African kimberlites are elevated.

Green (1985) extended the zeta method to minerals other than zircon, namely apatite and sphene. Values of zeta for all three of these minerals were determined with respect to three different U glass dosimeters and reduced to a common scale using precisely determined conversion factors for the relative fission track response of each glass dosimeter. All the determinations demonstrated a high level of consistency. On a scale appropriate to SMR 612, overall weighted mean zeta (OWMZ) values of 381.8 ± 10.3 for zircon, 353.5 ± 7.8 for apatite and 320 ± 12.4 for sphene were obtained. These unexpected differences between the three minerals remain unexplained. However, the study illustrated the need for all fission track age determinations to be derived from a calibration scale based upon the use of age standards. Consistent fission track counting must be attained in controlled experiments before reliable ages may be determined on unknown samples.

It has been found that fission tracks are usually stable for long time intervals at room temperature, but fading occurs when the temperature is elevated because the process promotes curing of the relevant radiation

Table 12.3 Zeta values for U-bearing dosimeter glasses irradiated with Zr standards of known ages

Mineral standard (Zircon)	*Zeta (ζ) value of dosimeter glasses*		
	NBS SRM 612	*Corning Glass*	
		CN1	*CN2*
Bishop tuff	326 ± 19.0	107.1 ± 6.2	110.5 ± 6.3
Fish Canyon tuff	$335.1 + 3.0$	111.3 ± 2.3	118.9 ± 2.6
Tardree rhyolite	340.0 ± 8.4	111.9 ± 3.8	120.7 ± 4.3
Kimberlites, S. Africa	358.7 ± 8.5	122.2 ± 4.0	132.0 ± 4.1
Weighted means	339 ± 5.0	113 ± 2.6	121 ± 3.6

damage. Naturally, the rates of fading vary between various minerals and glasses, and therefore two different samples subjected to the same higher temperature over the same time span may yield discordant fission track dates. Nevertheless, these may give useful information regarding the thermal history of the cogenetic samples.

The following equation is applicable to the description of the fading of fission tracks in minerals and glasses as a function of rise in temperature and length of heating:

$$t = \mathrm{A}e^{\mathrm{u}/kT},$$

where t is the annealing time for a specific reduction in fission track density, A is a constant, U is the activation energy in units of kcal mol^{-1} or eV, k is the Boltzmann constant (1.380658 JK^{-1}) and eVK^{-1} is the absolute temperature (in K). Taking natural logarithms on both sides, the following expression results:

$$\ln t = \ln \mathrm{A} + \mathrm{U}/kT.$$

the equation of a straight line in coordinates of ln t and 1/T with a positive slope of U/k and an intercept on the y-axis of ln A. The linear relationship between ln t and 1/T allows laboratory measurements derived at high temperatures and short time intervals to be extrapolated to lower temperatures and short time intervals. Naeser and Faul (1969) provided empirical data from which it emerged that fission tracks in apatite and sphene, common accessory minerals in igneous and metamorphic rocks, fade at quite different rates. For instance, they calculated that, over a heating period of 1 million years, apatite starts losing fission tracks at *c.* 50°C and is wholly annealed at *c.* 175°C whereas fission tracks in sphene begin fading only when the temperature reaches 250°C and complete annealing is attained only at 420°C. Consequently, fission track dates of apatite may be completely reset by periodic heating under conditions suggested by its track fading curve for 100% track loss. Under similar conditions, sphene does not lose any fission tracks and hence its fission track date is unaffected. It may be inferred that, when concordant fission track dates are derived from samples of these two minerals collected from the same rock, the latter must have cooled rapidly and did not undergo reheating at a later stage. By contrast, when discordant dates are recorded, i.e. the sphene date exceeds the apatite date, the rock either cooled slowly or was reheated to a temperature sufficient to produce fission track fading in the apatite, but not in the sphene.

The mineral apatite appears to be especially valuable as an indicator of the cooling history of a rock since it retains fission tracks only at temperatures which are appreciably below the blocking temperatures of the Rb–Sr or K–Ar geochronometers in coexistent micas, the precise temperature at which all such tracks are retained depending upon the actual cooling rate. If this rate is high, the time of cooling is short and of course complete track retention takes place at a higher temperature than is the case where the cooling rate is low. At a particular cooling rate, the retention goes from zero to 100% as the temperature falls and therefore the fission track density is a measure of the time interval that passed after the temperature dropped through the value at which 50% of the fission tracks are retained. Therefore, it is possible to interpret such track dates as the time which has elapsed since the rocks cooled through their 50% retention temperatures, which are also termed closing temperatures, in parallel with the blocking temperatures germane to the retention of radiogenic ^{40}Ar and ^{87}Sr.

Figure 12.1 records closing temperatures of selected minerals at different rates of cooling and incorporates data from Haack (1977) Sharma *et al.* (1980) and Bal, Lal and Nagpaul (1983). From this it may be seen that there is a wide variation in closing temperatures, and that they increase with more rapid rates of cooling. Consequently minerals of plutonic igneous or high-grade metamorphic rocks which cool slowly demonstrate a sequence of dates at which the temperature passed

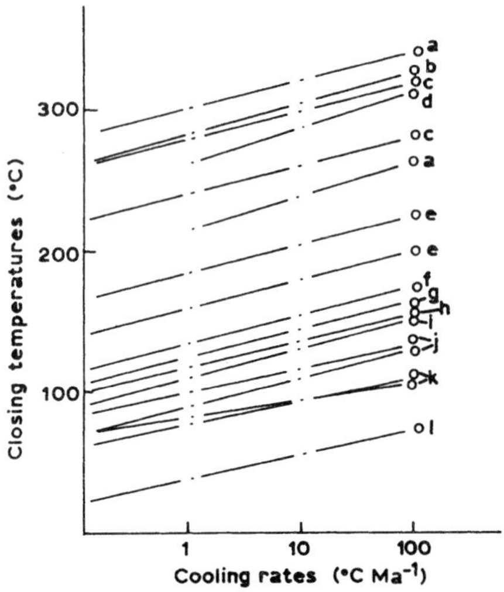

Fig. 12.1 Closing temperatures for the retention of fission tracks for minerals which cooled at different rates. The various closing temperatures for the different minerals reflect differences in the etching procedures used to show partially annealed fission tracks. a = sphene; b = garnet; c = epidote; d = zircon; e = allanite; f = vesuvianite; g = muscovite; h = hornblende; i = phlogopite; j = vermiculite; k = apatite; l = biotite.

through their respective closing temperatures. It may be inferred that, taking two cogenetic minerals such as apatite and sphene, differences between their fission track dates show the length of time necessary for the temperature to decline from the closing temperature of sphene to that of apatite. In this manner, using different minerals, the fission track dates and closing temperatures may be applied to reconstructing the cooling histories of both igneous and metamorphic rocks.

Some large-scale discrepancies in closing temperatures, e.g. for epidote and vermiculite, may arise from the different conditions of etching employed to expose tracks during the annealing experiments as a result of which the closing temperatures were assessed. Thus, for instance, HCl, NaOH and other etching agents are not equally efficient in the retrieval of fission tracks which have commenced fading because of annealing. Consequently the empirically determined closing temperatures in fact depend upon the etching solution, its temperature and the duration of the exposure. Additionally, any variations in the chemical composition of the mineral undergoing etching can influence its response to any particular etching agent. It may be seen that fission track dating of minerals ought to be carried out by employing the same etching procedures as those used to determine their closing temperatures. In this way, errors in cooling rates obtained from such data can be avoided.

Of course, the differences in the closing temperatures of mineral samples do not depend upon the cooling rates: this is demonstrated in Fig.12.1, which shows that the relevant lines are almost parallel. Once the cooling rate has been ascertained, the temperatures corresponding with the fission track dates of the minerals can be determined from the relationship between the cooling rate and the closing temperature.

Fission track dating can be used in conjunction with estimated geeothermal gradients in order to produce information regarding the rate of uplife and erosion. This approach has indicated an average uplift rate of 500 m Ma^{-1} for the Alps during the past 10 Ma and 800 m Ma^{-1} in the Himalayas during the past 500 ka. Crough (1983) gave an average fission track date of 12.5 Ma for three apatites collected from the Huayna Potosi batholith in the Bolivian Andes, the data showing that these minerals cooled through their closing temperature at about that time. The batholith itself is stated to be 211 Ma old. Data provided in Fig.12.1 show that the closing temperature of this mineral at a cooling rate of $10^\circ C\ Ma^{-1}$ is approximately 95°C and hence the average time-integrated cooling rate will be

$$\frac{dT}{dt} = \frac{95-10}{12.5} = 6.8^\circ C\ Ma^{-1},$$

if it is assumed that the average annual temperature in the Andes in

Bolivia at 4625 m elevation is 10°C. The geothermal gradient here is between 20 and 40°C km^{-1} (average *c.* 30°C km^{-1}) and so the mean uplift (or erosion) rate may be obtained from

$$\frac{dx}{dt} = \frac{dT/dt}{dT/dx} = 6.8/30 = 0.226 \text{km Ma}^{-1} (226 \text{ m Ma}^{-1}),$$

in which T is the temperature (°C), t that time (Ma) which has passed since fission track retention and x the distance (km) below the surface. The Andean result indicates a slightly lower uplift rate than that manifested in other young mountain chains.

Finally, reference may be made to FT-plateau dating. This concept is aimed at obtaining correct fission track dates for fading, and involves the fact that a damage trail which is left by a charged particle anneals initially at its ends, i.e. the etched tracks become shorter. Techniques have been developed so as to shorten induced fission tracks by annealing under controlled conditions; e.g. the work of Mehta and Rama (1970) regarding muscovite. Now it is possible to employ the noted distribution of lengths of natural fission tracks in minerals to refine the interpretation of dates given by them, minerals with well-marked unimodal track length distributions providing reliable cooling ages. The partial annealing of tracks is apparent from the skewness of the length distribution and results in a fission track date that is too low (because some of the tracks may be irrecoverable by etching). Minerals which underwent reheating after cooling could have bimodal track length distributions, and therefore give systematically erroneous dates.

The plateau method of fission track dating tries to correct fission track dates for partial fading (cf. Storzer and Poupeau, 1973). The basis is that induced fission tracks start annealing either at lower temperature or in a shorter time than natural tracks which commenced fading in the environment; consequently there may be a change in the ratio of these two types of tracks when a mineral sample is annealed by stepwise heating in a laboratory. It is implicit throughout that all fission tracks, whether natural or induced, commence fading above a specific temperature θ_1. If the original material cooled rapidly and was not reheated, then the natural fission tracks commence fading at the same temperature as the induced ones and the ratio between them will be constant with increasing temperature of annealing. Likewise, the fission track date calculated from the ratio of track densities will also remain constant. On the other hand, when natural fission tracks are partly annealed initially, then their density diminishes more slowly than that of induced tracks so that the density ratio increases with increasing temperature. At a higher temperature θ_2, this ratio approaches a constant value so that dates calculated from it constitute a plateau. The plateau date so derived includes a correction for partial fading of the natural fission tracks during slow cooling and hence represents that time which elapsed subsequent to the cooling of the sample through the zero track retention temperature. The procedure used to date an aliquot is to anneal it completely and then irradiate it with thermal neutrons so as to produce induced fission tracks. Both treated and untreated grains are afterwards heated together for 1 h at stepwise rising temperatures and, after each such step, the grains are mounted in epoxy resin, polished, etched and both the natural and induced tracks are determined. The maximum temperature attained is that at which all the tracks are lost in 1 h. It is possible also to employ an alternative technique in which mineral samples are heated at a fixed temperature, but for increasing time intervals. This is an isothermal approach in which the results from both natural and induced fission tracks give a response similar to that obtained by the previous isochronal technique. Natural glass has been dated by the plateau method and so have some minerals.

Some recent advances of interest may be cited. First, the fission track evidence for Late Triassic oceanic crust in the French Occidental Alps was given by Carpena and Gaby (1984). They examined five zircons and an apatite from plutonic rocks of the ophiolitic sequence from the Piedmontese zone, obtaining ages which showed that the minimum age of the Alpine oceanic crust here is between 192 ± 6 Ma and 212 ± 8 Ma, i.e. Late Triassic to Liassic. They reached the conclusion that, at least in the external part of this zone, the temperature – duration conditions of the high-pressure Alpine metamorphism were inadequate to reset the fission track clock of the zircons significantly, which suggested a short-lived metamorphic event. Breitzman, Birnie and Johnson (1983) studied fission track ages of the Chagai intrusives in Baluchistan, Pakistan. Again, zircon and apatite crystals were used and fission track dating by the external detector method gave Oliogocene and Miocene ages. Although the younger fission track results may represent uplift ages associated with this terrane (previously believed to be Cretaceous to Palaeogene in age), the field evidence is stated to support strongly the interpretation that they represent ages of emplacement. Two phases of intrusive activity occured, one about 35 Ma ago and the second some 20 Ma ago. The composition of rocks from both of these phases ranges from granite to tonalite and there are areas of hydrothermal mineralization connected with both of them. Fission track dating of rocks west of the Chagai Hills (including the Saindak porphyry copper deposits) gave an age of *c.* 12 Ma.

As has been mentioned earlier, a recurrent problem in geochronology is how to interpret apparent ages representing a mixture of an original formation age and a partial overprint due to a later thermal event. The consequent age will be intermediate between the original one and that of the overprinting event in question. The difficulty in interpretation arises from the fact that an essentially similar pattern of ages could also derive from slow cooling from the original formation age without any necessity to postulate a later thermal event. The application of the rival overprinting and slow cooling models mentioned has stimulated much discussion; below about 350°C (the only range in which fission track dating gives useful information, in any case), it is in principle feasible to differentiate between these models, using the distribution of fission track lengths. As has been seen, their utility derives from three properties of spontaneous fission tracks:

1. each has a relatively constant initial length controlled by the energy liberated in the fission decay and the nature of the recording medium;
2. fission tracks become progressively shorter during thermal annealing;
3. each track is actually of a different age and has undergone a different fraction of the total thermal history of the mineral sample.

Consequently, in a U-bearing mineral which underwent a complicated thermal history, the lengths of the different groups of fission tracks will vary according to when they formed. In principle, therefore, the track length distribution, in conjuction with the apparent fission age, can be of great value in reconstructing the variation of temperature through time. Gleadow *et al.* (1986) reported on the distributions of such track lengths in apatite samples from deep drill holes where temperatures are now within the annealing zone for this mineral. The samples came from three petroleum exploration wells in the Otway Basin of south-eastern Australia which are up to approximately 3 km deep. It was found that, at subsurface temperatures of *c.* 100° C and over, the length distribution characteristically becomes very broad without a pronounced peak. The lengths of the longest tracks remain essentially constant at all stages of natural annealing observed and the pattern is similar to that found in laboratory annealing of spontaneous fission tracks in apatite from an outcrop sample of the same formation. The results showed that confined track lengths can be used as indicators of the type of thermal history that a sample experienced in the temperature zone of increasing track stability. In fact, the length distributions reported provide a basis for interpreting fission track ages which might be ambiguous otherwise. An apatite age resulting from a uniform slow cooling thermal history will have a broad, negatively skewed length distribution, while a bimodal distribution gives clear evidence of a two-stage history involving partial annealing by a discrete thermal event. The apparent fission track age associated with a bimodal length distribution will be a 'mixed' age intermediate between the original age and that of the later event.

12.3 PLEOCHROIC HALOES

Pleochroic haloes are bands or zones of discoloration surrounding U- and Th-bearing inclusions in minerals which manifest such radiation damage effects because of the interaction of α-particles with atoms in the relevant crystal lattices to produce either displacement of the atoms or their electrons from their equilibrium positions.

Such coloured aureoles were first described as occurring in accessory minerals over a century ago, by A. Michel-Levy in 1882, more than a decade prior to the discovery of radioactivity in 1896 by H. Becquerel. Interestingly, the phenomena occur not only in such coloured minerals as biotite, tourmaline and chlorite, but also in colourless ones such as muscovite and fluorite. The inclusions involved are centrally located with respect to the haloes and commonly comprise zircon, apatite or sphene. The correlation with radioactivity was established by J. J. Joly and O. Mügge in 1907.

Pleochroic haloes are composed of concentric rings which have radii ranging from 10 to 50μm. They are produced by specific α-particles emitted by the radionuclides occurring in the central inclusions; the radii record the kinetic energies of the corresponding α-particles. Rings form as a result of the fact that the ionizing power of such α-particles maximize near the end of their ranges in particular media. Possibly the optimal mineral in which to study them is biotite because of its perfect cleavage and also because radiation damage discoloration is easily seen in it. Although many attempts have been made to identify which radionuclides produce which rings with specific radii in this mineral, the matter is still not completely solved even though many examples are known. Controversy exists over giant haloes with diameters which may reach 110μm. Some of these are caused by energetic alpha particles emitted by isotopes of Po, for instance some with radii of *c.* 54μm may be produced by rather uncommon 10.5MeV α-particles of ^{212}Po, a daughter of Th. Haloes with greater radii have speculative origins. Thus Hashemi-Nezhad, Fremlin and Durrani (1979) suggested that they resulted from the diffusion of radioactive daughters of U and Th. Alternatively, they may arise due to the occurrence of residual primordial ^{244}Pu which emits α-particles with energies far exceeding 15MeV during spontaneous fission. Yet another idea is that the giant haloes are the result of the decay of super-heavy elements with atomic numbers up to 127.

Pleochroic haloes have been used to date minerals and rocks. This method depends upon the fact that the intensity of their colours increases with the alpha dose to which they have been exposed. The relationship between the two may fall into any one of three categories. In the first place, a normal stage exists in which colour increases linearly with radiation dose. Then there is a saturated stage wherein the colour attains a maximum density and does not show any further increase with increasing radiation dose. Third, there is an inversion stage in which coloration decreases with intensity of radiation dose – this result from extreme radiation damage. As might be anticipated, only those materials exhibiting the first stage can be used in age determinations. A number of assumptions have to be made: one is that the coloration really is a linear function of the alpha dose. A second is that concentrations of radionuclides emitting alpha particles have not altered except through decay processes. Consequently, in order to date a mineral by this method, it is necessary to measure the above-mentioned alpha activity of the inclusion at the centre of the pleochroic halo and also to assess the intensity of the colour of this feature. In addition, it is essential to know the relationship between the α dose and the colour intensity in the mineral to be dated. It is possible to establish this experimentally by irradiation of the mineral sample with alpha particles from an artificial source or by examining haloes in other samples of the mineral which are of known age.

The dating of minerals by examination of pleochroic haloes is therefore perfectly possible, but there are a number of drawbacks. One is that the haloes are to some extent erased by heating of the mineral during metamorphism. Second, the α activities in the inclusions are difficult to measure. Finally, there is some ambiguity surrounding the exact relationship between the α dose and coloration in biotite which, as noted above, is the optimum mineral for age measurement using pleochroic haloes. Consequently, the technique is practically unused nowadays.

12.4 THERMOLUMINESCENCE

When ionizing radiation impinges upon a dielectric solid, electrons may be stripped off the parent atoms of the latter and some can become trapped in metastable states which are associated with crystal lattice defects, thus retaining the excess energy received from the external source. Heating such solids to temperatures below incandescence liberates this energy as light emission because the trapped electrons resume their stable ground states. Such light emission constitutes thermoluminescence, which is independent of the normal incandescence shown by solids thermally activated to attain higher temperatures. In addition, it increases linearly with the radiation dose until saturation is reached. Once electron traps are drained by heating, the phenomenon is lost unless the specimen is irradiated again. Most thermoluminescence is attributable to $CaCO_3$ and $MgCO_3$ minerals which emit yellow to orange light and to Na and K feldspars which emit blue light.

In thermoluminescence dating, a sample is heated at increasing temperatures and its light output is recorded as a glow curve, the area under which is a measure of the radiation dose to which the sample was subjected since it was last heated to *c.* 450°C. The annual radiation dose is determinable from the concentrations of U, Th and K in the rock sample concerned, and the time which has passed since the last heating is given by

$$t = \frac{\text{accumulated radiation dose}}{\text{annual radiation dose}},$$

where t is the radiation exposure age, the accumulated dose being given by the glow curve. There is an alternative means of assessing the annual radiation dose, involving irradiation of the sample and observation of the consequent thermoluminescence. The quantitative application of the method became feasible through the use of a photomultiplier and amplifier by Daniels and Saunders (1950) to record the phenomenon in natural samples. Zeller, Wray and Daniels (1955) suggested its utilization in dating geological materials and showed that it can be induced by pressure and the recrystallization of crystalline solids. In 1957 Zeller, Wray and Daniels tried to use it for dating limestones, but with limited success because other (then better) isotope dating techniques competed with theirs. Subsequently, archaeological materials such as ceramics, glass, shells and flints have been dated by this method and even Pleistocene sediments have given valuable data.

Among important researches those of Valladas *et al.* (1986) must be mentioned. They investigated the lower rock shelter of Le Moustier in the Dordogne, France, bearing in mind that the chronology of the Neanderthal cultures, usually termed Mousterian, flourishing between ~100 ka and 40 ka BP is as yet not well established. This particular locality comprises perhaps the best stratigraphic sequence, incorporating ten main Mousterian layers and two early Upper Palaeolithic ones, the stratigraphy of which is defined in terms of the regional climatic chronostratigraphy obtained from the alpine glacial sequence. A recent proposal was to correlate this with the stratigraphy based on oceanic oxgyen isotope ratios, but the lack of any absolute datings of the Mousterian settlements made this comparison uncertain for the time interval in question. Valladas *et al.* obtained dates by thermoluminescence measurements of 34 specimens of burnt flint recovered from the upper 3m of the Mousterian deposits and also from the superimposed Upper Palaeolithic layer. The results ranged from 56 to 40 ka BP and permitted correlation of the cold sediment deposits of Le Moustier with the cold intervals of the oxygen isotope records of the Mediterranean Sea during their isotopic stage 3. It was concluded that TL (thermoluminescence) dating of burnt flints ought to make it possible to secure close correlation between the 'floating' chronology of the Mousterian epoch based upon stratigraphic evidence and the more absolute chronology derived from the isotope record of sea cores.

Other interesting work was reported by Wintle and Westgate (1986). They obtained TL dates for loess deposits above and below the Old Crow tephra at the Halfway House site near Fairbanks, Alaska, USA, these being found to limit the age of the tephra to 86 ± 8 ka. This was an acceptable result because previously obtained radiocarbon dates on wood and peat less than 6 m above Old Crow tephra indicated that this must be >60 ka old. The loess contains quartz and feldspar; the feasibility of using these for dating this aeolian sediment was considered by them. The TL signal is produced when the mineral grains are heated; of course, thermal untrapping and later recombination of electrons within the crystal lattice occur, the electrons themselves being originally produced by the effect of ionizing radiation (α, β, γ and cosmic) in the crystalline environments resulting from the decay of radionuclides in the U and Th decay chains as well as by ^{40}K.

Interestingly, the electrons can also be untrapped by exposure to light and it is this zeroing mechanism that permits the TL signal to be utilized in dating Quaternary sediments. For dating, it is necessary to use laboratory measurements to obtain a radiation dose equivalent to that dose which the sample received since it was last exposed to sunlight, i.e. when it was deposited. This is the so-called equivalent dose, ED, which is then combined with the annual dose rate (calculated from measurements of the radioactive content of the sample) to derive the actual age thus:

$$\text{Age(ka)} = \frac{ED(\text{grays})}{\text{annual dose rate (grays ka}^{-1})}.$$

The gray (Gy) is the derived SI unit of absorbed dose of ionizing radiation, i.e. the energy in joules absorbed by 1 kg of irradiated material.

So far, unfortunately, there is no unique laboratory methodology for obtaining the ED because of the multifarious grain sizes and mineral species involved in the individual sediments.

The mineralogical and sedimentological analyses of the loess were effected on two samples from above and below the tephra. Magnetic separation of the fine sand component was carried out by means of a Franz Isodynamic Separator, giving the percentage of magnetic fractions (mainly muscovite). Quartz predominated, but proved to be less suitable for dating than feldspar because its TL signal saturates at a lower dose level and also it is easily bleached by exposure to light. Luckily, feldspars are at least 20 times more TL-sensitive than quartz (Berger, 1984). The feldspar signal can be enhanced by employment of a Schott UG-11 ultraviolet filter in the detection system.

The ED value, for TL age evaluation, was obtained by using the above-mentioned ultraviolet filter and also a 16h preheat at 150°C to isolate the geologically stable part of the glow curve (Wintle, 1985). The total bleach method described by Singhvi, Sharma, and Agrawal was used. Figure 12.2 illustrates the natural TL glow curve resulting from radiation in the environment of the sample since the sediment was deposited and the glow curves given by identical samples which underwent additional laboratory irradiation. The reduced TL level obtained by optically bleaching the sample with an unfiltered sunlamp for 450 min is also shown (Fig.12.2a). In Fig.12.2b, the TL response at the 320°C glow-curve temperature is plotted as a function of the applied radiation dose, using the curves from Fig.12.2a. The extrapolation of this plot back to the bleached level provides the laboratory dose equivalent to the radiation received since deposition, i.e. the ED. In this particular case the sample has an ED of 347 ± 29 Gy.

If the laboratory bleaching it too severe, overbleaching of the TL signal may cause an overestimation of the age of the sample. To avoid this, a partial bleach method, as described by Wintle and Huntley (1980), can be used. In work described by Wintle (1973) this was applied to one sample, QTL 71H, and a bleaching time of 18 h was used, a Corning 3.74 filter being placed between the sunlamp and the sample to remove photons with wavelengths of less than 550 mm. The ED obtained was the same as that derived using the total bleach method, confirming that the loess was well bleached prior to deposition. No anomalous fading was exhibited by the samples. The TL ages were obtained by dividing the ED by the annual radiation dose rate, calculated from measurements

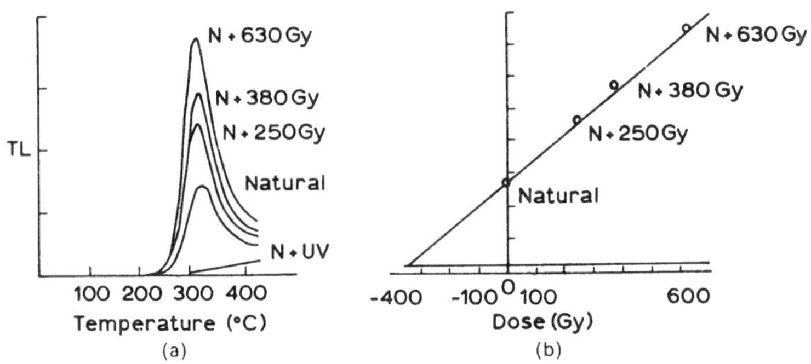

Fig. 12.2(a) TL glow curves for sample obtained from below the base of the Old Crow tephra which show the effect of additive doses using UG-11 filter and preheat treatment. One is natural, resulting from radiation in the environment since deposition. (**b**) TL growth curve for 320°C glow curve temperature taken from (a).

of the radioactive content of each sample. Bulk K content was measured by atomic absorption spectrometry and the quantity of ^{40}K calculated from it by use of the natural abundance ratio. The beta and gamma dose rate produced by the decay of ^{40}K can be calculated.

The U and Th decay chain contributions are assessed by thick-source α counting. The bulk α activity can be directly converted to dose rate if it is assumed that the decay chains are in secular equilibrium. The likeliest cause of disequilibrium is Rn loss: this can be checked by counting the samples twice, the second time sealed so as to prevent escape of the gas from the counting chamber. In fact, no increase in count rate was observed. The dose rate contribution due to alpha particles requires that lower TL sensitivity relative to beta and gamma radiation be measured. A cosmic ray contribution of 0.14 Gy ka^{-1} was also included. A correction factor was applied for water absorption of part of the radiation produced in the sediments. The value employed for the fractional water content, Δ = weight of water/dry sediment weight, was 0.20–0.05, the mean annual water content range for loess in temperate regions today. For the sample QTL 71A overlying the tephra, the TL age was 85 ± 7.9 ka. For those underlying it, QTL 71B gave a TL age of 86.1 ± 8.2 ka and two others provided dates statistically indistinguishable from the preceding ones. This implies relatively rapid deposition of the loess. The error is dominated by the uncertainty in past mean water content, the quoted value producing an error in the age of $\pm 5\%$. The age estimate determined accords with previous work and indicates that the tephra were deposited subsequently to the last interglacial. Also, it suggests that a palaeomagnetic excursion and reversal recorded below the tephra at the same locality represent the Blake Event. These researchers concluded that the TL method is suitable for late Pleistocene loess.

An investigation of the emission spectra of meteorites during thermoluminescence was made by Strain *et al.* (1986). Four ordinary chondrites of petrological type 5 were involved: a very broad emission structure, peaked near 470 nm and extending from 300 to at least 700 nm, was found for the glow peaks from *c.* 150 to 250° C. The higher temperature peak, 300–500°C, showed a different character with a narrow emission band near 400 nm. A tentative model for a luminescence site was proposed. Since the spectra of all four samples were in close accord, it is probable that both the trapping levels revealed at 170 and 350°C and the luminescence centres were not significantly perturbed by either their shock or thermal histories. These researchers made a tentative assignment of the luminescence site producing the lower-temperature broad emission feature as arising from Mn^{2+} ions in the oligoclase feldspar.

12.4.1 Induced Thermoluminescence Technique

Batchelor and Sears (1991) suspected that induced TL could be useful for investigating the metamorphism of eucrites since the technique has proved valuable in studying the metamorphism of chondritic meteorites. Some have called basic plutonites occurring among the Tertiary basic to ultrabasic complexes in Scotland eucrites, but this name is properly that of a type of meteorite. The Scottish rocks are mixed in composition; they are neither clearly defined gabbros nor norites, but combine the characteristics of both. Perhaps hypersthene-gabbro or augite-norite would be more suitable names. Induced TL has proved valuable in understanding the metamorphism of chondritic meteorites, these being essentially 'cosmic sediments', according to Batchelor and Sears (1991). In them the feldspar (which is responsible for the TL) is almost wholly of metamorphic origin. But in eucrites the feldspar is of igneous origin, although many of the meteorites are surface breccias and impact working may have converted some of the feldspathic material to glass. Possibly metamorphism could have caused recrystallization of the feldspars (depending on its timing in relation to the brecciation and the relative degrees of intensity of the two processes). If this happened, then the metamorphism can be surveyed using induced TL of the feldspar.

Batchelor and Sears (1991) selected interior chips of mass ~ 160mg from 18 eucrites (duplicates being taken where feasible) and these were homogenized by manual crushing in an agate mortar. Three 4 mg aliquots were put into shallow copper pans and drained of their natural TL by briefly heating to 500°C. That TL induced by a 15 TBq ^{90}Sr particle source was measured by heating at 7.5° C s^{-1} in TL apparatus equipped with blue and infrared bandpass filters (Corning 7–59 and 4–69). The data for three aliquots were averaged and the H3.8 ordinary chondrite was used as a normalization standard and instrument check. The intensity of the TL at maximum light production, i.e. the TL sensitivity, was plotted for each eucrite. Other researchers had divided the eucrites into 'equilibrated' and 'unequilibrated' groups and suggested that the former were metamorphosed, the latter not. A third group of workers had divided them into six types according to metamorphism. However, Batchelor and Sears found TL sensitivities to spread uniformly over a 17-fold range with no indication that the eucrites formed in two discrete environments, but rather in one which allowed a range of metamorphic conditions. They proposed a schema based on TL sensitivities and pyroxene characteristics under which eight types covered the observed range. While some of the differences between them no doubt reflect the mineral heterogeneity of the rocks, clearly there may be differences in metamorphism undergone by different clasts in a meteorite. The inference is that brecciation postdated metamorphism as, for instance, in the related howardite meteorites. Two of three cumulate eucrites in the Batchelor and Sears study appeared to have relatively young radiometric ages either due to formation in separate magma bodies or through particularly slow cooling. Increase in TL sensitivity in response to parent body metamorphism was ascribed to the increase in the degree of crystallinity of the feldspar. Cathodoluminescence petrography demonstrated that the feldspars in the Juvinas type 5 eucrite are brighter and more uniform in their luminescence than those in the Pasamonte type 2 eucrite. Surficial processes on the parent body may cause some of the feldspar to convert to glass, metamorphism reversing this. Of course, other mechanisms involving trace elements cannot be excluded. And with the howardites, changes in TL sensitivity were noted as correlating with bulk Al content. But these meteorites comprise mixed breccias, whereas the eucrites constitute breccias of one type of material.

Extra information regarding the metamorphic conditions undergone by the eucrites can be obtained from the temperature at which TL light production attains a maximum and relevant TL 'glow curves' (light production as a function of temperature as the sample is heated) for two representative eucrites were derived as well as one for an unusual eucrite. There were two peaks at *c.* 113°C and *c.* 192°C and these were taken to relate indirectly to ordered and disordered feldspar, respectively. Because all eucrites studied (except one) had a dominant peak at 113°C or had two peaks of comparable height, it was concluded that most eucrites cooled slowly below 800°C and that the exception had been heated to >1000°C. Ar–Ar data for this latter showed evidence of a major degassing event at *c.* 1 Ga ago and there were petrographic observations indicating that a shock event had affected it.

Batchelor and Sears (1991) have shown that induced thermoluminescence work could have wider implications than previously thought and that this approach is valuable in comprehending metamorphism in chondrites. Their feldspar has a novel history and origin. On the other hand, the eucrites were regarded as igneous rocks resembling terrestrial and lunar basalts. Obviously the success of their induced TL studies in assessing the thermal histories of eucrites implies that the technique could be applied to many types of basalt in which dry, low-level metamorphism occurred.

13

Cosmogenic radionuclides

The cosmic radiation which bombards the Earth causes various reactions within the atmosphere and, if it penetrates through, also in the hydrosphere and lithosphere. Of chief importance are the atmospheric reactions to which belong, for example, the production of the protecting ozone layer. These atmospheric reactions are a steady process and, although they depend possibly on changes in solar activity, the production and decay of cosmogenic radionuclides attains an equilibrium.

Cosmogenic radionuclides play an important part in tracing and dating water circulation above and below ground as well as in the oceans. Sediments can also be dated by means of cosmogenic radionuclides, which have been used in dating both groundwater and continental ice sheets. The most significant radionuclides involved in such researches are listed in Table 13.1

Cosmogenic tritium (T) is produced by several processes in the atmosphere, for instance by high-energy neutrons reacting with nitrogen ($^{14}N(nT)\ ^{12}C$). Tritium has a relatively short half-life of 12.3a, producing the stable ^{3}He atom by beta decay. The T inventory on Earth is estimated to be 1.8 kg of which about 11g are in the atmosphere. Basically, T could serve for dating in the same manner as ^{14}C but is mainly used for tracing methods. As T (and other radionuclides) is also produced by thermonuclear explosions and nuclear reactors, the T amount is naturally affected by these activities. Mainly used as a tracer, the T content of precipitation has been utilized to estimate the input of the radionuclide to groundwater systems and as well to investigate circulation patterns in water bodies, flow rates, vertical mixing and residence times of water in the oceans.

The element beryllium (named after the precious stone beryl and also called glucinum) was discovered by Vaquelin in 1798 and Arnold (1956) first detected ^{10}Be in marine sediment and noted its beta activity in BeO.

Table 13.1 Half-lives of some radionuclides and their appplication in geosciences

Radionuclide	$T_{1/2}(a)$	$\lambda(a^{-1})$	*Application in dating*
^{10}Be	1.5×10^6	0.462×10^{-6}	Marine sediment, manganese nodules, glacial ice, quartz in exposed rocks, terrestrial age of meteorites, petrogenesis of island arc volcanics
^{14}C	5730 ± 40	0.1209×10^{-3}	See Chapter 9
^{26}Al	0.716×10^6	0.968×10^{-6}	Marine sediment, manganese nodules, glacial ice, quartz in exposed rocks, terrestrial age of meteorites
^{32}Si	172(4)	0.4×10^{-2}	Biogenic silica, glacial ice
^{36}Cl	0.301×10^6	2.25×10^{-6}	Glacial ice, exposed volcanics, groundwater, terrestrial age of meteorites
^{39}Ar	269	0.257×10^{-2}	Glacial ice, groundwater
^{53}Mn	3.7×10^6	0.187×10^{-6}	Extraterrestrial dust abundances in ice and sediments, terrestrial age of meteorites
^{59}Ni	8×10^4	0.086×10^{-4}	Extraterrestrial dust abundances in ice and sediments, terrestrial age of meteorites
^{81}Kr	0.213×10^6	3.25×10^{-6}	Glacial ice, cosmic–ray exposure age of meteorites

Geochemically, Be has an atomic number of 4 and a valence state of +2. Eight isotopes exist with mass numbers ranging from 6 to 14 of which only ^{9}Be is stable. Two unstable isotopes, namely ^{7}Be and ^{10}Be, occur naturally and the result of cosmic ray-induced high energy spallation reactions with atmospheric nuclei in the upper atmosphere such as ^{14}N(n, ^{8}Li) ^{7}Be or ^{16}O(n, 4p3n) ^{10}Be. The cosmogenically produced Be reacts to BeO or $Be(OH)_2$ (both being toxic) and is incorporated in sediments as well as in continental ice sheets following the hydrologic cycle (Fig.13.1).

The application of ^{7}Be and ^{10}Be to dating was proposed by Arnold (1956) who also indicated that the ^{10}Be/^{9}Be ratio could be measured by means of solid-source mass spectrometry. Once formed, the cosmogenic Be isotopes are quickly removed from the atmosphere by rainfall, vertical diffusion, gravitational and various other precipitation processes. ^{7}Be undergoes radioactive decay by electron capture and produces the stable isotope ^{7}Li, the relevant half-life being 53.29 d. This fast decay is an asset in the examination of atmospheric processes, surface water in the oceans and the mixing of sediment in offshore marine and lake environments. ^{10}Be ($T_{1/2} = 1.5 \times 10^{6}$ a) decays to stable ^{9}Be by β-emission. Reaction cross-sections for ^{7}Be and ^{10}Be formation by spallation reactions between high-energy protons and O and N atoms have been assessed. The global production rate of ^{10}Be is estimated at between 1.5×10^{-2} and 3×10^{-2} atoms cm^{-2}s^{-1}, although a lower value has been derived from ^{10}Be concentrations in Antarctic snow by Raisbeck *et al.* (1978) and Reyss, Yokoyama and Guichard (1981). Nishizumi *et al.* (1989) determined absolute production rates for ^{10}Be, including the production due to nega-

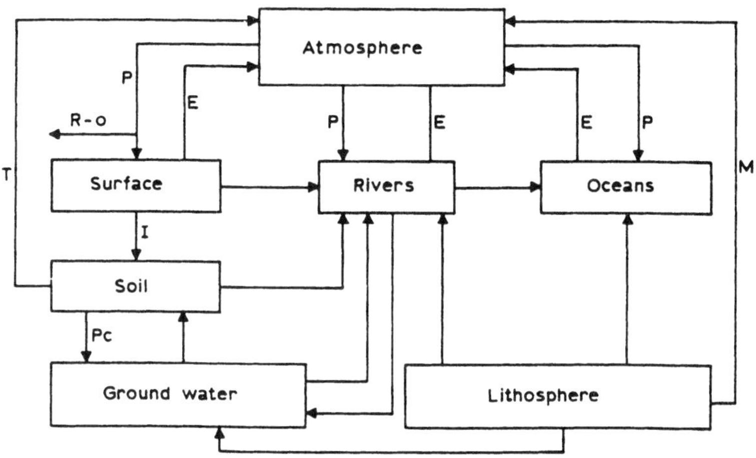

Fig. 13.1 Components of the hydrologic cycle (R-o = runoff, P = precipitation, E = evaporation, T = transpiration and Pc = percolation).

tive muons as $160(\mu^-, 3p3n)\ ^{10}Be$, using accelerator mass spectrometry (AMS) and quartz samples of exposed glacially polished rocks. The average production, also depending on such factors as altitude, and geomagnetic latitude was calculated from the measurements with 61.9 ± 3 atoms $g^{-1}a^{-1}$. The authors also calculated the values for ^{26}Al on this basis as 373.6 ± 27.8 atoms $g^{-1}a^{-1}$.

However, problems arose in using ^{10}Be for sediment dating because it has a low level of activity and its disintegration rate has been observed to vary sporadically with depth in cores since the associated abundances depend not just on time since deposition, but also on the rate of deposition and secular variations in their rates of production. The detection of ^{26}Al is even more difficult. Nowadays, the tandem accelerator mass spectrometry has facilitated ion counting and, together with cyclotrons, stimulated researches now being effected on smaller samples at greater levels of accuracy and more rapidly (Southon *et al.*, 1982).

The discovery of aluminium is attributed to F. Wöhler in 1827; the name derives from Latin *alumen* (alum) and *lumen* (light). The element Al has an atomic number 13 and 15 isotopes from $A = 23$ to $A = 37$, of which only ^{27}Al and ^{26}Al occur in nature. The former is stable, but ^{26}Al is radioactive and results from spallation of Ar in the atmosphere by cosmic ray protons, by interaction with neutrons as $^{28}Si(n, p, 2n)\ ^{26}Al$ and by reaction with negative muons as $^{28}Si(\mu^-, 2n)\ ^{26}Al$. The relevant reaction cross-section gives a production rate of $(1.1 \pm 0.3) \times 10^{-4}$ atoms $cm^{-2}s^{-1}$. ^{26}Al decays by positron emission (85%) and electron capture (15%) to stable ^{26}Mg, the total half-life being 0.716×10^6 a. Part of the ^{26}Al and ^{10}Be on the planetary surface may be associated with cosmic dust irradiated in deep space before it was deposited on Earth, but measurements of the disintegration rates of both in deep sea sediments and glacial ice from Greenland strongly imply that the fraction involved must be extremely small (Reyss, Yokoyama and Tanaka, 1976). As aforementioned, the production rates for the cosmogenic radionuclides vary with altitude, latitude and time (due to alterations in solar activity and the terrestrial magnetic field which affect the cosmic ray proton flux). Consequently, the abundance of any one of these radionuclides in a terrestrial reservoir is controlled by these and other factors, among which are the rate of sedimentation in the seas (if this is high, it will reduce the concentrations of radionuclides), bioturbation phenomena and the magnetic latitude as well as the time interval since deposition. The effects of bioturbation can be compared with those arising from the flow deformation of ice layers in glacial sites and were discussed by Krishnaswami and Lal (1982). ^{7}Be, ^{10}Be and ^{26}Al enter the seas in ionic form and of course mix with the stable isotopes ^{9}Be and ^{27}Al; however, since they have high ionic potentials (charge-to-radius ratios), many become absorbed on both inorganic and organic particles; the significance of scavenging is clearly indicated in information recorded by Measures and Edmond (1982). Concentrations increase from 4 pmol kg^{-1} at the surface to 28 pmol kg^{-1} at depths in excess of 4000 m. From this the average oceanic residence time of Be was calculated to be approximately 800 years. In the Pacific, estimates of the residence times for ^{10}Be in both surface and deep waters were made by Raisbeck *et al.* (1978, 1979). In surface waters the estimate was *c.* 15.6 a, perhaps too long because the possible upwelling of extra ^{10}Be from the depths was neglected. This short residence time influences the content of the radionuclide in deep sea sediment and is similar to the horizontal mixing time in the surface layer in question. Thus differences in the input of radionuclide into oceans tend to be smoothed out; in any case, since all of it enters through the surface, the content of deep sea sediment should be more or less independent of latitude. In fact, two cores from high latitudes (in the Arctic Ocean) yielded unusually low concentrations of ^{10}Be, but this may well relate to the presence of ice cover there which could also be coupled with a low quantity or absence of scavengers in the waters. Raisbeck *et al.* (1980) obtained concentrations of 2200 atoms g^{-1} in water from depths of 800 and 3140 m in the Mozambique Channel and calculated from these a residence time of 630 a, a value comparable with the estimated mixing time of the oceans (10^3a), this also implying that there should be a practically uniform deposition rate for the ^{10}Be radionuclide throughout the planetary seas. It could mean that secular variations in its production rate on such a timescale should be observable in deep sea cores.

Interestingly, sea water has very low concentrations of ^{10}Be compared with river waters, as well as the low surface seawater concentrations compared with deep water in the oceans. The inference is that the residence time of the scavenged Be is less than the value calculated from chemical analysis, and that the residence time of its ionic and molecular species is longer. With reference to this question, it is important to mention that Yokoyama *et al.* (1978) tried to assess the residence times of both ^{10}Be and ^{26}Al by assuming that their specific disintegration rates at the surface of a Mn nodule from the Tuamoutu Archipelago are the same as those of the radionuclides in ionic solution in sea water, and they derived a value of 1600 a for ^{10}Be and 1400 a for ^{26}Al. These long residence times are useful for dating authigenic minerals which may incorporate the same radionuclides from ionic solution in the oceans.

^{32}Si is produced by cosmic ray spallation of Ar in the upper atmosphere and later mixes with surface water. As a radioactive tracer, it is applicable to studies as measurement of ocean–water mixing and currents, glacier formation, sediment accumulation and the stability of groundwater on a timescale of hundreds of years. Lal and Somanyajulu (1984) claimed that the radionuclide can be used also in examination of biogeochemical cycles of Si in the oceans by measurements on siliceous sponges.

Naturally, the value of ^{32}Si in all such work depends upon knowing its half-life accurately. While its significance was indicated in the 1960s, no reliable estimates of this parameter had by then been made. A number based upon nuclear reaction cross-sections for producing ^{32}Si ranged from 60 to 700 a. It was suggested that a ^{32}Si source appropriate for a half-life measurement could probably be made through the ^{30}Si(t, p) ^{32}Si reaction, using the triton beam facility newly installed on the BNL 3.5MV van de Graaff accelerator. ^{32}Si decays 100% by beta emission to the ground state of ^{32}P with $Q_{\beta-} = 225$ keV. ^{32}P decays 100% by beta emission to the ground state of ^{32}S with $Q_{\beta-} = 1710$ keV and $T_{1/2} = 14.28$ d. A ^{32}Si was produced and some years later, around 1976, mass spectrometric techniques with tandem van de Graaff accelerators were developed and the proposal was made to use them for solving the ^{32}Si $T_{1/2}$ problem. A result of $T_{1/2} = 101 \pm 18$ a was obtained. A further independent accelerator measurement resulted in a value of 108 ± 18 a. Clearly, there is a great difference between the low-value accelerator and most cross-section results on the one hand and high-value geophysical determinations on the other. Hence Alburger, Harbottle and Norton (1986) reported a half-life for ^{32}Si obtained by direct decay measurement. Beta rays from a ^{32}Si ^{32}P source produced in 1968–1969 through the ^{30}Si(t, p) ^{32}Si reaction using a van de Graaff beam at $E = 3.4$MeV were counted with an end-window gas flow proportional counter system which included an automatic precision sample changer. Comparison counts were taken on beta rays from a ^{36}Cl source. The measurements were commenced in February 1982 and carried out at roughly 4 week intervals, each comprising 40 h of counting on each sample. The decay rate was assessed from the ^{32}Si/^{36}Cl ratio of counts. Small periodic annual deviations of the data points from an exponential decay curve were noted, but their origin is not certain and they seem not to have influenced the results. Based upon the analysis of 53 points taken over 48 months, these researchers produced the value $T_{1/2} = 172 \pm 4$a and adopted this for the half-life of ^{32}Si. The result in question is substantially greater than the previously reported measurements of 108 ± 18 a and 101 ± 18 a, but is lower than values based upon geophysical evidence.

The geochemistry of Si is interesting because the cosmic abundance of the element itself is only 0.05% by weight, according to Cameron (1973). Curiously, however, in meteorites and terrestrial planets, it is the most abundant element following oxygen. This can be explained on the basis of condensation processes in the early history of the solar system; these paralleled the ones occurring in interstellar space and leading to the formation of dust clouds. Infrared studies have established that silicate grains are the main component of interstellar dust clouds. The high chemical reactivity of Si with other major elements such as O_2, Ca, Mg, and Fe promotes the formation of a vast variety of silicate minerals and, of course, on Earth Si exists primarily in rock-forming minerals of which quartz and the feldspars are the most common. Despite its abundance, quartz contributes little to the global silicon cycle because of its very low solubility. Weathering of igneous silicates produces the clay minerals, mostly by hydrolysis.

There is only one known dissolved form of Si, i.e. monomeric silicic acid, H_4SiO_4, with concentrations in fresh waters and the oceans of 0.1–15 and 10–70 ppm, respectively. These quantities are actually lower than those anticipated taking the solubility into account. The control mechanism which governs the dissolved concentrations in question appears to be the reaction between dissolved silica and the clay minerals. In lakes and oceans, biogenic processes cause opaline silica to form; this is a main component of diatoms, radiolaria, silicoflagellates, etc., so that the dissolved silicon cycle is seen to be controlled by biological processes as well as the detrital–soluble interactions alluded to above.

The element chlorine, discovered by C.W. Scheele in 1774, has an atomic number of $Z = 17$ and 14 isotopes with a range of mass numbers from $A = 32$ to $A = 45$. Only the stable isotopes ^{35}Cl and ^{37}Cl occur naturally, together with the radionuclide ^{36}Cl. ^{36}Cl is valuable in both hydrological and glaciological work because it has properties making it appropriate for the study of confined groundwaters:

- its relatively long half-life $T_{1/2} = (3.01 \pm 0.04) \times 10^5$ a.
- the fact that it is not removed from solution by mineral interaction of secondary mineral formation allows estimation of very long groundwater residence times.
- the very high solubility of most natural chloride species guarantees that ^{36}Cl remains in solution.

Cosmic ray production of the radionuclide takes place both in the stratosphere and in the troposphere, the main production reaction being proton-induced spallation of ^{40}Ar with, in the troposphere, the neutron-induced reaction $^{36}Ar(n,p)^{36}Cl$ predominating. Approximately two-thirds of the production is stratospheric, and is latitude dependent because of the effect of the terrestrial magnetism upon cosmic ray intensities. Nevertheless, stratospheric mixing is rapid and the concentration of ^{36}Cl is uniform. The tropospheric production does not vary appreciably with latitude. ^{36}Cl atoms attach to atmospheric aerosols in the submicron size range and these are progressively removed by precipitation in the lower troposphere. Entry of the radionuclide from the stratosphere into the troposphere varies with the season and takes place through the tropopause. Thus the fallout rate is greater in middle latitude than at the poles or the equator. The fallout rate at Stripa (Sweden at 57°N) should be *c.* 14 ± 3 atoms $m^{-2}s^{-1}$, according to data from Lal and Peters (1967).

^{36}Cl has also been produced by neutron irradiation of oceanic aerosols following nuclear bomb testing at low altitudes, thus increasing the fallout of the radionuclide which rose during the period 1953 to 1964 to 70000 atoms $m^{-2}s^{-1}$ according to Bentley, Phillips and Davis (1986). As a result of the fact that nuclear testing over oceans started earlier than the major set of high-altitude detonations, the ^{36}Cl bomb peak predates the peak in the thermonuclear tritium by about 10 years. Elmore *et al.* (1982) gave a residence time in the atmosphere for Cl^- of less than 3a, so the bomb pulse of ^{36}Cl was transient and its fallout rates had dropped back to the natural level by the early 1970s, when weapons testing at low altitudes had terminated. As with tritium, the bomb pulse can be employed in studying the migration of water through the unsaturated zone into the groundwater system as well as the stratification of waters in unconfined aquifers.

In groundwaters, the ^{36}Cl content can be calculated if it is assumed that the average fallout is incorporated in the total annual precipitation and allowance is made for concentration consequent on evapotranspiration. The appropriate functions were given by Carlsson *et al.* (1983).

A subterranean production of ^{36}Cl must also be considered. World average Cl contents of granite and basalt are *c.* 50ppm and 200ppm, respectively, sedimentary rocks having highly variable Cl contents which range from *c.* 10ppm in sandstones to 20000ppm in deep-sea limestones (Parker, 1967). Of course, surfaces of rock outcrops are exposed to the cosmic neutrons flux incident at their localities and altitudes, and this means that some ^{36}Cl will be produced due to neutron capture by ^{35}Cl. At depths greater than a few metres, however, this flux is attenuated to ineffectiveness. Nevertheless, some ^{36}Cl production will take place because of the occurence of the relevant reaction due to the *in situ* neutron flux present in rock matrices through (α,n) reactions triggered by alpha particles from U and Th decay systems. This *in situ* neutron flux was recorded as being of the order of $10^{-4}\ cm^{-2}\ s^{-1}$ by Kuhn *et al.* (1984).

As regards *in situ* production of ^{36}Cl in rock matrices, some muon-induced spallation of Ca, K and Ar nuclei may well produce ^{36}Cl in the near-surface part of the granite, but these reactions are negligible compared with the neutron capture reaction, ^{35}Cl(n,α) ^{36}Cl. To obtain useful results for ^{36}Cl/Cl ratios and groundwater residence time, there are a number of prerequisites, which include the assumptions that a correct estimate of the ^{36}Cl input of cosmogenic ^{36}Cl at recharge is possible and that changes in the ^{36}Cl total or specific activity are attributable to decay of the initial ^{36}Cl or, alternatively, to a growing in of this radionuclide as a consequence of the *in situ* neutron irradiation of chloride in the migrating groundwater.

Of course, it is taken for granted that the groundwater constitutes a closed system with respect to this chloride, which must have been neither lost nor gained from the relevant rock matrix. Unfortunately, this situation is not normally achieved because groundwaters are observed to show increases in their salinities during aquifer evolution. The number of ^{36}Cl atoms, $^{36}N_t$, occurring in a groundwater at a time *t* after recharge took place is given by the following expression:

$$^{36}N_t = {}^{36}N_0(e^{-\lambda t}) + {}^{36}N_{eq}(1 - e^{-\lambda t})$$

in which the first term on the right-hand side refers to the radioactive decay of the cosmic input and the second to the aforementioned growing in of ^{36}Cl as a result of *in situ* neutron irradiation of chloride in solution (in closed systems).

Where the *in situ* neutron flux for the aquifer is small, the equation reduces to just the decay term; a simplified model of this type was applied to a confined aquifer in the Great Artesian Basin in which the ^{36}Cl activity decreases downdip over a distance of 700km from the recharge area (Airey *et al.*, 1983). In the case where the *in situ* neutron flux is high and the groundwater is closed to chloride, the in-growth term controls the ^{36}Cl activity change with time for the dissolved chloride. In cases in which the groundwater content of chloride rises due to interaction with the rock matrix involved, the specific activity of ^{36}Cl in solution must attain the equilibrium specific activity of ^{36}Cl in that matrix during the progress of solution. Here, no inferences regarding groundwater residence time can be made, although the relevant admixture of chloride from various rock matrices or with input chloride can be assessed.

Cosmogenic Ar has been recorded in atmospheric Ar and could be a useful tool in dating groundwater and ice from Greenland and Antarctica. The element Ar has an atomic number of 18 and 15 isotopes with mass numbers which range from 32 to 46. Three are stable, namely ^{36}Ar ^{38}Ar, and ^{40}Ar, and are found mostly in the atmosphere. The presence of ^{39}Ar in the atmosphere was established by Loosli and Oeschger (1968); it results mainly from the stable ^{40}Ar by a number of nuclear reactions of which the most important is ^{40}Ar(n, 2n)^{39}Ar, the daughter decaying by beta emission to the stable isotope ^{39}K with a half-life of 269 a. The concentration of ^{39}Ar in the atmosphere is 0.112 ± 0.010 dpm l^{-1}, of which less than 5% is due to atmospheric nuclear tests, according to Oeschger (1982). The radionuclide has a long residence time which both damps out fluctuations in its rate of production and contributes to probable constancy of its concentration in terms of time and latitude. It is precisely this which makes the ^{39}Ar potentially valuable for the dating of glacial ice as well as groundwater.

At about 70m depth, névé is converted to glacial ice and air bubbles are entrapped; such air is recoverable by *in situ* melting or crushing samples of ice in a vacuum chamber. It can then be analysed in order to assess rates of disintegration for ^{39}Ar and, for that matter, of radiocarbon and ^{81}Kr as well, all relevant to the dating of the ice sample(s). Dates reported from a borehole at Station Crete, Greenland, agree rather well with those obtained by the counting of annual layers shown by seasonal variations in the isotope composition of oxygen in the ice. Also, the CO_2 concentration may be determined in order to investigate time-dependent variations of this in the atmosphere. Such studies have shown that in both Greenland (at Camp Century) and Antarctica (at Byrd Station) the CO_2 content of the air trapped in ice cores rose significantly about 10 to 12 ka BP, probably by degassing of the surface layer of the oceans due to climatic warming when the last ice age came to an end. The main problem with the use of ^{39}Ar as a dating tool is partly due to its chemical inertness, which means that it does not interact with solids. In addition, it and other noble gas radionuclides are found mostly in the atmosphere, which is indeed their principal source. Hence the ^{39}Ar concentration of groundwater is dependent on the time interval which has elapsed since the water was in contact with the atmosphere. Also, it is necessary to analyse for the isotope by using very large volumes of water. Finally, excess ^{39}Ar may arise from the nuclear reaction ^{39}K (n, p) ^{39}Ar in rocks which contain measurable quantities of U and Th.

13.1 DATING WITH COSMOGENIC RADIONUCLIDES

Following the law of radioactivity, the disintegration rate of any cosmogenic radionuclide varies chronologically. The matter is discussed in detail elsewhere in this book and by Bowen (1988).

In the case of ^{10}Be dating, marine sediment is leached with hot concentrated HF so as to provide solutions containing both ^{10}Be and ^{26}Al. Amin, Kharkar and Lal (1966) calculated rates of sedimentation for Pacific and Indian Ocean cores, describing the initial disintegration rate of ^{10}Be as the ratio of its cosmogenic rate of production P and the rate of sedimentation a:

$$^{10}\mathrm{Be} = (P/a)\, e^{-\lambda h/a},$$

with the assumption that both P and a were constant during the time interval of deposition. The calculated sedimentation rates ranged from 1.2 to 6.4 mm 10^3 a^{-1}. However, it is probably inappropriate to propose that rates of sedimentation in the oceans at any particular locality were constant for geologically significant time intervals of millions of years; Inoue and Tanaka (1976) recorded an irregular variation of the relevant rates of disintegration which was attributed to changes in the rate of sedimentation caused by the deposition of volcanic detritus as demonstrated by the presence of glass shards.

Of course, it is also possible to examine ^{10}Be concentrations on land, for instance in soils. Nevertheless, there are few data available because of the low activity levels prevailing which necessitate the collection of large samples. A recent study by Brown *et al.* (1981) involved the use of a van de Graaff accelerator and demonstrated that sediment from the Atchafalaya River, Louisiana, USA had an average ^{10}Be concentration of $(4.42 \pm 0.55) \times 10^8$ atoms g^{-1}. They corrected for the presence of sand, which adsorbs about one-third of the available ^{10}Be, discovering that such adsorption increases with decreasing grain size, independently from the mineral composition of the relevant particles. A close similarity exists between this sediment and soil rich in clay from ground around the Department of Terrestrial Magnetism in Washington, DC, which implies that ^{10}Be concentrations may well be relatively constant in continental sedimentary deposits. The materials mentioned possess approximately ten times less of this radionuclide than average deep-sea sediment, for which a value of $(5.7 \pm 3.2) \times 10^9$ atoms g^{-1} was given. Interesting ^{10}Be data were also derived from tektites. Seven specimens from Thailand, the Philippines and Australia of about 700 ka of age yielded a value of *c.* $(1.06 \pm 0.26) \times 10^8$ atoms g^{-1}, according to Pal *et al.* (1982). It was concluded that the ^{10}Be value arose from sediment on the terrestrial surface. A couple of older tektites, actually moldavites from Czechoslovakia with an age of 15 Ma, had a content under 0.02×10^8 atoms ^{10}Be g^{-1}, probably because decay occurred after their formation by meteoritic impact on Earth, which confirms their terrestrial origin.

Both ^{10}Be and ^{26}Al radioactive decay systems have also been applied to the determination of the rate of growth of Mn nodules (Guichard, Reyss and Yokoyama 1978).

The application of ^{10}Be in volcanics is further of interest, since magma extruded by volcanoes originates partly by melting of descending oceanic crust during subduction and this includes pelagic sediments. It may be expected that recent lava flows ought to be endowed with both ^{10}Be and ^{26}Al. The presence of ^{10}Be was demonstrated in such materials from the Aleutian Islands, and also in Central America by Brown *et al.* (1982) who gave average concentrations of $(4.8 \pm 2.0) \times 10^6$ atoms g^{-1} in four volcanic rocks. Nine basalts and andesites from the Aleutians yielded values of $(4.3 \pm 1.3) \times 10^6$ atoms g^{-1}. As well as pelagic sediments, other sources may contribute to such concentrations: these include contamination from precipitation, some nuclear reactions caused by the radioactive decay of U, Th and their daughters, e.g. ^{7}Li(α, p)^{10}Be, and contamination with soil during the process of extrusion. As it has been suggested that the rather large and fairly uniform concentrations of the radionuclide exclude such secondary sources, the content in the lava flows of island arcs must have been the result of the incorporation of deep sea sediment into the magmas described above. If so, this would be a useful validation of the hypothesis that such magma arises, to some extent, within the descending oceanic crustal block. Anomalies remain to be explained, however, and one of these is potentially damaging to the hypothesis in question. This is the unusually high concentration of ^{10}Be (1.0×10^6 atoms g^{-1}) for a basalt sample from the Columbia River plateau, the rock having an age of 14 Ma. This antiquity means that almost all of the ^{10}Be originally incorporated from the initiating magma decayed away and, concomitantly, that the ^{10}Be existing now must have derived from alternative sources, either one of those already mentioned or perhaps through irradiation by cosmic ray neutrons and mesons.

The production rate of ^{10}Be and ^{26}Al is, however a prerequisite for geological studies. Clayton (1994) stated that the high abundance of radioactive ^{26}Al in meteoritic matter is produced by heavy cosmic rays in molecular clouds and accelerator mass spectrometry (AMS section 3.1.4(b)) has turned out to be a useful tool for high-sensitivity measurements of nuclides produced by cosmic rays. Nishiizumi *et al.* (1989) used the AMS to study the cosmic ray production rates from ^{10}Be and ^{26}Al concentrations in quartz from glacially polished rocks. They found average production rates of 61.9 ± 3.0 atoms $g^{-1} a^{-1}$ for ^{10}Be and 373.6 ± 27.8

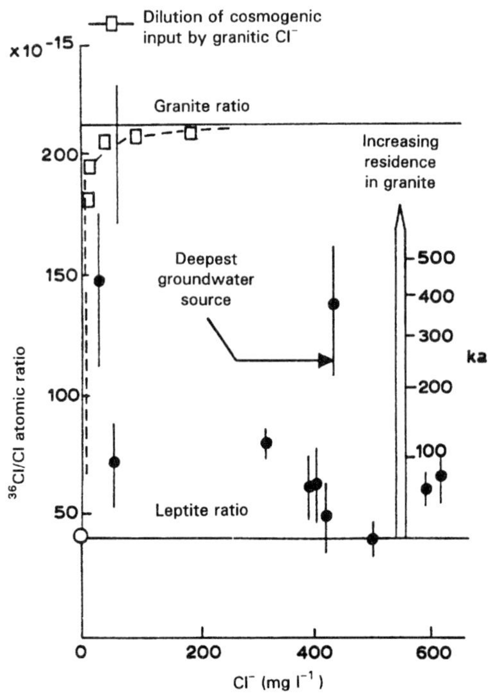

Fig. 13.2 $^{36}Cl/Cl$ atomic ratio of dissolved chloride in Stripa groundwaters as a function of chlorinity. The equilibrium ratios for the *in situ* neutron fluxes in the granite and leptite are shown and the dilution of the cosmic ^{36}Cl content in precipitation with granite chloride is represented by the broken line. Andrews et al. 1986.

atoms $g^{-1}a^{-1}$ for ^{26}Al. Errors are possibly $>10\%$ as snow covering, geometrical factors and particularly the exposure date lack exact determination.

The nuclides are further applied to the dating of ice sheets and groundwaters. The vast continental ice sheets of Antarctica and Greenland comprise sinks for cosmogenic radionuclides, including ^{10}Be and ^{26}Al as well as ^{36}Cl and ^{32}Si. As these have short residence times in the atmosphere, they fall out onto the ice within a year or so of their production. There is a second set with much longer half-lives such as ^{39}Ar, ^{3}H and ^{14}C which are associated with air trapped in the ice during its conversion from snow. These have rather long atmospheric residence times, being either isotopes of noble gases or forming gaseous compounds; therefore their abundances are not especially sensitive to fluctuations in their rates of production

Andrews *et al.* (1986) analysed ^{36}Cl in Stripa groundwaters; from these, Fig. 13.2 shows the ^{36}Cl/Cl atomic ratio and the ^{36}Cl concentration (atoms l^{-1}) plotted against the Cl^- content, and the changes in the former for various modes of groundwater evolution. A groundwater deriving its Cl^- only from the granite ought to evolve towards the right along the line marked 'granite ratio' as further Cl^- dissolves. Rainfall with the cosmic input of ^{36}Cl would have a ^{36}Cl/Cl ratio of $\approx 40\times10^{-15}$. Either way, however, the input will undergo rapid dilution as Cl^- is leached from the rock matrix. The broken line in Fig. 13.2 illustrates the change for the cosmic input. Groundwaters possessing salinities exceeding a few parts per million of Cl^- must originate in deeper situations or so, for Cl^- contents of *c.* 60 mgl^{-1}; the influence of rock-derived ^{36}Cl predominates. Samples plotting on the left-hand side are near the equilibrium values for granite chlorides. No noted ^{36}Cl contents of ^{36}Cl/Cl ratios are high enough to imply a bomb pulse input. Many samples are close to the leptite ratio line, from which it may be inferred that their ^{36}Cl derived mostly from the leptite which borders the granite. The alteration in their ^{36}Cl/Cl ratios to slightly higher values no doubt results from Cl^- in the groundwater arising in the leptite and residing in a higher neutron flux in the granite fracture system for up to 100 ka. Clearly, the groundwater residence times would be anything up to 100 ka greater than their transit times through the leptite. In the present case, two Cl^- sources, namely the leptite and the granite in question, are involved, and the concentration of ^{36}Cl atoms in groundwater obtaining its total salinity, W mgl^{-1}, from them is given by

$$^{36}\mathrm{N} = W_1S_1 + (W-W_1)S_2 \text{ atoms } l^{-1},$$

in which W_1 is the chlorinity derived from source 1 and S_1, S_2 are the specific concentrations (atoms mg^{-1}) of ^{36}Cl in the two sources. The ^{36}Cl atomic ratio R for the mixture is given by

$$R = [W_1R_1 + (W-W_1)R_2]/W$$

where R_1 and R_2 are the ^{36}Cl atomic ratios for two Cl^- sources. Using these equations, Andrews *et al.* (1986) were able to show that the granite contribution to the chlorinity is the minor one for the samples near the leptite ratio. The model assumes that there has been no significant groundwater residence within the granite and also that the granitic ^{36}Cl influence is due solely to the removal of Cl^- from the granite (and is not attributable to ^{36}Cl production within fracture fluid Cl^- owing to the higher neutron flux in the granite).

The origin of the ^{36}Cl in the Stripa groundwaters can be explained on the strength of two hypotheses. Firstly, Cl^- may originate from groundwater interaction deriving Cl^- from both the leptite and the granite, the residence times of groundwater in the latter being short compared with the half-life (301 ka) of ^{36}Cl. Secondly, the Cl^- of the deeper-lying groundwater may have been neutron-irradiated within the leptite with a later entry of the saline fluids into the granite fracture system where they underwent a higher neutron flux. The residence times in the granite would then be >100 ka for most of the borehole samples near the leptite ratio in Fig. 13.2. In both hypotheses, the Cl^- entering from the leptite could be either matrix Cl^- or Cl^- from an external source which resided in the rock in question for at least 1.5 Ma. The correct answer probably occurs somewhere between the two, i.e. some shallower mine waters on the left of the figure probably obtained most of their Cl^- by interaction with the granite and their flow paths probably did not contact leptite at all. The deepest groundwater source is not in equilibrium with the granite although it has the highest ^{36}Cl content of the groundwaters. Its hydrological situation implies that it possesses the longest flow path from leptite to granite and its ^{36}Cl content necessitates a subsequent residence time of *c.* 300 ka in the granite or, alternatively, some 50% of its chlorinity to have been derived from the granite. Perhaps the most interesting aspect of the Stripa researches has been the attempts made to use the data to derive generalized groundwater flow regimes. One (A) shows groundwater evolution within a granite, salinity resulting from interaction with the rock matrix and the $^{36}Cl/Cl$ atomic ratio of the dissolved Cl^- characterizing the rock. A second (B) has the groundwater recharging into a formation with lower radioelement contents (and thus neutron flux) than the granite, so that a lower $^{36}Cl/Cl$ atomic ratio is acquired as chlorinity increases. If the groundwater later enters the granite and chlorinity goes on rising due to interaction with the rock matrix, then the $^{36}Cl/Cl$ atomic ratio increases as a result of admixture of granite Cl^-. But if the chlorinity evolution ceased before the groundwater reached the granite, the ratio would increase with increasing residence time in the granite. The residence time in the granite fracture system could thereafter be estimated from the additional ^{36}Cl in growth within the granite. A third regime (C) has chlorinity assumed to have been derived from a source with a low ^{36}Cl/Cl atomic ratio. This could be evaporitic. Later infiltration into a formation with higher radioelement contents produces ^{36}Cl ingrowth in the dissolved Cl^-. In the event that the groundwater resides long enough in such a formation, the Cl^- will reach the saturation ratio for the *in situ* neutron flux. Later flow into a granitic fracture system would result in a further time-dependent rise in the ^{36}Cl/Cl ratio, provided that no further matrix Cl^- is dissolved.

^{36}Cl deposited in snow and ice sheets of Greenland and Antarctica may be used either in dating ice or in detecting variations in its production rate in the atmosphere. The ^{36}Cl/^{10}Be ratio may further serve to date glacial ice according to Nishiizumi *et al.* (1983), although results are not yet satisfactory.

13.2 ^{81}Kr DATING

13.2.1 Geochemistry

Krypton is an element with relative atomic mass of 83.80 and atomic number 36. An inert noble gas, it is found in the atmosphere (1 part in 670 000) and its boiling point is $-152.3°C$. It is used in some lasers and has 25 isotopes, mostly short-lived and radioactive. The six stable ones (78, 80, 92, 83, 84 and 86) comprise 100% of the abundance, but the one of interest, ^{81}Kr is unstable with a half-life of 2.3×10^5a. It has been used as a dating tool for meteorites, among other samples, and a typical instance may be cited.

13.2.2 ^{81}Kr Terrestrial Ages of Eucrites

Miura, Nagao and Fujitani (1993) analysed Yamato eucrites in order to establish their terrestrial ages and groupings using noble gas and chemical composition. The usual approach to determining the terrestrial age of a meteorite involves such cosmogenic radionuclides as ^{14}C, ^{36}Cl and ^{81}Kr. Clearly, the last one is applicable only to those with terrestrial ages around its half-life value. In the 5940 Yamato meteorite collection in Japan, there are some 50 eucrites, mostly polymict breccias from Antarctica, and these are fragments not from a single fall, but rather from at least two and perhaps as many as five separate falls. Elucidating this matter was one object of the Miura, Nagao and Fujitani (1993) investigation. Others were to discover the terrestrial ages for nine Yamato eucrites and to study their possible groupings. The noble gas analyses followed methods outline by Miura and Nagao (1991).

Electron probe microanalysis (EPMA) and instrumental neutron activation analysis were also used. In order to calculate the ^{81}Kr–Kr age using the isotopic ratios of cosmogenic Kr, the trapped and fission components were subtracted from the measured Kr isotopes. Fissiogenic ^{86}Kr can be evaluated from the Xe isotopic ratios and, to achieve this, the cosmogenic Xe isotopes had to be subtracted assuming that both ^{126}Xe and ^{130}Xe are the composite of trapped and cosmogenic isotopes. Xe isotopic ratios were derived from values given for Pasamonte, Angra dos Reis and St Severin. Trapped Xe was assumed to be atmospheric. A plot of $^{134}Xe/^{130}Xe$ against $^{136}Xe/^{130}Xe$ was obtained after correction for cosmogenic Xe. The isotope ratios lay on a mixing line between ^{244}Pu fission Xe and atmospheric Xe, which implies that ^{244}Pu existed in the eucrites when Xe retention started. Any contribution of Xe from the spontaneous fission of ^{238}U was thought to be too small to be noticeable in the plot. Subtracting atmospheric Xe as the trapped component gave fission Xe spectra which accorded with that of ^{244}Pu fission. Uranium concentrations of 190 and 150 ppb respectively were measured for two samples so that the upper limit for ^{136}Xe concentrations derived from ^{238}U could be estimated, assuming that the Xe retention age was 4.5 Ga. For the remaining eucrites a value of 150 ppb U was assumed. Calculated concentrations of ^{238}U-derived ^{136}Xe were less than 13% of the excess ^{136}Xe and the remaining excess ^{136}Xe derived from ^{244}Pu fission. The ^{244}Pu abundance of each eucrite was calculated as indicated by Eugster, Eberhardt and Geiss (1967). Fissiogenic ^{86}Kr abundances from U and Pu were obtained by using their concentrations and the fission yields. In evaluating the cosmogenic Kr isotope ratios, the concentration of trapped ^{83}Kr was calculated by utilizing a cosmogenic ratio $^{86}Kr/^{83}Kr = 0.015$ and the concentrations of ^{83}Kr and ^{86}Kr corrected for fissiogenic ^{86}Kr. The contribution made by fissiogenic ^{86}Kr to the measured ^{86}Kr was $< 4\%$. The cosmogenic Kr isotopic ratios presented were obtained by correcting trapped Kr (for which terrestrial atmospheric composition was assumed). Production rate ratios were calculated by using the cosmogenic ratios $^{80}Kr/^{83}Kr$ and $^{82}Kr/^{83}Kr$ and were in good agreement with others obtained with the formula using cosmogenic $^{78}Kr/^{83}Kr$. ^{81}Kr,–Kr ages were calculated using formulae from Marti (1967) and Eugster, Eberhardt and Geiss (1967). They were used as an apparent exposure age in calculating the terrestrial age. The latter was calculated by combining the apparent exposure age T_a and the true exposure age T_e using the following formula (e.g. Freundel, Schultz and Reedy, 1986):

$$T_t = (1 - \lambda_{81})\ 1\text{n}\,(T_a/T_e)$$

where the decay constant for ^{81}Kr (λ_{81}) is 3.25×10^{-6} a. The true exposure age was calculated from the ^{38}Ar production rate and the concentration of cosmogenic ^{38}Ar. The cosmic ray exposure ages based on this radionuclide for nine eucrites are within the range 8.5 to 70 Ma. Six of the nine have similar exposure ages which range from 58 to 70 Ma. They are members of a group with the longest exposure ages among the

Yamato eucrities. For them, Miura, Nagao and Fujatan (1993) noted that they are part of a group the calculated terrestrial ages of which are too short to be determinable using ^{81}Kr. Their terrestrial ages are less than 0.08 Ma, so the maximum radioactive decay of ^{81}Kr in them after they fell to Earth cannot exceed *c.* 20%. Therefore, the mean value of ^{81}Kr concentrations of $0.188 \pm 0.023\,(1\sigma) \times 10^{-12}\,\text{cm}^3\,\text{STP}\,\text{g}^{-1}$ shows a saturation activity of $0.226–0.188 \times 10^{-12}\,\text{cm}^3\,\text{STP}\,\text{g}^{-1}$ for these particular eucrites. This is actually somewhat high in the range $(0.129 - 0.252) \times 10^{-12}\,\text{cm}^3\,\text{STP}\,\text{g}^{-1}$ which has been recorded for a number of non-Antarctic eucrites (Macibini, Stannern, Bouvante, Juvinas, Sioux County and Millbillillie). This relatively high saturation level of ^{81}Kr might well be due to high concentrations of target elements in the six eucrites involved. For their samples Y-792511, Y-793164 and Y-794043, these authors gave terrestrial ages of 0.14, 0.11 and 0.28 Ma, respectively. These fall within the range of terrestrial ages previously observed for *c.* 40 Yamato meteorites of which over half have ages lower than 0.07 Ma. As for ^{244}Pu in eucrites, the Xe isotope ratios imply its presence when eucrites formed. The ^{244}Pu/Nd ratio is approximately constant through high chemical coherence. The weight ratio of ^{244}Pu/Nd seen in Angra dos Reis ADOR is 1.5×10^{-4} when its parent body formed (*c.* 4.55 Ga ago). The concentrations of REEs in most of the reported eucrites turn out to be higher by a factor of five to ten than those in chondrites. This is perhaps because of the differentiation of chondritic materials in a parent body without significant fractionation. Interestingly, negative Eu anomalies occur and in three eucrites there are positive Ce anomalies. The latter are especially significant because Ce can have valence of +4 in aqueous conditions and hence behaves differently from other REEs during storage of eucrites in ice. Regarding previously reported Yamato eucrite data, Miura, Nagao and Fujitani (1993) required >1 fall(s) for a monomict and >5 falls for polymict eucrites during the past 0.3 Ma. The grouping of Yamato eucrites was also considered.

Part Three
Stable isotopes in the biosphere

14

Relevant stable isotopes in nature

Life is an offensive, directed against the repetitious mechanism of the universe.

Alfred North Whitehead, 1861–1947 (Adventures of ideas)

The biosphere is part of the terrestrial ecosphere and it is generally defined as that space where any organism can exist. Therefore it will be obvious to the geo- or bioscientist that the atmosphere, the hydrosphere and the Earth's surface and soil, as well as marine sediments belong to the living space. During Earth history, life has conquered the terrestrial and lower atmospheric regions, starting from the ocean with organic molecules about 4.0Ga ago. Through all the evolutionary processes and despite all available elements within the biosphere, life mainly involved only a few elements, chiefly H, O, C, N, S and P for its basic purposes. These few essential organic elements are, amongst others, those which can be more or less incorporated by organisms and may be necessary or toxic. Those of interest for isotope geoscientists will be discussed in the following paragraphs.

The atmosphere, as known today, started at some time in the Palaeozoic finally to reach its present O_2 content. Now it comprises N_2 (78.09%), O_2 (20.95%) and noble gases (0.96%) and is a product of volatiles released from the solid material of the Earth. It has changed through chemical and biological processes in the course of time. Minor components are free H, partially ionized and only occurring in the upper stratosphere, ozone, H_2O, CO_2, CH_4 and, if exhalated by volcanic processes, H_2S. Reactions of participating compounds can for instance produce acids, aggressive to buildings and biological systems such as lakes or, in the case of the so-called artificial chlorofluorocarbons (CFCs) attack the ozone layer. The greenhouse gas CO_2 causes problems by increasing the average global temperature. Calculations carried out at the Max Planck Institute of

Meteorology, Hamburg, utilizing a high-speed Computer Cray C 69 seem to confirm strongly that the greenhouse effect is due to anthropogenic influences. However, that was already known but often neglected, because economic interests still dominate governmental actions. More than 20 Gt of CO_2 are released by mankind every year, and one of the buffers, the forests, is destroyed at a rate of 20 million hectares per year.

Basically, the atmosphere turned out to be very stable over long periods of time during Earth history. A well-balanced buffering system always kept it in a life-supporting state of equilibrium. Today, the influence of humanity on a nearly overpopulated planet disturbs the natural balance and, because the world cannot be divided nationally with respect to the ecosystem, vital problems have to be faced by all. Dividing rationally would obviate many, but is too much to hope for.

Despite the relatively simple chemical composition of the atmosphere, the isotope fractionation in rain and snow, the exchange of isotopes with organisms through photosynthesis and equilibrium processes at the atmosphere/hydrosphere boundary are mandatory for the understanding of preserved isotope signals. From a chemical point of view, it can be assumed that the mixing ratios of gases within the homosphere are constant and independent of gravitational influences. Vertical changes of components, i.e. the occurrence of ozone within the stratosphere, whereas the formation of radioactive elements by cosmic radiation becomes important. Relevant processes are for instance the formation of radioactive C and ^{3}H through the following reactions:

$$^{14}N + n \rightarrow {}^{14}C + H^{+} \text{ and } {}^{14}N + n \rightarrow {}^{12}C + {}^{3}H\,.$$

Cosmogenic radionuclides of geological importance include ^{39}Ar, ^{81}Kr and ^{85}Kr. They are discussed, with others, in Chapter 13.

The atmosphere is in contact with the Earth's surface, of which 71% or 3.61×10^{B} km^2 is covered with water, to comprise part of the hydrosphere with an average thickness of 3.8 km, a volume of 1.37×10^{18} m^3 and a density of 1.03 g cm^{-3}. Its mass (43×10^{24} g) represents only 0.7% of the total mass of the planet. Oxygen and H (relative abundances 85.9 and 10.6 wt% respectively (Clarke values)) are the most abundant elements in the hydrosphere and dominate all other elements occurring in sea water. Both atmosphere and hydrosphere are in contact with terrestrial and marine soil or sediments and living organisms (plants and animals), broadly speaking. The physical and chemical conditions however are different, not just due to the gaseous state of the atmosphere, the liquid state of the hydrosphere and also solid or biochemical active matter. The soil in both cases is also a living space for active organisms, either terrestrial or aquatic, where they are in contact with the atmosphere or hydrosphere.

As these remarks show, numerous different physical, chemical and biological processes have to be taken into account if the very complex and sensitive ecosystem of the Earth is to be understood. Isotopes provide a relevant tool and the development of knowledge within the last 50 years has been staggering.

14.1 OXYGEN AND HYDROGEN ISOTOPES

With 46.6 mass % and 55.1 atom %, oxygen is the most abundant element in the terrestrial crust. With the atomic number of 8, it has nine isotopes ranging in mass from 13 to 21 of which only the three stable isotopes ^{16}O, ^{17}O and ^{18}O are of geoscientific interest. Oxygen has a valence state of −2 and the three stable isotopes have the following relative abundances: $^{16}O = 99.763\%$; $^{17}O = 0.0375\%$; $^{18}O = 0.1995\%$.

Free O_2 occurs in air and water but most of the element is bound in various compounds like water molecules, siliceous and other minerals or in organic matter. The O_2 molecule was discovered *c.* 1770 independently by C. W. Scheele and J. Priestley. Oxygen is necessary for the respiration of nearly all organisms except anaerobic bacteria. Now being practically indispensible to life, its absence in the primaeval atmosphere of the Earth was mandatory for the development of the first organic compounds. Another oxygen molecule called ozone (O_3) (from the Greek *ozo* meaning 'I smell') is also important for life. It is produced in the upper stratosphere by electrical discharge and through ultraviolet radiation, the latter being highly absorbed by the ozone, of which the concentration is very large within the stratosphere (although very small compared to the O_2 concentration). The so called ozone layer, mainly depleted through the influence of CFCs, protects life from dangerous ultraviolet radiation. It is not clear how long the ozone layer has existed, but it may be assumed that ozone production increased parallel to the increasing O_2 content of the atmosphere. An effective protection through the ozone concentration is discussed for the Devonian. Again it was necessary for life that this protection did not exist within the earliest atmosphere but does now. The loss of the protection (which may well have occurred several times during Earth history, sometimes due to supernovae) will have been arduous for all organisms whether on the planetary surface or in shallow water either because of increasing ultraviolet radiation or through the shift of ozone production to the lower atmosphere.

Ozone is a highly reactive gas, whereas O_2 is more inert due to the great energy of the binding (494.0 kJ mol^{-1}). Most O_2 reactions occur at temperatures significantly greater than room temperature. Within the O cycle (as in the C cycle) biological reactions are of great importance. A large amount of CO_2 is used by plants during photosynthetic reactions and the O_2 produced serves for respiration and oxidation of organic matter. According to Holland (1978), the residence time for O_2 in the atmosphere is about 5500 a. As the organic C is partially deposited in the form of elemental C, it follows that the O_2 content of the atmosphere should increase, because of the lack of this matter for oxidation. However, this is buffered by the weathering of magmatic rocks and the oxidation of its content of C, S and Fe ions. Biological and mineralogical reactions together obviously maintain the balance of the existing atmospheric O_2 content.

The element hydrogen ($Z = 1$), which was discovered by H. Cavendish in 1776, possesses two stable isotopes ^{1}H (protium) and ^{2}H (deuterium or D) and the radioactive ^{3}H (tritium or T). Its valence state is $^{+}1$ (seldom $^{-}1$) and the abundances of H and D are 99.9852% and 0.0148% respectively. Tritium exists on Earth only in traces with a total amount of *c*. 1.8 kg. Free H is very rare and occurs only in the higher parts of the atmosphere. Hydrogen compounds in atmosphere, hydrosphere and crust contain only 1% by mass of this element. The main reservoir of H_2 is water molecules and important are also hydrocarbons (oil, gas) and some minerals and hydrates. Together with C and O (and some other elements) it is a main constituent of organic matter.

Hydrogen is almost insoluble in water and its isotopes are of interest since they have the greatest relative mass difference among all elements. This means that the H/D ratio of terrestrial samples can vary by up to as much as 70%, the maximum possible among stable nuclides. In extraterrestrial matter, even greater variations have been observed. Measurements of O_2 and H_2 isotopes are performed on double-inlet gas mass spectrometers using either CO_2 for the $\delta^{18}O$ ($\delta^{13}C$) ratios or H_2 for the D/H ratios. The usual standard is Vienna SMOW (V-SMOW) which replaced the former SMOW (Standard Mean Ocean Water), which was exhausted in 1976. The isotopic composition of V-SMOW relative to SMOW is $\delta D = +0.20‰$; and $\delta^{18}O = +0.04‰$. Another standard offered by the International Atomic Energy Authority (IAEA) for δD and $\delta^{18}O$ measurements is the designated SLAP (Standard Light Antarctic Precipitation), which has the following values relative to V-SMOW: D(SLAP) $= -428‰$; and $^{18}O(SLAP) = -2.2‰$. V-SMOW and SLAP are standards on a natural level. For the measurement of enriched isotopes and for the calibration of mass spectrometers the standards IAEA-302 (D/H) and IAEA-304 ($^{18}O/^{16}O$) are available.

Within the atmosphere O and H are of interest only in compounds, while free H_2 is luckily very rare, occurring in the protosphere from where it escapes into the space. If the Earth did not have such a good temperature balance, the planet would already have lost all H_2 and have suffered a fate similar to that of our neighbour Venus. Heated up with the consequence that all the water is vaporized, the water vapour would be dissociated by ultraviolet radiation and the H_2 would gradually disappear. That this might happen to the Earth in the far future might be due to increasing solar activity and the life span of the biosphere is indeed a matter for scientific 'prophecy', although the model of Caldeira and Kasting (1992) leaves probably another billion years.

The isotope composition of marine O_2 has also varied during the Phanerozoic, but the changes do not correlate well with those of the ^{32}S. They are the result of changes in the flux of O_2 entering and leaving oceans; this O_2 and S arise together from erosion of marine sulphate minerals.

Output of O_2 from oceans takes place by reduction of sulphate to CO_2 and H_2S, and also by the crystallization of sulphate minerals during evaporite formation. Oxidation of sulphide minerals permits O_2 to enter the S cycle, reduction of sulphate ions letting O_2 escape. Of the four atoms employed in making new sulphate, three come from the atmosphere and one from meteoric water. The average ^{18}O value of sulphate formed by the oxidation of sulphide is *c.* +2‰. The CO_2 that is liberated in the course of the reduction of sulphate is depleted in ^{18}O by *c.* +10‰ compared with the residual sulphate, so an increase in the sulphate reduction rate can enrich sulphate remaining in the oceans in the heavy stable isotope. On the other hand, an increase in sulphide oxidation can lower the ^{18}O of marine sulphate because gypsum favours ^{18}O by *c.* 3.5‰. The effect of erosion of marine sulphate minerals on ^{18}O of sulphate in the oceans is a function of the age of the deposit. The isotope composition of O_2 in marine sulphate is unaffected by isotopic equilibrium with water of the oceans or with water of hydration released during the dehydration of gypsum as it is buried. Additionally, O_2 in sulphate is not altered isotopically by exchange reactions with silicate minerals in the oceanic crust or with $CaCO_3$.

14.2 NITROGEN ISOTOPES

The atmosphere-friendly element nitrogen has a volume of 78.08% and weighs 75.5%, mostly consisting of the N_2 molecule, the major constituent of the atmosphere, which is also the biggest N reservoir. It occurs too as the oxides NO, NO_x and N_2O, of which the latter is a greenhouse gas of primarily biogenic origin. Within the stratosphere it is either photochemically split by the reaction $N_2O \rightarrow N_2 + O$ or it may be the source of ^{14}C. Within the range of seven N isotopes, ranging from $A = 12$ to $A = 18$ with $Z = 7$, the two stable isotopes ^{14}N and ^{15}N have relative abundances of 99.64% and 0.36%. Strongly depleted on Earth and rare in minerals and rocks, N is of great biological importance.

The atmospheric N_2 molecule cannot normally be assimilated by living cells because of its strong chemical bond and the resulting high energy of 942 kJ mol^{-1} which is neccessary to split it. Therefore nitrifying bacteria in the soil have a great importance in producing ammonia and nitrate through several steps from organic N_2 compounds. In soil, organic matter and sea or fresh water, the element N_2 with valence states between +5 and −3 occurs in the form of the above-mentioned oxides and also as nitrate NO_3^-), nitrite (NO_2^-), ammonium (NH_4^-), ammonia (NH_3) and in proteins and amino acids. Detailed information about N_2 isotopes, mass balances and and exchange reactions was summarized by Hübner (1986).

The isotope composition of N_2 is expressed in the usual δ notation, where

$$\delta^{15}N = \frac{(^{15}N/^{14}N)_{sa} - (^{15}N/^{14}N)_{std}}{(^{15}N/^{14}N)_{std}} \times 1000\ [‰],$$

and standard atmospheric N_2 is used as both a working and reporting standard. The former NBS-14 gas sample, however, had to be substituted because of the lack of material and has therefore been replaced by, for example, the ammonium sulphate standards IAEA-N-1, IAEA-N-2 and the potassium nitrate one IAEA-NO-3, the latter also of use in O_2 isotope analysis. Measurements of N isotope ratios are effected on gaseous N_2 with the best precision on GMS (gas mass spectrometers) with a double inlet system and usually a triple collector or a five-cup CNOS collector. Nitrogen compounds therefore have to be converted to N_2 (e.g., Bremner and Keeney, 1966; Ross and Martin, 1972).

The isotopic fractionation of N_2 isotopes leads to considerable differences in the $\delta^{15}N_2$ values due to the various fractionation processes which are mainly biological reactions and some equilibrium exchange reactions. Variations occur in the range of $\delta^{15}N < -20‰$ up to $+30‰$ in natural compounds as seen in Fig. 14.1 documented by Hoefs (1987). Compared to atmospheric N_2, the the isotope fractionation caused by nitrification and denitrification is large compared to the relative mass difference of the stable isotopes $^{15}N/^{14}N$. A detailed description of fractionation processes and different fractionation steps is given by Hübner (1986) and Hoefs (1987), including the summarized relevant fractionation factors.

Atmospheric N (N_2) occurs also as the major N_2 compound in the oceans and is in equilibrium with the atmosphere. As measurements of air samples from different locations and altitudes have shown (Sweeney, Liu and Kaplan, 1978; Mariotti, 1983), the $^{15}N/^{14}N$ ratio of atmospheric N_2 is more or less constant with $\delta^{15}N$ values close to zero and the residence time in the atmosphere is $> 10^7$ a according to Hattori, Campbell and Krouse (1983).

Differences in $\delta^{15}N$ values of trace gases are mostly dependent on their sources. The trigonal pyramid of the NH_3 molecule in pure air shows $\delta^{15}N$ isotope ratios of about $-10‰$ whereas measurements from barnyards were extremely positive with $+24.9‰$, both with a standard deviation of about 3‰. This particularly applies to N_2O, whose mass can be the same as the mass of the CO_2 molecules in air, depending on the combination of isotopes, which must be taken into account if CO_2 is extracted from air. The N_2O content in the atmosphere is small with an average of 5×10^{-6} vol%. However, corrections have to be made, as described for example by Mook and van der Hoek (1983), in order to achieve correct $\delta^{18}O$ and $\delta^{13}C$ values from CO_2 in air, although the concentration of CO_2 in air is much greater (0.035 vol%) and the N_2O/CO_2 ratio is only about 0.9×10^{-3}. If the N_2O is removed through combustion, as suggested by Mook and Jongsma (1987), a correction by adding $\delta^{13}C$ (0.23‰) and $\delta^{18}O$ (0.33‰) still remains neccessary because the combustion influences the isotope signals. As the main source for N_2O in the atmosphere is the production by bacteria in the soil, the concentrations of N_2O in the air and the $\delta^{15}N$ ratio may vary depending on the localities. The total average of the N_2O content of the atmosphere is probably the equilibrium state between the rates of photochemical destruction and biochemical production.

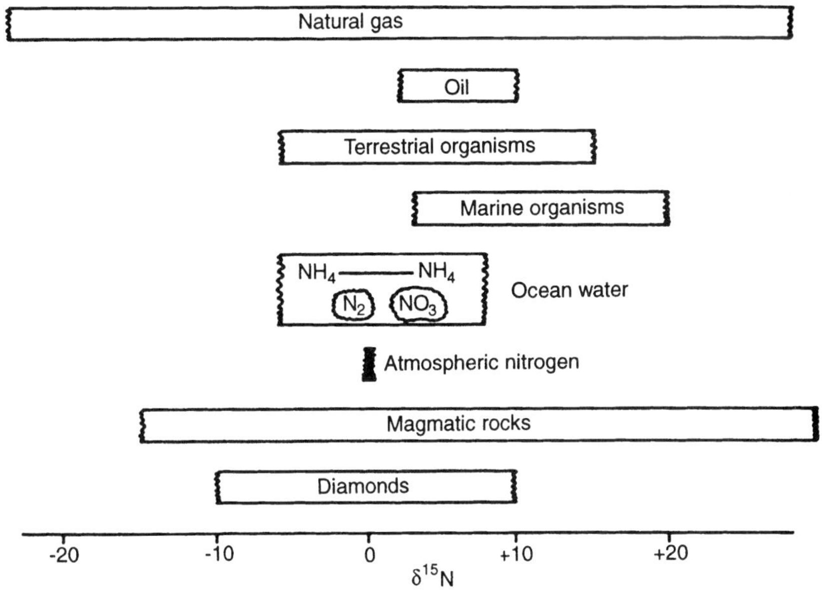

Fig. 14.1 Variations of $\delta^{15}N$ values in natural materials (after Hoefs, 1987).

14.2.1 Nitrogen in Natural Waters

In sea water, most N_2 occurs in equilibrium with the atmosphere. Its concentration depends on the solubility of atmospheric N_2 at the sea surface, which decreases with increasing temperature and chlorinity. The N_2 concentration remains constant within the sinking waters, whereas the amount of the other compounds which occur, i.e. NH_3, NO_2^-, and NO_3^+, depends upon location, season, depth and the biological activity. They are involved in a N_2 cycle, where the nitrate is incorporated in photoplankton in the photic zone. The decomposition of the plankton then releases amino acids and proteins which are transformed into NH_3.

The latter is therefore more concentrated in the upper layers of the water column. Bacterial activity converts the NH_3 into NO_2^- which is further oxidized to NO_3^+ and enriched in deeper layers, where plant activity is impossible due to the lack of light. Depending on the location, the NO_3^+ is brought back into the cycle by convection, preferentially occurring in oceans of the northern or southern hemisphere, thus causing an increase in nutrition within these areas. The NO_3^+ contents tend to rise with depth with average values of 1.2 mg L^{-1} in the Atlantic Deep Water and 2.5 mg L^{-1} in the Pacific Ocean due to the lack of vertical exchange in subtropical or tropical areas.

The oceans contain particulate organic matter (POM) of which the C/N ratio increases with depth since N is lost more quickly than C during POM degradation. It has been shown that the ^{15}N values of POM vary systematically with depth. The decrease of this parameter in particulate organic N (PON) in surface water is attributable to isotope fractionation during uptake of dissolved NO_3^+ by phytoplankton which prefer $^{14}NO_3$. Consequently surface-water NO_3^+ is enriched in $\delta^{15}N$, as was observed in the eastern part of the North Pacific Ocean. The degradation of dead phytoplankton down to *c.* 500 m depth is shown by the increase of $\delta^{15}N$ up to $+13‰$ and the decrease in concentration of PON. In greater depths the residual POM is stable and indeed gradually sinks to the ocean bottom. POM which originates from land sources has an appreciably lower ^{15}N than POM in the oceans; for example, an average value of $+2.5‰$ was observed in samples analysed by Sweeney, Liu and Kaplan (1978).

This difference provides a tool for studying the mixture of water masses in coastal regions as shown by Mariotti, Lancelot and Billen, (1984) who recorded that, in April 1974, the ^{14}N values of POM in the Scheldt estuary of the Netherlands rose towards the North Sea from an average of $+1.5 \pm 0.2‰$ *c.* 80 km inland from the coast to a mean value of $+8.0 \pm 1.8‰$ in the sea. It must be borne in mind, however, that seasonal growth of microorganisms produces important seasonal variations in the ^{15}N values of POM in both the North Sea and the freshwater part of the Scheldt estuary. The lowest values of ^{15}N are found in POM which is collected in the spring and early summer (April to June), when of course microorganism blooms take place. Altough a decrease in the concentration of dissolved NO_3^+ could be expected, it has not been demonstrated. In summer and autumn, North Sea POM showed high $\delta^{15}N$ values up to $+11.5‰$.

Regarding POM sinking to ocean bottoms deposits together with inorganic mineral particles, measurements show that the ^{14}N values of marine sediments fall around an average of about $+6.8 \pm 4.1‰$, demonstrating that the N in sediment derives from sources other than POM (although including POM) or perhaps is in some way modified isotopically after deposition. These values are of course lower than those of the North Sea (average $= 0.8 \pm 1.8‰$), which may be due to the fact that larger particles such as faecal pellets sink faster and hence are not enriched in ^{15}N by the oxidative degradation of nitrogenous compounds. Saito and Hattori (1980) started that particulates collected on sedimentation traps in the northern North Pacific Ocean possessed low ^{15}N values ($+2.9$ to $+4.4‰$) and inferred that the organic fraction of sediment may be a mixture of slowly descending particles with high ^{15}N values and fast-sinking ones with low values of ^{15}N. Measurement of POM from sediment traps and also from filtration, considered together with data on N_2 in samples of sediment, may permit an estimate of their relative contributions to sedimentation in various parts of the oceans to be made.

14.3 CARBON ISOTOPES

As Sir James Jeans, the English mathematician and astronomer (1877–1945) stated in his *The Mysterious Universe*, life exists in the universe only because the carbon atom possesses certain exceptional properties. In fact, carbon provides the basis for terrestrial life and is omnipresent in the universe and on Earth. Chemically, the element has an atomic weight of 12.011, an atomic number of 6 and a melting point of 3550°C. Of 12 C isotopes, the two stable isotopes ^{12}C and ^{13}C with abundances of 98.89% and 1.11% respectively and the radionuclides ^{11}C and ^{14}C are of geochemical interest. The latter are either produced artificially through several reactions as, for instance, ^{11}C or naturally through cosmic radiation. The allotropic varieties of C include graphite and diamonds with relative densities of 2.25 and 3.51, respectively.

The C isotopes are fractionated by incorporation into living systems and/or biochemical processing due to thermodynamic and kinetic effects imposed on the main metabolic pathways. Since biochemical reactions are mostly controlled by enzymes, and living systems usually constitute dynamic states undergoing rapid cycles of anabolism and catabolism, probably most biological fractionations are due to kinetic rather than to equilibrium effects. Such biological fractionations are basically retained when organic matter becomes incorporated into sediments. Thus, these effects are introduced into the rock cycle of the geochemical cycle, where their signatures remain on the crustal inventories not only of C, but also of S and other elements. Continuous biological processing in the exogenic exchange reservoir of the atmosphere and the oceans of carbon is responsible for the entry of the fluxes into the crust caused by photosynthetic C fixation. Biogenic phases preferentially concentrate the light isotope ^{12}C leaving the heavy species ^{13}C to accumulate in the residual inorganic phase. The circulation of C within our biosphere constitutes the C cycle shown in Fig. 14.2.

As CO_2 is a greenhouse gas, an important question is how to estimate the possible temperature changes caused by the shifting of C compounds within the global CO_2 reservoirs. The natural balance of the Earth's buffering system has been shown to work well without anthropogenic influence, but since the release of CO_2 from fossil fuel, the CO_2 content of the atmosphere increases. The C isotope ratio is usually expressed relative to the PDB standard after *Belemnitella americana* from the Cretaceous Pee Dee Formation, South Carolina. As this standard is exhausted, but is referred to in the literature, the isotope ratios are measured against other standards and related to PDB as

$$\delta_{S-PDB} = \delta_{S-STD} + \delta_{STD-PDB} + \delta_{S-STD} + \delta_{STD-PDB} \times 10^{-3}.$$

The above relative isotope ratios of sample, standard and PDB are as usual expressed in per mil, although the ratios are without a dimension and ‰ is not a physical unit. There are various C standards available from the IAEA which in some cases are also O_2 standards. Since PDB is virtually non-existent, the Vienna-PDB (V-PDB) was introduced in 1985 with NBS19 as a reference material. It has values of $\delta^{13}C = +1.95$ ‰ versus V-PDB and $\delta^{18}O = -2.20$‰ versus V-PDB.

Usually the standard inlet of a GMS is connected with a gas reservoir which serves as the internal laboratory standard, for instance industrial CO_2 or SO_2. The comparison standards are mainly used for the control of the measurements on the sample side. If industrial gases are used, they have to be calibrated and a gas should also have appropriate isotope ratios, not too far from the samples measured. Thus, it might be worth mentioning that gaseous standards can be produced by gravimetric methods and are available through Messer Griesheim, Krefeld, Germany. The isotope ratios can be chosen by the customer within a range of ‰: −4 to −8; −20

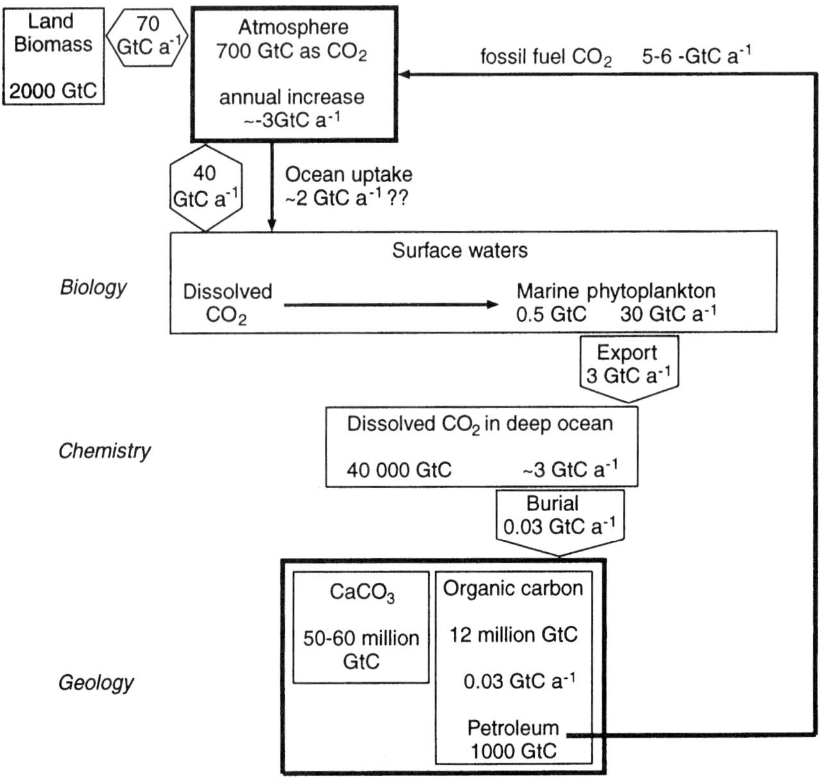

Fig. 14.2 Major reservoirs and C fluxes (1 GtC = 1 gigaton C = 10^{15} g C; after DeBaar and Suess, 1992).

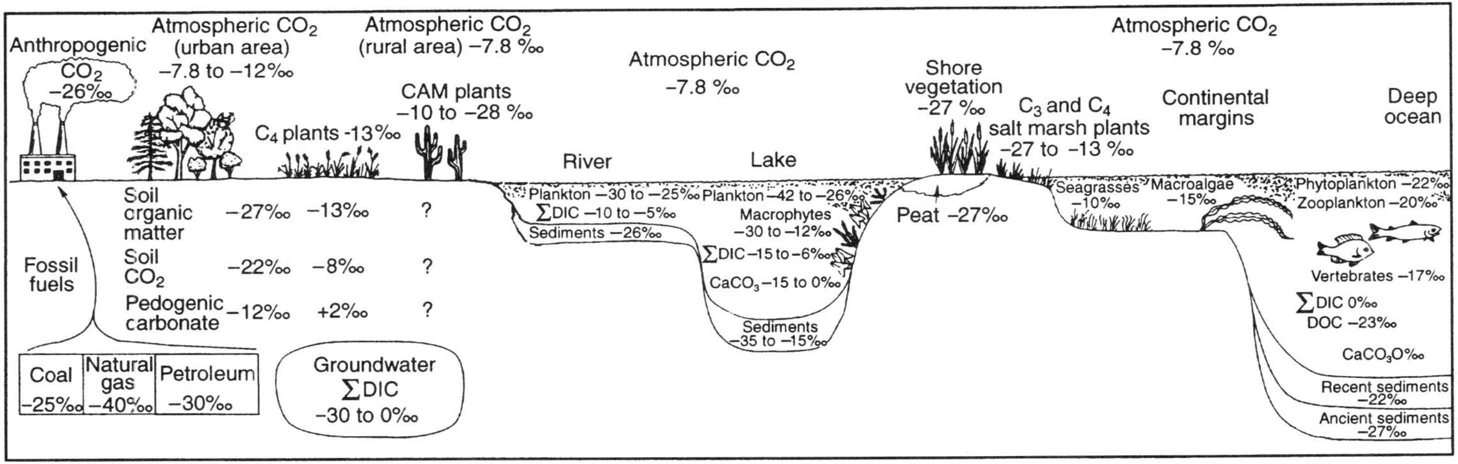

Fig. 14.3 Isotope ratios of various components from terrestrial, marine and freshwater environments (after Boutton, 1991b).

to −40 versus PDB for C. These standards, called ISO-TOP, are also available for H_2 and S.

14.3.1 Biospheric C Isotopes

The main biospheric C reservoirs are shown in Fig. 14.2 and the isotope ratios of inorganic C and organic C differ due to the reaction mechanisms involved. Thus the various compounds show C isotope ratios which are characteristic for the environment. Figure 14.3 shows some typical isotope ratios of major components from different ecosystems. The total range in ^{13}C observed in biologic- or organic-derived matter has shown values between +20 and −110‰ (versus PDB).

14.3.2 Atmospheric C

The components of the present atmosphere have already been mentioned and it is especially interesting for geoscientists to obtain information about the isotopic composition of atmospheric constituents and possible fractionation during, for example, respiratory processes, sedimentation or weathering of rocks and exchange at physicochemical boundaries, i.e. air–water. Additionally, the knowledge of such processes might provide information about the past atmosphere and its CO_2 content.

The main C compound in the atmosphere is CO_2, despite its low concentration which has increased since pre-industrial times from *c.* 290 ppm to 323 ppm in 1972 (Farmer and Baxter, 1974 a) and has now reached a value of 345 ppm (Friedli *et al.*, 1986). The average $\delta^{13}C$ value is *c.* 7.8‰, but it varies locally and seasonally. Minor components are CO and CH_4, of which the latter has been found to increase in the range 1 to 2% per year. Craig *et al.* (1988) stated a shift from 0.7 ppm to 1.7 ppm for the last 200 a, based on measurements from polar ice cores. The $\delta^{13}C$ ratios have in due course changed from *c.* 50‰ to a present-day value of 47.7‰. In addition to the usual CH_4 production due to bacterial activity, the increased burning of fossil fuel and gas as well as the digestive CH_4 production through large numbers of cattle might influence the $\delta^{13}C$ ratios. The amount of CO within the atmosphere is small with concentrations of 0.1 to 0.2 ppm, but shows a local variability with higher values (*c.* 0.5 ppm) in the southern hemisphere. The C isotope ratios range from −22‰ to −32‰ according to Stevens and Clayton (1971) and the observed average increase of *c.* 1% a^{-1} might as well be due to the burning of fossil fuel not least in car engines, whose emitted CO has mean $\delta^{13}C$ ratios of −27.4‰.

The importance of atmospheric CO_2 is due to several reasons. As a greenhouse gas, the increasing CO_2 level causes an increase of the mean global temperature and thus has severe consequences for society. The greenhouses gases, which also include CH_4 and other molecules with three

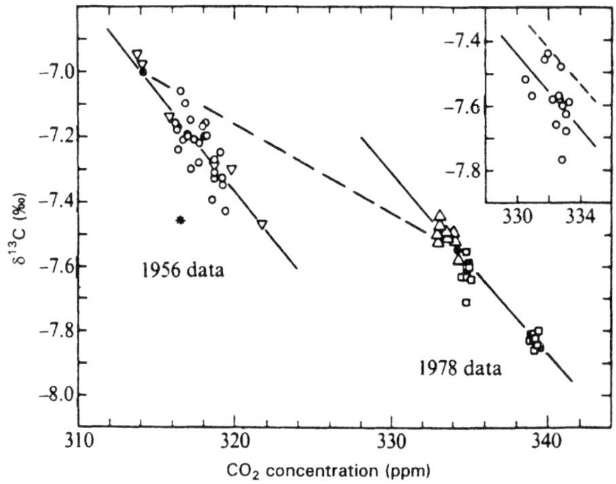

Fig. 14.4 Change of $\delta^{13}C$ ratios related to the concentration of CO_2 in the atmosphere, observed over a time span of 22 years. The mean shift is shown by the dashed line. (After Keeling, Mook and Tans, 1989.)

or more atoms, absorb infrared light (heat radiation) which is emitted from the Earth. Thus the heat balance is disturbed, and secondary processes such as global ice melting and connected sea-level changes occur. Consequently, scientific research aims at prognosis for the future, partially based on data from the past. Approaches have been made in order to estimate the CO_2 concentrations of the palaeoatmosphere, and a relationship between the atmospheric CO_2 level and the $\delta^{13}C$ in CO_2 has been observed.

Keeling (1985) and Keeling, Mook and Tans (1979) recorded results from measurements of atmospheric CO_2 enabling an estimate to be made if the $\delta^{13}C$ of tree rings mirror the usually varying atmospheric concentration of CO_2. Figure 14.4 shows the observed relationship between the atmospheric level of CO_2 and its $\delta^{13}C$ ratios. According to these observations, the $^{13}C/^{12}C$ ratio of the atmosphere decreased by *c.* 0.6‰ between AD 1956 and 1978, which fits well with the results derived from ice cores by Friedli *et al.* (1986), who found a decrease of approximately 1.2‰ for the last 130 years, thus implying that the changes became more rapid in the most recent period. This is obviously due to the progress of technology and rising population.

Freyer and Wiesberg (1973) and Farmer and Baxter (1974a) already implied that the $\delta^{13}C$ record in tree rings reflects changes in the CO_2 level of the atmosphere, and complications were discussed by the authors in due course (Freyer and Wiesberg, 1973; Farmer and Baxter, 1974b). In any case, the material available for research only provides data over a small time span, although the $\delta^{13}C$ values in tree rings can possibly provide a tool for the estimation of past CO_2 levels.

A main source for the reconstruction of the palaeoatmosphere comprises air inclusions trapped in continental and polar ice and, as mentioned before, much information derives from these investigations, e.g. Lorius *et al.* (1985) and Barnola *et al.* (1987), who presented a climatic and CO_2 record over *c.* 150 ka from the Antarctic Vostok ice core. Genthon *et al.* (1987) carefully concluded that climatic changes would be triggered by insolation changes with the relatively weak orbital forcing being amplified by possibly orbitally induced CO_2 changes. However, orbital, inclination and procession parameters should usually remain constant and, although cyclic variations (i.e. Milankovitch) can be identified by Fourier analysis, these explanations alone do not satisfy the observations. More information on the subject has been given by Bowen (1991) and thus will not be repeated in this chapter.

Naturally, the time span covered by polar and continental ice sheets is limited. However, the data obtained through this research have immensely increased knowledge of C and O_2 isotope variations during the last glacial/interglacial cycles and have also shown shifts in the O_2/N_2 ratio of the atmosphere (cf. a method developed by Sowers, Bender and Raynaud, 1993).

Another interesting method to estimate the pCO_2 of the paleoatmosphere is based on the fact that the C isotope fractionation during photosynthetic activity reflects the concentration of dissolved CO_2 and therefore allows estimation of the ancient atmospheric CO_2 concentration. A further mineralogical source for the estimation of pCO_2 has recently been described by Yapp and Poths (1992), based on earlier research, e.g. Yapp and Poths (1986), showing a relationship between the CO_2 pressure of the ancient atmosphere and the C content in goethites. The mineral goethite, α-FeOOH, contains small amounts of $Fe(CO_3)OH$, which preserves a record of the atmospheric CO_2 level from the time of mineral formation. The data derived from goethites to ironstones from the Upper Ordovician (Neda Formation, Wisconsin) suggest a CO_2 level 16 times higher than today. The subaerial formation of the goethites offers two sources for the CO_2, the atmosphere and the oxidation of organic matter. The mixing equation given by the authors is

$$\delta^{13}C = (\delta^{13}C_A - \delta^{13}C)\ [X(A)][1/X] + \delta^{13}C_o,$$

with C_A and C_O being the atmospheric and organic-derived C and X as the mole fraction of the $Fe(CO_3)$ OH. The measurments led to the following linear regression:

$$\delta^{13}C = 0.0162\,[1/X] - 20.1$$

with a high correlation coefficient of 0.98. Further calculations considering, among other things, the temperature of goethite formation and using the following equation:

$$\log pCO_2 = \log X + 6.04 - 1.57\ T$$

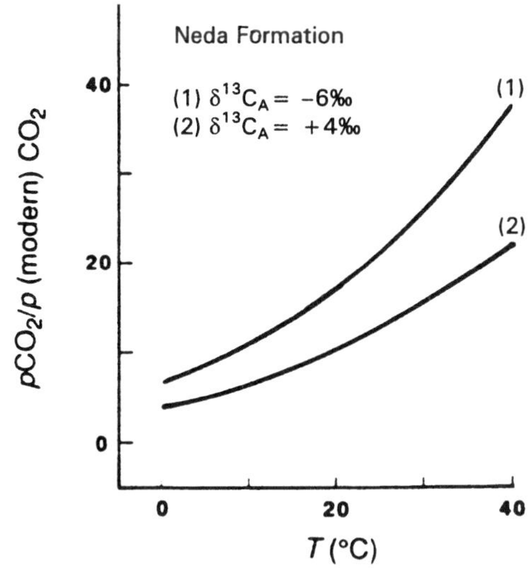

Fig. 14.5 Calculated values of ancient CO_2 (atmospheric) versus the temperature of goethite formation. Curves 1 and 2 assume a 10% range of atmospheric CO_2.

enabled a calculation of the pCO_2 from measured data (Fig. 14.5). This approach may be of interest as goethite of varying ages can be found.

Generally, the processes controlling the atmospheric composition need a far better understanding because not only may the CO_2 amount vary, but also the concentrations of O_2 and N_2. Bendes *et al.* (1994) analysed firn samples from various depths at the Vostok Station, Antarctica, for their $\delta^{15}N$ and O_2/N_2 ratios as well as for the CO_2 content, based on a method developed by Sowers, Bendes and Raynaud (1993). They found that gravitationally corrected CO_2 decreased in older (deeper) samples which are less influenced anthropogenically, whereas the corrected O_2/N_2 increases with depth due to anthropogenic O_2 consumption in the course of fossil fuel burning. Keeling and Shertz (1992) calculated a decrease of 0.018‰ a^{-1} for the O_2/N_2 ratio. As is obvious, the O_2 content of the atmosphere may also be influenced by the destruction of O_2 producers.

A possible uptake of anthropogenically produced CO_2 by the oceans is still being debated. Boyle (1988) presented a model for calculating the late Quaternary CO_2 variability, in which the vertical rearrangement of chemicals in the water column is a basic factor controlling the CO_2 uptake. Other studies, for example Schwartzman and Volk (1989), stressed the role of biotic influence on the chemical weathering of silicate rocks and Berner (1990, 1992) and Brady (1991) presented models to calculate the CO_2 level up to the Phanerozoic, also based on weathering processes of silicates. As a consequence of these and other pCO_2 estimations (e.g. Tans, Fung and Takahashi (1990), Brady (1991) we have some idea of CO_2 fluctuations during Earth history and know that maximum levels (when there were several times more CO_2 in the atmosphere than at present) have occurred, for instance in the Cretaceous. However, a largely neglected fact, the possible influence of the CO_2 pressure on organisms, seems to be worth a few further considerations.

Although there is no doubt that CO_2 is not a mutagen, Fajer, Bowers and Fakhri (1989) found CO_2 enrichment having an effect on plant–insect interactions. Larvae of herbivore insects raised in a 700 ppm CO_2 enrichment grew more slowly and showed an increased mortality. An even greater effect on microorganisms such as bacteria and viruses as discussed, for example, by Hämmerle and Inacker (1990), Suttle, Chan and Cottrell (1990), Beckmann and Klopries (1991) and Beckmann *et al.* (1995) might, because of the rapid generation changes of the microorganisms, lead to new qualities to which higher plants and animals are not adapted.

14.3.3 Marine C

The oceans are the largest C reservoir within the C cycle and contain both organic and inorganic C. The $\delta^{13}C$ of these differ, due to the reaction processes involved, and the usual expression for both parts is DIC and DOC, namely dissolved inorganic C and dissolved organic C. Most of this carbon is present as the HCO_3^- ion and some as the $CO_3^=$ ion, although the word 'dissolved' is not quite correct because the ions are a result of a dissociation of a molecule.

The dissolved part is the CO_2 which is dissolved in the sea water and is usually less than 1% of the total DIC. More correct are the abbreviations as TOC and TIC for the total organic or inorganic C because they neglect the production mechanism.

It should be mentioned that the acidity of sea water is determined by the H^+ ion concentration (as usual as for all solutions) which can be an acid or a base. The concentration is expressed as the *p*H value, which is the negative logarithm of the H^+ concentration and thus, $[H^+] = 10^{-pH}$. With a pH value of *c.* 8, the ocean water is a slightly basic solution and, as the acidity of the oceans is of importance for biological and geological processes, it might be asked if there is a relationship between the atmospheric CO_2 level and the pH value of sea water. In fact such a relationship has been calculated but the derivation, based on different reactions, will not be given here. However, it appears that, at temperatures of 25° C, shifts of atmospheric CO_2 will only result in small pH changes. Anyway, it is not yet clear whether, under normal conditions, the CO_2 pressure controls the pH value or vice versa. Regarding the anthropogenically emitted CO_2, a natural balance seems to be influenced, as might have happened also during some global events, for instance at the K–T boundary, when global wildfires could have caused the emission of CO_2.

Additionally, the acidity of the oceans might be buffered by the reaction of H^+ ions with clay minerals produced by weathering. The average $\delta^{13}C$ of the TIC is about 0‰, although surface waters show variations in a range of 1 to 3‰ for C isotopes (Anderson and Arthur, 1983). The average $\delta^{13}C$ value for the surface waters of central ocean areas is *c.* 2‰, as found in an extensive study by Kroopnick (1985), and is assumed to have decreased since pre-industrial times by *c.* 0.5‰, due to the anthropogenic production of CO_2. Quay Tilbrook and Wong (1992), on the other hand, report a $\delta^{13}C$ decrease of 0.4‰ for dissolved inorganic C of Pacific surface water for the much shorter time span between 1970 and 1990.

The marine limestones and calcareous shells of marine organisms generally reflect the C isotope ratios of the water in which they formed, supposing that non-equilibrium reactions (biological fractionation = vital effects) do not occur, but are usually enriched in heavy C up to 2‰. The temperature dependence for the equilibrium fractionation is fairly small (0.035‰ K^{-1}; Emrich, Ehalt and Vogel, 1977) and thus cannot account for the observed differences in the C isotope ratios. Also, the $\delta^{13}C$ of the surface water is a function of the rate of upwelling of intermediate deep waters as well as of temperature and associated rates of CO_2 invasion and retreat.

Light $\delta^{13}C$ values at high southern latitudes and at the equator are in part a consequence of the upwelling of ^{13}C-depleted cooler water, which is also capable of absorbing more CO_2, at the circumpolar and equatorial divergences. The main process controlling the $\delta^{13}C$ ratio of the sea water might be the biological production and decomposition of organic matter. Photosynthesis by marine organisms uses the Calvin cycle of the C_3 plants, the light ^{12}C isotope being preferentially incorporated in the organic matter produced, thus leaving the residual water within the photic and deeper surface zone of the oceans enriched in heavy C. The opposite effect occurs in deeper waters, if the dead organisms sink down and the ^{12}C is released again. Additionally, salinity changes and, in near shore environments, mixing may cause shifts of the $\delta^{13}C$ ratios.

For organic C in marine phytoplankton, the $\delta^{13}C$ values range from −10 to −31‰, but the average for the warm-water plankton lies between −17 and −22‰. On the other hand, a relationship between the temperature and the $\delta^{13}C$ ratio is not clear, because the data show no significant decrease in heavy C between 30 and 14°C but a rapid decrease between 10 and −2°C. Also, the rate of the decrease is not strictly linear, but varies in the range from 1.5‰ K^{-1} to 0.5‰ K^{-1}.

14.3.4 Freshwater and Terrestrial C

Carbon compounds in freshwaters such as lakes and rivers, usually dissolved HCO_3^-, exhibit $\delta^{13}C$ values lower than those from marine environments. Keith and Weber (1964) first compared isotope ratios of marine and freshwater molluscs showing variations in $\delta^{13}C$ of more than 15‰, similar to the range from −18 to 0‰ found by Baertschi (1957) for freshwater samples of different origin. The $\delta^{13}C$ of the total inorganic C is strongly depleted in heavy C due to the input of isotopic light CO_2 from decomposed terrestrial organic matter and this depletion is reflected in the isotope ratios of shells from freshwater molluscs, thus being distinguished from that of marine species. Other factors controlling the isotope ratios of C are similar to the marine environment, including the equilibrium with atmospheric CO_2, photosynthetic activity and the dissolution of carbonate rocks.

Quade, Cerling and Bowman (1989) analysed pedogenic carbonates of three profiles from the semiarid southern Great Basin, USA, for their C and O isotope ratios to elucidate the relationship between the isotopic composition and factors as climate, parent material and soil depth. They found that for all three profiles, the surface carbonates showed $\delta^{13}C$ values close to the equilibrium with atmospheric CO_2. The carbonates from a soil of depth >0.5 m further showed a total variation of *c.* 12‰ over an elevation range of c. 2500 m with decreasing $\delta^{13}C$ ratios parallel to increasing elevation (Fig. 14.6). The variations found are attributed to lower soil respiration rates at lower elevations as well as from variations in the occurrence of plant types (C_3, C_4, and CAM). Naturally, if the climate is a controlling factor, the isotopic composition of soil carbonate will differ with the environment. In arid areas the contribution of the flora might be neglected, whereas in humid areas the influence of plants and rain or snow dominates. Additionally, there is a seasonal sensitivity and variations in the concentration of soil CO_2, repiration rates and biomass production.

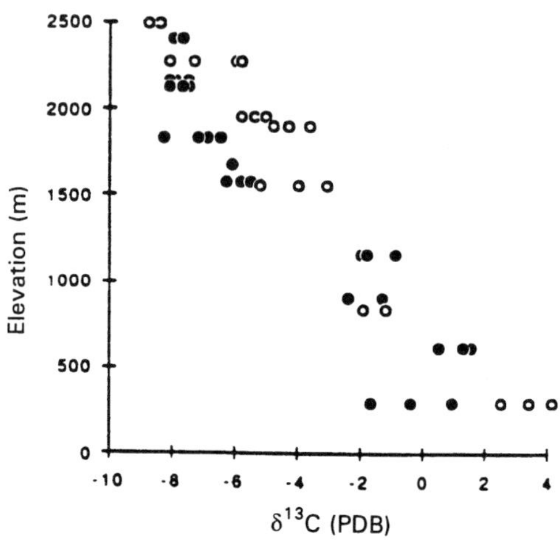

Fig. 14.6 Carbon isotope composition of pedogenic carbonates from depths > 0.5 m versus elevation (• = non-calcareous parent material; ○ = calcareous parent material; after Quade, Cerling and Bowmar, 1989).

Also depending on the seasons and meteorological conditions is the rate of C trapped or exported.

14.4 SULPHUR ISOTOPES

The element sulphur was known already at the time of Homer. It has valence states of +6, +4 and −2 and an atomic number of 16. Sulphur has 16 isotopes of which four are stable. The latter have the following abundances: $^{32}S = 95.02\%$; $^{33}S = 0.75\%$; $^{34}S = 4.21\%$; $^{36}S = 0.02\%$. The element is of minor importance within the atmosphere, where it occurs in the compounds SO_2 and H_2S, mainly emitted anthropogenically or in volcanic areas during eruptions or post-volcanic activity. Elemental S can also be found in these regions. The anthropogenic SO_2 derives mainly from the burning of fossil fuels (coal and crude oil), and the isotope technique might be helpful in future to estimate the degree of artificial pollution. Mineralogic compounds containing S are mainly sulphides and sulphates, the latter being easily soluble in water except for the sulphates of Pb, Ca, Sr and Ba. In sea-and freshwater, S occurs dissolved in the form of the SO_4^{2-} ion. The average concentration of SO_4^{2-} in ocean water as given by Turekian (1976) is 2.712 g kg^{-1}, and a residence time of 7.9×10^6 a has been estimated. The total amount of S in the oceans is *c.* 1.3×10^{15} t, in the form of SO_4^{2-} ions. Its isotope composition is nearly constant with a value of *c.* $^{34}S = +20‰$.

The input of SO_4^{2-} ions into the oceans mainly occurs through the discharge of fresh water. The SO_4^{2-} is derived from weathering sulphide-bearing sedimentary rocks (shales and carbonates), marine-derived evaporites and volcanic plus primary igneous rocks. The cycle is completed by removal of S from oceans both by the formation of evaporitic rocks and by the bacterial reduction of SO_4^{2-} to S^- succeeded by precipitation as pyrite or marcasite. The rates of such processes as well as the isotope compostion of S vary chronologically and reflect such factors as rates of weathering and erosion, and intensity of global igneous activity. Although the erosion of black shales containing light sulphides can lower the ^{34}S value of oceanic SO_4^{2-} the greater reduction of SO_4^{2-} by bacteria and abstraction of light S^- from oceans may increase this value. Interestingly, evaporite formation reduces the concentration of SO_4^{2-}, but does not alter the ^{32}S value.

Sulphur is a basic constituent of organic compounds and, as a part of many proteins, is vital for higher organisms. In agriculture, S compounds are used as fertilizers.

The fractionation factor for the chemical equilibrium:

$$^{32}SO_4^{2-}{}_{(aqueous)} + H_2\,{}^{34}S_{(gaseous)} \Leftrightarrow {}^{34}SO_4^{2-}{}_{(aqueous)} + H_2{}^{32}S_{(gaseous)}$$

has been given by Tudge and Thode (1950) as $\alpha = 1.075$ at 25°C. While this equilibrium cannot be reached by chemical systems at low temperature, it may be approached by enzyme-catalysed sulphate reduction in bacteria.

At high temperatures, the isotope exchange in the system $SO_4 - H_2S$ has been investigated by Thode *et al.* (1971).

The isotope composition of S is expressed by the common δ-notation in terms of $\delta^{34}S$ and defined in the usual manner as:

$$\delta^{34}S = \frac{(^{34}S/^{32}S)_{sa} - (^{34}S/^{32}S)_{std}}{(^{34}S/^{32}S)_{std}} \times 1000\ (‰),$$

$\delta^{34}S$ measurements are performed on SO_2 gas using a double inlet mass spectrometer with a double or multiple collector system. The $\delta^{34}S$ values are usually reported relative to the Cañon Diablo Standard (CDT), and the material is troilite – which the T in the abbreviation stands for – of the Cañon Diablo iron meteorite. Troilite is a hexagonal variety of pyrrhotite, a sulphite, with the ideal FeS formula, whereas most other pyrrhotites have in reality a composition of Fe_nS_{n+1}. Troilite occurs in thin veins or nodular masses in different types of meteorites and the standard has a $^{34}S/^{32}S$ ratio of 22.22‰. This isotope composition is close to that of mafic igneous rocks and hence the $\delta^{34}S$ value of a particular sample of terrestrial S represents a measure of the alteration which occurred in its isotope composition after its introduction into the crust of the Earth. The standard itself is not available, similarly to the PDB standard, but the IAEA distributes different intercomparison samples, for instance NBS–122 and NBS–123 (both sphalerite, Zns), NBS–127 (which is a Ba sulphate) and the standards IAEA–S–1 and IAEA–S–2 (Ag sulphide) or IAEA–S–4, which is S . All these standards are on a natural level. The NBS–123 sphalerite has for instance a ^{34}S of 17.32‰ and the Ba sulphate NBS–127, which is also appropriate for $\delta^{18}O$ measurements has δ-values of 20.58‰ (^{34}S) and 9.31‰ (^{18}O) versus CTD (V-SMOW).

The natural variation of the S isotope ratios can be very large due to the fractionation processes which occur within the different environments. In fact, the greatest range of ≈ -50 to $+100$‰ can be found in sedimentary rocks, whereas the sulphate in seawater turned out to have relative homogenous $\delta^{34}S$ values of about 21‰ (Rees, 1978).

Any variations in the isotope composition of S result from reduction of O_4^{2-} ions to H_2S by some anaerobic bacteria to cause enrichment of the H_2S in ^{32}S, and also by isotope exchange reactions between S-bearing ions, molecules and solids through which ^{34}S is concentrated in those compounds possessing the highest oxidation state of S or greatest bond strength. Biogenic fractionation occurs extensively in nature. The anaerobic bacteria, for example, *Desulfovibrio desulfuricans,* live in sediments deposited in oceans and lakes. They separate O_2 from SO_4^{2-} ions, excreting H_2S enriched in ^{32}S relative to the SO_4^{2-} . The degree of fractionation varies and depends upon certain rate-controlling steps in the reactions by means of which S is metabolized. In the case of inorganic systems, the amount of isotope fractionation during the reduction of SO_4^{2-} to H_2S is a function of the differing rates at which the S–O bonds are broken. Therefore, the first H_2S

produced through the inorganic reduction of SO_4^{2-} is enriched in ^{32}S by *c.* 22‰ compared with the SO_4^{2-}. The fractionation associated with bacterial reduction of SO_4^{2-} can be explained as a two-stage process comprising the entrance of SO_4^{2-} into the cell giving a small isotope shift, and the break-up of S–O bonds which produces a large change (up to 22‰) in the isotope composition and controls the process. Kaplan and Rittenberg (1964) achieved ^{32}S enrichments in H_2S of as much as 46‰ relative to SO_4^{2-}. Also, sulphide minerals in newly deposited sediment can be enriched in ^{32}S by *c.* 50‰ compared with associated marine sulphate. It is apparent that isotope fractionation in nature is more extensive than would be inferred from the laboratory experiments and is not, in fact, limited by the fractionation connected with the chemical reduction of SO_4^{2-} ions to H_2S.

In microorganisms, the extent of isotope fractionation is inversely proportional to the rate of SO_4^{2-} reduction, which itself is controlled by the temperature and/or the SO_4^{2-} concentration. It was found that the rate of this SO_4^{2-} reduction as well as the magnitude of the fractionation depends on the nature of the electron donor, ethanol giving the largest observed effect. The isotope fractionation of S by sulphate-reducing bacteria apparently became significant only after about 2.35 Ga ago, i.e. after photosynthetic sulphur-oxidizing bacteria had increased the S concentration of the oceans sufficiently. Hence, photosynthesis apparently preceded bacterial sulphate reduction, which was succeeded by oxygen respiration in Late Proterozoic time. However, as a result of the early evolution of green or purple sulphur-oxidizing bacteria, the presence of sulphate deposits in Precambrian rocks of Archaean age is not admissible evidence for O_2 being present in the early Precambrian atmosphere. It is interesting that sedimentary–volcanic rocks from Isua in West Greenland laid down over 3.7 Ga ago have average $\delta^{34}S$ values close to zero, from which it is clear that no sulphate-reducting bacteria then existed on Earth. According to Monster *et al.* (1979) bacterial sulphate reduction started between 3.2 and 2.8 Ga ago, although this may have to be revised in view of results obtained by Fripp, Donelly and Lambert (1979). These researchers obtained an average δ^{34} S value of $+0.6$‰ for 31 samples from banded Archaean iron formations from Zimbabwe. Cameron (1983) analysed sulphides from seven such iron formations from all over the world, and indicated that their $\delta^{34}S$ value of sulphides in Precambrian shales in South Africa were close to 0‰ until *c.* 2.35 Ga ago, when they shifted to negative values. This may have been triggered by an increase in the SO_4^{2-} concentration in the oceans which enhanced isotope fractionation by sulphate-reducing bacteria, these becoming significant participants in the terrestrial S cycle *c.* 2.3 Ga ago rather than in the 3.2–2.8 Ga time interval of Monster *et al.* Strauss and Schieber (1990) studied the pyrite genesis from samples of the Mid-Proterozoic Newland Formation, Montana, and found $\delta^{34}S$ values ranging between -14 to $+18$‰ (CTD). The samples with negative values contained a high amount of early diagenetic pyrite due to a formation by bacterial reduction of contemporaneous sea water. The $\delta^{34}S$ increases with a decreasing content of the early diagenetic pyrite, indicating a shift from an open system to a closed system with a lack of sulphate supply (Chanton, Martens and Goldhaber, 1987).

The S isotope ratios of sulphate minerals from Phanerozoic evaporites vary as a function of age and many values have been assessed in order to ascertain details of the history of the isotope composition of marine sulphate. Relevant results are shown in Fig. 14.7a, embodying results of Claypool *et al.* (1980). They demonstrate that the $\delta^{34}S$ values of marine sulphates probably went down from *c.* +30‰ during the Cambrian period to *c.* +10‰ at the end of the Permian, thereafter increasing in an irregular manner during the Mesozoic Era and today reaching +20‰.

Because it is clear that S isotopes are not markedly fractionated by precipitation of gypsum from sulphate-bearing brines ($\delta^{34}S$ gypsum-brine = +1.65 ± 0.12‰), it is assumed that the $\delta^{34}S$ values of marine sulphate minerals represent the isotope composition of the brines in which they were precipitated. However, the $\delta^{34}S$ values of sulphate minerals during any particular geological time interval usually vary within limits which makes it hard to choose a representative value for the oceans during that interval of time. This contemporaneous variability may be due to a number of possible causes. For instance, isotopically light sulphate may be introduced by rivers, or isotopic fractionation may take place while minerals precipitate from brines. The result is that $\delta^{34}S$ values from ancient oceans are

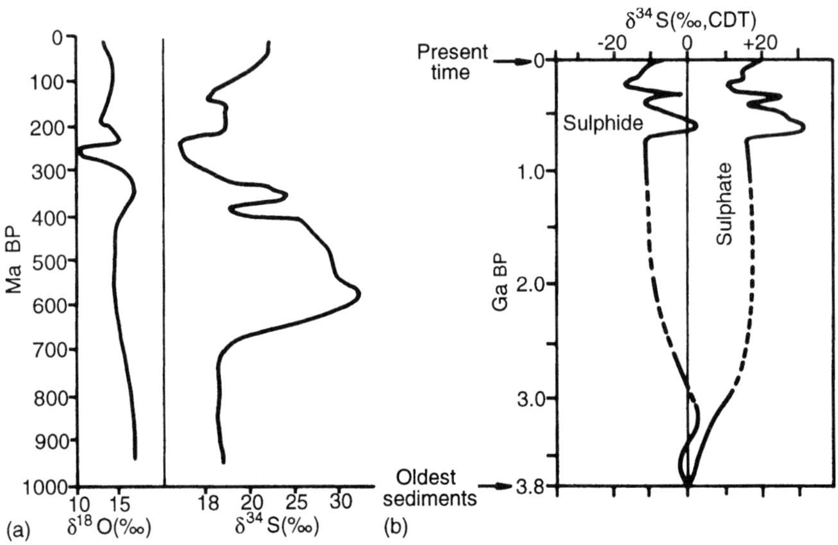

Fig. 14.7 The S isotope record (a) after Claypool *et al.* (1980), (b) after Schidlowski (1988).

assessed mainly on the general evidence, aberrant values being rejected since they may derive from non-marine origin of the particular deposit. Despite this, the $\delta^{34}S$ curve in Fig. 14.7a is believed to be substantially valid and the curve in Fig. 14.7b presented by Schidlowski (1982, 1988) is similar for the given time span. Additionally, secular variations must be considered (Holland, 1984) and further research, possibly investigating zonations as well, might help to answer open questions.

14.5 STRONTIUM ISOTOPES

The isotope composition of Sr in the hydrosphere is dependent upon the $^{87}Sr/^{86}Sr$ ratios of rocks interacting with water either on or close to the terrestrial surface. Strontium liberated into solution homogenizes isotopically by mixing while being transported, until the oceans or closed basins are reached. Thereafter the element may once again enter the rock cycle, mainly by coprecipitation with $CaCO_3$. Consequently, the sedimentary carbonate (and evaporite) rock record contains details of the alteration in the isotope composition of Sr in the oceans and on continents from the Proterozoic Eon onwards.

Continental chemical weathering contributes Sr into solution in rivers, lakes and groundwater, its isotope composition reflecting ages and Rb/Sr ratios of parent rocks as well as relative solubilities of the various minerals present. Of course, the isotope composition of Sr in the rocks will not be the same as that of Sr liberated by weathering because the Rb-bearing minerals are more resistant to this process. Thus there is preferential removal of Sr and the $^{87}Sr/^{86}Sr$ ratio of the Sr which enters solution is usually lower than that of the unweathered rocks. Consequently, chemical weathering increases the Rb/Sr ratios of rocks and enriches also weathered products in radiogenic ^{87}Sr. Brass (1975) quantified this chemical weathering effect, using an n-parameter defined as

$$n = \frac{1 - K_C}{1 - K_R},$$

in which K_C is the fraction of common Sr anf K_R the fraction of radiogenic Sr remaining in the rock in question. Usually, K_R exceeds K_C because minerals which resist weathering better contain more radiogenic ^{87}Sr than those containing common Sr.

Through chemical weathering, Sr contained in water in a drainage basin reflects the geology of the basin, so that the element may be employed to trace rivers which flowed across different rock types, and to investigate groundwater movement. A Sr isotope study of the Susquehanna River carried out by Fisher and Stuebar (1976) showed a variation in the $^{87}Sr/^{86}Sr$ ratio along its length between values of 0.7090 and 0.7154, resulting mainly from the mixing of water from tributaries. These drain a variety of different rock types – from carbonates, with lower $^{87}Sr/^{86}Sr$ ratios and higher Sr concentrations, to sandstone, shale, schist and granite gneiss. Despite their low area of *c.* 10%, the carbonates dominate the isotope composition and concentration of Sr in the river water. Actually, there is also a component of groundwater involved as well as direct run-off, but the isotope composition and concentration of Sr in these probably parallel values in the tributaries.

In brine lakes Sr concentrations probably increase greatly as a result of the evaporation and also of the precipitation of Sr-excluding minerals. Nevertheless, the isotope composition of Sr remaining in the brine is unaffected and related solely to the geology of the drainage basin. Therefore, the $^{87}Sr/^{86}Sr$ ratio can be used to ascertain whether the Sr derived from the rocks outcropping in a drainage basin entered a brine lake within it, or whether the lake in question obtained its Sr from sea water or geothermal brines of volcanic origin. Both groundwater and subsurface brines also contain Sr, the isotope composition of which depends upon that of the rock types with which they interacted. Following this principle, the $^{87}Sr/^{86}Sr$ ratios of oilfield brines have been employed in order to find out details of the origin of such brines as well as their migration histories; see for instance the Sr isotope study of Smackover brines and related solids in southern Arkansas, carried out by Steuber, Pushkar and Hetherington (1984). Here, and in other such work, these ratios were different from those of sea water at the time of deposition of the relevant host rocks; this is because the Sr derived mainly from the rocks with which the brines interacted. Of course, influence will also be exerted by ion-exchange relations with clay minerals, and it must always be remembered that such subsurface brines may represent a mixture of components from different sources. The Sr incorporated in evaporites and non-marine carbonates will have the same isotope composition as the parent brines in terms of the $^{87}Sr/^{86}Sr$ ratio, since Sr isotopes do not undergo fractionation during this process. Also, the $^{87}Sr/^{86}Sr$ ratios in such rocks do not vary geochronologically because both gypsum and calcite exclude Rb. Consequently, rocks of this type record the value of the original $^{87}Sr/^{86}Sr$ ratio in the brine from which they originally deposited. Any variations in $^{87}Sr/^{86}Sr$ ratios must be ascribed to alterations in the geology of the basin in which deposition took place.

The isotope composition of Sr in marine carbonates in Precambrian times is not well known, mainly because of the comparative rarity of such rocks compared with their frequent occurrence in the Phanerozoic Eon, but also as a result of the higher probability that the isotopic and other characteristic features of these rocks have been altered. Additionally, these rocks represent a far greater time interval than their Phanerozoic successors and are much more difficult to date, both because isotope dating methods are either inapplicable or imprecise and because index fossils are lacking. In point of fact, it is hard to distinguish marine from non-marine depositional basins in the earliest parts of the history of the Earth, at which time the oceanic salinity could have been much lower than it is now. Available information shows that, in the Archaean, the $^{87}Sr/^{86}Sr$ ratios of carbonates were close to those in the mantle, implying that the isotope geochemistry of the element of the terrestrial surface was then different from that of later times. This could have been related to more intense volcanism in the Archaean, and to the youth of surface rocks then. It could also be that the surface rocks possessed lower Rb/Sr ratios than later-formed igneous ones, so that the $^{87}Sr/^{86}Sr$ ratios of surface water resembled values then existing in the mantle. A final point is that carbonate rocks were then far from abundant and hence could not have moderated fluctuations in the $^{87}Sr/^{86}Sr$ ratio of the oceans as they did in the Phanerozoic.

In the early Proterozoic, i.e. between 2.5 and 2 Ga ago, $^{87}Sr/^{86}Sr$ ratios in marine carbonates may have risen appreciably above this ratio of mantle Sr, perhaps because radiogenic ^{87}Sr entered oceans from continental crustal rocks, which by then had become old enough to attain elevated $^{87}Sr/^{86}Sr$ ratios. This may reflect an increased erosion of ancient cratonic rocks induced by continental uplift accompanied by decreasing volcanism and sea-floor spreading. Then too, the $^{87}Sr/^{86}Sr$ ratio of the oceans could have been enhanced by continental glaciations such as that recorded by the Gowganda Formation of the Huronian Supergroup of northern Ontario in Canada and discussed by Card (1984). McCulloch (1994) reported primitive $^{87}Sr/^{86}Sr$ ratios of Archaean barite with an age of 3.45 Ga from the Pilbara Block of Western Australia. This initial ratio of 0.700 502 is similar to values gained from minerals of the Onverwacht Group basalts of the same age and implies a relatively high $^{87}Sr/^{86}Sr$ ratio of 0.700 50 for the early Archaean mantle.

Turning to the Phanerozoic, the isotope composition of Sr is now rather uniform in the oceans with $^{87}Sr/^{86}Sr = 0.709\,06 \pm 0.000\,03$, based on numerous analyses and corrected for interlaboratory bias to $^{87}Sr/^{86}Sr$ ratio of 0.708 00 for the Eimer and Amend $SrCO_3$ isotope standard. This value was substantiated by Burke *et al.* (1982). who derived $0.709\,10 \pm 0.000\,04$ relative to the same standard from 42 global modern carbonate samples.

In sea water, this ratio is influenced by the intermingling of three isotope varieties coming from young volcanic rocks, ancient sialic materials of the continental crust and Phanerozoic marine carbonate rocks. Hence the $^{87}Sr/^{86}Sr$ ratio of sea water indirectly reflects the rock types subjected to chemical weathering on continents and in ocean basins. It could be that the ratio in the oceans varied during geological time as a response to the types of rocks which were exposed to such weathering. The $^{87}Sr/^{86}Sr$ ratios of marine carbonate minerals are taken to be the same as those of sea water at the time of deposition if they are unaltered by diagenesis, dolomitization or regional metamorphism. Faure (1982) calculated average ratios in 20 Ma increments through the Phanerozoic and concluded that they varied systematically through this extensive time interval. Much additional evidence exists to support this view and short-term fluctuations have also been established. Nevertheless, it seems likely that oceanic Sr was isotopically homogeneous through out this time interval: this is feasible, bearing in mind the long residence time of the element in the oceans (*c.* 5 Ma) compared with the mixing time for these bodies of *c.* 1000 a. Incidentally, the concentration of Sr in the oceans is much higher at 7.7 $\mu g\ g^{-1}$ than in river water at 0.68 $\mu g\ g^{-1}$. Remembering the isotopic homogeneity of Sr in the oceans, short-term fluctuations in the $^{87}Sr/^{86}Sr$ ratio since the Cambrian simply reflect the varying proportions of Sr contributed by the young volcanics, ancient sial and marine carbonates mentioned above. During the Palaeozoic era the ratio underwent repeated fluctuations between the limits 0.7091 (Late Cambrian) and 0.7068 (Late Permian). By the Middle Triassic the ratio had risen to 0.7077, thereafter declining to 0.7068 by the Late Jurassic. Subsequently, it rose again through the Cretaceous and Tertiary, with only small fluctuations in the Middle Cretaceous and Early Tertiary time intervals.

The isotope geochemistry of Sr in the oceans is quite well understood, but the reason for any particular fluctuation of the $^{87}Sr/^{86}Sr$ ratio is unclear, although continental glaciation has been invoked, as also has orogeny. In addition, decrease of the ratio during the Mesozoic Era has been attributed to increase in the rate of sea-floor spreading and volcanism connected with the opening of the Atlantic Ocean and the rifting apart of Gondwana and Pangaea. However, major rapid fluctuations may have taken place solely when a decrease in the contribution of Sr from one source coincided with an increase in the other, i.e. the $^{87}Sr/^{86}Sr$ ratio of the oceans may have been decreased only when the input from the mantle increased and that from crustal sources declined, and vice versa.

14.6 CHLORINE ISOTOPES

The element chlorine derives its name from the Greek *chloros* which means yellow-green coloured. It was discovered in 1774 by C. W. Scheele and occurs mainly in sea water as Cl^- ion or bond in halogenous compounds and another main reservoir is the upper mantle. With an atomic number of 17, Cl has 12 isotopes of which two occur naturally. These are the stable isotopes ^{35}Cl and ^{37}Cl with relative abundances of 75.52% (^{35}Cl) and 24.471% (^{37}Cl) (Boyd, Brown and Lounsbury, 1955). Also the radionuclide $^{36}Cl(T_{1/2} = 3.01 \times 10^5 a)$ is of interest.

Chlorine belongs to the less commonly investigated isotopes, although Urey (1947) calculated fractionation factors for some Cl compounds. As Cl mainly occurs in the valence state 1 in the Cl^- ion, the natural range of isotope ratios is small. Additionally, the precision of $^{37}Cl/^{35}Cl$ measurements has not been satisfying. However, older investigations by Baertschi (1953) and Owen and Schaeffer (1955) indicated that the range of Cl isotope ratios in sea water is limited to 0.5‰. In due course, several authors measured Cl isotope ratios and confirmed this observation. Although some results might be questioned if a small isotope range was measured with a relatively low precision, it turned out that the δ^{37}C1 ratios mainly vary between +1 and −1‰ in both natural waters and mineralogical compounds.

Recent investigators such as Xiao and Zhang (1992), Long *et al.* (1993) and Magenheim *et al.* (1994) have described methods increasing the precision of δ^{37}C measurements. They used different methods, as the isotope ratios of Cl can be determined by both GIRMS and TIMS (N-TIMS) if other methods, e.g. electron microprobe, are neglected. The authors described the former methods as well, so that the following will be restricted to the new results. Long *et al.* (1993) applied the GIRMS for their analysis and achieved a routine precision of 0.09‰, using positive ions of methyl chloride gas (CH_3Cl^+) for the measurements. The method, described in detail by the authors, is summarized briefly for basic information.

Depending on the isotopic composition, the CH_3Cl has the masses 52 and 50. The 52/50 ratio is attributable to the ^{37}Cl ratio and thus the $\delta^{52}(CH_3Cl)$ is expressed as

$$\delta^{52}(CH_3Cl) = [(52.50)_{sa}/(52/50)_{std} - 1] \times 10^3 ‰,$$

with

$$CH_3Cl_{SMOC} = \delta^{52}{}_{sa} - \delta^{52}{}_{SMOC},$$

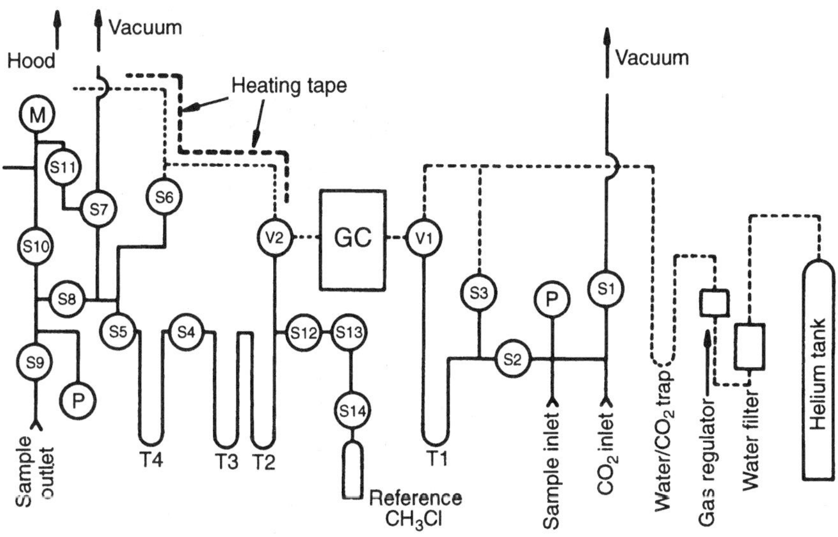

Fig. 14.8 Diagrammatic scheme for the separation of CH_3Cl. Solid line = vacuum line; dotted line = helium line; CF = cold finger; GC = gas chromatograph; M = manometer; P = pressure gauge; S = stopcock; T = cold trap; V = valve. (After Long *et al.*, 1993).

and SMOC as the international chloride standard (Standard Mean Ocean Chloride). Chloride solutions are prepared from solid sample by dissolving soluble matter or hydrolysing Cl-containing minerals. At least 3mg e^{-1} Cl^- ions are required for precise measurement. The Cl^- is then transformed into AgCl which is subsequently converted to CH_3Cl, as described by Long *et al.* (1993), through reaction with CH_3I:

$$AgCl + CH_3I \Rightarrow AgI + CH_3Cl$$

The sample preparation is quite time consuming and a further step is necessary to separate CH_3Cl and CH_3I from the reaction product, which is achieved by a vacuum apparatus shown in Fig. 14.8. Less sample material is required for the TIMS. Magenheim *et al.* (1994) used $2\mu g$ of Cl for the $^{37}Cl/^{35}Cl$ measurement of silicate and aerosols with low Cl levels. They achieved an external reproducibility of 0.25‰, which is also satisfactory for the observed total natural range of *c.* 3.5‰ for the $\delta^{37}Cl$ ratios. This is particularly important, as the minimal amount of Cl required allows $^{37}Cl/^{35}Cl$ determinations from a wide range of samples.

14.7 SILICON ISOTOPES

Elemental silicon (*silex* = pebble) was discovered by J. J. Berzelius in 1822. Silicon is the second most abundant element in the Earth crust and has the three stable isotopes ^{28}Si, ^{29}Si and ^{30}Si with the relative abundances (Bainbridge and Nier (1950) of 92.27%, 4.68% and 3.05% respectively. Slightly different values were given by Reynolds (1953). Among several radioactive isotopes, e.g. ^{26}Si, ^{27}Si, ^{31}Si, ^{32}Si, the cosmic ray-produced ^{32}Si is of geoscientific interest.

The $\delta^{30}Si$ value is expressed in the usual manner as the ratio of $^{30}\delta Si/^{28}Si$. Silicon isotopes are not widely used in geoscience because the $\delta^{30}Si$ variations are small and fractionation processes are limited, due to the nature of Si compounds. Usually bound to oxygen, as in siliceous organisms and minerals, the oxygen isotope ratios supply more ecological or geochemical information. Geochemical properties of stable Si isotopes have been discussed by Douthitt (1982) and observed isotope ratios vary within a range of 6‰. How far the stable Si isotopes can be of geoscientific interest in biogenic systems or in mineralogic processes is a question which may be answered if more data become available. As Si occurs in animals and plants, a biological fractionation during SiO_2 formation in skeletons may be possible, although source matter may show relatively limited $\delta^{30}Si$ values.

14.8 ISOTOPES OF NOBLE GASES

About 1% of the Earth's atmosphere consists of the noble gases He, Ne, Ar, Kr, Xe and Rd, of which Ar is the most abundant with 0.94 vol%. Only a few isotopes are of geoscientific interest apart from applications in radioactive dating. Amongst these are ^{3}He, ^{37}Ar, ^{39}Ar, ^{81}Kr, ^{85}Kr and ^{133}Xe. The abundance of the noble gases, except Ar and Rn, which only exist in traces, ranges from 9×10^{-6} (Xe) to 2×10^{-3} (Ne) vol%.

Helium has two stable isotopes, ^{4}He and ^{3}He, the latter contributing 1.3×10^{-4}% of the naturally ocurring helium (from *helios* = sun). A very interesting modification is helium-II, which can only be derived from ^{4}He. Its heat conductivity is better by a factor of 1000 than that of Ag and it has virtually no viscosity. An early dating method using the α-emission-produced particles which are seen as ^{4}He atoms was proposed by Rutherford in 1905. This U–Th–He method will not be discussed here, but is based on the assumption that the element He, produced by α-decay may be preserved in the minerals. As the α-particles, consisting of two protons and two neutrons, are emitted from the radioactive atom with a speed of about 16 000 km s^{-1}, not all He might be retained in a mineral, depending on the position of the mother nuclide within the crystal. This led to underestimated ages and the method has hardly been used since the 1950s and 1960s.

Mention should also be made of the ^{3}He–^{3}H method, which might replace the tritium method in determining the turnover times in ground water systems. A problem arising in this method is the identification of tritiogenic ^{3}He in ground waters. More information is given by Maloszewski and Zuber (1983).

The element argon (from the Greek *argos* which means 'the inert,' has a melting point of –189°C, a boiling point of –185°C and a density of 1.78 g l^{-1} under normal conditions. The atomic number is 18 and 15 isotopes with mass numbers ranging from 32 to 46 exist. Three of the isotopes are stable, namely ^{36}Ar, ^{38}Ar and ^{40}Ar, and are found mostly in the atmosphere. The unstable radiogenic ^{40}Ar is a decay product of ^{40}K and is discussed in Chapter 8. Of the radioactive isotopes, both ^{37}Ar and ^{39}Ar are of geoscientific interest. Whereas the latter occurs mainly naturally, ^{37}Ar is also of anthropogenic origin. The activity of ^{39}Ar in the atmosphere was first measured by Loosli and Oeschger (1968), i.e. 1 year after the initial measurements of ^{37}Ar. Both isotopes can be produced by several nuclear reactions, of which the most important are resulting from the decay of ^{40}Ar: ^{40}Ar (n, 2n) ^{39}Ar and ^{40}Ar (n,α) ^{37}Ar. The radioactive Ar isotopes have half-lives of 269 a (^{39}Ar) and 35.1 d (^{37}Ar) and decay either by β-emission as for ^{39}Ar ($^{39}\text{Ar} \rightarrow {}^{39}\text{K} + e^{-}$) or electron capture as for ^{37}Ar ($^{37}\text{Ar} + e^{-} \rightarrow {}^{37}\text{Cl}$). Both isotopes occur in the atmosphere, in groundwater which is in contact with the atmosphere and in bubbles of air preserved in glacial ice. At about 70 m depth, névé is converted to ice and air bubbles are entrapped. This offers a tool for dating ice samples of such different provenances as, for instance, Antarctica or Greenland or glaciers elsewhere. In fact, the first use of ^{39}Ar isotopes was to date ice samples by measuring the activity of the trapped Ar compared to recent atmospheric ^{39}Ar. As the half-life of ^{37}Ar is short and the concentration in the atmosphere small compared to ^{39}Ar (0.112 ± 0.010 dpm 1^{-1} Ar), ^{37}Ar is not useful for such measurements. It is necessary to collect large volumes of water or ice in order to analyse the rates of disintegration for ^{39}Ar and, for that matter, of radiocarbon and ^{81}Kr as well. Oeschger (1982) reported that data from a borehole at Station Crete, Greenland, agree well with those obtained by the counting of annual layers in the ice cores, shown by seasonal variations of $^{18}O/^{16}O$ ratios. Agreement of the measured ages has also been found with data derived from ^{32}Si dating.

A major problem in using ^{39}Ar as a dating tool is partly due to its chemical inertness. Unlike ^{14}C, ^{32}Si and ^{36}Cl, noble gases do normally not interact with solids and a first real reaction with the noble gas Xe was reported in 1962. In addition, Ar and the other noble gases and their radionuclides are found mostly in the atmosphere. Hence their concentration in groundwaters is dependent on the time interval which elapsed since the water was in contact with the atmosphere. Finally, some radionuclides such as ^{39}Ze, ^{37}Ar and ^{85}Kr are partially of anthropogenic origin, as ^{133}Xe, which does not occur naturally. Up to 5% of ^{39}Ar might be due to atmospheric nuclear tests and an increase of the ^{37}Ar activity has to be attributed to underground bomb tests.

15

Isotopes in palaeoclimatology and palaeoecology

The importance of isotope analysis for the reconstruction of palaeoclimates and the achievement of palaeoecological information has already been mentioned in previous chapters. Since the basic work of Urey and his associates in the 1950s, extended studies have been carried out taking into account new data and problems arising from the further development of the methods. The estimation of palaeotemperatures by measuring the oxygen isotope ratios of $CaCO_3$ from marine fossils had great consequences for interpreting the history of climatic change of Earth, although it has turned out that palaeotemperatures have to be interpreted carefully if reliable data shall be obtained. Unrealistic results can derive from various factors such as salinity, diagenetic alteration of the samples, ice volume effects, mixing with fresh water, life habits, biological fractionation or mineralogical structure of the $CaCO_3$. Although these parameters might influence and limit the value of temperature determinations, each one of them gives opportunities for deriving more palaeoecological information, especially if comparable data gained through faunal or floral analysis are available.

15.1 MARINE ISOTOPE PALAEOTHERMOMETRY

Since the estimation of palaeotemperatures of ancient oceans by measuring the oxygen isotope distribution between $CaCO_3$ and water derived from an original suggestion of H.C. Urey, the isotope tool has become very important for interpreting the history of climatic change on Earth.

The first carbonate–water palaeotemperature scale was based upon the fact that the isotopic composition of oxygen from $CaCO_3$ differs from that of the water and depends on the temperature when the carbonate in question is precipitated under equilibrium conditions. This difference arises from an isotope exchange reaction that can be represented by

$$\frac{1}{3}CaC^{16}O_3 + H_2{}^{18}O \rightleftharpoons \frac{1}{3}CaC^{18}O_3 + H_2{}^{16}O\,,$$

for which the equilibrium constant is

$$K = \frac{(CaC^{18}O_3)^{1/3}(H_2{}^{16}O)}{(CaC^{16}O_3)^{1/3}(H_2{}^{18}O)} = \frac{([CaC^{18}O_3]/[CaC^{16}O_3])^{1/3}}{[H_2{}^{18}O]/[H_2{}^{16}O]}$$

this latter being the $^{18}O/^{16}O$ ratio in the carbonate sample divided by the $^{18}O/^{16}O$ ratio in the water. The fractionation factor α is the quotient of both ratios expressed as

$$K = \alpha = \frac{R_c}{R_w}\,,$$

with R_c and R_w representing the $^{18}O/^{16}O$ ratios of carbonate and water respectively. In the system calcite–water the fractionation factor found is $\alpha = 1.0286$ at 25°C, so that the calcite precipitated in equilibrium will be enriched in ^{18}O to 2.86% or 28.6‰ in δ-notation. The fractionation factor is temperature-dependent with values near 1.0 and carbonate can either be enriched or depleted in heavy O isotopes. Of course, the isotope composition of sea water is unaffected since the mass of this water is larger by many orders of magnitude that that of the $CaCO_3$ equilibrated with it. Epstein *et al.* (1953) determined empirically the temperature dependence of the ^{18}O of calcite in equilibrium with water, the δ-values of these being later modified by Craig (1965). The revised expression for the temperature, T °C, is

$$T = 16.9 - 4.2(\delta_c - \delta_w) + 0.13(\delta_c - \delta_w)^2\,,$$

in which δ_C is the $\delta^{18}O$ of CO_2 obtained from a sample of carbonate by reaction with 100% H_3pO_4 at 25°C with respect to the mass spectrometer working standard gas and δ_w is the corrected ^{18}O of CO_2 equilibrated isotopically at the same temperature with the water from which the carbonate was precipitated, again measured against the same mass spectrometer working standard gas. Temperature equations for palaeotemperature determinations are generally based on empirical data either obtained through growth experiments under controlled conditions, for instance those of Epstein *et al.* (1953), or by inorganic precipitation of minerals, for example by Tarutani, Clayton and Mayeda (1969). Temperatures are calculated using a quadratic equation, describing a nearly linear relation of $\delta^{18}O$ values and temperature within a restricted temperature interval. A list of different palaeotemperature equations for carbonates is given, for example, by Wefer (1985). The phosphate–water temperature scale has been revised by Longinelli and Nuti (1973) but is rarely used.

Depending on the mineralogical structure of shell material precipitated by marine organisms, different palaeotemperature scales are in use. The following equation for molluscan aragonite has, for instance, been established by Grossmann and Ku (1986):

$$T = 21.8 - 4.69(\delta_{\text{aragonite}} - \delta_w)\ .$$

In marine environments, palaeotemperature determinations are mainly carried out on calcitic or aragonitic shells of such suitable organisms as bivalves and planktic or benthic foraminifera. Generally, all organisms precipitating their shell material in equilibrium with the surrounding sea water could be used if the original isotope signal has not been altered. Problems arise, however, from the interplay of various parameters which might influence the isotope signal. In fact, the temperature studies only supply reliable information when the other parameters are known, which in the case of palaeostudies is mostly not so.

15.2 ISOTOPIC COMPOSITION OF SEA WATER

Palaeoecological research based on the equilibrium fractionation of oxygen isotopes with the surrounding water obviously necessitates knowledge of the isotopic composition of sea water. The assumption of a more or less constant isotopic composition is generally accepted although sea water is in fact far from being homogenous, as a δ_w value used in palaeotemperature equations might imply. The isotopic composition of sea water has been studied extensively by Craig and Gordon (1965) and Broecker (1974) and the $\delta^{18}O$ and δD values of sea water are close to zero at a normal marine salinity of 35‰. Figure 15.1 shows the $\delta^{18}O$/salinity relationship of Atlantic surface waters and North Atlantic Deep Water (NADW). As can be seen, the oxygen isotope ratios increase or decrease parallel to the salinity, as already stated by Epstein and Mayeda (1953). Apart from this, glaciation binds light isotopes in ice masses or melting releases them. The main processes changing isotope ratios of sea water are therefore evaporation and precipitation on the one hand and oscillations of the ice masses on the other. As is obvious, these processes depend on the climate, mainly the temperature, and vary with the latitude and other geographical factors, for instance, with the altitude of mountain ranges on continents which can cause rain or snowfall to varying extents. Meltwater from polar regions is, on the other hand, distributed by ocean currents.

The relationships between δD and $\delta^{18}O$ are of the following type:

$$\delta D = M\delta^{18}O,$$

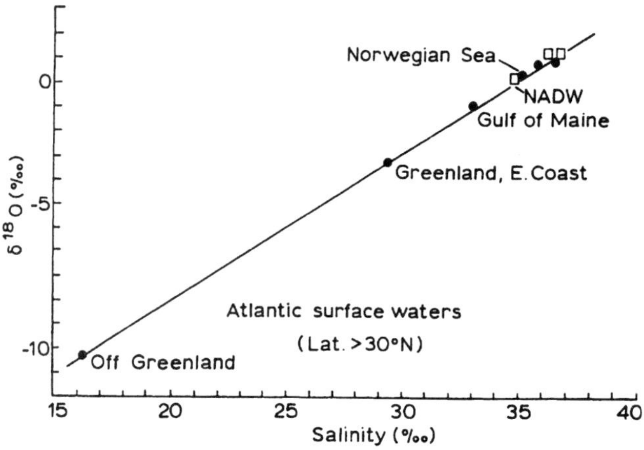

Fig. 15.1 Oxygen isotope ratios versus salinity for Atlantic surface waters and NADW (after Craig and Gordon (1965)).

in which M represents the slope of linear trajectories which decreases with increasing ratios of evaporation to precipitation in an area. In the North Atlantic Ocean M is $= 6.5$, compared with 7.5 in the North Pacific Ocean and 6 in the Red Sea. Of course, this preferential loss of water containing ^{16}O from ocean surfaces is accompanied by increasing salinity. Therefore, $\delta^{18}O$ and δD are correlated with salinity as shown in Fig. 15.1. All these parameters are modified at high latitudes by the mixing of ice- and snow-derived melt water with sea water. It has been shown also that average bottom water in the Pacific Ocean (PDW = Pacific Deep Water) deviates appreciably from surface water, and this refers to the influence of freezing of sea water on salinity and isotope composition. The ^{18}O isotope fractionation factor for ice in equilibrium with water is 1.002, and therefore the $\delta^{18}O$ value for sea ice is *c.* $+2‰$ relative to the sea water from which it arose. Consequently, although sea ice formation causes a significant increase in salinity, the $\delta^{18}O$ value of the water remains almost unchanged. Thus the density of this water increases and promotes sinking. Dense bottom water occurs near Antarctica, thereafter flowing northwards along the ocean bottom. The North Atlantic bottom water has a salinity and an oxygen isotope composition which lie on the surface water trajectory for the Atlantic Ocean. However, bottom waters in the other two oceans are displaced towards higher salinity or lower $\delta^{18}O$ relative to their surface waters, while surface water from the Weddell Sea, which is situated next to Antarctica, is compatible with both South Pacific surface water and NASW (North Atlantic Surface Water). In the Weddell Sea, freezing increases salinity but does not appreciably alter the $\delta^{18}O$ value. Antarctic bottom water enters the Pacific and Indian Oceans, at the same time mixing with Atlantic bottom water. Consequently the $\delta^{18}O$ values and salinities of Indian and Pacific bottom waters represent mixtures or roughly equal amounts of Antarctic and Atlantic bottom waters. The deviation to the left of the mixing line shown by Indian and Pacific bottom waters shows that a third component is involved, which has either a lower salinity or higher $\delta^{18}O$ value, or both, in comparison with Indian and Pacific bottom waters. This component has not been identified, but could be intermediate water in the Indian and Pacific which has an isotopic composition of $\delta^{18}O = -0.10‰$ and a salinity of 34.45‰.

As the seas contain O-bearing ions other than water, e.g. sulphate (SO_4^{2-}) and bicarbonate (HCO_3^-), these must be considered as well. Sulphate has a $\delta^{18}O$ of $+9.7‰$. The fractionation factor for the isotope exchange of oxygen between water and SO_4^{2-} ions is not known with certainty, but a theoretical value of 1.024 was given by Urey.

There is considerable uncertainty regarding the isotopic composition of pore waters. With regard to the fact that many organisms are shallow or deep burrowers, the isotopic milieu within the burrowing depths, usually not more than 1 m, needs more attention.

15.3 RELEVANT ^{18}O-DETERMINING PARAMETERS

Several problems arise in making palaeotemperature determinations utilizing the above equations. The most basic assumption to be made when palaeotemperatures are studied is the oxygen isotope composition of the ocean water. Firstly, the ^{18}O content of palaeo-oceans is not known. The δ_w value used in the palaeotemperature equations is a mere approximation based on present-day values and the knowledge of interaction between water and rocks. In earlier work, it was assumed that these had a practically constant isotopic composition very close to that of today, but this still left the matter of palaeosalinity, i.e. it is necessary to know whether the fossil organism to be analysed lived in ocean water with a salinity of about 35‰ or not. Urey (1948) posed the question, 'Has the ocean kept the same ratio of oxygen isotopes throughout all geological time?', and assumed no significant change for the last 60 Ma. This is in accordance with estimates of Popp *et al.* (1986) and Veizer, Fritz and Jones (1986), since a change of 2 or 3‰ is not really significant from a global point of view and for the time interval from the Palaeozoic to the Recent. A review of marine isotope geochemistry through time has also been given by Fallick and Hamilton (1989).

On the other hand, small variations such as these mentioned would affect the temperature determinations in the range 8°C to 15°C, depending on the palaeotemperature equation used. Extensive studies of marine waters by Craig and Gordon (1965) gave a mean value of $-0.28‰$ for the modern oceans. The value corresponds to the present ice volume and shifts to $-1.2‰$ for ice-free phases according to Shackleton and

Kennett (1975), if the light ^{16}O isotopes bound in polar regions are released by melting. The accumulation of ice would shift the isotope ratios of sea water to more positive values.

Attempts have been made to elucidate the isotopic composition of palaeo-oceans. For instance, the slope of the ^{18}O-salinity relationship for the Red Sea implies that a change in the salinity of 1‰ is associated with a difference of 0.29‰ in ^{18}O, this altering the isotopic temperature by *c.* 1°C. It is believed that during Pleistocene glaciations, the value rose to +0.9‰. Shackleton (1968) estimated that the ^{18}O of the oceans altered by 1.0–1.4‰ during the glacial and interglacial stages of this epoch. An interesting approach to the effects of continental glaciation involves the analysis of pelagic and benthic foraminifera deposited, with the concomitant assumption that the ocean bottom temperature is constant and variations in the ^{18}O values from tests of benthic species reflect variations in the ^{18}O content of sea water. Assuming a specific temperature, the ^{18}O values from such tests enable calculation of the ^{18}O values of sea water (Kahn, Oba and Ku, 1981). As early as 1955, Emiliani (1955a, b) outlined the most important features of Quaternary glacial-interglacial climatic functions from oxygen isotope determinations on planktonic foraminifera from Caribbean and equatorial Atlantic deep-sea cores. Shackleton and Opdyke (1973) established the chronology of climatic functions using isotopic data in conjuction with palaeomagnetic data derived from a western equatorial Pacific core.

Ocean water with high salinity has a higher ^{18}O content, ^{16}O being preferentially concentrated in the vapour phase during evaporation. On the other hand, water of lower salinity possesses a lower ^{18}O content since it is diluted by fresh water. It has been calculated that a variation in salinity as small as 1‰ would be accompanied by a 1°C error in temperature determinations in a non-glacial time interval in Earth history. Another difficulty arises from the metabolic effects which may influence carbonate precipitation, which are far from completely understood. It is known that some organisms such as molluscs and brachiopods deposit calcite or aragonite in isotope equilibrium with ocean water, but others certainly do not, e.g. echinoderms, crinoids and asteroids. It is possible that there may be an isotopic exchange between respiratory CO_2 and dissolved HCO_3^- at or near the site of skeletal deposition, which would constitute an adequate explanation of the phenomenon. In cases where non-extinct species of organisms are used in palaeothermometry, it is assumed that their depth habitats have remained unaltered through time.

Another question is whether the $CaCO_3$-secreting organisms grow their shells during only part of the local temperature range or throughout it. Epstein and Lowenstam (1953) demonstrated that for most species skeletal growth does not occur throughout the whole year; for instance, most pelecypods appear to grow mainly in the warm-temperature range, in contrast to gastropods which show both winter and summer growth. Actually, some do retain shell growth at low temperatures.

The isotope composition of oxygen in an aragonite or calcite shell is believed to remain unaltered until diagenesis causes its material to dissolve and recrystallize; indeed, some of the characteristics by means of which unchanged specimens can be identified have been discussed by Lowenstam (1961).

As mentioned above, the mineralogy of the shells is also significant for the isotopic composition of carbonates, as Sharma and Clayton (1965) pointed out. Thus the ^{18}O content of aragonite at 25°C is *c.* 0.6‰ greater than in coexisting calcite according to Tarutani, Clayton and Mayeda (1969). In addition, Rubinson and Clayton (1969) noted that the ^{13}C content of aragonite is enriched by 1.8‰ relative to calcite. For similar reasons, it is usually not feasible to determine palaeotemperatures from fresh water carbonates since there is a great variability in the isotope composition of meteoric waters (Fritz, Anderson and Lewis, 1975).

A further aspect regarding the isotope ratios of mollusc shells seems to be connected with the life habit. As a comparative study of different molluscs has shown, a relationship between burrowing depths and isotope ratios cannot be precluded (Attendorn, 1993). A systematic variation of C and O isotopes strongly suggests an alteration of the bottom water isotope signal with increasing burrowing depth of the organism. The effect (Fig.15.2) was found for different taxa and is thought to be due to isotopic differences within the pore water as well as to changing physical properties at the water–sediment interface, limiting, for example, the temperature and water exchange. Thus, there is a presumption, as pointed out by Attendorn , Bowen and Strauch (1997), that organisms exchange isotopes through the entire mantle area and at the surface of the siphon. Therefore the isotope ratio of shells is not only determined by the bottom water, which incorporated through the siphon and so, in order to attain reliable palaeoecological information, epifaunal species seem to be most suitable for isotope investigations.

Additionally, false interpretations of O isotope values may be due to diagenetic alteration of the original shell material, to a change of life habit if fossil species are compared to Recent ones, changes of depth habits during the life span of individuals (as found for foraminifera) and to the effects of bioturbation.

Most of the above-mentioned parameters influence the C isotope ratios as well, but this may be neglected here since the palaeotemperatures are calculated on the basis of the O isotope composition.

15.4 PALAEOTEMPERATURE DETERMINATIONS

When the isotope thermometer became a common tool in the 1950s, numerous investigations were carried out by the pioneers. Among them

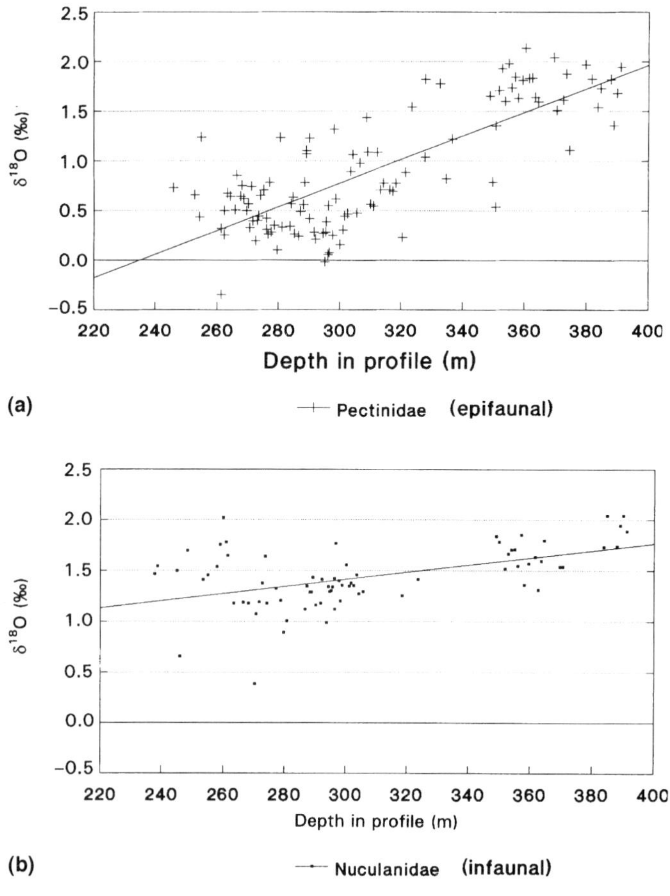

Fig. 15.2 Differences in the slope of O isotopic profiles for (a) epifauna and (b) shallow/deep burrowers versus depth in the Chattian. The samples from the Niederrheinische Bucht, north-west Germany, show similar isotopic trends for 16 investigated taxa but the different slopes indicate an influence of burrowing depth on the O isotope ratio.

were Emiliani (1953, 1954, 1955a, b, 1956, 1957, 1964), working on Cenozoic foraminifera, and Bowen (1961a–d, 1962, 1963a–c, 1964) who worked on Mesozoic belemnites. They also soon discovered relationships between the isotope ratios and salinity (Epstein and Mayeda, 1953).

It is now generally accepted that variations in the oxygen isotope composition of Quaternary foraminifera relate to climatic episodes; the degree to which changes in temperature and ice volume are represented is still undecided. Obviously ^{18}O maxima and minima act as stratigraphic markers for correlation between relevant Quaternary deep-sea sediment cores. Very detailed investigations on Tertiary isotope palaeoclimatology commenced in the late 1960s when sediment cores of this age were recovered and provided by the Deep Sea Drilling Project (DSDP). The good record that exists of oxygen isotope changes in planktonic and benthonic foraminifera from Pacific-derived Tertiary sediment shows, on the basis of results from the benthonic foraminifera, that the temperature of bottom waters, and therefore high-latitude surface temperatures, decreased through the Tertiary, but that the decrease was irregular and sometimes interspersed with episodes of temperature increase. The pattern of variations in high-latitude planktonic foraminifera accords with this phenomenon. By contrast, the isotope composition of low-latitude planktonic foraminifera demonstrates relatively little alteration (except for low values in the Oligocene), which implies that surface temperatures at these latitudes did not vary very much at all. There are two significant decreases in the isotopic temperatures of benthic foraminifera during Middle Miocene times and near the Eocene–Oligocene boundary. The so-called Middle Miocene event has been interpreted as the commencement of widespread glaciation in Antarctica, perhaps initiated by the establishment of Antarctica circumpolar circulation (Savin, Douglas and Stehli, 1975). The Eocene–Oligocene boundary event seems to have influenced low-latitude isotopic temperatures in some areas as well as in the actual benthic record (Keigwin, 1980). Two possible explanations have been presented. The first is the sudden evolution of cold, thermohaline circulation (Kennett and Shackleton, 1976). The second is the injection of low-salinity waters from the Arctic Ocean leading to climatic deterioration and the later production of cold bottom water (Thierstein and Berger, 1978).

Evidence for isotopic temperature variations across the Cretaceous–Tertiary (K–T) boundary has been discussed by Savin and Yeh (1981). In most places, there was a slight decrease in isotopic temperatures of both planktonic and benthic foraminifera, followed by a recovery just prior to the boundary in question. Of course, this decline was much smaller than those of the Tertiary events mentioned above. For the Cretaceous, older data provided by Lowenstam and Epstein (1959) and Bowen (1961a–d, 1963a–c, 1964a, b) demonstrated Albian and Coniacian–Santonian temperature maxima, although a lower temperature was indicated for the Santonian by Teis, Chupakhin and Naidin (1957). However, researchers from the former USSR used a standard with a different $^{18}O/^{16}O$ ratio and there is also a distinct possibility that their 'Santonian' is not chronostratigraphically equivalent to that employed by the Western workers. In addition, Naidin, Teis and Chupakhin (1956) suggested that the Russian 'Santonian' belemnites analysed probably passed their lives in deep waters, i.e. waters with a lower-temperature regime than the water where they died, i.e. shelf water. For the Jurassic, Bowen and Fritz (1963) found a temperature maximum in the Lias with a later gradual decline up to the Bajocian. In East Greenland the maximum again occurred in the Kimmeridgian, but took place in Europe during the Oxfordian.

Today, the value of palaeotemperature determinations is often questioned, for the number of parameters which might influence the results is quite large. It is obvious that a critical interpretation of temperature results is necessary but it would go too far to say that it is impossible to obtain reliable information. Thus, the careful choice of samples and the accumulation of environmental data is mandatory for palaeotemperature investigations. As an updated summary of climatic research has been given recently by Bowen (1991) together with a discussion of the pros and cons. Some recent – in the wider sense palaeoecologic –results will be presented in the following section, without summarizing former and initial research.

15.5 FORAMINIFERA

Since the foraminifera are ideal species occuring in nearly all deep sea drilling cores, they are widely used for palaeoecological research. It is clear that planktonic and benthic species represent different temperature regimes i.e. surface and bottom waters. However, if surface temperatures change and oxygen isotope ratios depend on climate, the bottom-water temperatures are more or less uninfluenced by annual temperature variations but show, on the other hand, the influence of ice volume effects. Oxygen isotope records from benthic foraminifera are therefore used as both indicators of glaciation and stratigraphic tools (see also section 16.4).

Pujol and Grazzini (1989) studied, amongst other species, the stable isotope ratios of the dominant planktonic foraminifera (*Globigerina bulloides* and *Uvigerina peregrina*) from the Alboran Sea (Western Mediterranean). They recognized nine ecozones different to those of Atlantic cores near Gibraltar which represent two phases, the first from the last glacial maximum to 9500 a BP, and the second from then to the present.

Environmental changes are reflected by the oxygen isotope ratios (Fig.15.3) and the interval from 18000 to 9500 a BP is mainly characterized by the subarctic, cold-water assemblage (*N. pachyderma, G. bulloides* and *G. quinqueloba*). This is in accordance with the $\delta^{18}O$ values which, expressed in terms of temperature, indicate low but increasing temperatures towards 9500 a BP. In any case, temperatures derived from both oxygen isotope ratios and faunal assemblages must be interpreted carefully. Similar to the oxygen isotope ratio, the percentage of a species does not depend only on the temperature. For the second interval (postglacial), oxygen isotope ratios and the co-occurence of subtropical foraminifera indicate warmer conditons and today's climate is supposed to begin at about 4000 a BP.

An interesting aspect was described by D'Hondt, Zachos and Schultz (1994). It is, for example, known that some foraminifera precipitate their

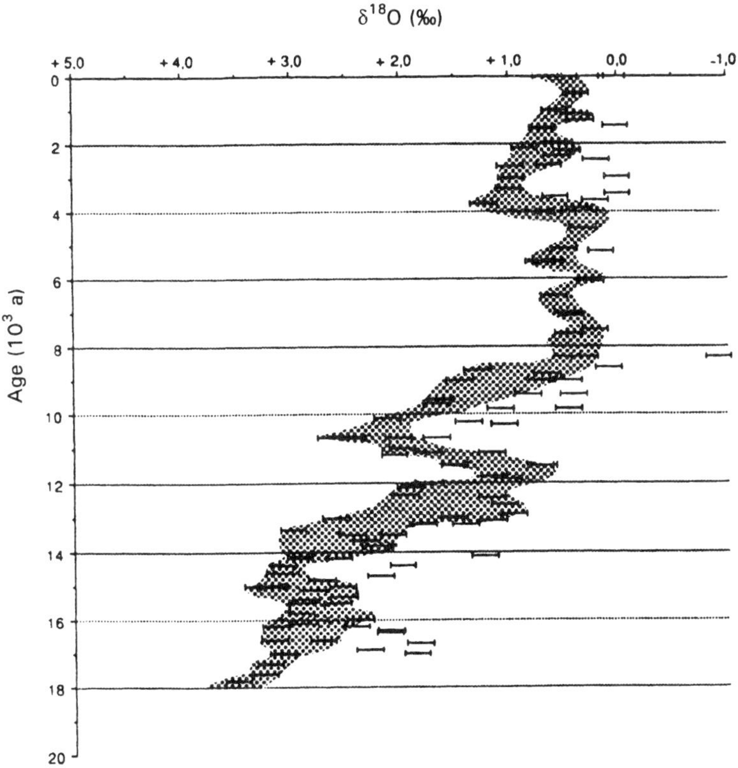

Fig. 15.3 Oxygen isotope ratio versus age for *G. bulloides* from four cores of the Alboran sea (after Pujol and Grazzini, 1989).

shells in isotopic disequilibrium and, assuming this to be a constant factor it is corrected for oxygen isotopes. This behaviour of the organism can be regarded as a vital effect and further biological influence on C isotope ratios may be due to photosymbiosis as, for instance, found among corals. The researchers studied planktic foraminifera from the late Palaeocene (*Acarinina nitida* (Martin), *Morozovella subbotinae* (Morozova), *Morozovella velascoensis* (Cushman) and *Subbotina triangularis* (White) of DSDP Site 758A. The samples showed a positive correlation between $\delta^{13}C$ values and test mass and a negative correlation for $\delta^{18}O$, resembling trends of Recent planktonic foraminifera in photosymbiosis with dinoflagellates. Results are shown in Figure 15.4.

An increase of C isotope ratios and a decrease of oxygen isotope ratios can be observed with different magnitudes. How far the depth habitats may also influence the isotope values is not clear, but assuming a strong reliance on the symbionts for nutrition, depth variations must be small, being restricted to the photic zone. Such correlations have been observed

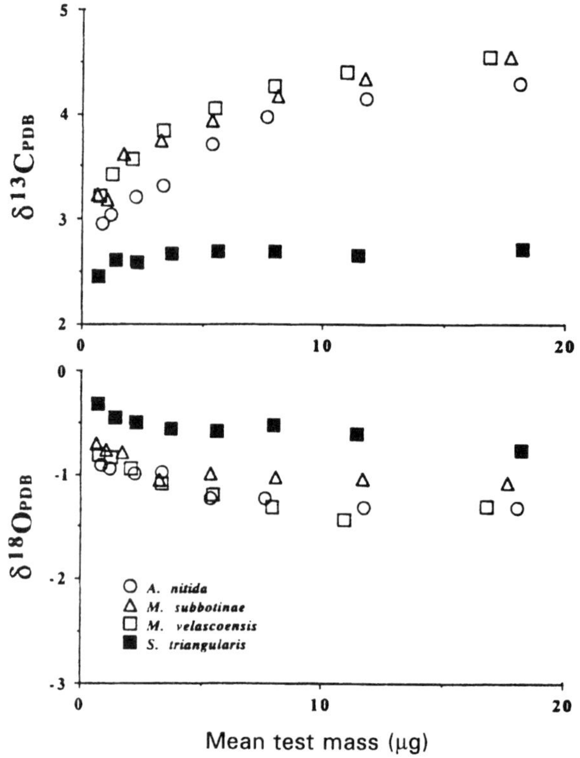

Fig. 15.4 (a) $\delta^{13}C$ and **(b)** $\delta^{18}O$ ratios versus mean test mass for planktonic foraminifera (from D'Hondt, Zachos and Schultz, 1994).

by several authors, for example Shackleton, Corfield and Hall (1985), D'Hondt and Zachos (1993) and Pearson, Shackleton and Hall (1993). Culturing studies (Spero and Lea, 1993) could explain the correlation by relating isotope signals of adult specimens and test size to the activity of photosymbionts. The results are important with respect to the evolution of planktonic foraminifera, although the possible $\delta^{13}C$ shift caused by photosymbiosis can complicate the interpretation of palaeoecological records.

Apart from palaeotemperature studies, ice volume estimations and the above-mentioned research, the foraminiferal records obtained from worldwide drillings enable stratigraphic correlations to be made (section 16.4) and provide further palaeoecological information. Hilbrecht, Hubberten and Oberhänsli (1992) studied Late Cretaceous planktonic foraminifera from North-West Europe and discussed palaeotemperatures and biogeographic and depth distribution of such genuses as *Hedbergella, Rotalipora Praeglobotruncana.* The regional patterns of C isotope variations found indicate mixing of waters from different areas and are

Table 15.1 Comparison of O and C isotope ratios for samples of Recent and Holocene specimens (from Beveridge and Shackleton, 1994)

Species	$\delta^{18}O_R{}^a$	$\delta^{18}O_H{}^a$	$\Delta\delta^{18}O$	$\delta^{13}C_R$	$\delta^{18}C_H$	$\Delta\delta^{18}C$
G. ruber	−0.77	−0.87	−0.10	1.06	0.53	−0.53
G. bulloides	−0.34	−0.50	—0.16	−0.90	−1.52	−0.62
O. universa	−0.35	−0.44	0.09	1.84	1.57	−0.27
G. inflata	−0.08	0	−0.08	0.26	0.06	−0.20

[a] R = Recent, H = Holocene; isotope ratios relative to PDB (‰).

interpreted as indicators of higher organic productivity in upwelling zones. Recent planktonic foraminifera have been used by Beveridge and Shackleton (1994) to estimate the $\delta^{13}C$ change caused by the release of ^{13}C-depleted anthropogenic CO_2, which causes a decrease of C isotope ratios in surface waters. The researchers compared the Recent specimens *G. ruber* and *G. bulloides* with older Holocene ones and found that they revealed a decrease in $\delta^{13}C$ values compared to the older samples. The different species show significantly variable amounts of decrease as shown in Table 15.1. A relevant decrease can be observed for the C isotope ratios, whereas $\delta^{18}O$ differences are small and sometimes similar to the older samples. The equation

$$\Delta\delta^{13}C_{(p-o)} = \frac{\Sigma CO_{2_a} \times (\delta^{13}C_a - \delta^{13}C_p)}{\Sigma CO_{2_p} - \Sigma CO_{2_a}}$$

gives the calculated difference $\Delta\delta^{13}C_{(p-o)}$ in $\delta^{13}C$ of the total CO_2 (ΣCO_2) between present and pre-industrial time with ΣCO_{2p} and $\delta^{13}C_p$ as DIC of the present-day mixed layer and with ΣCO_{2a} and $\delta^{13}C_a$ as anthropogenically added CO_2. However, if this is valid for foraminifera (and it may be proved by more data), it should in principle facilitate pCO_2 estimates of the palaeoatmosphere as well. $\delta^{13}C$ of CO_2 in the atmosphere is currently *c.* −7.8‰ and the oceans (which mainly determine the atmospheric C isotope ratios) have a $\delta^{13}C$ of *c.* 1.5‰ for DIC in surface waters, thus being *c.* 9‰ enriched compared to atmospheric values (Mook, 1986). Consequently, the burning of fossil fuel must not only have increased the pCO_2 from approximately 280ppm to 345ppm since pre-industrial times, but must also have caused a negative shift in $\delta^{13}C$ of the atmospheric CO_2 and, thus also in the oceans; this should possibly be reflected by trees as well. Friedli *et al.* (1986) estimated a decrease of *c.* 1.2‰ for atmospheric CO_2 over the last 130a, and thus $\delta^{13}C$ values ranging from *c.* −6.6‰ to −7.8‰ would have caused such observed $\Delta\delta^{13}C$ shifts in shells of foraminifera.

Benthic foraminifera records, on the other hand, supply information regarding environmental changes at greater depths. Annual temperature variations and salinity changes usually do not affect the isotope ratios of both C and O but ice volume effects and water circulation can be monitored. Mackensen *et al.* (1993) studied the isotopic composition of the total dissolved inorganic carbon $\delta^{13}C_{\Sigma CO2}$ from water and surface sediment samples (Eastern Atlantic Ocean and Weddell Sea) and compared these to the $\delta^{13}C$ values of live specimens of the benthic foraminifer *Fontbotia wuellerstorfi* and related genera. The deep and bottom water circulation patterns were found to be in agreement with those from other tracers but circumpolar deep water was depleted in $\delta^{13}C$ by *c.* 0.3‰ and the bottom water from the Weddell sea was enriched by 0.7 to 0.8‰. The decrease found in CPDW (Circumpolar Deep Water) coincides with nutrient maxima whereas the enrichment in heavy C indicates a decoupling of the dissolved phosphate signal from the $\delta^{13}C_{\Sigma CO_2}$ signal, possibly due to temperature-dependent fractionation processes during CO_2 exchange between atmosphere and surface waters.

Grobe *et al.* (1990) studied the isotope records from samples of four late Quaternary cores located near the Antarctic continental margin, among others, with respect to the sedimentation rates. The O isotope ratios evidence the influence of isotopically light meltwater released at the beginning of an interglacial. The sedimentation rates are then high due to an input of ice-rafted debris and they decrease with increasing distance from the continent. The isotope signal is also associated with increasing biological productivity (opal and foraminiferal content, bioturbation) as a result of higher input of light after melting of the sea ice coverage.

Naqvi, Charles and Fairbanks (1994) presented C and O isotope records of benthic foraminifera from intermediate-depth cores of the north-east Indian Ocean dating back to the last glacial. Results from the Andaman Sea show a $\delta^{13}C$ shift from the last glacial to the Holocene of 0.43‰, similar to the ocean mean value for deep ocean cores of 0.46‰, as reported by Curry *et al.* (1988). The deep-sea values usually show a more positive shift compared to intermediate depths and therefore it is concluded that the upper Indian Ocean has experienced a $\delta^{13}C$ shift greater than the global mean. The isotope ratios presented in Figure 15.5 show a $\delta^{13}C$ increase at core depths of *c.* 300 and 100 cm and the $\delta^{18}O$ increase is about 0.26‰ greater than the expected ice volume effect, thus indicating a cooling of *c.* 1.1°C with the assumption of constant salinity. The $\delta^{13}C$ increase is supposed to reflect fluctuations in the NADW (North Atlantic Deep Water) production.

Generally, the C and O isotope records obtained from foraminifera supply a useful tool for paleoecological studies of the marine environment, although results may sometimes reflect local rather than global changes.

15.6 BIVALVES

Among the molluscs, bivalves are most commonly used for C and O isotope investigations. It is generally accepted that bivalves precipitate

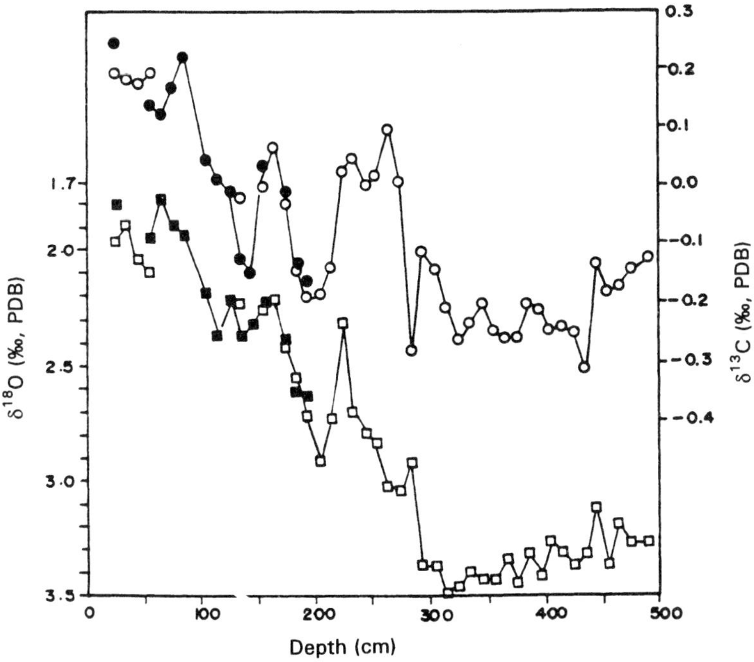

Fig. 15.5 Carbon (upper curve) and O (lower curve) isotope ratios from benthic foraminifera *Planulina wuellerstorfi* (solid symbols) and *Cibicidoides pachyderma* (open symbols). (After Naqvi, Charles and Fairbanks, 1994).

their shell carbonate in equilibrium with the surrounding sea water and therefore are especially suitable for palaeoecological studies. However, the assumptions to be made include the δ_w of ocean water, which has been discussed in detail by Craig and Gordon (1965) and the salinity, which may especially vary near continental margins. Isotope ratios may be measured by separating material from single growth layers, the so-called seasonality studies, or on whole specimens, in order to obtain average values for the life span of a mollusc. However, it must be pointed out that temperature determinations using fossil shells are a mere approximation, usually underestimating the temperatures, except where dilution with isotopically light fresh water has occurred. Supposing a correct δ_w and a salinity of 35‰, the average temperature over a growth period will not give the maximum temperatures. Unfortunately, very little is known in most cases about the annual time of growth, the duration of a growth period, the number of growth increments added per period and the physiological constraints influencing growth, for example nutrition or age-dependent growth deceleration. Further complications may be due to life habits and the mineralogical structure of the shell.

Barrera *et al.* (1994) investigated the shell microstructure of *Laternula elliptica* from Antarctica and found that $\delta^{18}O$ values of samples from the exterior surface are similar at different distances from the umbo (average *c.* 4.5 ×), thus indicating that bulk samples would provide reliable palaeoecological information as well. However, other shell regions showed variations of both C and O isotope ratios which are attributed to vital effects. The authors also found an enrichment of $\delta^{18}O$ in the aragonitic shell of *L. elliptica* relative to the calcitic *Adamussium colbecki* from the same location and time at temperatures below 0°C. In fact, differences have been found for the equilibrium fractionation of C and O in the systems calcite– water and aragonite–water (Rubinson and Clayton, 1969; Tarutani, Clayton and Mayeda, 1969).

This fractionation applies to both organically and inorganically precipitated aragonite, which has been found to be enriched by 0.6‰ at 25°C in O and 1.8‰ in C, relative to calcite. Alas, biogenic aragonite has also been found to be depleted (Tarutani, Clayton and Mayeda, 1969; Horibe and Oba, 1972) and therefore the matter does not become clearer. The application of the palaeotemperature equation for aragonite–water, given by Grossman and Ku (1986) may also lead to wrong temperature estimations. A comparative study of C and O isotopes including 11 taxa from different orders of Bivalvia (Attendorn, 1993) showed similar trends for the profiles of upper Oligocene samples from the Niederrheinische Bucht (north-west Germany). The shell material was diagenetically unaltered and ideally it could be expected that samples from the same location and depth have similar isotope ratios. However, the isotope ratios for both O and C showed differences which cannot be due to temperature and the salinity; these must be the same for all taxa. Temperatures calculated from these samples could be compared with other results from faunal and floral analysis which indicate a warm temperate to subtropical climate. These results were, however, only indicated by epifaunal *Pectinidae* which precipitated an originally calcitic shell. Oxygen isotope ratios from infaunal species often lead to much lower temperature results (Fig. 15.6), exceeding possible corrections for aragonite and variability due to the family. These results indicate that palaeotemperature determinations must be evaluated critically, although the general trend of increasing temperatures toward the Miocene is reflected in all profiles.

As isotope ratios of bivalves are also interpreted with respect to other environmental changes, it might sometimes be impossible to estimate the contribution of the various parameters to the $\delta^{13}C$ and $\delta^{18}O$ values. Krantz, Williams and Jones (1987), Romanek *et al.* (1987), Romanek and Grossman (1989) and Krantz (1990) studied, for example, environmental changes indicated by the O and C isotope ratios of bivalves. The authors analysed annual and minor internal growth increments and discussed

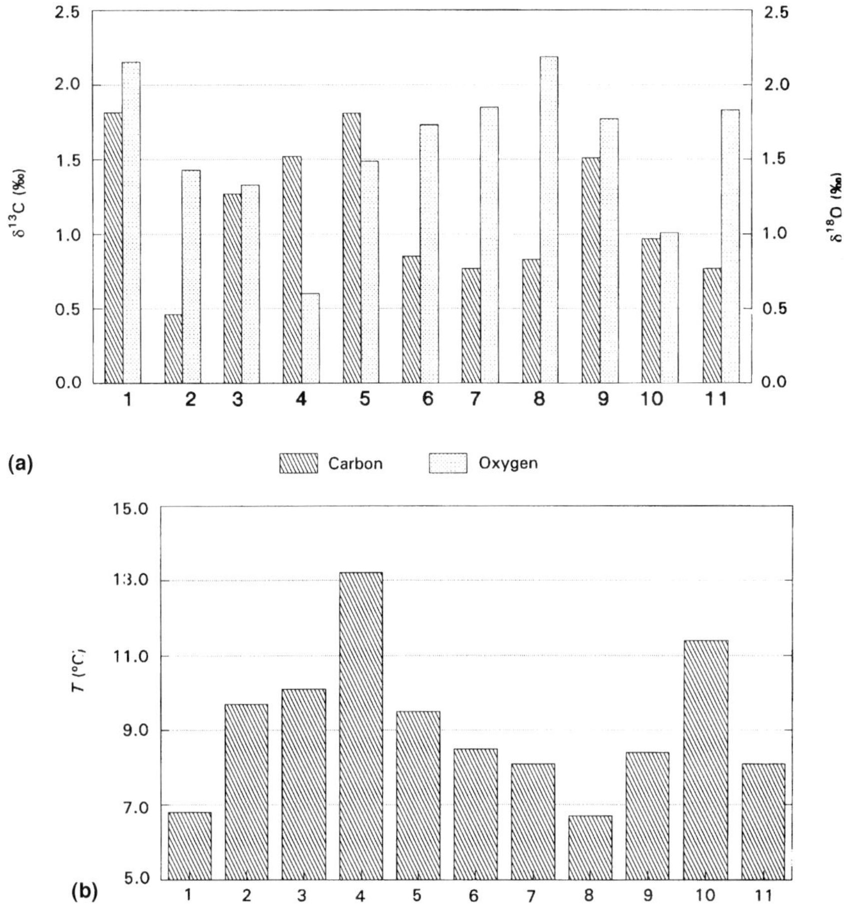

Fig. 15.6 (a) Carbon and oxygen isotope ratios and **(b)** calculated average temperatures for Bivalvia from the Upper Oligocene (Chattian) of the Niederrheinische Bucht (northwest Germany). 1 = *Nucula* sp.; 2 = *Nuculana westendorpi*; 3 = *Pecten* sp.; 4 = *Palliolum hausmanni*; 5 = *Glycimeris obovata, L imopsis retifera* and *L aurita*; 6 = *Astarte gracilis*; 7 = *Astarte pygmaea*; 8 = *Astarte falsopropinqua*; 9 = *Digitaria koeneni*; 10 = *Laevicardium tenuisulcatum*; 11 = *Carbula gibba*. (After Attendorn, 1993.)

the relation of the isotope records to, for example, seasonal hydrographic changes and temperature variations, water depth, biological productivity, ice volume offsets during the late Pleistocene, etc. (Krantz, Williams and Jones 1987). An interesting aspect is illustrated by the growth curves of *Tridacna maxima* (Fig.15.7). A correlation of O isotope ratios and annual cycles was made by using recorded temperatures and the decreasing cycle length indicates slower growth at an age of 10 years, probably related to the onset of sexual maturity. The decrease of

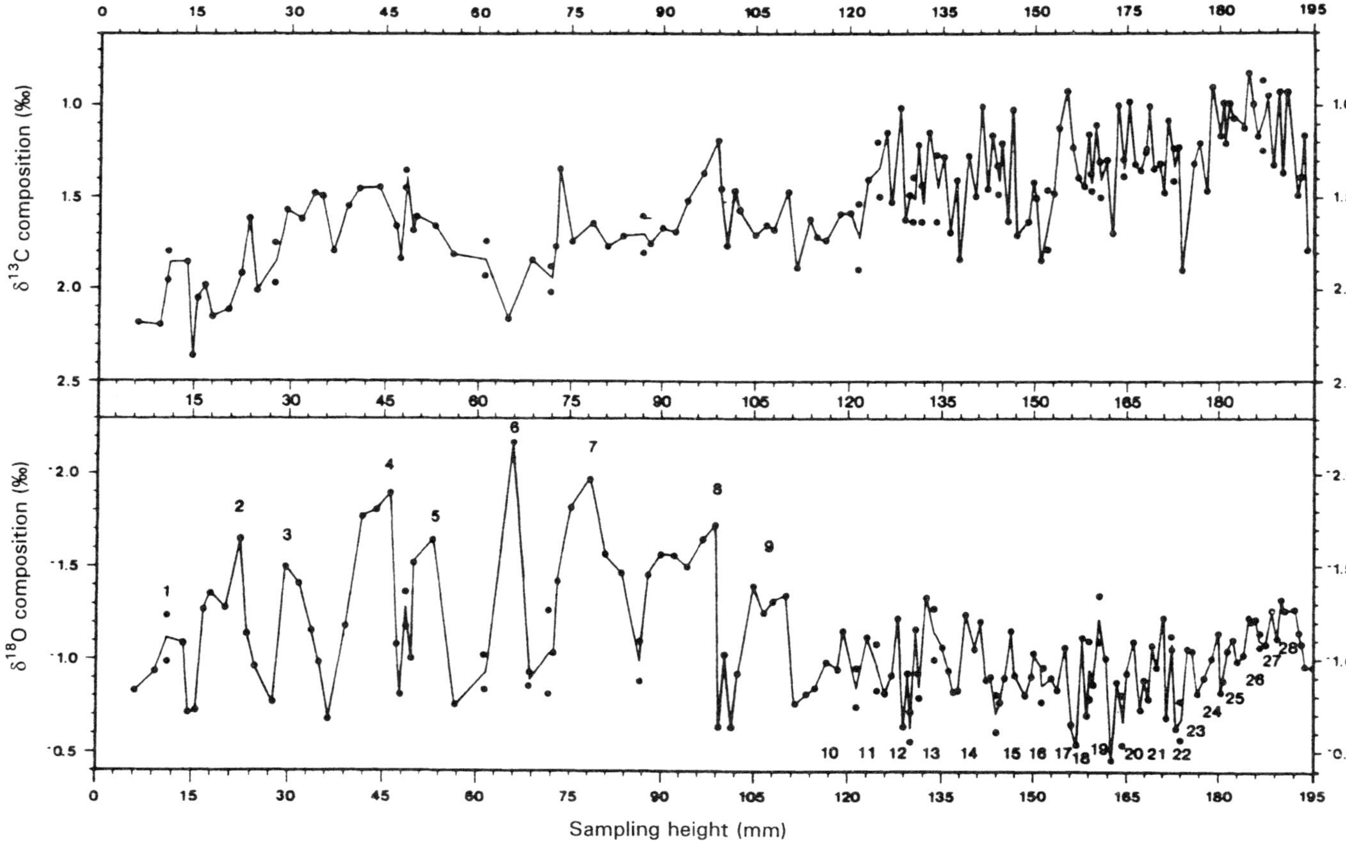

Fig. 15.7 Carbon and oxygen isotope ratios of *T ridacna maxima* from the Rose Atoll (south-west Pacific). Annual cycles and thus the ages of the specimen are given by the numbers (after Romanek *et al*. 1987).

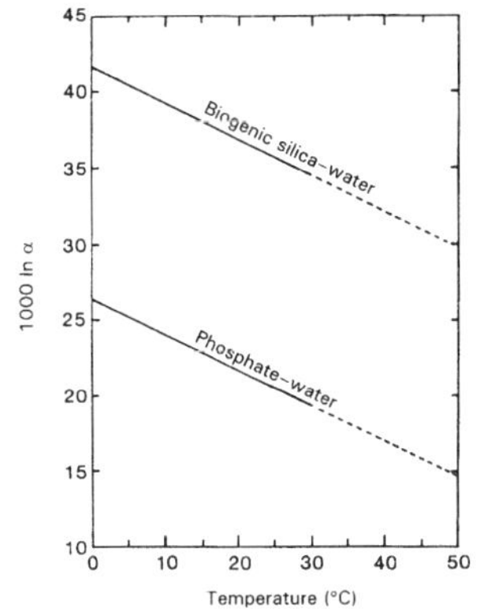

Fig. 15.8 $\delta^{18}O$ isotope fractionation (1000 1n α) versus temperture for biogenic silica–water and phosphate–water. Dashed lines represent the extrapolated extension.

the amplitudes during later growth is attributed to either shell precipitation, recording equilibrium conditions only in the winter months or isotopic disequilibrium, if the growth period lasts for the whole year. Generally, such investigations might supply information regarding shell growth and ontogenetic changes, but a restriction is placed by the necessary shell size. However, there is a lack of physiological knowledge and further studies are necessary.

Bivalves may further indicate depth habitat, water mass (mixing) and salinity as discussed by Israelson *et al.* (1994) on the base of $\delta^{13}C$ and $\delta^{18}O$ ratios from shell increments of bivalves from East Greenland. Basically, there is a vast amount of environmental information preserved in mollusc shells and both O and C isotopes may well indicate salinity changes, (Attendorn and Strauch, 1994), O_2-deficient waters (Shimoyama and Hamano, 1988) and palaeotemperatures (Cornu *et al.*, 1993). Nevertheless, the more parameters are known, the more reliable will be the interpretation.

15.7 OTHER TAXA

Various other taxa have been used to achieve palaeoecological information. Among them are other molluscs such as nautiloidea, gastropods and scaphopods, and also brachiopods and corals, ostracods, algae or fish. Some of them are known to precipitate their hard parts in isotopic disequilibrium, for example some algae, which either prefer to incorporate the light or heavy O isotope. Apart from calcitic shells, skeletons or otoliths, silicate and phosphate may be of interest and the temperature/fractionation relations are shown in Fig. 15.8. However, O isotope ratio investigations from these systems are rare compared to that of carbonate material. This is due, on the one hand, to lack of knowledge of processes controlling the fractionation and, on the other hand, to the complicated sample preparation methods for O from silicates. In addition, the more or less rare occurrence of some taxa prevents them from being an appropriate tool for isotope investigations.

An extensive study of $\delta^{13}C$ and $\delta^{18}O$ isotope ratios from skeletons of calcareous organisms has been carried out by Wefer (1985). Nevertheless, it is not yet clear whether all organisms which are supposed to fractionate O isotopes in equilibrium with the surrounding seawater really reflect equilibrium conditions or, in addition, mirror physiological processes.

Corals, for example, do not neccessarily precipitate their $CaCO_3$ skeleton in equilibrium with the surrounding sea water. The isotope record preserved may give information about climatic variations but may also

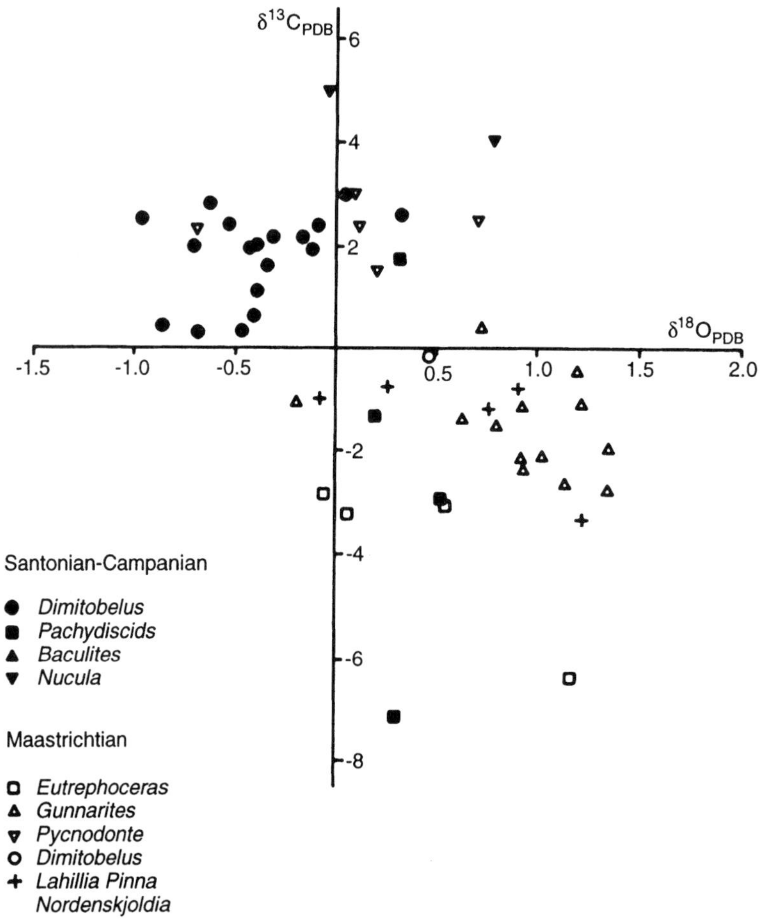

Fig. 15.9 Carbon and O isotope ratios from Late Cretaceous Antarctic molluscan fossils (after Pirrie and Marshall, 1990a).

depend on the photosynthetic activity of symbionts ($\delta^{13}C$) and metabolism ($\delta^{18}O$). As demonstrated by Cole, Fairbanks and Shen (1993), a comparison of the $\delta^{18}O$ ratios from annual layers of corals with historic records of El Niño events indicates a coupling of coastal South American climate with the Pacific El Niño/Southern Oscillation. The samples were taken from increments along the axis of maximum growth and a resolution of 16 samples per year was achieved. Intensive rainfall with low $\delta^{18}O$ values, due to increased isotopic distillation in warm tropical areas, causes salinity changes and, thus, decreasing O isotope ratios within the surface mixed layers. The lower O isotope ratios are preserved in the aragonite skeletons of corals from shallow habitats which therefore reflect either rainy or dry conditions. Extremely small samples of micrometer thickness and microgram weight, cut by using modern microtomes, naturally improve such high-resolution investigations. Gagan, Chivas and Isdale (1994) presented, for example, a 6–year O and C isotope record (1978-1984) of *Porites lutea* from Pandora Reef (Central Great Barrier Reef), a long-lived and globally distributed species. Their results indicate that the species investigated accurately records the timing of monsoonal runoff from the nearby Burdekin River as well as anomalous sea surface temperatures.

Pirrie and Marshall (1990a,b) studied Late Cretaceous samples from James Ross Island, Antarctica including bivalves, ammonites and belemnites. Using only diagenetically unaltered calcitic and aragonitic shell material, they calculated average temperatures of 13.6°C ($\delta^{18}O = -0.23‰$) for the Santonian–Campanian and 11.7°C ($\delta^{18}O = 0.66‰$) for the Maastrichtian (Fig.15.9), thus indicating cool temperatures in polar regions with broad climatic zonation during the Late Cretaceous. The authors also found differences in the O isotope ratios of the nektonic *Dimitobelus* and the epifaunal *Inoceramus* (Pirrie and Marshall 1990 b), which is attributed to a more intense diagenetic alteration of the epifaunal bivalve, possibly due to pore waters expelled from underlying volcanoclastic sediments, and modifying the isotopic composition of the bottom water. Interestingly, the coarse prismatic shell layer of *Inoceramus* showed variable alteration associated with the boundaries of the prisms, which is possibly connected with the destruction of the original organic cover. Middle Jurassic ammonites from Poland (Lukow) were studied by Brand (1986) and differences in the degree of diagenetic alteration between the outer shell and the septa were detected, with the consequence that the less-influenced shell material from the outer layers is supplying the most reliable palaeocological information. Palaeotemperatures calculated from this material ranged from 11.0°C to 14.5°C for the Polish Callovian sea.

It is generally accepted that variations in the O isotope composition of fossils relate to climatic episodes, although the degree to which changes in temperature and ice volume are represented is still undecided. Obviously $\delta^{18}O$ maxima and minima act as stratigraphic markers for correlation (section 16.4) and detailed investigations on palaeoclimatology, especially for the Tertiary, commenced in the late 1960s when sediment cores of this age were recovered and provided by the Deep Sea Drilling Project (DSDP). The good record that exists of oxgyen isotope changes in planktonic and benthonic foraminifera from Pacific-derived Tertiary sediment shows, on the basis of results from the benthonic foraminifera, that the temperature of bottom waters, and therefore high-latitude surface temperatures, decreased through the Tertiary, but the decrease was irregular and sometimes interspersed with episodes of temperature increase. The pattern of variations in high-latitude planktonic foraminifera accords with this phenomenon. By contrast, the isotope composition of low-latitude planktonic foraminifera demonstrates relatively little alteration (except for low values in the Oligocene), which implies that surface temperatures at these latitudes did not vary much at all. There are two significant decreases in the isotopic temperatures of benthic foraminifera during Middle Miocene times and near the Eocene–Oligocene boundary. The so-called Middle Miocene event has been interpreted as the commencement of widespread glaciation in Antarctica, perhaps initiated by the establishment of Antarctic circumpolar circulation (Savin, Douglas and Stehli, 1975). The Eocene–Oligocene boundary event seems to have influenced low-latitude isotopic temperatures in some areas as well as in the actual benthic record (Keigwin, 1980). Possible explanations include the sudden evolution of cold, thermohaline circulation (Kennett and Shackleton, 1976) and the injection of low-salinity waters from the Arctic Ocean, causing climatic deterioration and later production of cold bottom water.

Evidence for isotopic temperature variations at the K–T boundary has been discussed by Savin and Yeh (1981). In most places, there was a slight decrease in isotopic temperatures of both planktonic and benthic foraminifera, followed by a recovery just prior to the boundary in question. This decline was much smaller than those of the Tertiary events mentioned above. For the Cretaceous, data provided by Lowenstam and Epstein (1959) and by Bowen (1961a–d, 1963a–c and 1964a,b) demonstrated Albian and Coniacian–Santonian temperature maxima. For the Jurassic, Bowen and Fritz (1963) found a temperature maximum in the Lias with a later gradual decline up to the Bajocian. In East Greenland the maximum again occurred in the Kimmeridgian, but took place in Europe during the Oxfordian.

16

Other applications of biospheric carbon and oxygen

16.1 CARBON ISOTOPES IN ORGANIC MATTER

In this field (C_{org}) belong also the isotopes in plants and animals which are discussed in sections 16.2 and 16.3, but some other aspects will briefly be summarized in the following. If one divides organic matter into groups one could easily find three distinctive trends. A large group is the living matter to which belong the fauna and flora both terrestrial and marine. The second group would be dead matter, originated by the first group and sedimented sometimes in amounts sufficient to create deposits of coal, oil or gas. Furthermore, there are organic compounds, molecules consisting of organic elements, also deriving from the first group or sometimes even found as parts of amino acids in extraterrestrial matter. Several $\delta^{13}C$ isotope ratios for organic matter were given, for example, by Bowen (1988) and Hoefs (1987).

Carbon isotope investigations made on organic matter usually aim at (1) identifying the source material, (2) achieving more information about the carbon cycle, (3) determining the effect of secondary processes and (4) obtaining ecological data.

Progress in elucidating the carbon cycle has been made by analysing organic compounds by means of compound-specific analysis (CSIA; Matthews and Hayes, 1978; Hayes *et al.*, 1990), a method combining gas chromatography and isotope ratio mass spectrometry (Fig.16.1). An extension of this method towards $\delta^{15}N$ and $\delta^{18}O$ analysis has recently been described by Brand, Tegtmeyer and Hilkert (1994) and the 'tool' may lead to a far better understanding of biogeochemical processes.

Jasper and Gagosian (1990, 1993) combined analyses of $^{13}C_{org}$ and N/C ratios in a core (DSDP 619) from the Pigmy Basin in the Gulf of Mexico.

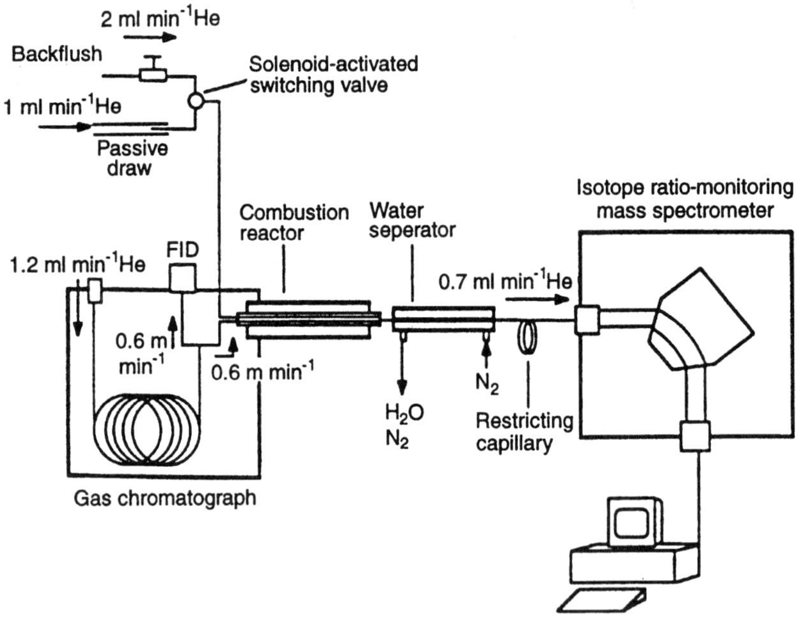

Fig. 16.1 Schematic diagram of a GC–IRMS arrangement (after Hayes *et al.*, 1990).

The results of this sequence from the Late Quaternary (Fig.16.2) show a mixed influence of terrigenous organic matter with negative $\delta^{13}C$ ratios (– 25.6‰), probably from C_3 plants, during glacial stages and more enriched marine organic matter (– 24.2 and – 22.9‰) during the interglacial (see also Jasper and Gagosian, 1989). Typical values for the N/C atomic ratios are given for the dominance of either marine or terrigenous organic matter as N/C = 0.13 – 0.20 and 0.01 – 0.05 respectively. Similar results yielded the analysis of sedimentary organic matter from the Norwegian – Greenland Sea by Botz *et al.* (1991). The samples belong to a part of core 23059 – 3, covering the last 150 ka. The *n*-alkane chain length and the $\delta^{13}C$ of the TOC from samples showed parallel variations in parts of the core and the ratio $C_{19} - C_{22}/C_{26} - C_{29}$ (marine short-chain *n*-alkanes/terrigenous long-chain *n*-alkanes) indicate a change in the relative amount of terrestrial to marine organic matter in the sediments. Rieley *et al.* (1991) compared carbon isotope of *n*-alkanes from the leaves of lakeside trees with the $\delta^{13}C$ values of the lake sediments. The *n*-alkanes from the leaf samples ranged from – 30.1‰ to – 38.7‰, depending on the species, whereas the *n*-alkanes from the lake sediments exhibited carbon isotope ratios between – 30.1‰ and – 35.9‰, thus indicating that they derive from a mixed input of deciduous leaf waxes. Such investigations may elucidate the origin of organic matter found in sediments. As there are different sources for the organic compounds to

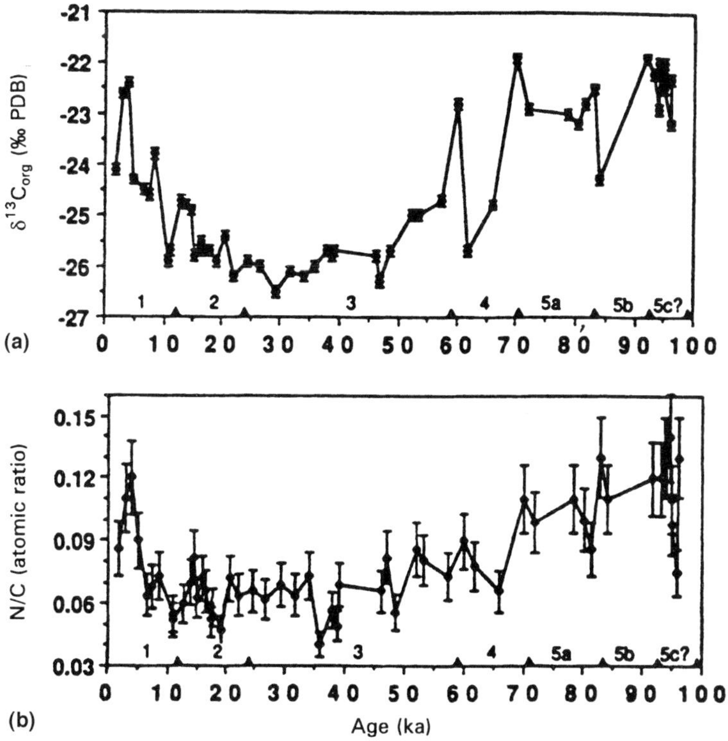

Fig. 16.2 (a) $\delta^{13}C$ and **(b)** N/C ratios versus time for DSDP Leg 619 in the Pigmy Basin (after Jasper and Gagosian, 1990).

which belong primary and secondary producers as well as consumers, the compound-specific analysis of organic matter potentially can reconstruct biogeochemical processes. Analyses of the organic compounds from oil shales were, for example, presented by Simoneit *et al.* (1993) who found an unusual $\delta^{13}C$ enrichment in cyclic hydrocarbons from a Permian deposit in Tasmania. The bulk kerogen carbon showed $\delta^{13}C$ values of *c.* $-16.6‰$ compared to other Permian C_{org} which usually varies between -25 to $-30‰$. The enrichment could reflect local conditions with a high biological production rate and a CO_2 drawdown. Carbon isotope ratios found in kerogen from the Triassic Serpiano oil shale by Chicarelli *et al.* (1993) were more usual with an average of $-30.8‰$.

An approach to reconstructing sea surface temperatures and salinity has been made by Rostek *et al.* (1993). The method is based on the abundance ratios of unsaturated alkenones in phytoplankton algae which showed in culture experiments that the degree of unsaturation in synthesized ketone depends on the growth temperature.

16.1.1 Isotopes in Fossil Fuel

As coal, petroleum, gas and oil shales derive from organic material, the carbon is strongly enriched in ^{12}C. Coal usually has $\delta^{13}C$ values similar to that of modern land plants not showing much variation with either the age or rank of coal involved. Average values of $\delta^{13}C = 25‰$ (PDB) reflect the absence of any appreciable isotopic fractionation during coalification as well as indicating that the source plants had an isotopic composition similar to that of modern plants.

Carbon isotope values of petroleum range from -18 to $-34‰$ with an average of *c.* $-28‰$, similar to the petroleum standard NBS-22, an oil with a $\delta^{13}C$ value of $-29.4‰$ relative to PDB. Petroleum is probably derived from marine and/or terrestrial organic matter and shows an enrichment in ^{12}C by *c.* 10‰ relative to modern aquatic plants, perhaps as a result of isotopic fractionation processes arising during the conversion or by preferential accumulation of fractions enriched in the light ^{12}C isotope. For example, lipids have been found to be enriched in ^{12}C by *c.* 8‰ (Degens *et al.* 1968) compared with other biogenic compounds. In addition, the lipid fraction is the most stable of all plant constituents as well as being insoluble in water and containing hydrocarbons which most closely resemble petroleum in composition. As proposed by Stahl (1979), petroleum may have formed from kerogen as the result of thermal effects associated with burial of the source materials. Kerogen is the insoluble organic C fraction in sedimentary rocks which arises from the diagenetic alteration of humid acid complexes and, as mentioned before, $\delta^{13}C$ values of kerogen are close to those of petroleum.

Natural gas is much richer in ^{12}C than the petroleum with which it is associated. Methane, for example, might be bacterially produced or by thermal influence on organic matter. The biologically produced CH_4 is enriched in ^{12}C by 20 to 40‰ compared to thermically produced CH_4 and may show $\delta^{13}C$ values up to $-100‰$. The heavier fractions were shown to be enriched in ^{13}C relative to CH_4 in the following order to increasing heavy carbon enrichment: ethane, propane, butane.

The H and N isotope ratios in natural gases can exhibit considerable variations. Gas from wells in Kansas has yielded molecular H and N, and one particular case is extreme. Gas from the Scott 1 well near Junction City contained as main components 33.6% H_2, 64.8% N_2 and 1% O_2. The δD values ranged from $-740‰$ to $-826‰$, which constitutes the greatest D depletion ever recorded to 1983. $\delta^{15}N$ values were $-0.2‰$ to $-0.45‰$. Variations in δD up to 100‰ and more are usually observed, wheras $\delta^{15}N$ values commonly range between -15 and $+20‰$. However, extreme values have also been found for nitrogen. The S isotope ratios in coal also show a great variability and $\delta^{34}S$ values ranging from *c.* $+24$ to $-30‰$ have been reported. The S in coal occurs in a number of chemical forms such as pyrite and marcasite, organically bound S and sulphates (gypsum, barite). Most abundant are pyrite and organic S-containing compounds and the low-S coal, i.e. less than 1% of total S, vary within a very restricted, nearly homogeneous range of *c.* 4.6 to 7.3‰. With increasing content of total S (> 1%) increases the variability of the $\delta^{34}S$ values which often show a depletion in ^{34}S. Whereas the low-S coal seems to reflect the S isotopic composition of the original plant material high-S coal shows no systematics and may possibly result from post-depositional infiltration of marine or non-marine sulphate into the coal in conjunction with partial or total reduction of this sulphate to sulphide by bacteria.

16.2 ISOTOPES IN TERRESTRIAL PLANTS

Plants are a major biogenic reservoir for C in either marine or terrestrial environments. The isotope ratios in plants have been thoroughly investigated with respect to the palaeoecologic and palaeoclimatic information they might contain. Terrestrial plants fix atmospheric CO_2 during photosynthesis and the carbon is depleted in ^{13}C relative to the source CO_2. Of special interest, therefore, has been the fractionation of isotopes within the plants and the possible sources of isotopes in air, water or soil. Basically the isotopes of C, O and H are of importance. Within the great variety of terrestrial plants C_4, C_3 and CAM plants can be distinguished.

Numerous reports have been made on the fractionation of isotopes in plants since, in the 1950s, isotope variations in plants were observed. It was realized that the C isotope ratios found are different depending on the photosynthetic pathway and clearly distinguish C_3 and C_4 plants. The extent of C isotope fractionation represents a balance between the diffusion of CO_2 to chloroplasts and the carboxylation reaction, which is different due to the initial carboxylating enzymes of the different pathways (RuBP carboxylase for C_3 and PEP carboxylase for C_4 plants). Apart from these two groups, the CAM plants (crassulacean acid metabolism) show slightly different isotope ratios (Table 16.1).

The initial chemical product formed in the carboxylation reaction in most plant families is the three-carbon molecule phosphoglyceric acid. M. Calvin won the Nobel prize for chemistry (1961) for his work, showing

Table 16.1 Photosynthetic pathways, $\delta^{18}O$ and discrimination Δ (data from Ehleringer, 1991)

	$\delta^{13}C$ (‰)	Δ (‰)
C_3 photosynthesis	−21 to −35	13 to 22
C_4 photosynthesis	−10 to −14	4 to 8
Facultative CAM	−15 to −20	8 to 13
Obligate CAM	−10 to −14	4 to 8

that CO_2 is first incorporated in the C_5 ribulose-1, 5-diphosphate and the resulting C_6 molecule immediately disintegrates into two C_3 molecules (phosphoglyceric acid) under participation of the enzyme carboxydismutase. The so-called C_3 plants include mainly higher plants, whereas the C_4 plants are, for example, tropical grasses and CAM plants are the succulents growing in relatively water-limited environments. The scheme in Fig.16.3 shows the different pathways.

Sample preparation techniques will not be discussed in this context. They are described, for example, for the isotopes of H, C and O by several authors including Epstein, Yapp and Hall (1976) and Feng, Krishnamurthy and Epstein (1993).

However, different isotope ratios are found in different parts of plants. Schleser (1990) investigated the $\delta^{13}C$ values in leaves of *Fagus sylvatica L.* from Sicily and found that the $\delta^{13}C$ ratios for the leaf ribs were consistently about 1‰ higher than those for intercostal tissue. In addition, the isotope ratios from samples varied with the height in the tree (Fig.16.4), indicating the influence of different microenvironments. The data indicate that the microenvironment in which leaves grow influences the isotope ratios of C. Schleser, Bernhardt and Hunka (1989) also studied the morphological and anatomical variability of *Diplotaxis erucoides* populations from Sicily and found that populations growing during the summer months exhibit distinct xeromorphic features. The leaf area was

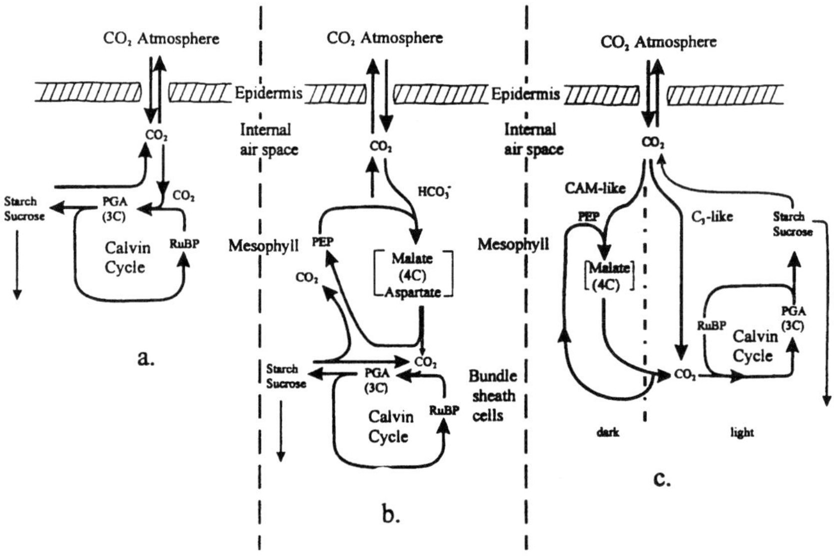

Fig. 16.3 Steps in CO_2 fractionation during photosynthesis. Possible pathways of C are indicated by the arrows (after Schleser, 1995).

strongly reduced and leaf thickness increased almost three times compared with winter populations. These differences are attributed to differenes in intercellular volume and partly due to alterations in mesophyll cell sizes. The organic material of the summer populations exhibited $\delta^{13}C$ values in the order of – 27‰ to – 28‰, while the corresponding values for the winter populations were in the order of – 31‰ to – 33‰. Analysis of *D. erucoides* populations from the transitional period revealed intermediate C isotope ratios.

Dawson and Ehleringer (1993) studied the H isotopic ratio (δD) of stem water and found deviations from source water values in young, nonuberized woody stems (Fig.16.5). The isotope ratios reflect an enrichment associated with cuticular water loss and/or transpiration. The extent of this enrichment is greater in trees which produce a new set of leaves annually than in trees which retain their leaves for 2 or more years. Stem-water enrichment is also greater in rapidly growing stem tissues than in mature, fully suberized or lignified stems which have ceased growing. Benner *et al.* (1987) reported large differences between the C isotope compositions of the polysaccharide and lignin components of a variety of vascular plants. They demonstrated that the C isotope composition of *Spartina* detritus gradually changed during biochemical processing. Polysaccharides are preferentially removed, leaving a material enriched in lignin-derived C and depleted in ^{13}C.

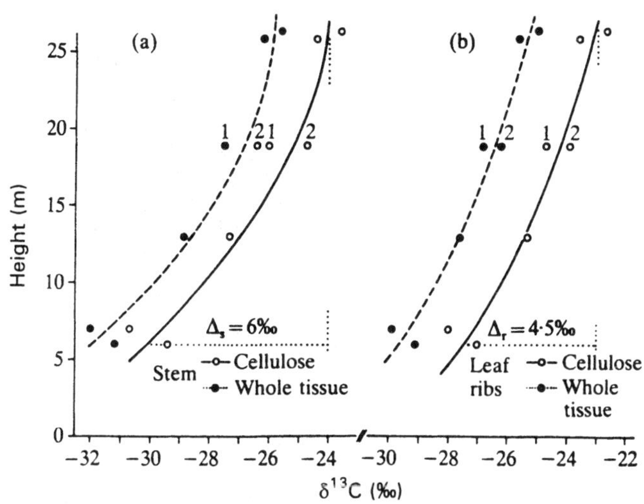

Fig. 16.4 Carbon isotope variations of (a) stems and (b) leaf ribs versus height in the tree for *Fagus sylvanatica* L. from Sicily. The indices 1 and 2 refer to parts of the tree which were shaded or exposed to sunlight.

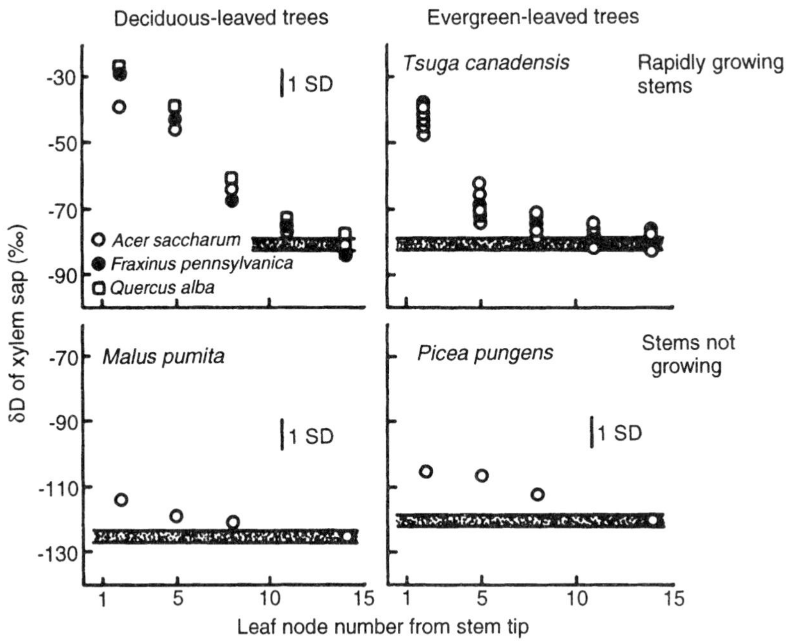

Fig. 16.5 δD ratio of xylem from coniferous tree species versus the number of nodes away from the stem tip. The bar represents the isotopic composition of source water. The upper and lower parts of the diagram compare rapidly growing stems with those either not growing or dormant.

The isotope ratios in plant matter also enable a stratigraphical correlation. Becker, Kromer and Trimborn (1991) presented a late glacial and Holocene tree-ring chronology. Similarly to the information derived from deep-sea sediment or polar ice cores, trees preserve information about past environments. The authors found changes in tree ring growth rates to be related to past climatic anomalies and to be reflected in the isotope composition of tree-ring cellulose. Assuming the isotopic composition in tree rings mirrors changes in the atmospheric and the hydrospheric composition, they established a 9 928 a absolutely dated dendrochronological record of Holocene oak (*Quercus robur, Quercus petraea*) – and a 1604 a floating Late Glacial and Early Holocene chronology of pine (*Pinus sylvestris*) from subfossil tree remnants deposited in alluvial terraces of south central European rivers. The pine sequence investigated provided records of dendro-dated ^{14}C, ^{13}C and ^{2}H patterns for the late Younger Dryas and the entire Preboreal (10000 – 9000 a BP). As a result the authors suggest that the Late Glacial/Holocene transition may be identified and dated by ^{13}C and ^{2}H tree-ring chronologies by means of radiocarbon age calibration, stable isotope analysis and dendrochronology. The stable C isotopes in the wood of palm trees were analysed by Wiesberg and Tavares (1987) in order to investigate the accumulation of fossil fuel-derived CO_2 in the atmosphere, but no significant trend was found for the last 100 a. However, the authors state that palm trees may give better isotopic records of atmospheric CO_2 than trees with rings, as indicated by a small 'biological noise' found in palm stems.

Stable C isotope compositions of different plant matter are used to trace the C flow in ecosystems and to obtain ecological information. However, parameters controlling the isotopic composition in plants are numerous and physiological processes still need a better understanding.

16.3 ISOTOPES AND DIET

DeNiro and Epstein (1978) investigated the effects of diet on the distribution of C isotopes in animals. They stated that the isotopic composition of the whole body of an animal is related to the isotopic composition of its diet, but the animal is enriched in ^{13}C on average by *c.* 1‰. In some cases which they examined, this is counterbalanced by the ^{13}C depletion arising from the respired CO_2. The isotopic relationships between the whole bodies of animals and their diets are similar for different species raised on the same diet and for the same species raised on different diets, but the ^{13}C values of whole bodies of individuals of a species raised on the same diet may differ by as much as 2‰. The relationships between the $^{13}C/^{12}C$ ratio of a tissue and that of the diet depends upon the type of tissue as well as the nature of the diet involved. Many of the isotopic relationships between major biochemical fractions (lipid, carbohydrate and protein) are qualitatively preserved as diet C becomes incorporated into the animal. However, it was found that the difference between the ^{13}C values of a biochemical fraction in an animal and in its diet can be as great as 3‰. The ^{13}C values of the biochemical constituents collagen, chitin and the insoluble organic fraction of shells, all of which are frequently preserved in fossil material, relate to the isotopic composition of the diet. DeNiro and Epstein (1978) stated that it is feasible to perform dietary analysis based upon the determination of the $^{13}C/^{12}C$ ratio of animal C, opining that analysis of the total animal C will usually give a better measure of diet than the analysis of individual tissues, biochemical fractions or biochemical components. The distribution of C isotopes from different tissues is shown in Fig.16.6 as summarized by Gearing (1991). As can be seen, the $\delta^{13}C$ ratios of the muscle and the whole animal are in good accordance. Shell and bone material is, however, not presented, as the inorganic values are much more positive.

As the $\delta^{13}C$ values can give information about ecosystems if recent organisms are studied, the isotope method of dietary analysis should also be applicable to fossil material providing that the original ^{13}C

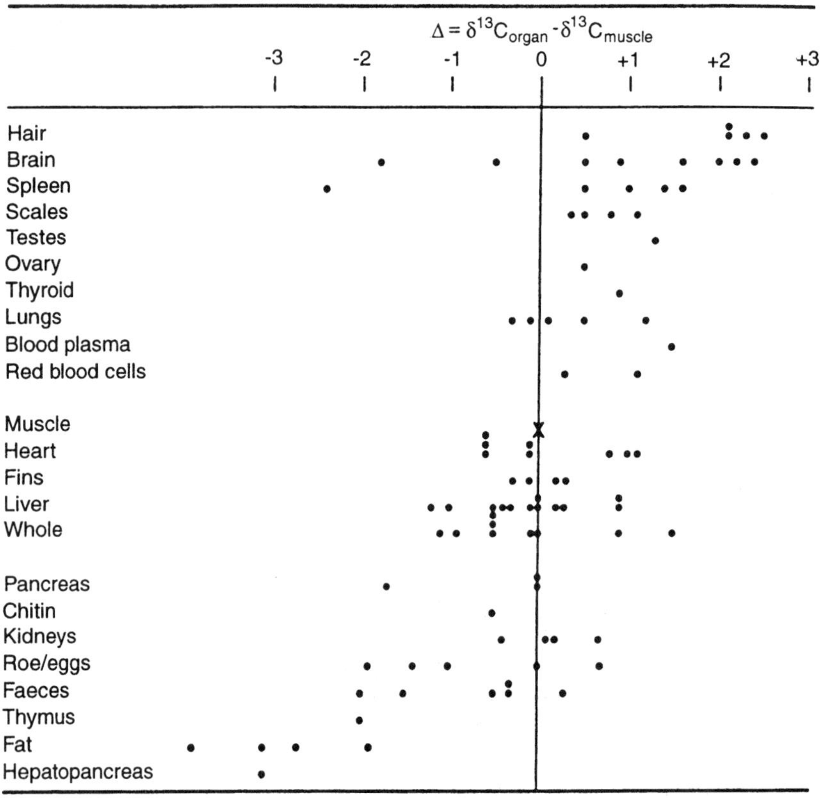

Fig. 16.6 Isotope ratios of tissues and other materials from different marine and terrestrial animals (mice, humans, gerbils, turtles, fish, shrimp, bivalves and insects) compared to that of muscle.

value of some component has been preserved. Often, the best-preserved animal C is included in the carbonate fraction of either invertebrate shells, vertebrate bones or teeth. In both marine and non-marine molluscs, the ^{13}C values of the shell carbonates reflect mostly those of the CO_2 dissolved in the water rather than the diet ^{13}C values and indeed aquatic carbonate-deposition organisms of other phyla may well obtain the C in their shell carbonate from this source too. In terrestrial snails, the C in their shell carbonate derives from a uncertain source. Even if there is some dietary influence involved, this will be hard to evaluate because of the possible contribution of carbon from the atmosphere and the ingestion of $CaCO_3$. Both of these have ^{13}C values far more positive than dietary sources, which would be swamped out in the ^{13}C value of the resultant shell carbonate. A similar situation is encountered in the case of vertebrate bone because the source of the carbon in the relevant carbonate fraction is again uncertain. However, there may be some dietary influence as analyses of bones obtained from mice raised on known diets have shown, and measurements of radiocarbon in fossil bones demonstrated that the carbon in the carbonate fraction is exchangeable with the CO_2 dissolved in groundwater and/or atmospheric CO_2 so that the original ^{13}C value of the carbonate fraction is unlikely to be preserved in fossil bone. Important to palaeoclimatology and palaeoecology is the observation that the relative amounts of potential diet sources ingested by an animal are determinable from the ^{13}C value of the C in the animal if these sources possess sufficiently different ^{13}C values. For instance, the ^{13}C values of aquatic plants and animals frequently do not overlap those of terrestrial organisms, and it should be feasible to utilize this difference in order to ascertain the relative contributions of these two kinds of organisms to the diets of animals living in near-shore environments. In addition, C_3 and C_4 plants act as potential diet sources. The ^{13}C values of most of the former range from -24 to $-34‰$ while those of the latter lie between -6 and $-19‰$, a difference sufficiently large to enable the relative quantities of C_3 and C_4 plants eaten by animals to be determinable from the ^{13}C value of its C. However, other isotopes such as O_2 and N_2 are also of importance if environmental information is required.

Marine palaeotemperatures have, for example, been studied amongst others by Kolodny and Raab (1988) from the O_2 isotope ratios of phosphates preserved in fish remains. They used the phosphate–water temperature scale calculated by Longinelli and Nuti (1973), thus assuming an equilibrium fractionation between the water and phosphate of the fish. The resulting temperatures were in good agreement with apatite from marine reptile skeletons, suggesting a warm climate in the Cenomanian/Turonian and a cooling trend towards the end of the Cretaceous. As the phosphate–water temperature scale produces reasonable temperatures, O_2 in marine phosphates is obviously in equilibrium with the surrounding seawater, whereas the diet is either of no influence or the dietary matter is also isotopically equilibrated. Oxygen isotope ratios from terrestrial mammals on the other hand require a more sensible interpretation. Diagenetic processes change the primary $\delta^{18}O$ values in bones and teeth, whereas the $\delta^{13}C$ ratios seem to be much less affected. Wang and Cerling (1994) have studied closed and open systems, showing that the original $\delta^{13}C$ signal of apatite structural carbonate is retained in closed diagenetic systems and that in open systems the diagenetic alteration of ^{13}C from biogenic apatite can be small if the porosity of the material is low, such as is the case in tooth enamel. Although the authors found the ^{18}O ratios may not be preserved, especially if temperatures and water/mineral ratios were high during diagenesis, oxygen isotopes may provide ecological information.

Chillón *et al.* (1994) analysed the O isotope ratios of bones and teeth from horses, mainly *Equus stenonis*, collected at different locations in south-eastern Spain. Samples covered the time from the Pliocene to the late Pleistocene and were found to differ by up to several per mil, even when from the same deposit. These results lead to the conclusion that, apart from diagenetic change in the $\delta^{18}O$ values, a strong influence of taphonomy might exist.

As mentioned above, marine organisms show phosphate O_2 isotope ratios in equilibrium with the sea water. A similar effect has been detected by Longinelli (1984) and Luz, Kolodny and Horowitz (1984) for the phosphate of bones and teeth from mammals. As the authors stated, there is a close relationship between the O_2 isotope composition of the organic matter and the ingested water. Thus the phosphate is precipitated in equilibrium with the body fluid which has normally a rather constant temperature for endothermic organisms. Empiricial data show a linear relation for the isotope fractionation as investigations in different groups of mammals indicate (Longinelli, 1984; Luz, Kolodny and Horowitz, 1984). Recent results from Bryant, Luz and Froelich (1994) lead to a linear regression for modern equids which is expressed as

$$\delta^{18}O_{PO_4}\,(\pm 2.46) = 22.9\,(\pm 1.13) + 0.6863\,(\pm 0.1614)\,\delta^{18}O_w.$$

The comparison of fossil horse teeth from Nebraska with the modern samples indicated diagenetic alteration for the fossil material which, however, can be detected by a decreased P concentration and increased crystallinity.

Other ecological information which might be of further importance for the reconstruction of palaeoenvironments has been described by Wang, Cerling and MacFadden (1994). Their study of fossil horse teeth has shown a discrepancy between the classic explanation of the assumed co-evolutionary change from brachydont to hypsodont tooth morphology due to the spreading out of savanna grasslands at the beginning of the Miocene. The hypsodont or high-crowned dentition has been interpreted as a result of the more abrasive silica-containing savanna grass. The C isotope ratios of 50 samples from horses, covering a time span from Eocene to Pleistocene, lead, however, to the conclusion that the hypsodont teeth appeared much earlier than the savanna grassland which is mainly dominated by C_4 grasses (Cole, 1986). Figure 16.7 shows the relation between the measured $\delta^{13}C$ ratios and age in comparision with data from MacFadden (1992). The data are in agreement with the isotopic composition of modern herbivore mammals which are fed on a C_3 diet. A similar result has been found for samples from Greece by Quade, Solounias and Cerling (1994). They studied carbonates from paleosols and found carbonate isotope ratios precluding savanna grasslands but indicating C_3 forests or grasses. Their data are also supported by the $\delta^{13}C$ values of teeth enamel from Samos (Greece) and Pasalar

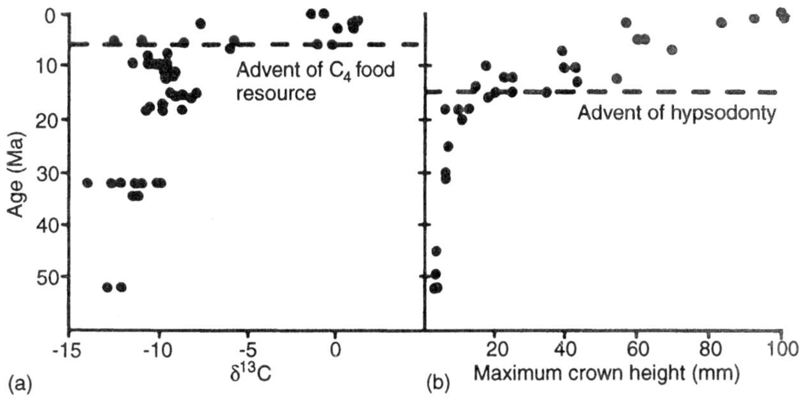

Fig. 16.7 (a) Carbon isotope ratios of tooth enamel of horses, showing the transition to hypsodont teeth and the occurrence of a C_4 diet (after Wang, Cerling and MacFadden, 1994) **(b)** Variation of maximum crown height with age for North American horses (data from MacFadden, 1992).

(Turkey). The prospects offered by isotope investigations on fossil mammals for solving ecological questions are quite good, and there has been an increase in research in the last few years, although the relation between food and $\delta^{13}C$ values had been recognized early by Craig (1953).

MacFadden *et al.* (1994) concluded from their research on C isotope ratios of South American mammals that further work on entire herbivore faunas will lead to a better understanding of 'dietary resource partitioning in ancient terrestrial communities'. They found $\delta^{13}C$ values ranging from -8.2 to $-5.5‰$ for the Late Miocene suggesting mixed plant communities, whereas evidence for C_4 plants are not clearly given until the Plio-Pleistocene.

Another interesting application regarding isotope ratios in mammals has been suggested by Sillen and Lee-Thorp (1994) who studied the Sr/Ca ratio together with $\delta^{13}C$ values. A connection between the Sr/Ca ratio and the trophic level of animals had already been proposed by Toots and Voorhies (1965). The isotope ratios may also perhaps provide information about predator–prey relationships. This is of special interest in the case of extinct species and could possibly elucidate the kind of nutrition favoured by the prey, without being restricted to anatomical and morphologial methods. The Sr/Ca ratio is biologically determined and decreases with increasing trophic level, whereas the $\delta^{13}C$ ratio of the apatite of both predator and prey is similar. Thus the vegetation incorporated is also expressed by the isotope ratio of the predator but the Sr/Ca ratio should be lower at the end of the food chain. Sillen and Lee-Thorp (1994) found that leopards prey mainly on C_3-nourished animals. Although there is a lack of comparative data, isotope investigations might be helpful with respect to this application, especially if accompanied by palaeontological data.

Among the isotope tools the nitrogen in the protein collagen, which is omnipresent in bones and teeth, is also of possible interest. Variations in $\delta^{15}N$ may be due to such factors as diet, trophic level and climate. Isotope studies of recent and fossil mammals have been carried out, e.g. by Cormie and Schwarcz (1994) and Bocherens, Fizet and Mariotti (1994). They have shown large variations of both carbon (9.7‰) and nitrogen (13.8‰) isotope ratios for freshly killed specimens on an inter-regional scale, whereas the samples of bone collagen from white-tailed deer showed a narrower range of 1.8‰ ($\delta^{13}C$) and 1.2‰ ($\delta^{15}N$) for nearby locations. The $\delta^{15}N$ ratio differs too between herbivores and carnivores, similarly to the $\delta^{13}C$ ratio, due to the trophic level, and also between browsers and grazers (Ambrose and DeNiro 1986 and 1989). The fossil samples from Pleistocene bears investigated by Bocherens, Fizet and Mariotti (1994) have, on the other hand, shown that the $\delta^{15}N$ ratios of tooth collagen are not reliable for reconstruction of the adult diet. They were 2‰ higher than the values for the bone collagen, which might be due to an enriched milk diet during weaning. The same reason might have lead to $\delta^{13}C$ ratios being depleted by *c.* 1‰.

It is obvious that all this work still requires an improved knowledge of the still incompletely understood relationships between diet and the related incorporation of isotopes into organic mineral phases or tissue. Additionally, it is most important to investigate unaltered material. However, further studies, not least of modern species used for comparison, might lead to results providing reliable ecological information regarding terrestrial environments.

16.4 ISOTOPES AS STRATIGRAPHIC TOOLS AND EVENT MARKERS

Within the field of geochronology, isotope signals can provide a stratigraphic tool on both small and global scales. The term isotope stratigraphy, however, seems to imply a dating method, either relative as in biostratigraphy or absolute as in the case of radioactive age determination. Actually the tool offers more comparative than absolute information, but nevertheless has useful possibilities. As is common in geology, lithological profiles may be correlated on the basis of physical and petrographic rock features (mainly in sedimentary rocks); isotope signals can be correlated as well. Uncertainties exist in both methods, for instance the horizontal extent of a significant marker in sediments is possibly limited and therefore might not be found in some profiles which represent different facies. This would make correlation difficult if the fossil content is poor. Generally, the isotope signal should still enable a correlation, for an isotopic shift is not influenced by the lack of, for instance, an ash layer. Global correlations of isotope signals can be made, as the results derived from DSDP (Deep Sea Drilling Project) drillings have shown. On a smaller or even global scale, annual isotope variations can be found in ice or in lake sediments and varves, enabling the correlation or the counting of periods similar to growth rings in dendrochronology.

It is obvious that the time span is limited for such investigations and that the lack of layers or very thin horizons will lead to errors. In addition, the interpretation of isotope signals or their correlation can be questionable if the peaks are within the range of normal oscillation. On the other hand, the isotope data also provide palaeoecological and palaeoclimatological information. In the case of $\delta^{13}C$ and $\delta^{18}O$ measurements, CO_2 can be released from carbonate sediments or calcite/aragonite fossil shell materials as well as from air bubbles trapped in ice.

As the temperature is a determining factor in regard to the isotope composition of O_2 and H_2 in snow laid down in polar regions and high mountain areas, extremely negative δ-values reflect the altitude and latitude effects of atmospheric precipitation which can be described by a Rayleigh distillation process and the seasonal fluctuations of temperature. Measurements in snow have shown variations in $\delta^{18}O$-values which are more positive in summer layers and more negative in winter snows, due to the lower air temperatures in winter. Of course, problems may result in regions of low precipitation such as Central Antarctica. For instance, snow can be shifted by deflation and redeposited in another location; consequently the thickness of annual layers may vary and hiatuses may occur.

Large ice-covered areas such as Greenland, Antarctica and possibly large glaciers are ideal for the development of snow and ice stratigraphy and much knowledge has been gained from investigations within these areas. Although the time span represented in ice is relatively small, it covers a part of Earth history characterized by great climatic changes. Reeh *et al.* (1991) reported a $\delta^{18}O$ record from Central West Greenland, covering a time span of 150 ka from the present interglacial back to the end of the previous glacial period, but using surface samples of the marginal area at Pakitsoq instead of samples from layers of ice cores. This is possible because the movement of the ice masses causes a rotation of the horizontal layers sedimented in the accumulation zone. This can be explained by the plastic behaviour of ice at higher pressure in deeper layers and the flow pattern demonstrates that drilling in the accumulation zone gives access to the same layers as surface sampling from the ablation boundary to the margin.

The age of the samples increases either with drilling depth or with distance from the ablation zone to the margin. If the isotope ratios of elements are to serve as event markers, the assumption must be made that the events in question influenced the physical or chemical conditions of the biosphere. Thus it follows that, at least on a global scale, events are recorded somehow by isotopic signals. There is no way, whatever is proposed, whether impacts, global wildfires or ice ages and biological production, that physical or chemical properties can change. It does not matter if, for instance, sea-level changes are caused by tectonic processes or ice accumulation/melting, or if temperatures change due to solar activity or the greenhouse effect. As usual, in dealing with isotopes the difficulty is to determine which of several processes, all causing similar isotope variations, has been responsible in this particular case. Even more complications can arise, for normally several processes may overlap in different directions. For example, high temperatures, leading to negative $\delta^{18}O$ ratios, can cause increased evaporation which leads to high $\delta^{18}O$ values. This kind of dilemma is common within isotope geoscience.

Mackensen *et al.* (1989) presented an isotope stratigraphic record from the Antarctic continental margin which dates back nearly 1 Ma (Fig.16.8).

Isotopic anomalies are observed in connection with global events, for example the mass extinction at the K–T boundary; if seen throughout Earth's history, about a dozen of these mass extinction events occurred and were accompanied by isotope anomalies on a global scale.

Hollander, McKenzie and Hsü (1993) observed a 3.0‰ decrease in the $\delta^{13}C$ of kerogen and a $\Delta\delta^{13}C$ increase of 1.2‰ in sediments of the K – T boundary (Fig.16.9) at Woodside Creek, New Zealand, consistent with the occurrence of a 'Strangelove Ocean', i.e. the dissolved CO_2 was enriched due to reduced plankton production. This reduced productivity seems to have lasted *c.* 1000 a and was followed by an increased production of plankton, shown by increasing $\delta^{13}C$ values (7.0‰) for kerogen and decreasing $\Delta\delta^{13}C$ values (5.4‰), indicating a reduced pCO_2, possibly due to a CO_2 depletion in surface waters caused by 'plankton blooms'. These conditions seem to have lasted for 1 Ma (Hsü and McKenzie 1991), and consequently have influenced the atmospheric pCO_2 and therefore the global climate. Similar results have been presented by Zachos, Arthur and Dean (1989) for samples below the K–T boundary obtained from the DSDP site 577 in the North Pacific (Shatsky Rise), although the time span of reduced biological productivity is much larger, *c.* 500 ka, during which the C isotope ratios are gradually re-established without indicating increased productivity.

The Permian–Triassic (P – T) boundary in the Meishan section of the Zhejiang Province in China, for instance, shows both a decrease of *c.* 6‰ in $\delta^{13}C$ and an increase of iridium up to 2.02 ppb, as high-resolution results of Dao-Yi and Zheng (1993) have demonstrated. These authors have suggested an extraterrestrial impact as a cause for the K – T

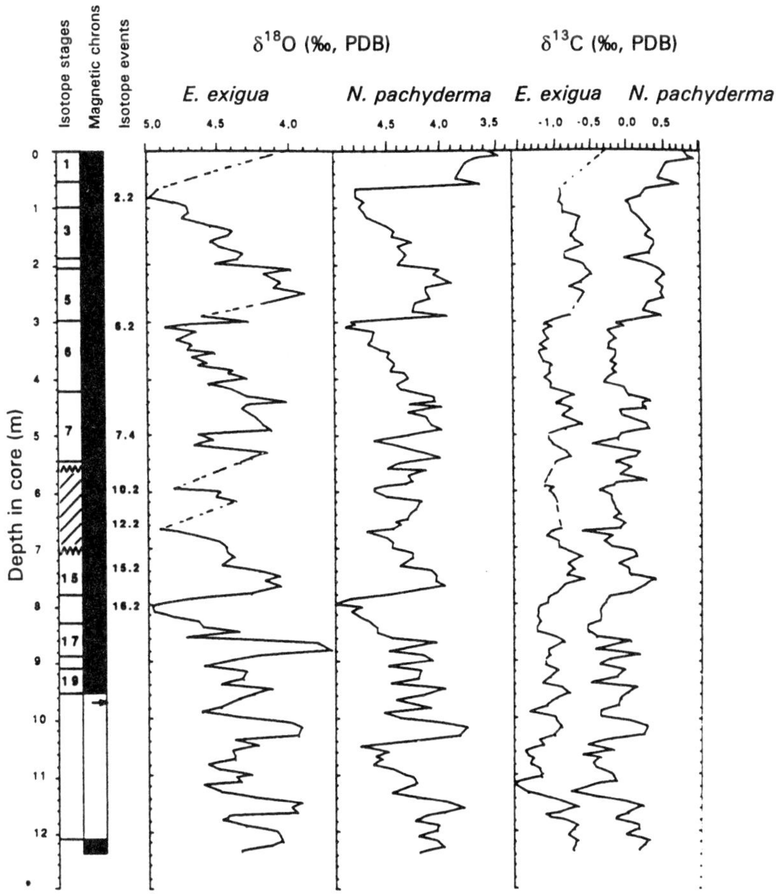

Fig. 16.8 Isotope stratigraphy (Antarctic continental margin) from benthic foraminifera of core PS 1388–3 (after Mackensen *et al.*, 1989).

boundary event. This accords with results from Holser *et al.* (1989), who found a parallel δ^{13}C and δ^{18}O drop of several per mil at the PTB in the Carnic Alps, Austria (Fig.16.10), accompanied by an Ir anomaly. The chemical changes and isotope shifts may be attributed to a regression of the sea and geochemical data are given by the authors. The δ^{18}O values may represent a temperature increase. Similar results were found for the Frasnian – Famennian boundary at the Luoxiu section in South China by Zheng, Hong-Fie and Lian-Feng (1993). A negative C peak of – 6.6‰ occurred in a 20cm thick layer (Fig.16.11), suggesting a short catastrophic event. The event is related to a small increase of δ^{18}O ratios,

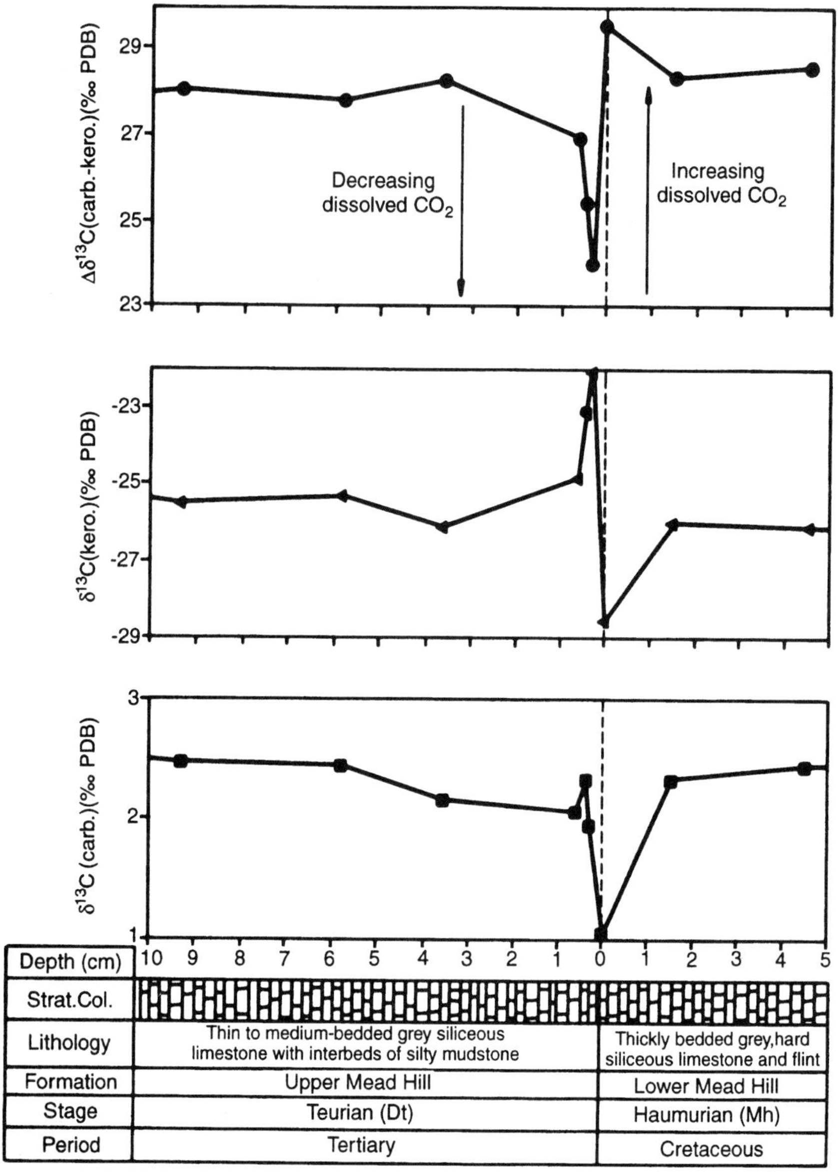

Fig. 16.9 $\delta^{13}C$ values of C and kerogen and $\Delta^{13}C$ of sediments below the K–T boundary (from Hollander, McKenzie and Hsü, 1993).

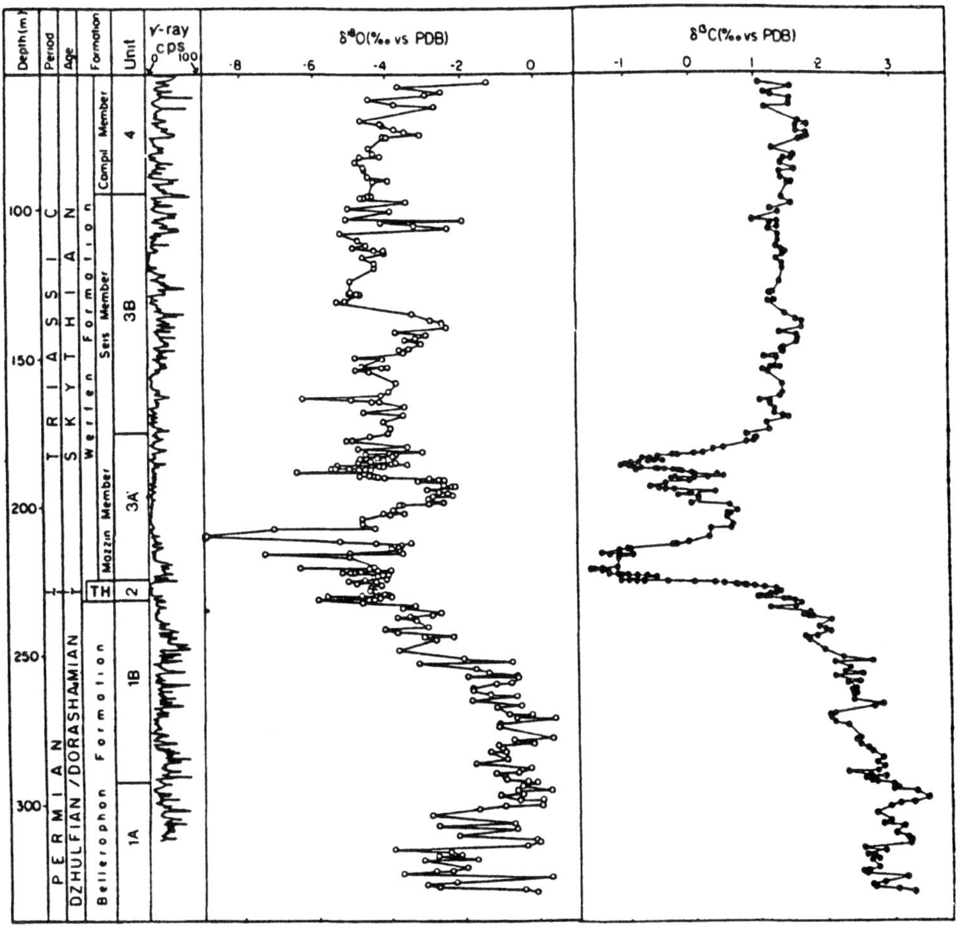

Fig. 16.10 Carbon and Oxygen isotope variations near the Permian–Triassic boundary (from Holser *et al.*, 1989).

which could be due to a temperature drop following the event. The decrease of the $\delta^{13}C$ values could be attributed to reduced organic productivity or bacterial decomposition of organic matter.

However, Joachimski, Buggisch and Anders (1994) found positive $\delta^{13}C$ excursions of up to $+3.1‰$ in a section through the Frasnian – Famennian boundary in the Carnic Alps in Austria and four locations in Germany, which must be due to a shift of $\delta^{13}C$ of the dissolved marine C.

From consideration of the oxygen isotope variations, the temperature may have either dropped or increased. Regarding the theories about the consequence of an impact, it is still unknown what really happened. Either change can be imagined, resulting in a kind of nuclear winter or

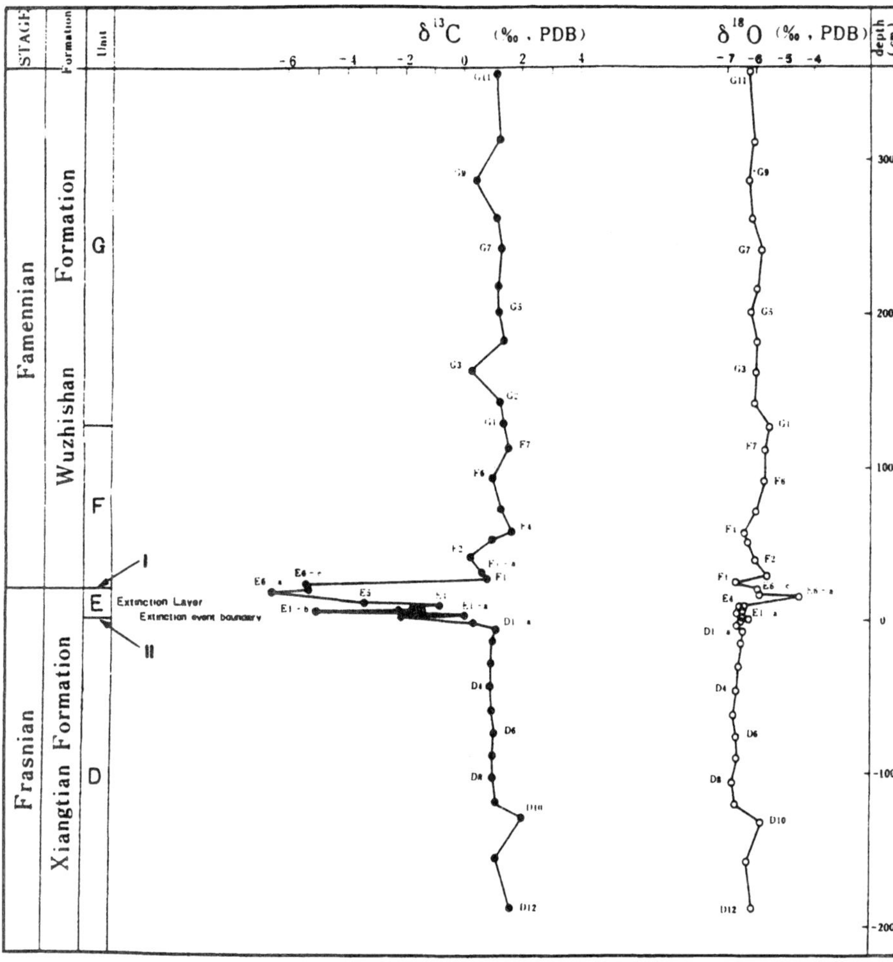

Fig. 16.11 $\delta^{13}C$ and $\delta^{18}O$ ratios at the Frasnian-Famennian boundary (after Zheng, 1993).

an increased greenhouse effect. The result might have been different depending on whether an impact occurred in the ocean or an the continent. More confusing still are the C isotope ratios, and it must be assumed that a global event may cause different reactions locally.

Additionally, isotope ratios could be altered by other effects. In any case, global events are not necessarily the result of impacts. Carbon and oxygen anomalies for the P-T boundary have also been recorded from brachiopod shells by Gruszczynski *et al.* (1989). At the Ordovician–Silurian boundary, Long (1993) compared the C and O isotope ratios to sea-level changes (Fig.16.12). Talent *et al.* (1993) distinguished nine to 12 extinction events during the Middle Palaeozoic on the basis of data derived from brachiopods, corals, conodonts and ammonoids.

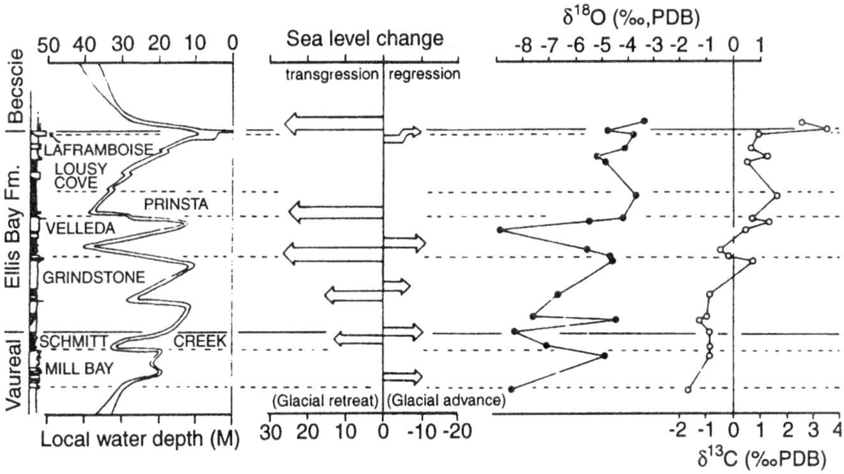

Fig. 16.12 Stratigraphy, $\delta^{13}C$ and $\delta^{18}O$ ratios compared to sea level at Anticosti Island, Quebec (after Long, 1993).

Part Four
Isotopes in the lithosphere

17

Extraterrestrial matter

A plausible impossibility is always preferable to an uncommitting possibility.

Aristotle, 384–322 BC (Poetics, 7)

A great variety of rock material is used for isotope ratio determinations. This includes rocks of extraterrestrial origin as lunar or meteoritic matter, but these are rare compared to the sedimentary, igneous or metamorphic rocks available from the lithosphere. However, isotope ratios obtained from measurements of meteoritic material or lunar rocks may yield a better understanding of the processes which formed our solar system. Investigations on isotopes of lithospheric rocks of either origin may, on the other hand, give information about rock formation, cycling of matter within the lithosphere and related processes such as temperature conditions and weathering.

The usual δ-notation is utilized to describe the isotope composition of the oxygen. The reference standard used is SMOW for all minerals except carbonates, for which the isotope ratios are expressed relative to PDB. The SMOW standard, however, is used for carbonates too, where they are compared with data from silicates when these coexist with the carbonates and the results are to be employed for geothermometric purposes. The usual preparation methods have been described in section 3.1.3.

Normally, the $\delta^{18}O$ values of minerals and rocks are positive, reflecting their enrichment in ^{18}O and the δD values are negative, showing a depletion of deuterium relative to SMOW. Different rock types and minerals will be discussed in Chapters 18 and 19 regarding their isotopic compositon and their geoscientific significance, e.g. geothermometry.

17.1 ISOTOPES IN METEORITES AND LUNAR ROCKS

The mineralogical composition of extraterrestrial matter is more or less similar to that of rocks from the Earth, but it is characteristic of the minerals that they are water free. Lunar rocks can be divided into three types:

- basaltic fine-grained rocks of different mineralogical composition, which are nearly undifferentiated in comparison to the Earth, are found in the maria;
- anorthositic rocks with doleritic texture which occur on the terrae;
- sedimentary cover called lunar soil or regolith consisting of mineral and rock debris with globular glass similar to the tektites on Earth.

In contrast, the meteorites show a greater variety. These stones which 'fell down from heaven' were first recognized as extraterrestrial in 1794 by the physicist E. Chladni. Despite the meteoritic input on Earth, which is estimated to average *c.* 1000 t d^{-1}, it consists mainly of small dust particles either from dust or from smaller meteorites (up to 100 kg), which lose their cosmic speed of 40 km s^{-1} in the upper atmosphere and are broken into small grains which become incorporated into sediments or are preserved in ice. Therefore the material for analysis is rare, although some of the meteorites are famous, at least amongst scientists; this is caused either by the destruction which followed the impact (Chicxulub and K–T boundary extinction) or by the speculations of science fiction lovers (Tungushka event). The holy stone in Mecca, the Kaaba, is also supposed to be a meteorite. In any case, relicts of meteorites with masses >100 t are not found because the energy of the impact is big enough to cause total evaporation of the material.

Meteorites are generally divided into stony and iron meteorites, of which the latter are the heaviest, comprising *c.* 40% of all meteorite finds. The stone meteorites mainly consist of siliceous mineral similar to the basic rocks on the Earth. Both are further divided with respect to their mineralogical composition, for example sideroliths or chondrites. The isotopic analysis of their components has now shown that the former assumption of homogeneity of matter in the solar system was wrong. Various isotopes show large variations and this leads to conclusions which will be discussed in the following.

The oxygen isotope ratios of stony meteorites and lunar rocks accord with those of terrestrial rocks; they can be accounted for by isotopic fractionation among the component minerals at varying temperatures. However, some carbonaceous chondrites, including Allende, are enriched in ^{16}O and this has been explained by postulating that they contain grains containing pure ^{16}O which might have formed by nuclear reactions in an exploding star that contributed matter to the solar nebula. This is confirmed because we know that the reaction cycles in a star can only produce elements up to the mass of Fe by nuclear processes, and this depends on the stellar mass (a star the mass of our Sun would only be large enough for He burning after H_2 has run out), the heavy elements in the solar system must derive from stellar collapse, produced by neutron capture processes. This shows that the solar nebula contains extra solar matter derived from former star generation and is not isotopically homogeneous. If the anomalous isotope compositions of oxygen in carbonaceous chondrites are plotted on a graph with coordinates of ^{17}O and ^{18}O, where ^{17}O is defined as the 'per mil' difference between the $^{17}O/^{16}O$ ratio of meteorites and that of SMOW, then the O of terrestrial and lunar rocks as well as that of most meteorites form a line with a slope of 0.5.

The magnitude of this slope is a result of the dependence of the fractionation of isotopes during chemical reactions upon the difference in their masses; so that the $^{16}O/^{18}O$ ratio is more affected (by a factor of two) than the $^{17}O/^{18}O$ ratio. This produces the situation in which the corresponding values lie along a line of slope 0.5 in coordinates of ^{17}O (y) and ^{18}O (x). However, in many carbonaceous chondrites, Allende included, oxygen isotope ratios fall on a line with a slope of 1 said to derive from the mixing of oxygen in the solar dust cloud with ^{16}O from another source. It is important to add that Clayton *et al.* (1984) found FUN-type O isotope anomalies in a number of exotic inclusions of Allende. The O in these deviates from the usual fractionation line defined by carbonaceous chondrites. This may be ascribed to a combined isotope exchange with an external oxygen reservoir and chemical fractionation. The differences are shown in Fig. 17.1.

As the 'strange' oxygen isotope ratios are due to matter possibly originating from another star and, in fact, Clayton found such mineral grains, the question arose as to whether this should not be possible for other ele- ments as well. Thus an extraterrestrial, or better an extrasolar,

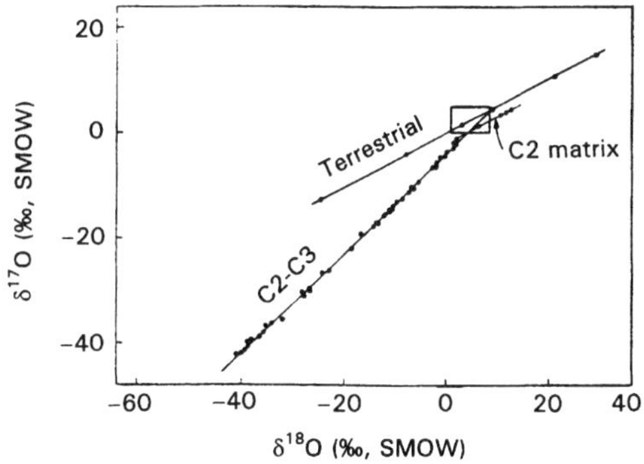

Fig. 17.1 Variations of O isotopes of terrestrial matter and carbonaceous chondrites compared to mass-dependent fractionation (after Clayton, Onuma and Mayeda, 1976).

origin of matter might also be an explanation for anomalies found for Mg and Xe. However, if the matter derives from supernova explosions of presolar age, very little is known about the isotopic composition of the Galaxy. As the only tool given is the spectral analysis of light, data must be looked at critically. Nevertheless, there seems to be a difference in the isotopic composition within our galaxy. Its centre has been found to be enriched in heavy isotopes compared to its plane (by a factor of 2 to 3) and Earth material is depleted relative to the Galaxy. Another explanation for oxygen isotope anomalies in meteorites has, for instance, been given by Navon and Wasserburg (1985). Self-shielding of UV radiation during photodissociation could cause non-mass-dependent isotopic effects. Their calculations, partially based on experiments, confirm that this effect can lead to anomalous isotope shifts. However, the arguments can only explain some of the isotope variations observed in meteorites, because neither the physicochemical environment in a solar nebula nor a necessary rapid isolation and trapping mechanism is known. Furthermore, the different mineral grains of meteorites show isotopic variations. The olivine grains of the Allende meteorite have been studied by Weinbruch *et al.* (1993), using the ion microprobe technique (section 3.1.4(d)), to investigate a possible relationship between the FeO content of different olivines and the O isotope ratios. They found that forsterite grains of <1wt% FeO have isotope ratios of *c.* $-12‰$ (δ^{17}O) and $-8‰$ (δ^{18}O), thus showing similar values to olivine grains with somewhat higher FeO content [$-12‰$ (δ^{17}O) and $-9‰$ (δ^{18}O)]; both indicate a formation not influenced by the crystallization of melt products. Fayalite-rich rims of the grains with a higher FeO content are, however, found to be less enriched then the forsterite cores [$-3‰$ (δ^{17}O) and $\pm 0‰$ (δ^{18}O)]. The reasons for these isotopic differences are still controversial and might be due to formation during metamorphism of a parent body. Generally, the causes of the oxygen isotope variations in meteorites remain unclear, although they enable a classification of meteorites into seven groups (Table 17.1, Clayton and Mayeda, 1978).

Among these groups, isotope ratios of oxygen or other elements may vary, depending on the type of meteorite or the mineralogic component.

Table 17.1 Classification of meteorites (after Clayton and Mayeda, 1978)

Group	*Meteorite type*	*Increasing* ${}^{16}_{7}$O
1	L and LL chondrites	
2	H chondrites	↓
3	E chondrites and aubrites	↓
4	Eucrites, howardites and diogenites	↓
5	Ureileites	↓
6	C2 carbonaceous chondrites	↓
7	C3 and C4 carbonaceous chondrites	↓

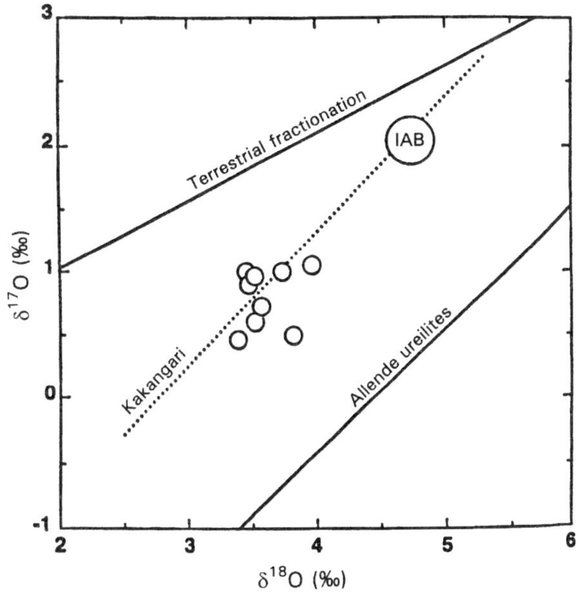

Fig. 17.2 ^{17}O versus ^{18}O for acapulcoites (circles) in comparison with other meteorite types (from Clayton, Mayeda and Nagahara, 1992).

Oxygen isotope relationships in primitive achondrites and among carbonaceous chondrites have previously been reported by Clayton, Mayeda and Nagahara (1992) and Prinz *et al.* (1993). Figure 17.2 shows the $\delta^{17}O$ and $\delta^{18}O$ isotopic compositions of several acapulcoites which are clearly distinct from those of the Allende ureilites or the IAB winonaites and terrestrial matter.

Apart from the oxygen isotope anomalies, Kerridge (1985) analysed whole-rock samples of 25 carbonaceous chondrites for their δD, $\delta^{13}C$ and $\delta^{15}N$ isotope ratios with results unfortunately yielding too few systematic trends and thus not enabling a classification to be made, but still indicating different origins or fractionation processes. In iron meteorites, for example, Prombo and Clayton (1993) found $\delta^{15}N$ ratios ranging from -90 to $+150‰$. This indicates that a well-mixed solar nebular N, reservoir, as concluded from data of bulk samples from stony meteorites with a $\delta^{15}N$ range $>1000‰$, must be questioned. As a consequence of the isotope variations in meteorites, different reservoirs and parent bodies must be assumed. Presolar components in meteorites have, for example, been found in the Allende meteorite and chondrites by Clayton *et al.* (1977) or in chondrites by Alexander *et al.* (1993). The latter authors have analysed 11 ordinary chondrites of the petrologic classes 3 to 6 for C and N isotopes by stepped combustion combined with gas source static mass spectrometry. They found the chondrites to contain isotopic anomalies in minerals and associated noble gas components (see also Alaerts, Lewis and Anders, 1979; Anders, 1988) similar to those of carbonaceous chondrites.

Observed isotope variations in meteorites, including H_2 and S, have been summarized in more detail by Hoefs (1987). Although an additional large amount of data has been gathered in the meantime by many investigations, it is still unclear which reservoirs and fractionation processes contribute to the measured isotope ratios. We do not really know if a homogeneity of the nebular matter can be assumed, and how fractionation processes have influenced the distribution of isotopes during formation of the solar system. Lavrukhina and Ustinova (1993) discussed, for example, an initial ^{53}Mn heterogeneity of the early solar system; further, the amount, origin and role of presolar matter are also not known. Whether isotopic compositions are regarded as 'unusual' (e.g. Amari *et al.* (1992) found δ^{13}C values of 2500‰ in interstellar SiC grains) is a question of the point of view and a consequence of the usual assumption that the terrestrial conditions can be compared to the often unknown astrophysical processes in space.

17.1.1 Impacts and Tektites

Big impacts are fortunately rare, although the Moon's surface might give an impression of their effects. It must be assumed that during its history the Earth has also been struck by many large meteorites, but the coupled endo- and exogenous processes have removed the traces. Thus only a few impacts and related rock formations can be the subject of isotope analysis. Tektites and impact glasses have a high silica content similar to obsidian. Tektites are not of meteoric origin, as formerly thought, but derive from terrestrial rocks melted and splashed around over great distances by the impact energy. Consequently, tektites from different events are products of different source rocks and might therefore be used as tracers. A problem arises, on the other hand, from the possible alteration of the original isotope signal of the source rocks during vaporization of the material and formation of tektites.

Taylor and Epstein (1969) considered the vapour fractionation of oxygen isotopes as such a process and, in fact, the δ^{18}O ratios of tektites has been found to increase if the SiO_2 content decreases. Additionally, meteoric water may be incorporated (Engelhardt *et al.* 1987) and therefore the porosity and related water content of target rocks may be factors controlling the oxygen isotope ratios of tektites, as pointed out by Chamberlain and Blum (1993). However, it is difficult to estimate the extent to which the various source rocks contribute to the isotope ratios of the finally formed tektites. Blum, Chamberlain and Hingston (1993) compared samples from the Manson Impact Structure, Iowa, with Haitian impact glasses from the K–T boundary and found that neither the silicate nor the carbonate rocks from the Manson impact, even if mixing is calculated, are appropriate source rocks for the Haitian impact glasses. The Si-rich (Ca-rich) impact glasses showed ^{87}Sr/^{86}Sr ratios of 0.70820 –

0.70878‰ (0.70796‰) and oxygen isotope ratios of 6.2–9.0‰ (13.1–14.5‰). These values are not comparable to the isotope ratios of source rocks which mostly have a higher ^{87}Sr content and are very different in $\delta^{18}O$ values. Tektites usually show oxygen isotope ratios in the range *c.* 9 to 12‰, although slightly greater $\delta^{18}O$ ratios of 12.8 to 14.6‰ have been reported by Taylor and Epstein (1966).

17.1.2 Exposure Ages of Meteorites

Analogous to the exposure age of different terrestrial rocks, cosmic rays cause spallation reactions on the surfaces of meteorites as well as upon those of asteroids and solar system planets unprotected by atmospheres and magnetic fields. Products accrete in a layer *c.* 1m thick as long as exposure continues. If , as has been suggested, meteorites are asteroidal fragments, this accretion will have commenced when they broke off from their parents, so the exposure age may be defined as the time interval separating such a break-off from the falling of the relevant meteorite upon the Earth. Of course, removal of this meteorite from space ends the production of cosmogenic nuclides, with an accompanying decline in their activities with their respective half-lives. Measurement of the activities of some long-lived cosmogenic radionuclides facilitates calculation of the terrestrial ages of such meteorites, i.e. the time intervals which have passed since they fell on Earth.

The number of radionuclides resulting from cosmic ray irradiation of meteorites is considerable; the abundances depend upon the compositon of these meteorites. Many stable nuclides are also produced by the spallation reactions involved. The activity of a cosmogenic radionuclide made by irradiation of a meteorite in a constant cosmic ray flux during a time t is

$$N_r\lambda = P_r(1 - e^{-\lambda t}) ,$$

in which N_r is the number of radioactive product atoms, P_r their rate of production, λ the decay constant of the product nuclide and t the duration of the irradiation. If the relevant half-life is short compared with the irradiation interval, λ_t is a large number so that $e^{-\lambda t}$ approaches zero as time goes on. Therefore, the decay rate of the product radionuclide comes to equal its rate of production:

$$N_r\lambda = P_r ,$$

meaning that the decay rates of radionuclides fulfilling these criteria are constant and independent of the time of exposure to cosmic rays. The abundance of stable nuclides produced by cosmic rays in meteorites rises with time:

$$N_s = P_s t ,$$

where the subscript s refers to the stable nuclide. The ratio of a stable to a radioactive nuclide resulting from cosmic ray irradiation of a meteorite is given by

$$\frac{N_s}{N_r}=\frac{P_s \lambda t}{P_r},$$

and the irradiation time (exposure age) of the meteorite in question is

$$t=\frac{N_s}{N_r}\frac{P_r}{P_s}\lambda-1,$$

where N_s/N_r is the atomic ratio of any stable to any unstable cosmogenic nuclide (a quantity which can be measured) in a unit weight of the meteorite. By contrast, the ratio of their respective rates of production (P_r/P_s) is much more difficult to evaluate.

It is apparent that P is determined by the size of the cosmic ray flux, the effective cross-section of all spallation reactions contributing to the formation of any particular product nuclide and the concentration of the target atoms in the meteorite. The cross-section in question involves a number of factors such as the intensity and energy spectrum of the cosmic rays to which a meteorite was exposed, the distance of a sample from the surface of a meteorite prior to its entry into the atmosphere (this determines the degree to which it is shielded from cosmic rays), the intensity and energy being distributed to a particular product integrated over the energy spectrum of the cosmic rays. If the ratio of the rates of production of two nuclides formed in a meteorite is taken, the intensity of the cosmic ray flux and the shielding corrections cancel each other out. If the two selected have similar mass numbers and formed from the same target nuclides in the meteorite, the ratio in question is approximated by the ratio of their respective cross-section, σ:

$$\sigma=P_r/\mathrm{P}_s \approx \sigma_r/\sigma_s,$$

and these cross-sections can be measured in the laboratory by bombardment of appropriate targets with high-energy protons or calculated from spallation theory.

Cosmic ray exposure ages for meteorites have been assessed using pairs of nuclides such as ^{4}He–^{3}H, ^{41}K–^{40}K and ^{80}Kr–^{81}Kr. It is assumed that such meteorites were totally shielded from cosmic rays while they constituted parts of a parent body before being liberated by collision, and also that the dimensions of these meteorites and their shapes underwent no changes during exposure to cosmic rays. Such cosmic ray ages show considerable variation from $<0.20\times10^{8}$a to *c.* 1.0×10^{9}a, all appreciably below the crystallization age of meteorites, which is *c.* 4.6×10^{9}a. Iron meteorites appear to have originated prior to stones, but the non-clustering of derived dates implies that impacts in the belt of asteroids probably took place continuously, not from time to time. Meteorites with exposure ages under 1Ma years are unsaturated with regard to ^{10}Be, ^{26}Al, ^{36}Cl and ^{53}Mn. Antarctic meteorites have been investigated and measurements of terrestrial ages made for samples collected near the Yamato Mountains and the Allan Hills. In the latter area, the ^{26}Al activities of 26 chondrites ranged from 11.2 ± 0.4 to 89 ± 2 dpm kg^{-1}, implying terrestrial ages of less than 100ka to 0.7Ma (Evans, Rancitelli and Reeves, 1979).

18

Isotopes in rocks and minerals

A great variety of rock material is exposed on the Earth's surface and, depending on the rock type, there is variation in the mineralogy as well as the conditions during rock formation and, not least, the age of the rocks. As it would lead too far to discuss the geological cycle of rock formation, especially since the resulting rocks are involved in different processes, thus causing various rock types, the isotope studies mentioned can only give an overview restricted to very common rocks.

18.1 ISOTOPES IN ROCK/MINERAL–WATER SYSTEMS

Fluids in a wider sense are omnipresent participants in mineralogical systems. Rocks and minerals interact with meteoric or hydrothermal waters and thus, the isotope ratios of either or both components, rock and fluid, experiences a shift in their δ-values. Furthermore, the liquid can be preserved in a mineral as fluid inclusions, later providing information of its original isotopic composition.

Various aspects considering the processes during rock–water interaction have been summarized by Hoefs (1987). The main parameters controlling the isotope exchange between the phases are the water/rock ratios and the temperature if a closed system is assumed. In nature, systems are usually open and fluids can either escape from or infiltrate the system. Considering that the temperatures can vary from very low (<100°C) to several hundred °C in magmatic systems and water may either be liquid or in a gas phase, the matter becomes far more complicated, depending on the nature of the fluid involved.

A breakthrough in this field came recently with the development of the laser preparation technique, described in section 3.1.3(c), which facilitates the investigation of zonation in single mineral grains. It can be shown that isotope zoning occurs in minerals due to the shift of isotope ratios in the fluids during crystallization. In any case, this had been expected from the physicochemical point of view at the time when only whole rock/mineral samples could be analysed. Presuming equilibrium conditions and neglecting, to some extent, the nature of the fluid, i.e. the pressure or the pH value, some recent studies will be presented in the following.

Chamberlain and Conrad (1991, 1993) studied isotope zonation in garnets and found nearly concentric variations (Fig. 18.1) in $\delta^{18}O$ values decreasing from the centre to the outer rim of the mineral grain. The zonation indicates a temporal isotope variation of the fluid and the magnitude from core to rim is *c.* 3‰ in $\delta^{18}O$ for a system which is supposed to be closed to infiltrating waters. It also seems from Fig. 18.1 as if the contact between two grain boundaries may have stopped growing further in one direction. Garnets from an open system showed, on the other hand, a minor decrease of *c.* 0.5‰. Fluid inclusions were measured separately and found not to be responsible for the gradient described. Data therefore may be useful to distinguish open and closed systems as well as to calculate time-integrated fluid fluxes (Chamberlain and Conrad, 1993). Additional information may be obtained regarding the origin of the fluids involved and their isotopic history. Regarding fluids, a problem should be stressed which affects all rocks, namely the access of fluids, which always depends on the permeability either supplied by channels and pores in sediments or by individual components in magmatic or metamorphic rocks. One might say, no access no change, and this is certainly true if mineral grains are considered and solid-state diffusion is considered to be a slower process.

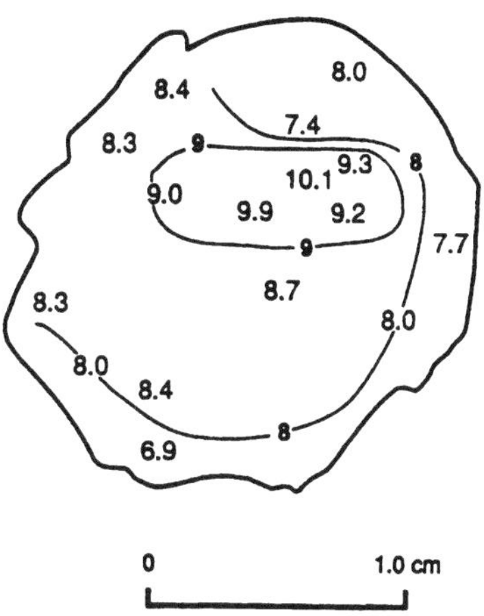

Fig. 18.1 Oxygen isotope zoning in a garnet grain from a regional metamorphic terrane (after Chamberlain and Conrad, 1991).

18.2 MAGMATIC ROCKS

As a variety of magmatic rocks from different genetic series exists and the evolution of magma generation in a series is also variable, the isotopes in magmatic rocks may provide information about the rock/melt history.

Oxygen isotope compositions of young volcanic rocks, together with data from radiogenic isotopes of Sr, Nd, Pb and H_2, evidence crustal contamination of magma given that certain assumptions are made. One assumption is that the magma was isotopically homogeneous when it formed, and another that the ^{18}O values of the rocks sampled did not undergo isotopic fractionation during crystallization. Also, these values must not have been later modified through isotopic exchange with water of variable isotopic composition over a range of temperatures. However, even if the magma is considered to be homogeneous, fractionation during crystallization and secular variations are almost certain. The new laser probe technique has facilitated the measurement of zones in single mineral grains and isotopic heterogenity has been observed in magmatic minerals (see section 18.5). Additionally, plutonic rocks often do not fulfil these formerly assumed criteria. They often have $\delta^{18}O$ values which were changed after crystallization as a result of re-equilibration of the oxygen with water. In the case of lava flows, the influence of fractional crystallization on their ^{18}O values is reduced in most cases because of their rapid rates of cooling which hinder the separation of high-temperature mineral phases. Neverthelesss, hydrothermal alteration of lava flows is to be expected and could change the ^{18}O values of the rocks in accordance with the extent of this phenomenon, the isotope composition of the oxygen in the water and the temperature.

In addition, difference in $\delta^{18}O$ values and also the abundances of stable radiogenic isotopes of Sr, Nd, Pb and Hf within a suite of volcanic rocks may be derived from heterogeneous source rocks, thus not necessarily showing crustal contamination. From this and other considerations, it is clear that when country rock is assimilated by a magma, the process is far from simple and also involves alteration in the composition of the magma by a simultaneous process of fractional crystallization. A mathematical model has been developed to describe the effects both of simultaneous fractional crystallization and country rock assimilation. It was used by Mensing *et al.* (1984) to explain the positive correlation of $\delta^{18}O$ values and $^{87}Sr/^{86}Sr$ ratios in flows of basalt in northern Victoria Land in Antarctica. Results showed that the contaminant is not collinear with the data array in coordinates of $\delta^{18}O$ and initial $^{87}Sr/^{86}Sr$, or in coordinates of $^{87}Sr/^{86}Sr$ and 1/Sr. The above-mentioned discussion of crustal contamination effects in volcanic rocks may be extended to the question of the origin of granitic magma. The effects of differences in the protoliths of such magmas should be reproduced in the oxygen isotope compositions of the resultant granitic rocks, and these differences have been found, for instance in the New England Batholith of Australia which comprises many plutons. The latter have been divided into four suites, two of which are thought to have formed from magmas derived by meeting of clay-rich sedimentary rocks and two arising from igneous protoliths, these falling into the so-called S and I types of Chappell and White (1974). O'Neil, Shaw and Flood (1977) showed that the S-type granitic suites are enriched in ^{18}O relative to the I-type ones, the distinction occurring at a $\delta^{18}O$ value of *c.* 10‰. The S-type range is from +10.4 to +12.5‰ and the I-type is from +7.7 to +9.9‰. In this particular case, the oxygen isotope composition is valuable in the distinguishing of granites stemming from different protoliths. Unfortunately, the isotopic compositions of H_2 seem to have altered through interaction with meteoric water and hence are not useful in identifying the protoliths in question. Later studies have found similar phenomena in other parts of the world, from which it may be inferred that sometimes the $\delta^{18}O$ values of individual plutons are practically constant. Normally, they correlate with initial $^{87}Sr/^{86}Sr$ ratios. In addition, plutons progressively enrich in ^{18}O with increasing distance from convergent plate margins. $\delta^{18}O$ values up to 16.6‰ for whole rocks and 17.8‰ for volcanic matrix have been found by Barth *et al.* (1993) in late Hercynian andesitic to rhyolitic samples from a volcanoplutonic complex in northern Italy. The oxygen isotope ratios, together with other geochemical data (Rb–Sr, Sm–Nd) indicate an open system as well as disequilibrium between quartz–feldspar pairs which suggests a secondary influence of medium–low temperature hydrothermal waters.

Of special interest in helping to estimate the history of degassing further may be the carbon isotope ratios of CO_2 trapped in vesicles or dissolved in basaltic glasses. Using stepwise combustion at increasing temperatures, C is released and often different isotope ratios are found. Blank, Delaney and Des Marais (1993) measured $\delta^{13}C$ ratios in the range −4.8 to −9.3‰ for different grain sizes of the crushed glass at the final combustion step (*c.* 1200°C). Only the last step is considered to release indigenous C, whereas C released at up to *c.* 600° is interpreted to be secondary. The C isotopic composition of the primary or indigenous C was found to be similar to that of ocean ridge basalts and related glasses which usually exhibit a $\delta^{13}C$ range between 4 and 8‰. Values lower than MORB are reported by Macpherson and Mattey (1994), who studied Central Lau Basin basaltic glasses with $\delta^{13}C$ ranging from −8.8 to −16.4‰.

A more or less exotic group of magmatic rocks has recently been studied by Hubberten, Katz-Lehnert and Keller (1988), who analysed carbonate phases from the Miocene Kaiserstuhl alkaline complex in Germany for their C and O_2 isotope ratios. The carbonatites, which are very rare magmatic rocks with *c.* 300 occurrences worldwide, originate from terrestrial subcrustal sources and crystallize from carbonate-rich magmas. $\delta^{13}C$ values in carbonate minerals of this rock type usually range from -2 to $-8‰$ and regional variations may indicate the isotopic composition of subcrustal C.

At the Kaiserstuhl, three groups can be distinguished. These are the subvolcanic sövite intrusions (among the host of available names commonly in use for calcite carbonatite), the alvikitic dykes and extrusive carbonatites. A great range of C and O_2 isotope ratios was found for the three groups (Fig. 18.2) and the results indicate that the isotope ratios of C and O_2 depend on the geological setting, the temperature and on the volatiles involved, and whether the system is open or closed.

As magmatic rocks are of manifold types so are the isotope studies of magmatic rocks. Some of the observed isotope ratios in magmatic rocks are therefore shown in Table 18.1.

In igneous rocks of mantle origin, the isotope composition of S is like that of meteorites, as has been shown from analyses of both continental and oceanic basaltic rock. Tholeiites from the Mid-Atlantic Ridge at 30°N yielded $\delta^{34}S$ values which range from $+0.3‰$ to $+1.6‰$ and, in granite rocks, sulphide minerals exhibit a greater variability, for example, in Europe the $\delta^{34}S$ values range from $+9‰$ to $-4‰$, whereas in Japan an even greater range ($+9‰$ to $-11‰$) has been found. In addition, some large gabbroic intrusives are strongly enriched in ^{34}S compared with meteoritic S. It may be inferred that these deviations relate to variability of the isotope composition of S in the mantle. Interestingly, the studies of Hubberten (1990) showed that the S isotopic

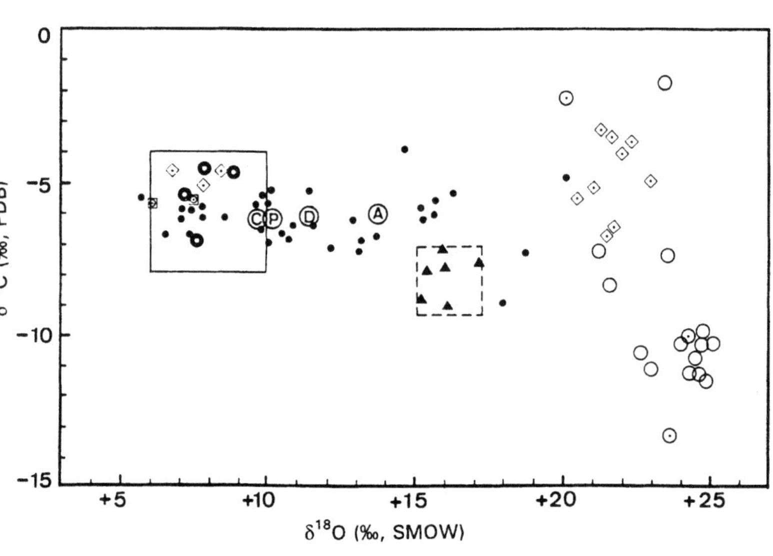

Fig. 18.2 $\delta^{13}C$ and $\delta^{18}O$ values for carbonatites from the Kaiserstuhl alkaline complex in Germany. ⊙ = sövitic intrusives; ● = carbonatitic dykes; ▲ = extrusive carbonatites; A = alvikites s.s.; C = comb-layer dykes; P = porphyritic carbonatites; D = average of all carbonatite dykes. (After Hubberten, Katz-Lehnert and Keller 1988.)

Table 18.1 Isotopic composition of various magmatic materials (data from Hoefs, 1987)

Origin	*$\delta^{18}O$ range*
Tholeiitic basalts from mid-ocean ridges	5.4–6.6
Tholeiitic basalts from oceanic islands	5.0–5.8
Tholeiitic basalts from the continent	6.4–9.0
Alkali basalts from oceanic islands	5.5–8.2
Alkaline baslats from the continent	5.9–8.2
Oceanic island arcs	5.0–9.7
Continental margin arcs	5.2–8.8

composition of oceanic basalts is in accordance with material from the upper mantle, because S is completely mobilized and no fractionation therefore occurs. Nor does the differentiation of basaltic magma cause a shift in the isotopic composition. Isotopic alteration is mainly due to intruding sea water which carries dissolved SO_4^{2-} and causes weathering and transport of dissolved components. Bacterial activity at the water–rock boundary may further result in sulphate reduction.

In the case of granite rocks, it is probable that the wide range of $\delta^{34}S$ values arises from the incorporation of fractionated S coming from sedimentary (country) rocks. The granitic magmas may show positive $\delta^{34}S$ values because they include S donated by marine sulphate during melting or assimilation of sedimentary rocks in a subduction zone. If they show negative values, this could be ascribed to their formation by melting of either sedimentary or metamorphic rocks containing S previously enriched in the light stable isotope ^{32}S by bacterial fractionation.

Some basalts have abnormal S isotope compositions, even in plutonic rocks of basaltic composition. If the mantle S is likely to have an isotope composition close to that of S in meteorites, alternative explanations must be sought for the anomalous cases: these must relate to alteration of the $\delta^{34}S$ values of mantle-derived rocks in some way. One possibility is that the basaltic magma may become contaminated with S originating in sedimentary rocks, and sea water can definitely cause hydrothermal alteration of basalts. Yet another scenario envisages isotopic fractionation occurring in the course of degassing of SO_2 prior to crystallization, but Hubberten (1990) found that the overall isotope ratios do not change during the eruption of lava in depths >500 m, due to the fact that the water pressure prevents volatiles from escaping. The contamination process, termed 'sulphurization', was described by Cheney and Lance (1967). They referred to mafic igneous intrusions enriched in ^{34}S, including the Palisade Sill in New Jersey, the Duluth Gabbro in Minnesota, the Stillwater Complex in Montana and the Sudbury norite in Ontario. All of these, together with the Insizwa Sill in South Africa, possess disseminated sulphide minerals which have $\delta^{34}S$ values tending to become progressively more positive near contacts with the associated country rocks. With reference to the Duluth Complex, Ripley (1981) demonstrated that, in the Dunka Road deposit, sulphide minerals such as pyrrhotite and chalcopyrite occur irregularly concentrated near the base of the gabbro where it intruded the Precambrian Virginia Formation. The relevant $\delta^{34}S$ values range from $+0.2$ to $+15.3‰$, with a mean of *c.* $+7.5‰$. In the altered Virginia Formation itself, the sulphide minerals have fairly similar values for the parameter, ranging from $+0.2$ to $25.8‰$. High positive values for $\delta^{34}S$ in both the Dunka Road ore bodies and sulphides in the Virginia Formation imply that most of the S in the ore came from pyrite in the country rock.

Sea water alteration has been shown to occur in oceanic serpentinites. In these, most of the S is in the form of sulphate with $\delta^{34}S$ values ranging form $+15.4$ to $+20.2‰$. However, coexisting sulphides are much less positive, having values of ^{34}S ranging from $+2.0$ to $+5.6‰$. The sulphate has become enriched in the heavy stable isotope ^{34}S and this has been interpreted as due to the transfer of S from sea water to the serpentinites. Supporting evidence from ultramafics and tholeiites as well as from lava flows and native sulphur deposits was given by Hubberten Nielser and Puchelt *et al.* (1975) and Puchelt and Hubberten (1980). Sakai Casadevall and Moore (1982), on the basis of an investigation into basalt flows from Kilauea volcano in Hawaii, found that submarine flows possess an average of 700 ± 100 ppm S with molar SO_4^{2-}/S ratios which range from 0.15 to 0.56. The average $\delta^{34}S$ value of the total S in these is $+0.7 \pm 0.1‰$, consistent with a mantle origin, but the sulphates are enriched in ^{34}S relative to the sulphides by *c.* $7.5 \pm 1.5‰$. The flows emerging above sea level possess about 150 ± 50 ppm S and are depleted in sulphate. The $\delta^{34}S$ value of the total S is $-0.8 \pm 0.2‰$ and O_4^{2-} is enriched in ^{34}S by a mere $3.0‰$ compared with the S^-. Of course, loss of SO_2 from a basaltic magma can either enrich or deplete the residual S in the melt in ^{34}S depending upon the partial pressure (fugacity) of O_2, fO_2. The relationship of $\delta^{34}S$ in the melt to fO_2 is a result of the fact that the SO_4^{2-}/S ratio in the melt increases with increasing fO_2, affecting the $\delta^{34}S$ value of the escaping S-bearing gases: where there are conditions of high fO_2, the SO_4^{2-}/S^- ratio in the melt increases, SO_2 is depleted in ^{34}S and the ongoing loss of SO_2 produces a progressive enrichment of the melt in the heavy stable isotope ^{34}S. Consequently, the outgassing of SO_2 from a basaltic melt under high-fO_2 conditions is responsible for moving the $\delta^{34}S$ value of S in the basalt in a positive direction, and vice versa. Clearly, the fugacity of O_2 the melt and the amount of S lost directly influence the $\delta^{34}S$ values of mantle-derived basaltic magma. The various interrelationships involved have been quantitatively described by appropriate equations by Sakai, Casadevall and Moore (1982).

The SO_4^{2-}/S^- ratio in a melt depends upon its oxygen fugacity, therefore, and this relative to the reaction

$$S^{2-} + 2O_2 \rightleftharpoons SO_4^{2-}.$$

The oxygen fugacity of a basaltic melt under conditions of equilibrium is governed by those mineral phases which crystallize from it, mainly fayalite, magnetite and quartz. Equilibrium among these phases at a given temperature becomes possible only at a specific value of fO_2. Although it is known that the fO_2 factor in the melt determines the effect of outgassing upon the $\delta^{34}S$ value of the magma, and the magnitude of the enrichment or depletion depends upon the fraction of S remaining and the fractionation factor, the dependence of the S isotope fractionation factor between SO_2 and silicate melts on fO_2 is not yet accurately established.

It should be further mentioned that the IAEA at its meeting held in Vienna on 1–3 December 1993 recommended adopting for the IAEA-S-1 standard the value of $\delta^{34}S = -0.30‰$ versus a defined, hypothetical V-CTD (Robinson, 1995). The CDT troilite is not only exhausted but it was also found that the original standard is not isotopically homogenous.

$\delta^{13}N$ values recorded from igneous rocks are usually positive, demonstrating their enrichment in the heavy stable isototope of N_2 relative to atmospheric N_2. In fact, an inverse correlation may exist between $\delta^{15}N$ and the relevant N_2 concentrations in the igneous rocks concerned. Data on nitrogen are rare, due to the usually low abundance of nitrogen and the difficult sample preparation from silicates. In recent years, the emphasis in such work has been placed more on meteorites and lunar rocks and has been extended to planetary atmospheres. Nitrogen measurements were, for example, carried out on crustal granitic rocks of the Cornubian batholith, south-west England, by Boyd, Hall and Pillinger (1993). They studied the release of ammoniacal N_2 from minerals, i.e. micas and feldspar and whole-rock samples by vacuum pyrolysis and static vacuum mass spectrometry (Wright *et al.*, 1988; Boyd and Pillinger, 1991). Micas were found to lose most of the N at 900° and 1000°C, possibly coupled with the loss of water and F, whereas no significant loss could be observed at temperatures below the melting point of feldspar. Such information may become important for temperature determinations and other petrographical problems such as tracing sediments and metasediments. The whole-rock samples of the granite showed a $\delta^{15}N$ range between $+5.1‰$ to $+10.2‰$, whereas the majority of samples ranged between $+8.4$ and $+10.2‰$, values usually found in metasediments from the amphibolite facies. The data suggest that the ammonium derives from organic matter incorporated in metasediments.

Generally, the laser probe and other techniques may, in addition to whole rock or mineral analysis, open a wide field for rock/mineral studies and mineralogical experiments in both closed and open systems with different mineral phases and under low or high temperature pressure conditions. Results can be expected to be promising with respect to volatiles and rock–mineral interactions. It may also provide more information about fractional crystallization and isotope fractionation between the melt and solid phases, which are supposed to be small at high temperatures, although it is largely unknown how far partial melting of rocks or mantle material may influence the isotope ratios.

18.3 SEDIMENTARY ROCKS

In detrital materials, the O_2 and H_2 isotope compositions are controlled by the respective isotope compositions of the constituent minerals and rock particles. Therefore such rocks are heterogenous from an isotopic point of view, and lack internal isotopic equilibria. During phenomena such as metamorphism, however, isotope re-equilibration may occur. In shales and other deep-sea sediments, the ^{18}O values embrace a range of $+5$ to $+25‰$, compared with δD values ranging from -30 to $-100‰$. Detrital quartz and feldspar do not equilibrate isotopically with sea water; most of the clay mineral fraction appears to belong to this category and thus is not in isotope equilibrium with the sea water. In such clay minerals, isotopic compositions of O_2 and H_2 are controlled by the water with which the clay minerals were in contact when they formed, and also by the relevant values for the isotope fractionation factors for O_2 and H_2 in clay–water systems and the environmental temperature. Thus the isotope composition of the source rocks plays no role. Some of the fractionation factors are listed in Table 18.2. These data demonstrate that clay minerals are enriched in ^{18}O relative to water, but depleted in D at temperatures of the Earth's surface.

Because the isotope compositions of clay minerals formed at these temperatures are governed by the isotope composition of meteoric

Table 18.2 Fractionation factors for clay–water systems at temperatures of the terrestrial surface

Mineral	*Oxygen*	*Hydrogen*
Montmorillonite	1.027	0.94
Kaolinite	1.027	0.97
Glauconite	1.026	0.93
Gibbsite	1.018	0.984
Illite	1.0234	–

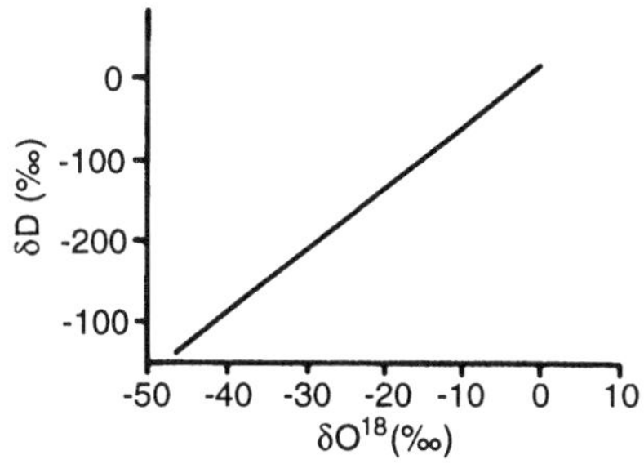

Fig. 18.3 Relationship between δD and δ^{18}O (SMOW) in samples of meteoric water. Isotopic compositions of both water and snow show a latitude effect reflecting the temperature of condensation.

water, and because the δD and ^{18}O values of the latter are related by the following equation: (the evaporation effect)

$$\delta D = 8\,\delta^{18}O + 10$$

which is plotted in Fig. 18.3. The isotope compositions of clay minerals formed at a particular temperature ought to be related by similar equations having the form $\delta D = A^{18}O + B$. In these, the slope of A is determined by the ratios of the fractionation factors and the intercept B is determined by their absolute values. Appropriate equations for some the clay minerals in at Earth-surface temperatures are given in Table 18.3. It has been found that, while many clay-rich soils lie close to the kaolinite line on a plot of δD against δ^{18}O, hydrothermal clays formed at higher temperatures plot nearer the meteoric-water line. This is stated to be due to the fact that clays which equilibrated isotopes at higher temperatures have δD and δ^{18}O values approaching those of water because the fractionation factors come closer to unity as the temperature rises. Consequently, it is possible to distinguish hypogene clay minerals formed at higher temperatures near hydrothermal ore deposits from the supergene clays which equilibrated at Earth-surface temperatures of 22°C.

Supergene clays in soils reflect those climatic conditions which existed when they formed since their δD and δ^{18}O values are linked with the

Table 18.3 Equations for calculation of fractionation factors for some clay minerals

Montmorillonite	$\Delta = 7.3\,\delta^{18}O - 260$
Kaolinite	$\Delta = 7.5\,\delta^{18}O - 220$
Glauconite	$\Delta = 7.3\,\delta^{18}O - 260$

corresponding values of the meteoric water, and those are related to the climatic conditions which prevailed. The clay minerals in soils covering much of the USA have been shown to have been formed in the Tertiary period under rather warmer climates than those of today.

In Recent marine limestones, $\delta^{18}O$ values range from +28 to +30‰ relative to SMOW. Lithification affects such values so that their $\delta^{18}O$ contents have been recorded as decreasing with geological age, dropping to about +20‰ in Cambrian rocks. On the other hand, the associated waters must have undergone a progressive enrichment in the heavy stable isotope. The effect is attributed to post-depositional recrystallization and oxygen isotope exchange with waters depleted in $\delta^{18}O$ compared with sea water. It has been observed in oil-field brines and formation water and was confirmed by Veizer and Hoefs (1976). In cherts a similar decrease in $\delta^{18}O$ values has been found, quantitatively from +32‰ now to *c.* +20‰ 3.4 Ga ago. Secular trends of this type led to the inference that Precambrian surface temperatures were higher than at present, but this may not be wholly justified. Thus, the post-depositional oxygen exchange may also have taken place in cherts (but more slowly, the exchange rate for quartz being much slower than that for calcite). It is also just feasible that the $\delta^{18}O$ of sea water increased through gelogical time, as Veizer *et al.* (1982) suggested. This may not be so, however, since there is some evidence that the isotope composition of sea water is buffered by deep hydrothermal convection along mid-oceanic ridge systems.

As fresh water is depleted to a greater or lesser extent in ^{18}O relative to sea water, the $\delta^{18}O$ values of fresh water limestones of Mesozoic to Recent age are usually lower and more variable than those of marine limestones of the same age. Carbon isotope ratios are also lighter and even more variable in δ-values, due to the different mainly ^{13}C-depleted C sources.

The origins of dolomite have been sought for many years. It was shown empirically that those formed at 25°C ought to be enriched in ^{18}O by 5–7‰ compared with coexisting calcite. In addition, the $\delta^{18}O$ of $CaCO_3$ alters continuously as Mg or Sr enters the structure of the calcite. The fractionation of O between dolomite and water is related to temperature within the range 250–300°C by the following expression:

$$10^3 \ln \alpha_W^d = 3.06 \times 10^6 T^{-2} - 3.24,$$

but the difficulty is that in most samples the difference in $\delta^{18}O$ values is appreciably lower than expected, maybe because the appropriate isotope fractionation factor for dolomite relative to calcite may be less than anticipated. Alternatively, dolomite may form by alteration of $CaCO_3$, in which case its $\delta^{18}O$ value depends upon that of the dolomitizing solution. If the latter hypothesis is valid, then coexisting calcite and dolomite cannot be in isotope equilibrium and the $\delta^{18}O$ value of dolomite in nature will vary in a quite unsystematic manner. If dolomitization occurs near the time of deposition, then the $\delta^{18}O$ enrichment may be small since the pore fluid will most probably resemble sea water. On the other hand, if dolomitization is deferred and then takes place by interaction with solutions with $\delta^{18}O$ values different from those of sea water, the consequent $\delta^{18}O$ values of dolomite will probably be both variable and unpredictable. Fossil dolomites often associate with evaporites and intertidal and supratidal dolomites of this type normally replace pre-existing carbonates; hence they are not primary deposits. However, they show relatively high $\delta^{18}O$ and also high $\delta^{13}C$ values. 'Bacterial' dolomites have the broadest spectrum of $\delta^{13}C$ values of any depositional carbonate, ranging from $-60‰$ to $+21‰$, according to Deuser (1970).

Precambrian sedimentary rocks are especially interesting because of the possible bearing of the results upon the elucidation of biological processes and evolution in the Archean Eon. It must be borne in mind that sulphate-reducing bacterial causing isotope fractionation of S did not exist then. In fact, the oxygen content of the atmosphere was not sufficient to permit the oxidation of sulphide to sulphate until about 2 Ga ago.

In view of these observations, it is unsurprising that sedimentary–volcanic rocks from Isua, West Greeland, deposited much earlier have an average of $\delta^{34}S$ value of almost zero. It was concluded by Monster *et al.* (1979) that bacterial sulphate reduction commenced between 3.2 and 2.8 Ga ago, on the basis of investigations in sedimentary rocks from the Aldan Shield of Siberia and also in iron formation in the Lake Superior area in North America.

Cameron (1982) showed that the $\delta^{34}S$ values of sulphides in Precambrian shales in South Africa were close to zero until some 2.35 Ga ago, after which they became negative; he attributed the phenomenon to an increase in the SO_4^{2-} concentration of oceans enhancing isotopic fractionation by sulphate-reducing bacteria. This suggests, sulphate-reducing bacteria began participating in the terrestrial S cycle around 2.3 Ga ago, i.e. at a much younger time as envisaged by Monster *et al.* (1979). That this may be valid is underlined by work by Hattori, Campbell and Krouse (1983) in which the authors found that 44 pyrite samples collected through the 65 m thick Matinenda Formation in the Quirke II Mine (which relate to the uraniferous conglomerates and sandstones of the Blind River region, Ontario) yielded $\delta^{34}S$ values with a rather narrow range from -1.0 to $+1.1‰$ and that even the pyrite which was intimately associated with organic matter demonstrated no evidence of isotopic fractionation. The absence of fractionation in the relevent deposit is apparent when the range in the parameter is compared with that found in Mesozoic and Tertiary uraniferous sandstones, namely $-50‰$ to $+29‰$. The inference is that there was no sulphate in the basin in which the formation deposited, implying that the atmosphere then lacked enough O_2 to convert sulphide into sulphate. The pyrite itself derived from Archaean gneisses, being transported as detrital grains around which authigenic pyrite later precipitated. However, since both the cores and overgrowths have similar $\delta^{34}S$ values (practically zero), the latter could not have been precipitated by the bacterial reduction of sulphate.

All this evidence makes a convincing case that O_2 did not become a significant atmospheric component until after some 2 Ga ago. In time intervals prior to that, sulphate could have been formed on the planet through volcanic emanations, and also by photosynthetic sulphur bacteria. If so, photosynthesis preceded sulphate respiration and the latter preceded oxygen respiration in the Late Proterozoic. This would accord with recorded $\delta^{34}S$ values of Archaean sulphates which are very nearly zero, different only to a slight extent from those of associated sulphide minerals. Perry, Monster and Reimer (1971) recorded results obtained from barytes of supposed sedimentary origin collected from the Swaziland System of the Barberton Mountain Land, South Africa, over 3 Ga in age. These gave an average $\delta^{34}S$ value of $+3.4‰$. Roughly contemporaneous sulphides gave a value of $+0.42‰$. The formation of the sulphides was believed to be due to oxidation of volcanic H_2S by green or purple photosynthetic sulphur bacteria.

Because of the rarity of Precambrian marine sulphate deposits and the problem of dating them accurately as well as the possibility of alteration in their isotope compositions, it is difficult to reconstruct the actual variation in their $\delta^{34}S$ values. Whelan and Rye (1978) found that two stratigraphically distinguishable anhydrite formations in the Grenville Series in New York show very different $\delta^{34}S$ values with the respective ranges of from $+7$ to $+10‰$ and from $+18$ to $+30‰$. Interestingly, these concord with data from sedimentary anhydrite in the Upper Roan Group, Zambia, for which a value of $\delta^{34}S = +17.5 \pm 3‰$ for marine sulphate during the time interval 1.3–1.0 Ga ago has been given. In the Bitter Springs Group (1000 – 800 Ma), Central Australia, and the Shaler Group (850 – 635 Ma), Northwest Territory, Canada, results point to a repressentative value for $\delta^{34}S$ of $+16.5 \pm 2‰$ for the Late Proterozoic. Whatever the deficiencies in the approach, it is clear that Late Proterozoic sedimentary anhydrite formations are enriched in ^{34}S because of the activities of sulphate-reducing bacteria. This became the basis for the complex pattern of variation of the ^{34}S parameter and also the $\delta^{18}O$ parameter in marine sulphates during the subsequent Phanerozoic time interval.

18.4 METAMORPHIC ROCKS

Metamorphism changes rocks within the Earth's crust by means of heat, pressure and chemically active fluids. Considering thermal metamorphism, mainly controlled by heat, dynamic metamorphism, due to pressure and a combination of both, and dynamothermal metamorphism, it becomes obvious that metamorphic rocks may have undergone a variety of conditions. There are also other types of metamorphism, and anatexis may occur as well, but this shall be neglected in the following. As heat, pressure and a fluid phase comprising water and carbon dioxide are present during metamorphism, the rock/fluid interaction is of importance as well as the temperatures under equilibrium conditions. Many aspects regarding formation processes and isotope ratios of metamorphic rocks have been summarized by Hoes (1987) and, generally, the majority of both igneous and metamorphic rocks possess $\delta^{18}O$ values in the range of $+5$ to $+15‰$, whereas the δD values fall between -50 to $-100‰$ with an average of $-85‰$. The analysis of single mineral grains may elucidate processes of fluid-rock/mineral interaction (see also section 18.1). Yui *et al.* (1990) analyzed, for example, serpentinites of the Rengtien ophiolite for hydrogen and oxygen isotopes to study the process of serpentinization. They found different values for lizardites ($\delta^{18}O = 3.6‰$, $\delta D = -48.0 - -49.0‰$), foliated antigorites ($\delta^{18}O = 3.5‰ - 5.8‰$, $\delta D = 45.0 - -69.0‰$), picrolites ($\delta^{18}O = 4.2‰ - 4.3‰$, $\delta D = -65.0 - -67.0‰$) and chrysotiles ($\delta^{18}O = 4.1‰ - 4.5‰$, $\delta D = -80.0$). The authors suggest that the lizardite may have formed in an oceanic environment, whereas the antigorite and chrysotile might have its origin in a continental environment.

Young and Rumble (1993) also found zonations in metamorphic garnet from the Gasset schist (Vermont, USA). The $\delta^{18}O_{(V\text{-}SMOW)}$ values varied in a millimeter scale exhibiting $\delta^{18}O$ values of $10.9 \pm 0.3‰$ (core) and 10.1 ± 0.2 (rim). Other minerals were found to show uniform isotope ratios within the analytical errors. Staurolite, muscovite and chlorite had, for example, average oxygen isotope ratios of $10.8‰$, $11.9‰$ and $10.3‰$ respectively. Quartz in garnet cores was found to be isotopically heavier ($14.8‰$) than vein quartz ($14.3‰$). Although these values are not as distinct as from the garnet in Fig. 18.1, the authors could distinguish two mineral parageneses on the basis of $\delta^{18}O$ values and mineral texture. It is also possible that not all minerals will reflect the metamorphic conditions.

18.5 ORE DEPOSITS

The isotope composition of water associated with the formation of hydrothermal ores and related rock-wall alteration can be assessed by analysis of water from fluid inclusions and by calculation of ^{18}O- and D-values of the aqueous fluid which equilibrated both O_2 and H_2 with minerals at a specified temperature. Whilst the first approach is direct, it has the disadvantage that the isotope composition of the water in question may have been altered through exchange with O_2 and H_2 in the host mineral. It might be argued that this host material does not invariably contain these elements, but then they may be present in inclusions or fractures. The second approach obviates this difficulty, being based only upon the validity of the appropriate mineral–water exchange equilibria and the isotopic geothermometric equations utilized in determining the temperature.

Unfortunately, although oxygen isotope equilibria have been extensively studied, the hydrogen equivalents between minerals and water are far from being fully known. Bearing this in mind, attempts have been made in order to assess the isotope composition of water which entered into the deposition of ores as well as accompanying wall-rock reactions. These studies demonstrated that there are various possible sources of hydrothermal fluids, and that different isotopic varieties of water were involved in ore deposit formation in any particular locality. Hydrothermal fluids may have started as meteoric water or as trapped sea water as well as in the shape of connate formation waters with variable ^{18}O and D values. Other possible origins include magmatic water and geothermal waters which could be enriched in ^{18}O. Finally, metamorphic water could be associated with such formation and both this and magmatic water cannot be studied at first hand but rather only through their effects on the isotopic compositions of minerals with which they equilibrated at a specified temperature. Metamorphic waters underwent such modification of their isotope composition by equilibration with O- and H-bearing minerals in the course of a metamorphic episode at temperatures ranging from 300 to 600°C. It appears that the range of δD in such waters is from -20 to $-65‰$ and that of $\delta^{18}O$ is from $+5$ to $+25‰$. Magmatic waters have their isotope compositions governed by equilibration with magmas which can range from 700 to 1100°C in temperature. It has been found that most plutonic and volcanic rocks possess fairly uniform ^{18}O- and D-values and this is reflected in the restricted values recorded from magmatic waters, namely from $+5.5$ to $+10.0‰$ for $\delta^{18}O$ and -50 to $-85‰$ for δD. In fact, the fundamental source of both metamorphic and magmatic waters cannot be ascertained from their respective ranges in isotope compositions. It is suspected, however, that this may well have been either meteoric water or sea water, or a combination of both.

Bacteria which live in freshly deposited sediments reduce sulphate and enrich the resultant H_2S in ^{32}S so that S which has undergone this process is enriched in the light stable isotopes relative to marine sulphate. However, S associated with mantle-derived igneous rocks has an isotope composition close to that found in meteorites, with $\delta^{34}S$ values near zero. These observations might suggest that the isotope composition of S could be useful in distinguishing sulphide ore deposits connected with igneous sources from those having a sedimentary origin: S from the former has a narrow range of $\delta^{34}S$ values grouped about zero, whereas biogenic S tends to have negative and more variable values. Nevertheless, attempts to employ this approach have been rather disappointing, partly because of the very complex geological histories of ore deposits which render it impossible in many cases to assign them into the categories of syngenetic (sedimentary) and epigenetic (igneous–hydrothermal). The latter often have a broad range of $\delta^{34}S$ values because they contain a set of generations of minerals which were sequentially deposited under different conditions. By contrast, a narrow range of these values can be found in ore deposits associated with sedimentary rocks because they formed from an infinite sulphate reservoir under constant environmental conditions. Of course, the isotopic composition of S in strata-bound sulphide deposits connected with sedimentary and volcanic rocks may be related to that of marine sulphate at the time of deposition; it has been shown that S in sedimentary types of strata-bound sulphide deposits is enriched in ^{34}S by 11.7‰ on average, while volcanic types show an enrichment of 17.5‰ compared with contemporaneous marine sulphate. This accords with the assumption that the S in them arose from marine sulphate through bacterial reduction. Therefore, the $\delta^{34}S$ values of strata-bound sulphide deposits of various ages may reflect also the time-dependent variation of the isotopic composition of marine sulphate.

An objection to this last conception was made by Ohmoto and Rye (1979), who indicated that the bacterial reduction of marine sulphate cannot explain the presence of sulphide materials in strata-bound deposits adequately since the host rocks usually possess low C contents and the depositional temperatures are too high (50–350°C) for such bacterial activity. They proposed instead that the sulphides form as a consequence of the inorganic reduction of the marine sulphate by Fe^{2+} in hot volcanic rocks in contact with sea water.

In the course of the precipitation of sulphide materials from aqueous solutions or their crystallization from sulphide liquids, isotopic equilibration among solids and between solids and liquids may produce slight differences in the ^{34}S values of cogenetic minerals: these have been recorded, for example, by Thode (1970). Taking bond strengths into account, the ^{34}S enrichment of common sulphide minerals ought to decrease in the following order: pyrite > sphalerite > chalcopyrite > galena, and this has been generally confirmed. Therefore, it may be that isotopic equilibrium is almost attained by these minerals in nature, and hence the $\delta^{34}S$ values of cogenetic sulphide minerals may be valuable in assessing the temperatures of equilibration if a suitable determination of the variation of $\delta^{34}S$ as a function of temperature be made, either empirically or by calculation. Fractionation factors for S isotopes in many S-bearing compounds as a function of temperature have been compiled (Ohmoto and Rye, 1979). The $\Delta^{34}S$ values of pyrite–galena, sphalerite–galena and pyrite–sphalerite have been calculated for the temperature range 27–527°C as a function of $10^6/T^2$, where T is the absolute temperature in Kelvin. The $\Delta^{34}S$ values for these mineral pairs decrease linearly with increasing temperature above *c.* 150°C; of the three pairs, pyrite–galena is the most sensitive geochronometer, but only rarely forms in equilibrium.

Many researchers have studied the fractionation of S isotope by sulphide minerals and obtained results confirming that the $\delta^{34}S$ value of mineral pairs are linearly related to the equilibrium temperature given as $10^6/T^2$. The experimental results can be expressed in equations taking the following form:

$$\Delta = \frac{A + 10^6}{T^2}, \tag{18.1}$$

in which A is a constant equal to the slope of straight lines through the origin and T is the absolute temperature in K. Some expressions obtained by solving this equation are given in Table 18.4.

Table 18.4 Sulphur isotope geothermometers based on the isotope compositions of coexisting minerals in equilibrium with a common S reservoir

Mineral pair	*Temperature* (K)[a]	*Uncertainty* (K)	
		1[b]	2[c]
Pyrite–galena	$\frac{(1.01 \pm 0.04)10^3}{\Delta^{1/2}}$	±25	±20
Sphalerite (pyrrhotite)–galena	$\frac{(0.85 \pm 0.03)10^3}{\Delta^{1/2}}$	±20	±25
Pyrite–chalcopyrite	$\frac{(0.67 \pm 0.04)10^3}{\Delta^{1/2}}$	±35	±40
Pyrite–pyrrhotite (sphalerite)	$\frac{(0.55 \pm 0.04)10^3}{\Delta^{1/2}}$	±40	±55

[a]The equations which relate the temperature to the $\Delta^{34}S$ value of the mineral pairs were derived by solving equation (18.1) for T.
[b]The uncertainty 1 of the calculated temperature results from the uncertainty of equation (18.1) at a temperature of 300°C.
[c]The uncertainty 2 is caused by the analytical error in $\Delta^{34}S$ of ±0.2‰.

So far, it has been assumed that the $\delta^{34}S$ value of sulphide minerals relates solely to the isotopic composition of S in the fluids from which they are derived and the temperature of isotopic equilibration. The matter is, however, more complicated because the $\delta^{34}S$ value of aqueous S_4^- ions relates to the relative proportions of H_2S, HS^- and S^{2-} which all fractionate S isotopes. Chemical equilibria involving H^+ ions govern the abundances of these ionic and molecular species so that the $\delta^{34}S$ values of the aqueous S^- ion and of sulphide minerals in equilibrium with it relate to the pH of the ore-forming fluid. In fact, the idea was extended to include the effects of a number of additional physicochemical parameters upon the $\delta^{34}S$ values of sulphide minerals which can precipitate out of any give ore-forming fluid: important ones were found to be the fugacities of both S and O_2, the total S content and the ionic strength of the ore-forming fluid involved. Sulphur-bearing species which are significant in ore-forming fluids below 500°C are H_2S, HS^-, S^{2-}, SO_2^{4-}, HSO_4^-, KSO_4^- and $NaSO_4^-$, They are interrelated by the following chemical equilibria:

$$H_2S_{(aq)} \rightleftharpoons H^+ + HS^-,$$

$$HS^- \rightleftharpoons H^+ + S^{2-},$$

$$2H^+ + SO_4^{2-} \rightleftharpoons H_2S_{(aq)} + 20_2,$$

$$HSO_4^- \rightleftharpoons H^+ + SO_4^{2-},$$

$$KSO_4^- \rightleftharpoons K^+ + SO_4^{2-},$$

and

$$NaSO_4^- \rightleftharpoons Na^+ + SO_4^{2-}.$$

Clearly, the activity of H^+ ions controls the relative proportions of the H_2S, HS^- and S^{2-} ions which can coexist, the O_2 fugacity controlling the abundance of $SO_4^{\,2-}$ relative to aqueous H_2S.

Large deposits of native sulphur are located in the cap rocks of salt domes, e.g. in the Gulf Coast region of the USA and in some evaporite sequences in the Mediterranean, for instance in Sicily. In the case of the Gulf Coast salt domes, petroleum may be entrapped along their flanks and the native sulphur is localized in the capping calcite and anhydrite rocks, which formed by the leaching of rock salt by groundwater. Anhydrite usually predominates and was originally disseminated, in quantities ranging from 5% to 10% in the rock salt concerned. Both in Texas and in Louisiana, the $\delta^{34}S$ values in anhydrite are similar, implying a common source which is thought to be the Louann salt (Jurassic in age). Some information from the Texan Boling salt dome is given in Table 18.5 from which it may be noted that the cap rock anhydrite is enriched in ^{34}S and has variable values, whereas the native sulphur is depleted in the heavy stable isotope and correspondingly enriched in the lighter ^{32}S.

Table 18.5 $\delta^{34}S$ isotope ratios from S in minerals in the Boiling salt dome

Material	$\delta^{34}S$ (‰)[a]
Anhydrite disseminated in halite	+16.2
Anhydrite, anhydrite cap rock	+17.9
Anhydrite, calcite cap rock	+26.3
Native sulphur, anhydrite cap rock	+3.0
Native sulphur, calcite cap rock	+2.6
Marcasite and pyrite in calcite rock	+5.8
Native sulphur with above-mentioned sulphides	+4.2
Sulphate, bleedwater	+25.2
H_2S bleedwater	+0.8

[a]Recalculated from $^{32}S/^{14}S$ ratios assuming a value of 22.22 for the $^{34}S/^{32}S$ ratio of Cañon Diablo.

This is probably due to isotope fractionation of S by bacteria-reducing SO_4^{2-} ions to H_2S gas and possibly living on the petroleum saturating the cap rocks. ^{32}S would then be preferentially incorporated and the residual sulphate would have been left enriched in the heavy stable isotope to a degree dependent upon the fraction of SO_4^{2-} actually converted into S^- in a particular part of the cap rock and also upon the metabolic rate of the bacteria. The H_2S must have been later ozidized by SO_4^{2-}.

Calcite in the upper portions of cap rocks resulted from the precipitation of CO_2 gas released by bacteria during the metabolic oxidation of petroleum. In the Boiling dome, the petroleum has a $\delta^{13}C$ value of -26.2‰ relative to sedimentary limestone, compared with a $\delta^{13}C$ value of -32‰ obtained from the calcite. Obviously, that latter mineral is enriched in the light stable isotope of C compared with normal marine limestones and resembles the associated petroleum in isotope composition. Perhaps this enrichment is the result of fractionation in the course of metabolic oxidation of the petroleum by the bacteria. The Ca of the calcite must have come from the anhydrite which provided SO_4^{2-} ions later reduced to H_2S by bacteria. In sum, therefore, the isotope composition of the C in cap rock calcite is in accordance with the probable origin of the native sulphur by bacterial reduction of sulphate to H_2S succeeded by oxidation to the native element.

19

Isotope geothermometers

Differences in the $\delta^{18}O$ values of cogenetic minerals can serve as geothermometers. As the temperature is the main parameter controlling the isotope fractionation, a set of studies of the equilibration of minerals with water over a variety of temperatures led to a temperature dependence which can be expressed by

$$\ln \alpha \approx 1/T^{2},$$

with α as the fractionation factor and T as the absolute temperature. A pressure dependence is mainly neglected, although Rumble (1982) stated that pressure changes influence the involved volatiles, which can cause a shift in the isotope fractionation. This is, in fact, true from the physical point of view, although this complicating effect is not yet quantified. However, experimental and empirical data have shown that, if equilibrium conditions occur, temperature determination for mineral pairs is possible.

If minerals equilibrated O_2 with the same reservoir at the same temperature, and no subsequent alteration in isotope composition took place, then in an igneous or metamorphic rock their differing $\delta^{18}O$ values reflect the fact that their respective fractionation factors at a given temperature have different values. This means that the difference in the $\delta^{18}O$ values of two coexistent minerals in a rock sample is dependent on the temperature of final equilibration. For example, in the case of the fractionation of oxygen in quartz–water and calcite–water systems, the following relationships obtain:

$$1000 \ln \alpha_{\mathrm{w}}^{\mathrm{q}} = 3.38(10^{6}T^{-2}) - 2.90\,\text{(quartz)} \tag{19.1}$$

$$1000 \ln \alpha_{\mathrm{w}}^{\mathrm{c}} = 2.78(10^{6}T^{-2}) - 2.89\,\text{(calcite).} \tag{19.2}$$

Subtracting equation (19.2) from (19.1) leads to the following expression:

$$1000 \ln \alpha_{\mathrm{w}}^{\mathrm{q}} - 1000 \ln \alpha_{\mathrm{w}}^{\mathrm{c}} = 0.60(10^6 T^{-2}).$$

This leads to

$$\delta q - \delta c = \Delta_{\mathrm{c}}^{\mathrm{q}} = 0.60(10^6 T^{-2}) \tag{19.3}$$

taking into account that $1000 \ln \alpha_{\mathrm{w}}^{\mathrm{q}} \approx \delta_{\mathrm{q}} - \delta_{\mathrm{w}}$ and $1000 \ln \alpha_{\mathrm{w}}^{\mathrm{c}} \approx \delta_{\mathrm{c}} - \delta_{\mathrm{w}}$. This simply embodies the fact that the difference in $\delta^{18}O$ values of coexistent quartz and calcite is inversely proportional to the square of the temperature at which the two equilibrated oxygen with the same reservoir. Equation (19.3) is in principle a geothermometer and similar geothermometers can be secured from other appropriate mineral pairs. Their sensitivities are dependent upon the differences in the slope of the fractionation lines for each particular mineral and also upon the temperature. As temperature rises, the Δ-value for any selected pair of minerals decreases towards the analytical uncertainty and hence the most sensitive thermometers employ mineral pairs of which the fractionation lines relative to water possess highly divergent slopes, e.g. quartz–magnetite. Some isotope fractionation equations of mineral–water pairs are listed in Table.19.1.

An early application of the approach was that of Anderson, Clayton and Mayeda (1971) which derived the following expression for the oxygen isotope fractionation between plagioclase and magnetite at temperatures higher than 700°C:

$$\Delta_{\mathrm{m}}^{\mathrm{p}} = (4.72 - 1.19\beta)\, 10^6 T^{-2},$$

in which β is the mole fraction of anorthite in the plagioclase. This was used to calculate temperatures for volcanic rocks of which the temperature of crystallization had been assessed by other techniques: their results were in good accordance with, for example, those measured directly with an optical pyrometer. These researchers gave an oxygen concordancy diagram for the plagioclase–pyroxene–magnetite assemblage from which it was clear that the Δ-values of mineral pairs in volcanic rocks are usually smaller than those recorded from equivalent plutonic rocks. By inference, the geothermometers showed that the latter equilibrated their oxygen isotopes at a lower temperature than volcanic rocks, their temperatures actually being subsolidus ones which do not correspond to the temperature of crystallization. The phenomenon is the result of the very different cooling rates of the two rock categories in question. Volcanic rocks cool faster and hence preserve the isotope composition which was established at the liquidus temperature. In plutonic rocks, on the other hand, cooling was slower and this enabled minerals to re-equilibrate oxygen as the temperature fell.

A general problem has been presented by Eiler, Valley and Baumgartner (1993). The interdiffusion between coexisting minerals affects all

Table 19.1 Temperature variation of isotope fractionation factors for some mineral–water systems

Mineral	*A*	*B*	*Temperature range* (°C)
Quartz–albite	0.50	0	400–800
Quartz–jadeite	1.09	0	400–800
Quartz–anorthite	1.59	0	400–800
Quartz–diopside	2.08	0	400–800
Quartz–wollastonite	2.20	0	400–800
Quartz–magnetite	6.11	0	400–800
Quartz–magnetite	7.8 ± 0.4	0	600
Quartz–magnetite	$6.1 + 0.2$	0	800
Quartz – calcite	0.50	0	500–700
Quartz–zoisite	1.56	0	500–700
Quartz–muscovite	2.20	-0.60	>500
Quartz–plagioclase	$(0.79 + 0.90\beta)$	$-(0.43 - 0.30\beta)$	400–500
Quartz–plagioclase	$(0.46 + 0.55\beta)$	$+(0.02 + 0.85\beta)$	500–800
Albite–jadeite	0.59	0	400–800
Albite–anorthite	1.09	0	400–800
Albite–diopside	1.58	0	400–800
Albite–wollastonite	1.70	0	400–800
Albite–magnetite	5.61	0	400–800
Albite–calcite	0.00	0	500–700
Albite–zoisite	1.06	0	400–800
Anorthite–diopside	0.49	0	400–800
Anorthite–wollastonite	0.61	0	400–800
Anorthite–magnetite	4.52	0	400–800
Plagioclase–amphibole	$(2.178 - 1.04\beta)$	-0.30	500
Plagioclase – biotite	$(2.27 - 1.04\beta)$	-0.60	500
Plagioclase – ilmenite	$(+4.32 - 1.04\beta)$	0	500
Plagioclase–magnetite	$(4.60 - 1.04\beta)$	0	500
Plagioclase – garnet	$(1.19 - 1.04\beta)$	-0.60	500
Plagioclase–muscovite	$(1.23 - 1.04\beta)$	-0.60	500
Plagioclase–olivine	$(2.90 - 1.04\beta)$	0	500
Plagioclase – pyroxene	$(1.70 - 1.04\beta)$	0	500
Diopside – wollastonite	0.12	0	400–800
Diopside–magnetite	4.03	0	400–800
Wollastonite–magnetite	3.91	0	400–800
Pyroxene–olivine	1.24	0	500
Pyroxene–garnet	0.21	0	500
Jadeite–anorthite	0.50	0	400–800
Jadeite–diopside	0.99	0	400–800
Jadeite–wollastonite	1.11	0	400–800
Jadeite–magnetite	5.02	0	400–800
Calcite–zoisite	1.06	0	500–700

Table 19.1 *contd.*

Mineral	*A*	*B*	*Temperature range* (°C)
$CaCO_3$	2.78	−2.89	0–500
$SrCO_3$	2.69	−3.24	0–500
$BaCO_3$	2.57	−4.23	0–500
Dolomite	3.20	−1.50	300–510
Anhydrite	3.88	−2.90	100–575
Dissolved SO_4^{2-}	3.25	−5.10	70–350
Dissolved SO_4^{2-}	2.88	−3.60	110–200
$BaSO_4$	3.0	−6.9	180–350
Quartz	3.38	−2.90	200–500
Quartz	2.51	−1.46	500–750
Quartz	4.10	−3.70	500–800
Quartz	3.55	−2.57	195–573
Quartz	3.23	−2.94	573 – 1000
Alkali feldspar	2.91	−3.41	350–800
Alkali feldspar	3.13	−3.70	500–800
Plagioclase	$(2.91 - 0.76\beta)$	$-(3.41 - 0.14\beta)$	350–800
Plagioclase	$(3.13 - 1.04\beta)$	−3.70	500–800
Muscovite	2.38	−3.89	400–650
Muscovite	1.90	−3.10	500–800
Rutile	4.10	−1.46	575–775
Magnetite	−1.47	−3.70	500–800

rocks and causes resetting and discordance of stable isotope geothermometers. It is commonly observed in slowly cooled igneous and metamorphic rocks and the fast grain boundary (FGB) model described the stable isotope fractionations and intracrystalline zonation which result from closed system interdiffusion (Eiler, Baumgartner and Valley, 1991, 1992). The model assumes that grain boundary diffusion is much faster than volume diffusion, and that it accounts for exchange among all minerals in a rock. The authors state that previous models of closure temperature violate mass balance restrictions and will be inaccurate in most rocks. Their modelling results are described for amphibolites and hornblende granites and gneisses; biotite granites, schists and gneisses; pelitic and semi-pelitic rocks; garnet peridotites; anorthosites, gabbros, pyroxenites and related rocks; and calc-silicate rocks. Given examples of mineral pairs and specific rock types that allow accurate stable isotope thermometry include plagioclase–pyroxene-bearing anorthosites and garnet–quartz in garnetiferous quartzites. Closed-system processes are obviously capable of producing a variety of patterns of stable isotope resetting, discordance, mineral zonation and fractionation reversals.

According to these results a reinterpretation of many studies of stable isotope thermometry is possible and necessary.

Marine carbonate rocks undergo alteration during metamorphism in the isotope composition of their oxygen and carbon, both parameters $\delta^{18}O$ (relative to SMOW) and $\delta^{13}C$ (relative to PDB) being lowered near an intrusive contact. On occasions, the process goes so far that metamorphosed calcite becomes indistinguishable from igneous calcite in carbonatites. Decrease in $\delta^{13}C$ values in carbonate minerals may be attributed to decarbonation reactions, the liberated CO_2 that emerges from them being correspondingly enriched in the heavy stable isotope. Such reactions are of the following type:

$$\text{calcite} + \text{quartz} \rightarrow \text{wollastonite} + CO_2,$$

However, contrary to this argument, fumarolic CO_2 has been recorded as possessing negative $\delta^{13}C$ values, even in locations where marine sedimentary rocks are present below the surface. Bottinga (1969) demonstrated that CO_2 gas in equilibrium with calcite at temperatures exceeding 200°C yielded an enrichment in $\delta^{13}C$ of only 2.8‰. It appears, therefore, that decarbonation reactions of the above type may not be the main source of fumarolic CO_2. Curiously, however, a ^{13}C enrichment of CO_2 evolved during decarbonation reactions was inferred by Shieh and Taylor (1969) as being greater than the value predicted by the theoretical calculations of Bottinga (1969). Such discrepancies are related to the degree to which the $\delta^{13}C$ values of carbonate rocks are reduced by increase in metamorphic grade.

Graphite is found in metamorphosed sedimentary and also in igneous rocks; it perhaps arose from progressive H_2 loss from organic compounds inducing recrystallization of residual atomic C. Graphites in such metamorphosed sedimentary rocks have been recorded as having $\delta^{13}C$ values compatible with a biogenic origin of the C, but there are also heavy graphites characterized by $\delta^{13}C$ values of from −2 to −8‰ in meteorites, ultramafic rocks and skarns, and it is unlikely that these could have been produced directly from biogenic C compounds. They may have arisen instead by the reduction of carbonates or CO_2 with molecular H_2, because it is known that the carbonate minerals can be reduced by H_2 in the temperature range 200−600°C, following the simple reactions

$$CaCO_3 + 4H_2 \rightarrow CH_4 + H_2O + Ca(OH)_2,$$

and the resultant CH_4 perhaps later decomposing at higher temperatures and forming graphite and H_2.

Both graphite and calcite are quite often found in marbles and may well be in isotopic equilibrium with each other. Calculation of the appropriate fractionation factors shows that graphite is probably depleted in $\delta^{13}C$ relative to calcite by amounts up to 18‰ at between 100 and 700°C. In some samples, however, analytically derived temperatures show a broad scatter, indicating that isotopic equilibrium between graphite and calcite was not attained.

If the isotope compositions of coexisting calcite and graphite are compared with the theoretically calculated fractionation curve of Bottinga mentioned above, it seems that equilibrium is sometimes not reached. In other cases, fractionations are larger than those predicted and this implies a partial isotope exchange. In yet other instances, particularly in high-temperature assemblages fractionations smaller than those predicted have been explained by assuming that the graphite was not derived from organic matter, but originated by the reduction of calcite, as was suggested by Wada (1977). On the other hand, Valley and O'Neil (1981) gave evidence that isotope equilibrium had been attained, but that the fractionations calculated by Bottinga are about 2‰ too large in the temperature range of interest.

The matter was further investigated by Kreulen and Van Beek (1983), who worked on samples from Naxos, the largest island of the Cyclades, Greece. This consists of a central migmatite dome which is surrounded by a series of schists and marbles. The temperatures of metamorphism inferred from mineral assemblages in various rock types decrease regularly from 700°C in the central part of the migmatite to 380°C in the south-eastern area of the island. Roughly half of the metamorphic rocks are marbles with dispersed graphitic materials in them; these become rare towards the dome, perhaps because of oxidation and combustion in the migmatite. Both graphite and calcite were separated from the same volume of rock material and analysed following the procedure outlined by Kreulen (1980). Results showed that the ^{13}C of the calcite ranges from − 3.9 to +5.8‰, most of the values grouping between 0 and +3‰. The graphites have $\delta^{13}C$ values ranging from − 19.8 to − 1.4‰; these correlate with the grade of metamorphism, low values occurring in the low-grade rocks and high values in high-grade rocks. With increasing temperature of metamorphism, fractionation between calcite and graphite decreases, as may be seen in Fig.19.1 in which the coordinates are fractionations of calcite–graphite pairs and temperatures of metamorphism.

In the lower-grade rocks, the fractionations are much more scattered than in the higher-grade ones. The graphite in the former (< 500°C) is present in a very fine dispersal, which implies a biogenic origin. In the higher-grade marbles, however, the mode of occurrence does not allow organic origins to be distinguished from inorganic ones. It may be inferred that the increasing rarity of the graphite towards the migmatic probably favours disappearance by oxidation rather than formation by reduction of calcite, but the isotope compositions involved do not provide a definitive solution to this question since only a single sample is

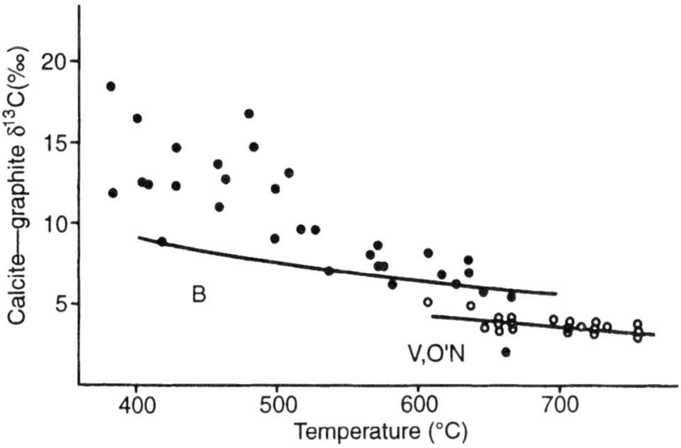

Fig. 19.1 Isotopic fractionation of calcite–graphite pairs versus temperatures of metamorphism. Fractionation curves are proposed by Bottinga, 1969 (B) and Valley and O'Neil, 1981 (V, O'N).

within the range of -20 to -25‰ characteristic of organic matter. The rest possess higher $\delta^{13}C$ values which reflect various degrees of isotope exchange with calcite. In Fig.19.1 the distribution of the calcite–graphite fractionations implies that equilibrium was approached from the upper side of the diagram, this being compatible with a biogenic origin from the graphite since fractionations between calcite and organic matter in non-metamorphosed sediments are usually large.

Clearly, because of isotope exchange with calcite, the $\delta^{13}C$ values of graphite have (with the single exception noted) become much higher than those of the precursor organic matter and the isotopic fractionations of calcite–graphite pairs are scattered widely for samples below 550°C but demonstrate a more restricted range at higher temperatures. The scattered fractionations suggest that isotopic exchange was often incomplete, which is perhaps illustrative of very low exchange rates at these temperatures. At higher temperatures, the fractionations show more uniform values to be expected in conditions of isotopic equilibrium. While isotope exchange is thought to occur at temperatures as low as 380°C, near-equilibrium values are believed not to be attained below 550°C; this view accords with that of Valley and O'Neil (1981). Kreulen and Van Beek (1983) proposed that a good criterion for such equilibrium is that, above 550°C, the fractionations appear to be almost independent of the $\delta^{13}C$ of the calcite (+5.8‰) and still normal. In Fig.19.1 the samples of Kreulen and Van Beek (1983) plot either above or on the fractionation curve of Bottinga. Over the whole temperature range, the smallest fractionations, i.e. the most complete isotope exchanges, coincide with Bottinga's curve: this could be a matter of chance or could accord with the possibility that, even at higher temperatures, equilibrium is frequently not completely reached. The conclusion must be reached that the data presented by Kreulen and Van Beek (1983) agree better with the theoretically derived fractionation curve of Bottinga than with the empirical fractionation curve proposed by Valley and O'Neil (1981). It was suggested that this inconsistency between the researchers could be, at least partly, a result of uncertainties in the temperature calibration. That of Kreulen and Van Beek (1983) was based on interpolation of isograd temperatures estimated by Jansen and Schuiling (1976). They studied mineral assemblages in many rock types such as pelitic schists, siliceous dolomites, amphibolites, ultramafic rocks and metabauxites. Some of the uncertainties stem from the calibration of individual mineral reactions, e.g. the Al_2SiO_5 triple point and X_{CO_2} of the metamorphic fluid (between 0.6 and 0.9) and this was taken into account only in part. However, the majority of the mineral assemblages provided fairly consistent temperature estimates, consequently yielding an accuracy to $\pm 50°C$ for the temperature calibration of the Naxos samples. Rye *et al.* (1976) applied the quartz–muscovite and quartz–biotite oxygen isotope thermometers which usually agree with mineral temperatures below 600°C, although at higher temperatures the isotope temperatures are generally below the mineral ones, possibly because of oxygen isotope retrogradation.

The conclusions derived from this discussion must be founded upon the various calibrations of the calcite–graphite isotope thermometer which, as noted above, differ by over 200°C in the temperature range of interest. Each is supported by data suggesting equilibrium. Now both cannot be correct, but the determination of which one is accurate does nothing to reduce doubts as to the applicability of the geothermometer. In addition, the relations between the slope of the fractionation curves and the scatter of the data points strongly imply that the thermometer is characterized by too low a sensitivity. A better comprehension of the system could ultimately give a valuable tool for testing equilibrium and tracing geochemical processes involved in the formation of graphite-bearing marbles.

References

Adams, S. (1994) No way back. *New Scientist, Inside Science*, **75**, 1–4.

Airey, P. L., Bentley, H., Calf, G. E. *et al.* (1983) Isotope hydrology of the Great Artesian Basin. Proc. Int. Conf. Groundwater and Man, Sydney, NSW. *Australian Resource Council, Conf. Ser.* **8**, 1–11.

Alaerts, L., Lewis, R. S. and Anders, E. (1979) Isotopic anomalies of noble gases in meteorites and their origins. III. LL-chondrites. *Geochim. Cosmochim. Acta*, **43**, 1399–1415.

Alburger, D. E., Harbottle, G. and Norton, E. F. (1986) Half-life of ^{32}Si. *Earth Planet. Sci. Lett.*, **78**, 168–76.

Alexander, C. M. O'D., Arden, J. W., Ash, R. D. and Pillinger, C. T. (1993) Presolar components in the ordinary chondrites. *Earth Planet. Sci. Lett.*, **99**, 220–9.

Alexander, E. C., Michelson, G. M. and Lanphere, M. A. (1978) A New ^{40}Ar/^{39}Ar dating standard. *US Geol. Surv. Open File Rep.*, **78–701**, 6–8.

Allègre, C. J. and Ben Othman, D. (1980) Nd–Sr isotopic relationship in granitoid rocks and continental crust development: a chemical approach to orogeneis. *Nature*, **286**, 335–42.

Allègre, C. J. and Condomines, M. (1976) Fine chronology of vocanic processes using ^{238}U–^{230}Th systematics. *Earth Planet. Sci. Lett.*, **28**, 395–406.

Allègre, C. J. and Luck, J. M. (1980) Osmium isotopes as petrogenic and geologic tracers. *Earth Planet. Sci. Lett.*, **48**, 148–54.

Allègre, C. J., Brick, J. L., Fourcade, S. and Semet, M. P. (1975) Rubidium –87/Strontium–87 age of Juvinas basaltic achondrite and early igneous activity in the solar system. *Science*, **187**, 436–8.

Allison, G. B. and Hughes M. W. (1976) Environmental tritium in the unsaturated zone: estimation of recharge to an unconfined aquifer. *Interpretation of Environmental Isotopes and Hydrochemical Data, Proc.* Vienna, IAEA, SM-182/4, pp. 57–72.

Alpher, R. A. and Herman, R. C. (1951) The primeval lead isotopic abundances and the age of the Earth's crust. *Phys. Rev.*, **84**, 1111.

Amari, S., Hoppe, P., Zinner, E. and Lewis, R. S. (1992) Interstellar SiC with unusual isotopic compositions. *Lunar Planet. Sci. XXIII* (abstract submitted to the 23rd Lunar and Planetary Science Conference), pp. 27–8.

Amari, S., Hoppe, P., Zinner, E. and Lewis, R. S. (1993) The isotopic compositions and stellar sources of meteoritic graphite grains. *Nature*, **365**, 806–9.

Ambrose, S. H. and DeNiro, M. J. (1986) Reconstruction of African human diet using bone collagen carbon and nitrogen isotope ratios. *Nature*, **319**, 321–4.

Ambrose, S. H. and DeNiro, M. J. (1989) Climate and habitate reconstruction using stable carbon and nitrogen isotope ratios of collagen in prehistoric herbivore teeth from Kenya. *Quaternary Res.*, **31**, 407–22.

Amin, B. S., Kharkar, D. P. and Lal, D. (1966) Cosmogenic ^{10}Be and ^{26}Al in marine sediments. *Deep Sea Res.*, **13**, 805–24.

Anders, E. (1988) Interstellar noble gases, C and N, in *Meteorites and the Early Solar System* (eds J. F. Kerridge and M. S. Matthews) Univ. Arizona Press, 927–55.
Anders, E. and Ebihara, M. (1982) Solar system abundances of the elements. *Geochim. Cosmochim. Acta*, **46**, 2363–80.
Anderson, A. T., Clayton, R. N. and Mayeda, T. K. (1971) Oxygen isotope thermometry of mafic igneous rocks. *J. Geol.*, **79**, 715–29.
Anderson, E. C. and Libby, W. F. (1951) Worldwide distribution of natural radio-carbon. *Phys. Rev.*, **81**, 64–9.
Anderson, E. C. Libby, W. F., Weinhouse, S. *et al.* (1947) Natural radiocarbon from cosmic radiation. *Phys. Rev.*, **72**, 931–6.
Anderson, T. F. and Arthur, M. A. (1983) Stable isotopes of oxygen and carbon and their application to sedimentologic and paleoenvironmental problems. *SEPM Short Course No.* 10, Dallas, 1983, 1/1–151.
Andrews, J. N., Fontes, J. -Ch., Michelot, J. -L. and Elmore, D. (1986) *In-situ* neutron–flux, ^{36}Cl production and groundwater evolution in crystalline rocks at Stripa, Sweden. *Earth Planet. Sci. Lett.*, **77**, 49–58.
Arnold, J. R. (1956) Beryllium-10 produced by cosmic rays. *Science*, **124**, 584–5.
Arnold, J. R. and Libby, W. F (1949) Age determinations by radiocarbon content: checks with samples of known age. *Science*, **110**, 678–80.
Aston, F. W. (1919) A positive ray spectrograph. *Phil. Mag.*, **38**, 707–14.
Attendorn, H. -G. (1993) Vergleichende Isotopenuntersuchungen des Kohlen-und Sauerstoffs an oberoligozänen Mollusken der Niederrheinischen Bucht. Dissertation, Münster, 157 pp.
Attendorn, H. -G. and Strauch, F. (1994) Die Kohlenstoff-Sauerstoff-Isotopenverhältnisse als Brackwasserindikatoren bei neogene Cardiiden der Westpeloponnes (Griechenland) *Münster. Forsch. Geol. Paläont.*, **76**, 353–64.
Attendorn, H. -G., Bowen, R. and Strauch, F. (1996) The influence of life habits on the carbon and oxygen isotope ratios of marine bivalvia. *Geology*, in press.
Avak, H. and Brand, W. A. (1995) The Finnigan MAT HDO-Equilibrator: A fully automated H_2O/gas phase equilibration system for hydrogen and oxygen isotope analysis. Finnigan MAT, *Application News* No. **11**, 1–13.
Baadsgaard, H., Nutman, A. P., Bridgwater, D. *et al.* (1984) The zircon geochronology of the Akilia association and Isua supracrustal belt, West Greenland. *Earth Planet. Sci. Lett.*, **68**, 221–8.
Baertschi, P. (1953) Über die relativen Unterschiede im $H_2{}^{18}O$-Gehalt natürlicher Wässer. *Helv. Chim. Acta*, **36**, 1352–8.
Baertschi, P. (1957) Messung und Deutung relativer Häufigkeits-variationen von O^{18} und C^{13} in Karbonatgesteinen und Mineralen. *Schweiz. Mineral. Petrogr. Mitt.*, **37**, 73 pp.
Baggott, J. (1994) New light on the cold chemistry of space. *New Scientist*, 1945, 18.
Bainbridge, K. T. and Nier, A. O. (1950) Relative Isotopic Abundances of the Elements. Prel. Rep. No. **9**, Nucl. Sci. Ser., Natl Res. Counc., USA, Washington, DC.
Baksi, A. K. (1987) Critical evaluation of the age of the Deccan Traps, India; implications for flood-basalt volcanoes and faunal extinction. *Geology*, **15**, 147–50.
Baksi, A. K., Hsu, V., McWilliams, M. O. and Farrar, E. (1992) $^{40}Ar/^{34}Ar$ dating of the Brunhes-Matuyama geomagnetic reversal. *Science*, **256**, 356–7.
Bal, K. D., Lal, N. and Nagpaul, K. K. (1983) Zircon and sphene as fission-track geochronometer and geothermometer: a reappraisal. *Contrb. Mineral. Petrol.*, **83**, 199–203.
Bar, M., Kolodny, Y. and Bentor, Y. K. (1974) Dating faults by fission track dating of epidotes –an attempt. *Earth Planet. Sci. Lett.*, **22**, 157–62.
Bard, E., Hamelin, B., Fairbanks, R. G. and Zindler, A. (1990) Calibration of the ^{14}C timescale over the past 30,000 years using mass spectrometric U-Th ages from Barbados corals. *Nature*, **345**, 405–10.
Barker, R., Millard, H. T. Jr, Hedge, C. E. and O'Neil, J. R. (1976) Pikes Peak Batholith: geochemistry of some minor elements and isotopes and implications for magma genesis, in *Studies in Colorado Field Geology* (eds R. C. Epis and R. J. Wimer), *Colorado School of Mines, Prof. Contrib.*, **8**, 44–56.
Barnola, J. M., Raynaud, D., Korotkevich, Y. S. and Lorius, C. (1987) Vostok ice core provides 160,000-year record of atmospheric CO_2. *Nature*, **329**, 408–14.

Barrera, E. (1994) Global environmental changes preceding the Cretaceous-Tertiary boundary: Early-Late Maastrichtian transition. *Geology*, **22**, 877–80.

Barrera, E., Tevesz, M. J. S., Carter, J. G. and McCall, P. L. (1994) Oxygen and carbon isotopic composition and shell microstructure of the bivalve *Laternula elliptica* from Antarctica. *Palaios*, **9(3)**, 275–87.

Barth, S., Oberli, F., Meier, M. *et al.* (1993) The evolution of a calc-alkaline basic to silicic magma system: geochemical and Rb-Sr, Sm-Nd, and $^{18}O/^{16}O$ isotopic evidence from the Late Hercynian Atesina-Cima d'Asta volcano-plutonic complex, northern Italy. *Geochim. Cosmochim. Acta*, **57**, 4285–300.

Batchelor, J. D. and Sears, D. W. G. (1991) Metamorphism of eucrite meteorites studied quantitatively using induced thermoluminescence. *Nature*, **349**, 516–8.

Baxter, M. S., Crawford, R. W., Swan, D. S. and Farmer, J. G. (1981) ^{210}Pb dating of a Loch Lomond sediment core by conventional and particle track methods and some geochemical observations. *Earth Planet. Sci. Lett.*, **53**, 434–44.

Becker, B. and Kromer, B. (1993) The continental tree-ring record-absolute chronology, ^{14}C calibration and climatic change at 11 ka. *Palaeogeogr. Palaeoclim Palaeoecol.*, **103**, 67–71.

Becker, B., Kromer, B. and Trimborn, P. (1991) A stable-isotope tree-ring timescale of the Late Glacial/Holocene boundary. *Nature*, **353**, 647–49.

Beckmann, G. and Klopries, B. (1991) CO_2 increase: questions beyond climatic change. *Energy*, **16**, 1317–1330.

Beckmann, G., Klopries, B., Hämmerle, H. *et al.* (1995) The chemical history of the atmosphere-self organizing processes and biological consequences. *Münster. Forsch. Geol. Paläont.*, **77**, 263–91.

Bell, K. and Powell, J. L. (1969) Strontium isotopic studies of alkalic rocks: the potassium-rich lavas of the Birunga and Toro-Ankole regions, east and central equatorial Africa. *J. Petrol.*, **10**, 536–72.

Bender, M. L., Sowers, T., Barnola, J. -M. and Chappellaz, J. (1994) Changes in the O_2/N_2 ratio of the atmosphere during recent decades reflected in the composition of air in the firn at Vostok Station, Antarctica. *Geophys. Res. Lett.*, **21**, 189–92.

Benner, R., Fogel, M. L., Sprague, E. K. and Hodson, R. E (1987) Depletion of ^{13}C in lignin and its implications for stable carbon isotope studies. *Nature*, **329**, 708–12.

Bentley, H. W., Phillips, F. M. and Davis, S. N. (1986) Chlorine-36 in the terrestrial environment, in *Handbook of Environmental Geochemistry*, Vol. **2** (eds P. Fritz and J. -Ch. Fontes), Elsevier, pp. 427–480.

Berger, G. W. (1984) Thermoluminescence dating of glacial silts from Ontario. *Can. J. Earth Sci.*, **21**, 1393–9.

Berger, W. H. (1981) Oxygen and carbon isotopes in foraminifera: an introduction. *Palaeogeogr., Palaeoclim., Palaeoecol.*, **33**, 3–7.

Bernat, M. and Allègre, C. J. (1974) Systematics in uranium-ionium dating of sediments. *Earth Planet. Sci. Lett.*, **21**, 310–4.

Bernatowicz, T. J. and Podosek, F. A. (1978) Nuclear components in the atmosphere, in *Terrestrial Rare Gases* (eds E. C. Alexander and M. Ozima), Scientific Society Press, Japan, pp. 99–135.

Berner, R. A. (1990) Atmospheric carbon dioxide levels over Phanerozoic time. *Science*, **247**, 1382–6.

Berner, R. A. (1992) Weathering, plants, and the long-term carbon cycle. *Geochim. Cosmochim. Acta*, **56**, 3225–31.

Bertram, C. J. and Elderfield, H. (1993) The geochemical balance of the rare earth elements and neodymium isotopes in the oceans. *Geochim. Cosmochim. Acta*, **57**, 1957–86.

Beveridge, N. A. S. and Shackleton, N. J. (1994) Carbon isotopes in recent planktonic foraminifera: a record of anthropogenic CO_2 invasion of the surface ocean. *Earth Planet. Sci. Lett.*, **126**, 259–73.

Bigeleisen, J. (1965) Chemistry of isotopes. *Science*, **147**, 463–71.

Bigeleisen, J. and Mayer, M. G. (1947) Calculation of equilibrium constants for isotopic exchange reactions. *J. Chem. Phys.*, **15**, 261–7.

Bingen, B. Demaiffe, D. and Hertogen, J. (1990) Evolution of feldspars ar the amphibolite-granulite-facies transition in augen gneiss (SW Norway): geochemistry and Sr isotopes. *Contrib. Mineral. Petrol.*, **105**, 275–88.

Birck, J. -L. and Allégre, C. J. (1988) Manganese-chromium isotope systematics and the development of the early solar system. *Nature,* **331**, 579–84.

Blank, J. G., Delaney, J. R. and Des Marais, D. J. (1993) The concentration and isotopic composition of carbon in basaltic glasses from the Juan de Fuca Ridge, Pacific Ocean. *Geochim. Cosmochim. Acta,* **57**, 875–87.

Blattner, P., Dietrich, V. and Gansser, A. (1983) Contrasting 180 enrichment and origins of High Himalayan and Transhimalayan intrusives. *Earth Planet. Sci. Lett.,* **65**, 276–86.

Blum, J. D., Chamberlain, C. P. and Hingston, M. P. (1993) Strontium and oxygen isotope study of M-1, M-3 and M-4 drill core samples from the Manson Impact Structure, Iowa: comparison with Haitian K-T impact glasses. *Lunar Planet. Sci. XXIV* (abstract submitted to the 24th Lunar and Planetary Science Conference), pp. 135–6.

Blum, J. D., Chamberlain, P. C., Hingston, M. P. *et al.* (1993) Isotopic comparision of K/T boundary impact glass with melt rock from the Chicxulub and Manson Impact Structures. *Nature,* **364**, 325–7.

Bocherens, H., Fizet, M. and Mariotti, A. (1994) Diet, physiology and ecology of fossil mammals as inferred from stable carbon and nitrogen isotope biogeochemistry: implications for Pleistocene bears. *Palaeogeogr., Palaeoclim, Palaeoecol.,* **107**, 213–25.

Bogard, D. D., Garrison, D. H., McKoy, T. J. and Keil, K. (1993a) $^{39}Ar/^{40}Ar$ ages of acapulcoites and lodranites: evidence for early parent body heating. *Lunar Planet. Sci. XXIV* (abstract submitted to the 24th Lunar and Planetary Science Conference), pp. 141–2.

Bogard, D. D., Garrison, D. H., Shih, C. -Y. and Nyquist, L. E. (1993b) $^{39}Ar/^{40}Ar$ dating of two granites: the age of Copernicus. *Geochim. Cosmochim. Acta,* 57.

Boltwood, B. B. (1907) On the ultimate disintegration products of the radioactive elements. *Am. J. Sci.,* **4(23)**, 77–88.

Bonani G., Balzer, R., Suter, M. *et al.* (1980) Radioisotope dating using an EN-Tandem accelerator. *Proc. Europhysics Conf. on Nuclear Physics Methods in Material Res.,* Darmstadt, pp. 357–9.

Bottinga, Y. (1969) Calculated fractionation factors for carbon and hydrogen isotope exchange in the system calcite-carbon dioxide-graphite-methane-hydrogen-water-vapour. *Geochim. Cosmochim. Acta,* **33**, 49–64.

Botz, R., Erlenkeuser, H., Koch, J. and Wehner, H. (1991) Analysis of sedimentary organic matter of a glacial/interglacial change (oxygen isotope stage 5/6) in the Norwegian-Greenland Sea. *Marine Geology,* **98**, 113–19.

Boudin, A. and Deutsch, S. (1970) Geochronology: recent development in the lutetium-176/hafnium-176 dating method. *Science,* **168**, 1219–20.

Boutton, T. W. (1991a) Stable carbon isotope ratios of natural materials: I. Sample preparation and mass spectrometric analysis, in *Carbon Isotope Techniques* (eds D. C. Coleman and B. Fry), Academic Press Inc., pp. 155–71.

Boutton, T. W. (1991b) Stable carbon isotope ratios of natural materials: II. Atmospheric, terrestrial, marine and freshwater environments, in *Carbon Isotope Techniques* (eds D. C. Coleman and B. Fry), Academic Press Inc., pp. 173–85.

Bowen, R. (1961a) Paleotemperature analyses of Mesozoic Belemnoidea from Germany and Poland. *J. Geol.,* **69**, 75–83.

Bowen, R. (1961b) Paleotemperature analyses of Mesozoic Belemnoidea from Australia and New Guinea. *Bull. Geol. Soc. Am.,* **72**, 769–74.

Bowen, R. (1961c) Paleotemperature analyses of Belemnoidea and Jurassic paleoclimatology. *J. Geol.,* **69**, 309–20.

Bowen, R. (1961d) Oxygen isotope Paleotemperature analyses of Mesozoic Belemnoidea from Europe, India and Japan. *J. Paleontol.,* **35**, 1077–84.

Bowen, R. (1962) Paleotemperature analyses of Jurassic Belemnoidea from East Greenland. *Experientia,* **18**, 438–41.

Bowen, R. (1963a) Measurement of paleotemperatures of the Upper Aptian of Mozambique, Africa and Middle Cretaceous paleoclimatology. *Am. J. Sci.,* **261**, 566–70.

Bowen, R. (1963b) O^{18}/O^{16} paleotemperature measurements on Mesozoic Belemnoidea from Neuquen and Santa Cruz provinces, Argentina. *J. Paleontol.,* **37**, 714–18.

Bowen, R. (1963c) Oxygen isotope paleotemperature measurements on Lower Jurassic Belemnoidea from Bamberg (Bavaria) Germany. *Experientia,* **19**, 401–3.

Bowen, R. (1964a) Faunal realms in Jurassic and Cretaceous belemnites. *Geol. Mag.,* **101**, 374–76.

Bowen, R. (1964b) Oxygen isotope paleotemperature measurements on Mesozoic Belemnoidea and their importance in paleoclimatic studies, in *Advances in Organic Geochemistry* (ed. U. Colombo,) Pergamon Press, Oxford, 1–13.
Bowen, R. (1988) *Isotopes in the Earth Sciences.* Elsevier Appl. Sci., 647 pp.
Bowen, R. (1991) *Isotopes and Climates.* Elsevier Appl. Sci., 483 pp.
Bowen, R. and Fritz, P. (1961) Oxygen isotope paleotemperature analyses of Lower and Middle Jurassic fossils from Pliensbach, Württemberg (Germany). *Experientia,* **19**, 461–70.
Boyd, A. W., Brown, F. and Lounsbury, M. (1955) Mass spectrometric study of natural and neutron-irradiated chlorine. *Can. J. Phys.*, **33**, 35–43.
Boyd, S. R. and Pillinger, C. T. (1991) Rubidium sulphate-ammonium sulphate solid solution: a standard for use during the determination of nitrogen abundance and isotopic composition at the ppm level by static vacuum mass spectrometry. *Anal. Chem.*, **63**, 1332–5.
Boyd, S. R. Hall, A. and Pillinger, C. T. (1993) The measurement of $\delta^{15}N$ in crustal rocks by static vacuum mass spectrometry: application to the origin of the ammonium in the Cornubian batholith, southwest England. *Geochim. Cosmochim. Acta*, **57**, 1339–47.
Boyle, E. A. (1988) The role of vertical chemical fractionation in controlling late Quaternary atmospheric carbon dioxide. *J. Geophys. Res.*, **93**, 15701–14.
Bradly, P.V. (1991) The effect of silicate weathering on global temperature and atmospheric CO_2. *J. Geophys. Res.*, **96**, 18101–6.
Brand, U. (1986) Paleoenvironmental analysis of Middle Jurrassic (Callovian) ammonoids from Poland: trace elements and stable isotopes. *J. Palentol.*, **60**, 293–301.
Brand, U. and Veizer, J. (1980) Chemical diagenesis of a multicomponent carbonate system –1: trace elements. *J. Sed. Petrol.*, **50**, 1219–36.
Brand, U. and Veizer, J. (1981) Chemical diagenesis of a multicomponent carbonate system–2: stable isotopes. *J. Sed. Petrol.*, **51** 987–97.
Brand, W. A., Tegtmeyer, A. R. and Hilkert, A. (1994) Compound-specific analysis: extending toward $^{15}N/^{14}N$ and $^{18}O/^{16}$. *Org. Geochem.*, **21**, 585–94.
Brannon, J. C., Podosek, F. A. and McLimans, R. K. (1992) Alleghenian age of the Upper Mississippi Valley zinc-lead deposit determined by Rb-Sr dating of sphalerites. *Nature*, **356**, 509–12.
Brass, G. W. (1975) The effect of weathering on the distribution of strontium isotopes in weathering profiles. *Geochim. Cosmochim. Acta*, **39**, 1647–53.
Breitzman, L. L., Birnie, R. W. and Johnson, G. D. (1983) Fission-track ages of the Chagai Intrusives, Baluchistan, Pakistan. *Bull. Geol. Soc. Am.*, **94**, 253–8.
Bremner, J. M. and Keeney, D. R. (1966) Determination and isotope ratio analysis of different forms of nitrogen in soils. III. *Soil Sci. Am. Proc.*, **30**, 577–82.
Broecker, W. S. (1963) A preliminary evaluation of uranium series inequilibrium as a tool for absolute age measurement on marine carbonates. *J. Geophys. Res.*, **68**, 2817–34.
Broecker, W. S. (1964) Radiocarbon dating: a case against the proposed link between river mollusks and soil humus. *Science*, **143**, 596–7.
Broecker, W. S. (1974) *Chemical Oceanography*, Harcourt Brace Jovanovich, New York.
Broecker, W. S., Thurber, D. L., Goddard, *J. et al.* (1968) Milankovitch hypothesis supported by precise dating of coral reefs and deep-sea sediment. *Science*, **159**, 297–300.
Brooks, C., James, D. E. and Hart, S. R. (1976) Ancient lithosphere: its role in young continental volcanism. *Science*, **193**, 1086–94.
Brown, G. E. (1988) High-density equation of state. *Nature*, **336**, 519–20.
Brown, L., Sacks, S., Tera, F. *et al.* (1981) Beryllium-10 in continental sediments. *Earth Planet. Sci. Lett.*, **55**, 370–76.
Brown, L., Klein, J., Middleton, R. *et al.* (1982) ^{10}Be in island-arc volcanoes and implications for subduction. *Nature*, **299**, 718–20.
Brown, M. (1993) *P-T-t* evolution of orogenic belts and the causes of regional metamorphism. *J. Geol. Soc. London*, **150**, 227–41.
Bryant, J. D., Luz, B. and Froelich, P. N. (1994) Oxygen isotopic composition of fossil horse tooth phosphate as a record of continental paleoclimate. *Palaeogeogr., Palaeoclim. Palaeoecol*, **107**, 303–16.
Burchart, J., Dakowski, M. and Galazka, J. (1975) A technique to determine extremely high fission track densities. *Bull. Acad. Polon. Sci., Ser. Sci. de la Terre*, **23**, 1–8.

Burke, W. H., Denison, R. E., Hetherington, E. A. *et al.* (1982) Variation of seawater $^{87}Sr/^{86}Sr$ throughout Phanerozoic time. *Geology*, **10**, 516–9.
Busson, P. (1991) Sampling error? *Nature*, **352**, 187.
Caldeira, K. and Kasting, J. F. (1992) The life span of the biosphere revisited. *Nature*, **360**, 721–3.
Cameron, A. C. W. (1973) Abundance of elements in the solar system. *Space Sci. Rev.*, **15**, 121–46.
Cameron, E. M. (1982) Sulphate and sulphate reduction in early Precambrian oceans. *Nature*, **296**, 145–8.
Cameron, E. M. (1983) Genesis of Proterozoic iron formation: sulphur isotope evidence. *Geochim. Cosmochim. Acta*, **47**, 1069–74.
Cameron, M., Collerson, K. D., Compston, W. and Morton, R. (1981) The statistical analysis and interpretation of imperfectly-fitted Rb-Sr isochrons from polymetamorphic terrains. *Geochim. Cosmochim. Acta*, **45**, 1087–97.
Card, K. D. (1984) Proterozoic Geology of the North Shore of Lake Huron, Ontario. *Field Trip 15, Geol. Assoc. Canada and Mineral. Assoc. Canada*, 49 pp.
Carlsson, L., Olsson, T., Andrews, J. N. *et al.* (1983) Geochemical and Isotope Characterization of the Stripa Groundwaters. SKBF/KBS, Prog. Rep., 83–01, Stockholm.
Carpena, J. and Caby, R. (1984) Fission-track evidence for Late Triassic oceanic crust in the French Occidental Alps. *Geology*, **12**, 108–11.
Carswell, R. F. (1989) Galactic chemical evolution. *Nature*, **341**, 692–3.
Chamberlain, C. P. and Blum, J. D. (1993) Oxygen isotopes as tracers of tektite source rocks: an example from the ivory coast tektites and Lake Bosumtwi crater. *Lunar Planet. Sci. XXIV* (abstract submitted to the 24th Lunar and Planetary Science Conference), pp. 267–8.
Chamberlain, C. P. and Conrad., M. E. (1991) Oxygen isotope zoning in garnet. *Science*, **246**, 403–5.
Chamberlain, C. P. and Conrad, M. E. (1993) Oxygen-isotope zoning in garnet: a record of volatile transport. *Geochim. Cosmochim. Acta*, **57**, 2613–29.
Chanton, J. P., Martens, C. S. and Goldhaber, M. B. (1987) Biogeochemical cycling in an organic-rich coastal marine basin. 8. A sulfur isotopic budget balanced by differential diffusion across the sediment-water interface. *Geochim. Cosmochim. Acta*, **51**, 1201–8.
Chappell, B. W. and White, A. J. R. (1974) Two contrasting granite types. *Pacific Geology*, **8**, 173–4.
Chappell, J. (1974) Geology of coral terraces. Huon Peninsula, New Guinea: a study of Quaternary tectonic movements and sea level changes. *Bull. Geol. Soc. Am.*, **85**, 553–70.
Cheney, E. S. and Lance, I. M. (1967) Evidence for sulfurization and the origin of some Sudbury-type ores. *Mineralium Deposita*, **2**, 80–94.
Chicarelli; M. I., Hayes, J. M., Popp, B. N. *et al.* (1993) Carbon and nitrogen isotopic compositions of alkyl porphyrins from the Triassic Serpiano oil shale. *Geochim. Cosmochim. Acta*, **57**, 1307–11.
Chillón, B. S., Alberdi, M. T., Leone, G. *et al.* (1994) Oxygen isotopic composition of fossil equid tooth and bone phosphate: an archive of difficult interpretation. *Palaeogeogr., Palaeoclim., Palaeoecol.*, **107**, 317–28.
Church, S. E. and Tatsumoto, M. (1975) Lead isotope relations in oceanic ridge basalts from the Juan de Fuca-Gorda ridge area, N. E. Pacific Ocean. *Contrib. Mineral. Petrol.*, **53**, 253–79.
Clauer, N. (1982) The Rb-Sr method applied to sediments: Certitudes and uncertainties, in *Numerical Dating in Stratigraphy*, 2 vols, John Wiley & Sons, New York, pp. 245–76.
Clauer, N., Caby, R., Jeanette, D. and Trompette, R. (1982) Geochronology of sedimentary and metasedimentary Precambrian rocks of the West African craton. *Precamb. Res.*, **18**, 53–71.
Claypool, G. E., Holser, W. T., Kaplan, I. R. *et al.* (1980) The age curves of sulfur and oxygen isotopes in marine sulfate and their mutual interpretation. *Chem. Geol.*, **28**, 199–260.
Clayton, D. D. (1994) Production of ^{26}Al and other extinct radionuclides by low-energy heavy cosmic rays in molecular clouds. *Nature*, **368**, 222–3.

Clayton, R. N. (1955) Variations in oxygen isotope abundances in rock minerals. PhD thesis, California Institute of Technology.
Clayton, R. N. (1961) Oxygen isotope fractionation between calcium-carbonate and water. *J. Chem. Phys.*, **34**, 724–6.
Clayton, R. N. and Degens, E. T. (1959) Use of carbon isotope analyses of carbonates for differentiating freshwater and marine sediments. *Bull. Am. Assoc. Petrol. Geol.*, **43**, 890–7.
Clayton, R. N., Mayeda, T. K. (1963) The use of bromine pentafluoride in the extraction of oxygen from oxides and silicates for isotopic analysis. *Geochim. Cosmochim. Acta*, **27**, 43–52.
Clayton, R. N. and Mayeda, T. K. (1978) Genetic relations between iron and stony meteorites. *Earth Planet. Sci. Lett.*, **40**, 168–74.
Clayton, R. N., and Mayeda, T. K. and Nagahara, H. (1992) Oxygen isotope relationships among primitive achondrites. *Lunar Planet. Sci. XXIII* (abstract submitted to the 23rd Lunar and Planetary Science Conference), pp. 267–8.
Clayton, R. N., Onuma, N. and Mayeda, T. K. (1976) A classification of meteorites based on oxygen isotopes. *Earth Planet. Sci. Lett.*, **30**, 10–8.
Clayton, R. N., Onuma, N., Grossman, C. and Mayeda, T. K. (1977) Distribution of the presolar component in Allende and other carbonaceous chondrites. *Earth Planet. Sci. Lett.*, **34**, 209–24.
Clayton, R. N. McPherson, G. J., Hutcheon, I. D. *et al.* (1984) Two forsterite-bearing FUN inclusions in the Allende meteorite. *Geochim. Cosmochim. Acta*, **48**, 335–48.
Clemens, S. C., Farrell, J. W. and Gromet, L. P. (1993) Synchronous changes in seawater strontium isotope composition and global climate. *Nature*, **363**, 607–9.
Clemens, S. C., Gromet, L. P. and Farrell, J. W. (1995) Artefacts in Sr isotope records. *Nature*, **373**, 201.
Cole, J. E., Fairbanks, R. G. and Shen, G. T. (1993) Recent variability in the southern oscillation: isotopic results from a Tarawa Atoll coral. *Science*, **260**, 1790–3.
Cole, M. M. (1986) *The Savannas: Biogeography and Geobotany*. Academic Press, London, 438 pp.
Coleman, M. L. (1971) Potassium-calcium dates from pegmatitic micas. *Earth Planet. Sci. Lett.*, **12**, 399–405.
Compston, W., Williams, I. S., Campbell, I. H. and Gresham, J. J. (1985) Zircon xenocrysts from the Kambalda volcanics: age constraints and direct evidence for older continental crust below the Kambalda-Norseman greenstones. *Earth Planet. Sci. Lett.*, **76**, 299–311.
Corless, J. T. (1968) Observations on the isotopic geochemistry of calcium. *Earth Planet. Sci. Lett.*, **4**, 475–8.
Cormie, A. B. and Schwarcz, H. P. (1994) Stable isotopes of nitrogen and carbon of North American white-tailed deer and implications for paleodietary and other food web studies. *Palaeogeogr., Palaeoclim., Palaeoecol.*, **107**, 227–41.
Cornu, S. Pätzold, J., Bard, E. *et al.* (1993) Paleotemperature of the last interglacial period based on δ^{18}O of *Strombus bubonius* from the western Mediterranean Sea. *Palaeogeogr., Palaeoclim., Palaeoecol.*, **103**, 1–20.
Craig, H. (1953) The geochemistry of the stable carbon isotopes. *Geochim. Cosmochim. Acta*, **3**, 53–92.
Craig, H. (1954) Carbon-13 in plants and the relationships between carbon-13 and carbon-14 variations in nature. *J. Geol.*, **62**, 115–49.
Craig, H. (1961) Mass spectrometer analysis of radiocarbon standards. *Radiocarbon*, **3**, 1–3.
Craig, H. (1965) The measurement of oxygen isotope paleotemperatures, in *Stable Isotopes in Oceanographic Studies and Paleotemperatures*, Spoleto, 26–27 July 1965, Cons. Naz. delle Richereche, Lab. di Geologia Nucleare, Pisa, pp. 161–182.
Craig, H. and Gordon, L. I. (1965) Deuterium and ^{18}O variations in the ocean and the marine atmosphere. *Stable Isotopes in Oceanographic Studies and Paleotemperatures*, 2nd Spoleto Conf. Proc., Pisa, pp. 1–122.
Craig, H., Chou, C., Welhan, *J. er al.* (1988) The isotopic composition of methane in polar ice cores. *Science*, **242**, 1535–9.

Creaser, R. A., Papanastassiou, D. A. and Wasserburg, G. J. (1991) Negative thermal ion mass spectrometry of osmium, rhenium and iridium. *Geochim. Cosmochim. Acta*, **55**, 397–401.

Crough, E. A. C. (1977) Fission product yields from neutron-induced fission, in *Atomic Data and Nuclear Data Tables*, Harwell Acad. Press.

Crough, S. T. (1983) Apatite fission-track dating of erosion in the eastern Andes, Bolivia. *Earth Planet. Sci. Lett.*, **64**, 396–7.

Crozaz, G. and Langway, C. C. Jr (1966) Dating Greenland firn-ice cores with Pb^{210}. *Earth Planet. Sci. Lett.*, **1**, 194–6.

Curry, W. B., Duplessy, J. C., Labeyrie, L. D. and Shackleton, N. J. (1988) Changes in the distribution of $\delta^{13}C$ of deep water $T\,CO_2$ between the last glaciation and the Holocene. *Paleoceanography*, **3**, 317–41.

Dalrymple, G. B. (1989) The GLM continuous laser system for $^{40}Ar/^{39}Ar$ dating: description and performance characteristics. *US Geol. Surv. Bull.*, **1890**, 89–96.

Dalrymple, G. B. and Duffield, W. A. (1988) High precision $^{40}Ar/^{39}Ar$ dating of Oligocene rhyolites from the Mogollon-Datil volcanic field using a continuous laser system. *Geophys. Res. L ett.*, **15**, 463–6.

Damon, P. E., Donahue, D. J., Gore, B. H. *et al.* (1989) Radiocarbon dating of the Shroud of Turin. *Nature*, **337**, 611–3.

Daniels, F. and Saunders, D. F. (1950) The thermoluminescence of rocks. *Science*, **111**, 462.

Dansgaard, W. (1964) Stable isotopes in precipitation. *Tellus*, **16**, 436–68.

Dao-Yi, X. and Zheng, Y. (1993) Carbon isotope and iridium event markers near the Permian/Triassic boundary in the Meishan section, Zhejiang Province, China. *Palaeogeogr., Palaeoclim., Palaeoecol.*, **104**, 171–6.

Dawson, T. E. and Ehleringer, J. R. (1993) Isotopic enrichment of water in the 'woody' tissues of plants. Implications for plant water source, water uptake, and other studies which use the stable isotopic composition of cellulose. *Geochim. Cosmochim. Acta*, **57**, 3487–92.

DeBaar, H. J. W. and Suess, E. (1992) Ocean carbon cycle and climate change-an introduction to the Interdisciplinary Union Symposium. *Global Planet. Change*, **8**, VII–XI.

Debuyst, R., Dejehet, F., Grün, R. *et al.* (1984) Possibility of ESR-dating without determination of the annual dose-rate. *J. Radioanal. Chem. Lett.*, **86**, 399–410.

Degens, E. T. and Ross, D. A. (1969) *Hot Brines and Recent Heavy Metal Deposits in the Red Sea.* Springer Verlag, New York, 600 pp.

Degens, E. T., Behrendt, M., Gotthardt, B. and Reppmann, E. (1968) Metabolic fractionation of carbon isotope in marine plankton: II. Data on samples collected off the coast of Peru and Ecuador. *Deep Sea Res.*, **15**, 11–20.

Deino, A. D. and Potts, R. (1990) Single crystal $^{40}Ar/^{39}Ar$ dating of the Olorgesaile Formation, southern Kenya rift. *J. Geophys. Res.*, **95**, 8453–70.

Dempster, A. J. (1918) A new method of positive ray analysis. *Phys. Rev.*, **11**, 316–25.

DeNiro, M. J. and Epstein, S. (1978) Influence of diet on the distribution of carbon isotopes in animals. *Geochim Cosmochim. Acta*, **42**, 495–506.

Denoyer, E. R., Kenneth, J. F. and Hager, J. W. (1991) Laser solid sampling for inductively coupled plasma mass spectrometry signals and their implications for optimizing trace analysis. *Anal. Chem.*, 63.

DePaolo, D. J. and Wasserburg, G. J. (1976) Nd isotopic variations and petrogenetic models. *Geophys. Res. Lett.*, **3**, 249–52.

DePaolo, D. J. and Wasserburg, G. J. (1977) The source of island arcs as indicated by Nd and Sr isotopic studies. *Geophys. Res. Lett.*, **4**, 465–68.

Deuser, W. G. (1970) Extreme $^{13}C/^{12}C$ variations in Quaternary dolomites from the continental shelf. *Earth Planet. Sci. Lett.*, **8**, 118–24.

DeVries, H. (1958) Variation in concentration of radiocarbon with time and location on Earth. *Proc. Koninkl. Ned. Akad. Wetenschap.*, **B61**, 94–102.

D'Hondt, S. and Zachos, J. C. (1993) On stable isotopic variation and earliest Paleocene planktonic foraminifera. *Paleoceanogr.*, **8**, 527–47.

D'Hondt, S., Zachos, J. C. and Schultz, G. (1994) Stable isotopic signals and photosymbiosis in Late Paleocene planktic foraminifera. *Paleobiology*, **20**(3), 391–406.

Dickin, A. F. (1987) La-Ce dating of Lewisian granulites to constrain the ^{138}La β-decay half-life. *Nature*, **325**, 337–8.
Dinçer, T. and Davies, G. H. (1967) Some considerations on tritium dating and the estimates of tritium input function. *Proc. 8th Cong. Instl Assoc. Hydrogeol.*, pp. 276–85.
Dodson, M. H. (1973) Closure temperature in cooling geochronological and petrological systems. *Contrib. Mineral. Petrol.*, **40**, 259–74.
Dodson, M. H. (1976) Kinetic processes and thermal history of slowly cooling solids. *Nature*, **259**, 551–3.
Dodson, M. H. (1979) Theory of cooling ages, in *Lectures in Isotope Geology* (eds E. Jäger and J. C. Hunziger), Springer-Verlag, Berlin, pp. 194–202.
Dodson, M. H. (1986) Closure profiles in cooling systems. *Materials Science Forum*, **7**, 145–54.
Dodson, M. H. (1989) Diamond dating anomalies. *Nature*, **337**, 207.
Doe, B. R. and Hart, S. R. (1963) The effect of contact metamorphism on lead in potassium feldspars near Eldora Stock, Colorado. *J. Geophys. Res.*, **68**, 3521–30.
Doe, B. R. and Stacey, J. S. (1974) The application of lead isotopes to the problem of ore genesis and ore prospect evaluation. A review. *Econ. Geol.*, **69**, 757–76.
Doell, R. R. and Dalrymple, G. B. (1966) Geomagnetic polarity epochs: A new polarity event and the age of the Brunhes-Matuyama boundary. *Science*, **152**, 1060–1.
Doell, R. R., Dalrymple, G. B., Smith, R. L. and Bailey, R. A. (1968) Paleomagnetism, potassium-argon ages and geology of rhyolites and associated rocks of the Valles caldera, New Mexico. *Mem. Geol. Soc. Am.*, **116**, 211–48.
Douthitt, C. B. (1982) The geochemistry of the stable isotopes of silicon. *Geochim. Cosmochim. Acta*, **46**, 1449–58.
Druffel, E. R. M. and Griffin, S. (1993) Large variations of surface ocean radiocarbon: evidence of circulation changes in the southwestern Pacific. *J. Geophys. Res.*, **98**, 20249–59.
Druffel, E. R. M. and Williams, P. M. (1990) Identification of a deep marine source of particulate organic carbon using bomb ^{14}C. *Nature*, **347**, 172–5.
Eberhardt, P., Geiss, G. and Houtermans, F. G. (1955) Isotopenverhältnisse von 'gewöhnlichem' Blei und ihre Deutung. *Z. Physik*, **141**, 91.
Edmunds, M. G. (1993) An elementary history lesson. *Nature*, **365**, 293–4.
Ehleringer, J. R. (1991) ^{13}C/^{12}C fractionation and its utility in terrestrial plant studies, in *Carbon Isotope Techniques* (eds D. C. Coleman and B. Fry), Academic Press, New York, pp. 187–200.
Eikenberg, J. Signer, P. and Wieler, R. (1993) U-Xe, U-Kr and U-Pb systematics for dating uranium minerals and investigations of the production of nucleogenic neon and argon. *Geochim. Cosmochim. Acta*, **57**, 1053–69.
Eiler, J. M., Baumgartner, L. P. and Valley, J. W. (1991) Numerical modeling of diffusive exchange of stable isotopes in the lithologic setting. *Geol. Soc. Am. Abstr. Prog.*, **23**, A447.
Eiler, J. M., Baumgartner, L. P. and Valley, J. W. (1992) Inter-crystalline stable isotope diffusion: a fast grain boundary model. *Contrib. Mineral. Petrol.*, **112**, 543–57.
Eiler, J. M., Valley, J. W., and Baumgartner, L. P. (1993) A new look at stable isotope thermometry. *Geochim. Cosmochim. Acta*, **57**, 2571–83.
Elmore, D., Tubbs, L. E., Newman, D. *et al.* (1982) ^{36}Cl bomb pulse measured in a shallow ice core from Dye 3, Greenland. *Nature*, **300**, 735–7.
Emiliani, C. (1953) Oxygen isotopes and paleotemperature determination. *Congr. Int. Quaternaire, 4e, Rome/Pisa, Actes*, p. 831.
Emiliani, C. (1954) Temperatures of Pacific bottom waters and polar superficial waters during the Tertiary. *Science*, **119**, 853–6.
Emiliani, C. (1955a) Pleistocene temperatures. *J. Geol.*, **63**, 538–78.
Emiliani, C. (1955b) Pleistocene temperature variations in the Mediterranean. *Quaternaria*, **2**, 87–98.
Emiliani, C. (1956) Oligocene and Miocene temperatures of the equatorial and subtropical Atlantic ocean. *J. Geol.*, **64**, 281–88.
Emiliani, C. (1957) Temperature and age analysis of deep sea cores. *Science*, **125**, 383–7.

Emliani, C. (1964) Paleotemperature analyses of the Caribbean cores A 254-BR-C and CP-28. *Bull. Geol. Soc. Am.*, **74**, 129–44.

Emirich, K., Ehalt, D. H. and Vogel, J. C. (1970) Carbon isotope fractionation during the precipitation of calcium carbonate. *Earth Planet. Sci. Lett.*, **8**, 363–71.

Encarnación, J. P., Musaka, S. B. and Obille, E. C. Jr (1993) Zircon U-Pb geochronology of the Zambales and Angat ophiolites, Luzon, Philippines: evidence for an Eocene back-arc pair. *J. Geophys. Res.*, **98**, 19991–20004.

Engelhardt, W. V., Luft, E., Arndt, *J. et al.* (1987) Origin of moldavites. *Geochim. Cosmochim. Acta*, **51**, 1425–43.

Epstein, S. and Lowenstam, H. A. (1953) Temperature-shell-growth relations of recent and interglacial Pleistocene shoal-water biota from Bermuda. *J. Geol.*, **61**, 424–38.

Epstein, S. and Mayeda, T. (1953) Variation of O^{18} content of waters from natural sources. *Geochim. Cosmochim. Acta*, **4:** 213–24.

Epstein, S., Thompson, P. and Yapp, C. J. (1977) Oxygen and hydrogen isotopic ratios in plant cellulose. *Science*, **198**, 1209–15.

Epstein, S., Yapp, C. J. and Hall, J. H. (1976) The determination of the D/H ratio of nonexchangeable hydrogen in cellulose extracted from aquatic and land plants. *Earth Planet. Sci. Lett.*, **30**, 241–51.

Epstein, S., Buchsbaum, H. A., Lowenstam, H. A. and Urey, H. C. (1953) Revised carbonate-water isotopic temperature scale. *Bull. Geol. Soc. Am.*, **64**, 1315–26.

Erez, J. (1978) Vital effect on stable-isotope composition seen in foraminifera and coral skeletons. *Nature*, **273**, 199–202.

Erlank, A. J., Smith, H. S., Marchant, J. W. *et al.* (1978) Hafnium, in *Handbook of Geochemistry* (ed. K. H. Wedepohl), Springer-Verlag, Berlin, section 72.

Eugster, O., Eberhardt, P. and Geiss, J. (1967) ^{81}Kr in meteorites and ^{81}Kr radiation ages. *Earth Planet. Sci. Lett.*, **2**, 77–82.

Evans J. C., Rancitelli, L. A. and Reeves, J. H. (1979) ^{26}Al content of Antarctic meteorites: implications for terrestrial ages and bombardment history. *Proc. 10th Lunar Sci. Conf.*, pp. 1061–72.

Evernden, J. F. and Richards, J. R. (1962) Potassium-argon ages in eastern Australia. *J. Geol. Soc. Australia*, **9**, 1–49.

Fajer, E. D., Bowers, M. D. and Fakhri, A. B. (1989) The effects of enriched carbon dioxide atmospheres on plant-insect herbivore interactions. *Science*, **243**, 1198–1200.

Fallick, A. E. and Hamilton, P. J. (1989) The isotope geochemistry of ocean waters through time. *Trans. Roy. Soc. Edinburg: Earth Sci.*, **80**, 177–81.

Farmer, J. G. and Baxter, M. S. (1974a) Atmospheric carbon dioxide levels as indicated by the stable isotope record in wood. *Nature*, **247**, 273–5.

Farmer, J. G. and Baxter, M. S. (1974b) Drs Farmer and Baxter reply. *Nature*, **252**, 757.

Faure, G. (1977) *Principles of Isotope Geology*, John Wiley & Sons, 464 pp.

Faure, G. (1982) The marine strontium-geochronometer, in *Numerical Dating in Stratigraphy*, Vol. 1 (ed. G. S. Odin), John Wiley & Sons, pp. 73–9.

Faure, G. (1986) *Principles of Isotope Geology*,. 2nd edn, John Wiley & Sons, 589 pp.

Faure, G. and Powell, J. L. (1972) *Strontium Isotope Geology*. Springer-Verlag, New York, 188 pp.

Feng, R., Machado, N. and Ludden, J. (1993) Lead geochronology of zircon by laser probe inductively-coupled plasma mass spectrometry (LP-ICPMS). *Geochim. Cosmochim. Acta*, **57**, 479–86.

Feng, X., Krishnamurthy, R. V., and Epstein, S. (1993) Determination of D/H ratios of nonexchangeable hydrogen in cellulose: A method based on the cellulose-water exchange reaction. *Geochim. Cosmochim. Acta*, **57**, 4249–56.

Fisher, D. E. (1977) Fission/alpha analysis of the U, Th families. *Nature*, **265**, 227.

Fisher, R. and Stuebar, A. M. (1976) Strontium isotopes in selected streams within the Susquehanna River basin. *Water Resources Res.*, **12**, 1061–68.

Fleischer, R. L. (1980) Isotope disequilibrium of uranium: alpha recoil damage and preferential solution effects. *Science*, **207**, 979–81.

Fleischer, R. L. and Price, B. P. (1964a) Decay constant for spontaneous fission. *Phys. Rev.*, **133**, B63–4.

Fleischer, R. L. and Price, P. B. (1964b) Glass dating by fission fragment tracks. *J. Geophys. Res.*, **69**, 331–9.

Fleischer, R. L., Price, P. B. and Walker, R. M. (1965) Spontaneous fission-tracks from extinct ^{244}Pu in meteorites and the early history of the solar system. *J. Geophys. Res.*, **70**, 2703–7.

Fleischer, R. L., Naeser, C. N., Price, P. B. *et al.* (1965). Fossil particle tracks and uranium distributions in minerals of the Vaca Muerta meteorites. *Science*, **148**, 629–32.

Freedman, W. L., Madore, B. F., Mould, J. R. *et al.* (1994) Distance to the Virgo cluster galaxy M100 from Hubble Space Telescope observations of Cepheids. *Nature*, **371**, 757–62.

Freundel, M., Schultz, L. and Reedy, C. (1986) Terrestrial ^{81}Kr-Kr ages in Antarctic meteorites. *Geochim. Cosmochim. Acta*, **50**, 1–11.

Freyer, H. D. and Wiesberg, L. (1973) ^{13}C-decrease in modern wood due to the large scale combustion of fossil fuels. *Naturwissenschaften*, **60**, 517–8.

Friedli, H. Lotscher, H., Oeschger, H. *et al.* (1986) Ice core record of the $^{13}C/^{12}C$ ratio of atmospheric CO_2 in the past two centuries. *Nature*, **324**, 237–38.

Friedmann, I. (1953) Deuterium content of natural waters and other substances. *Geochim. Cosmochim. Acta*, **4**, 89–103.

Fripp, R. E. P., Donelly, T. H. and Lambert, I. B. (1979) Sulphur isotope results for Archaean banded iron formation, Rhodesia. *Geol. Soc. S. Africa, Spec. Publ.*, **6**, 205–8.

Fritz, P. and Fontes, J. Ch. (1986) *Handbook of Environmental Isotope Geochemistry*, Vol. 2, Elsevier, 557 pp.

Fritz, P., Anderson, T. W. and Lewis, C. F. M. (1975) Late-Quaternary climate trends and history of Lake Erie from stable isotope studies. *Science*, **190**, 267–9.

Fryer, B. J., Jackson, S. E. and Longerich, H. P. (1993) *In-situ* U(Th)-Pb geochronology by laser ablation microprobe inductively coupled plasma mass spectrometry (LAM-ICP-MS). V. M. Goldschmidt Conference 1992, Program and Abstracts, p. A39.

Fujimaki, H., Tatsumotu, M. and Unruh, D. M. (1981) Lu-Hf systematics of the lunar and terrestrial basalts. *Lunar Planet. Sci. XIV* (abstract submitted to the 14th Lunar and Planetary Science Conference), pp. 225–6.

Fukuchi, T., Imai, N. and Shimokawa, K. (1986) ESR dating of fault movement using various defect centres in quartz: the case in the western Suth Fossa Magna, Japan. *Earth Planet. Sci. Lett.*, **78**, 121–128.

Gagan, M. K., Chivas, A. R. and Isdale, P. J. (1994) High-resolution isotopic records from corals using ocean temperature and mass-spawning chronometers. *Earth Planet. Sci. Lett.*, **121**, 549–58.

Galliker, D. Hugentobler, E. and Hahn, B. (1970) Spontane Kerspaltung von U-238 und Am-241. *Helv. Phys. Acta*, **43**, 593.

Gallino, R., Busso, M., Picchio, G. and Raiteri, C. M. (1990) On the astrophysical interpretation of isotope anomalies in meteoritic SiC grains. *Nature*, **348**, 298–300.

Gariepy, C., Allègre, C. J. and Xu, R. H. (1983) The Pb-isotope geochemistry of granitoids from the Himalaya-Tibet coallision zone: implications for crustal evolution. *Earth Planet. Sci. Lett.*, **74**, 220–34.

Garrison, D. H. and Bogard, D. D. (1993) $^{39}Ar/^{40}Ar$ dating of shockmelted phases of the Chico chondrite and implications for K-Ar dating of impact melts. *Lunar Planet. Sci. XXIV* (abstract submitted to the 24th Lunar and Planetary Science Conference), pp. 397–8.

Gaudette, H. E., Vitrac-Michard, A. and Allègre, C. J. (1981) North America Precambrian history recorded in a single sample: high-resolution U-Pb systematics of the Potsdam sandstone detrital zircons, New York State. *Earth Planet. Sci. Lett.*, **54**, 248–60.

Gearing, J. N. (1991) The study of diet and trophic relationships through natural abundance ^{13}C, in *Carbon Isotope Techniques* (eds D. C. Coleman and B. Fry), Academic Press, Inc., London, pp. 201–16.

Geiss, J. (1954) Isotopenanalysen an 'gewöhnlichem' Blei. *Z. Naturforschung*, **9A**, 218.

Genthon, C., Barnola, J. M., Raynaud, D. *et al.* (1987) Vostok ice core: climatic response to CO_2 and orbital forcing changes over the last climatic cycle. *Nature*, **329**, 414–8.

Gentner. W., Kleinmann, B. and Wagner, G. A. (1967) New K-Ar and fission-track ages of impact glasses and tektites. *Earth Planet. Sci. Lett.*, **2**, 83–6.

Gentner. W., Storzer, D. and Wagner, G. A. (1969) Das Alter von Tektiten und verwandten Gläsern. *Naturwissenschaften*, **56**, 255–60.
Gill, J. B., Morris, J. D., Johnson, R. W. (1993) Timescale for producing the geochemical signature of island arc magma: U–Th–Po and Be–B systematics in recent Papua New Guinea lavas. *Geochim. Cosmochim. Acta*, **57**, 4269–4283.
Gilligan, L. B. and Barnes, R. C. (1990) New England fold belt, New South Wales-Regional geology and mineralization, in *Geology of the Mineral Deposits of Australia and Papua New Guinea* (ed. F. E. Hughes), Vol. 2, Australasian Institute of Mining and Metallurgy, pp. 1417–23.
Gleadow, A. J. W. and Lovering, J. F. (1974) The effect of weathering on fission track dating. *Earth Planet. Sci. Lett.*, **22**, 163–8.
Gleadow, A. J. W., Duddy, I. R., Green, P. L. and Hegarty, K. A. (1986) Fission track lengths in the apatite annealing zone and the interpretations of mixed ages. *Earth Planet. Sci. Lett.*, **78**, 245–54.
Godfrey, J. D. (1962) The deuterium content of hydrous minerals from the East-Central Sierra Nevada and Yosemite National Park. *Geochim. Cosmochim. Acta*, **26**, 1215–45.
Godwin, H. (1962) Half-life of radiocarbon. *Nature*, **195**, 984.
Goldich, S. S. and Mudrey, M. G. Jr (1972) Dilatancy model for discordant U-Pb zircon ages, in *Contribution to Recent Geochemistry and Analytical Chemistry* (ed. A. I. Tugarinov), Nauk. Publ. Office, Moscow, USSR, pp. 415–8.
Goldstein, S. J., Perfit, M. R., Batiza, R. *et al.* (1994) Off-axis volcanism at the East Pacific Rise detected by uranium-series dating of basalts. *Nature*, **367**, 157–9.
Green, P. F. (1985) Comparison of zeta calibration baselines for fission-track dating of apatite, zircon and sphene. *Chem. Geol., Isotope Geosci. Sect.*, **58**, 1–22.
Green, P. F. and Durrani, S. A. (1978) A quantitative assessment of geometry factors for use in fission-track studies. *Nucl. Track Detect.*, **2**, 207–13.
Greenaway, P. (1971) Calcium regulation in the freshwater mollusc *Limnaea stagnalis* (L.) (Gastropoda: Pulmonata) II. Calcium movements between internal calcium compartments. *J. Exp. Biol.*, **54**, 609–20.
Gribbin, J. (1994) Sleeping giant at the centre of the galaxy. *New Scientist*, 21, October, 18.
Grobe, H., Mackensen, A., Hubberten, H.-W. *et al.* (1990) Stable isotope record and late Quaternary sedimentation rates at the Antarctic continental margin, in *Geological History of the Polar Oceans: Arctic versus Antarctic* (eds U. Bleil and J. Thiede), Kluwer Academic Publishers, Dordrecht, pp. 539–72.
Grosse, A. V. (1934) An unknown radioactivity. *J. Am. Chem. Soc.*, **56**, 1922–3.
Grossman, E. L., and Ku, T. L. (1986) Oxygen and carbon isotope fractionation in biogenic aragonite: temperature effects. *Chem. Geol.* **59**, 59–74.
Grün, R. (1985a) ESD-Dating without determination of annual dose: a first application on dating mollusc shells, in *ESR Dating and Dosimetry* (eds. M. Ikeya and T. Miki), Ionics Publ. Co., Tokyo, pp. 115–23.
Grün, R. (1985b) Beiträge zur ESR-Datierung. *Sp. Publ., Geol. Inst. Univ. Köln, Germany*, **59**, 157 pp.
Grün, R. (1989) ESR dating for the early Earth. *Nature*, **338**, 543–4.
Gruszczyski, M., Halas, S., Hoffman, A. and Malkowski, K. (1989) A brachiopod calcite record of the oceanic carbon and oxygen isotope shifts at the Permian/Triassic transition. *Nature*, **337**, 64–8.
Guichard, F., Reyss, J.-L. and Yokoyama, Y. (1978) Growth rate of a manganese nodule measured with ^{10}Be and ^{26}Al. *Nature*, **272**, 155–6.
Haack, U. (1977) The closing temperature for fission-track retention in minerals. *Am. J. Sci.*, **277**, 459–64.
Hall, E. T. (1983) Radiocarbon Dating by accelerator-a progress report, in *Proc. 22nd Symp. Archaeometry*, Bradford Univ., 10 March-3 April 1982, pp. 130–4.
Hames, W. E. and Hodges, K. V. (1993) Laser ^{40}Ar/^{39}Ar evaluation of slow cooling and episodic loss of ^{40}Ar from a sample of poly-metamorphic muscovite. *Science*, **261**, 1721–3.
Hamet, J., Nakamura, N., Unruh, D. M. and Tatsumotu, M. (1978) Origin and history of the accumulate eucrite, Moama as inferred from REE abundances, Sm-Nd and U-Pb systematics. Proc. 9th Lunar Sci. Conf., *Geochim. Cosmochim. Acta*, 1115–36.
Hamilton, P. J., Evensen, N. M., O'Nions, R. K. and Tarney, J. (1979b) Sm-Nd systematics of Lewisian gneisses: implications for the origin of granulites. *Nature*, **277**, 25–8.

Hamilton, P. J., Fallick, A. E., Macintyre, R. M. and Elliott, S. (1987) Isotopic tracing of the provenance and diagenesis of Lower Brent Group sands, North Sea, in *Petroleum Geology of North West Europe* (eds J. Brooks and K. Glennie), Graham & Trotman, 939–49.

Hamilton, P. J., O'Nions, R. K., Evensen, N. M. *et al.* (1978) Sm-Nd isotopic investigations of the Isua supracrustals and implications for mantle evolution. *Nature*, **272**, 41–3.

Hamilton, P. J., Evensen, N. M., O'Nions, R. K. *et al.* (1979a) Sm-Nd dating of Onverwacht Group volcanics, Southern Africa. *Nature*, **279**, 298–300.

Hammer, C. U., Clausen, H. B., Friedrich, W. L. and Tauber, H. (1988) Dating of the Santorini eruption. *Nature*, **332**, 401–2.

Hämmerle, H. and Inacker, O. (1990) Der Einfluβ des CO_2-Anstiegs auf die Biosphäre. *Energiewirtschaftliche Tagesfragen*, **9** 1990, 640–4.

Hamza, M. S. and Epstein, S. (1980) Oxygen isotopic fractionation between oxygen of different sites in hydroxyl-bearing silicate minerals. *Geochim. Cosmochim, Acta*, **44**, 173–82.

Harmon, R. S., Ku, T.-L., Matthews, R. K. and Sinart, P. L. (1979) Limits of U-series analysis: Phase 1 results of the Uranium-Series Intercomparison Project. *Geology*, **7**, 405–9.

Harrison, T. M. (1981) Diffusion of ^{40} Ar in hornblende. *Contrib. Mineral. Petrol.*, **78**, 324–31.

Harrison T. M., Duncan, I. and McDougall (1985) Diffusion of ^{40}Ar in biotite: temperature, pressure, and compositional effects. *Geochim. Cosmochim. Acta*, **49**, 2461–8.

Hart, S. R., Erlank, A. J. and Kable, E. J. D. (1974) Sea floor alteration: some chemical and Sr isotopic effects. *Contrib. Mineral. Petrol.*, **44**, 219–42.

Hart, S. R., Brooks, C., Krogh, T. E. *et al.* (1970) Ancient and modern volcanic rocks: a trace element model. *Earth Planet. Sci. Lett.*, **10**, 17–28.

Hashemi-Nezhad, S. R., Fremlin, J. H. and Durrani, S. A. (1979) Polonium haloes in micas. *Nature*, **278**, 333–5.

Hattori, K., Campbell, F. A. and Krouse, H. R. (1983) Sulphur isotope abundances in Aphebian clastic rocks: implications for the coeval atmosphere. *Nature*, **302**, 323–6.

Hayes, J. M., Freeman, K. H., Popp, B. N. and Hoham, C. H. (1990) Compound-specific isotopic analyses: A novel tool for reconstruction of ancient biogeochemical processes. *Org. Geochem.*, **16**, 1115–28.

Hebeda, E. H., Freudel, M. and Schultz, L. (1987) Radiogenic, fissiogenic and nucleogenic noble gases in zircons. *Earth Planet. Sci. Lett.*, **85**, 79–90.

Hemmann, A. (1989) Indikation für einen Umweltgeschädigten Mammutbaum mit Hilfe der stabilen Kohlenstoffisotope. Unveröffentl. Diplomarbeit, Univ. Köln, 54 pp.

Heumann, K. G. and Luecke, W. (1973) Calcium isotope ratios in natural carbonate rocks. *Earth Planet. Sci. Lett.*, **20**, 341–6.

Heumann, K. G., Kubassek, E., Schwabenbauer, W. and Stadler, I. (1979) Analytical method for the K/Ca age determination of geological samples. *Fresenius Z. Anal. Chem.* **297**, 35–43.

Hilbrecht, H., Hubberten, H.-W. and Oberhänsli, H. (1992) Biogeography of planktonic foraminifera and regional carbon isotope variations: productivity and water masses in late Cretaceous Europe. *Palaeogeogr., Palaeoclim., Palaeoecol.*, **92**, 407–21.

Hirt, B. and Epstein, S. (1964) Die Isotopenzusammensetzung von natürlichem Kalzium. *Helv. Phys. Acta*, **37**, 179.

Hirt, B., Herr, W. and Hoffmeister, W. (1963) Age determinations by the rhenium-osmium method, in *Radioactive Dating*, International Atomic Energy Authority, pp. 35–44.

Hodges, K. V., Hames, W. E. and Bowring, S. A. (1994) ^{40}Ar/^{39}Ar age gradients in micas from high temperature-low-pressure metamorphic terrain: evidence for very slow cooling and implications for the interpretation of age spectra. *Geology*, **22**, 55–8.

Hoefs, J. (1987) *Stable Isotope Geochemistry*, Springer-Verlag, Berlin, 241 pp.

Hoefs, J. and Emmermann, R. (1983) The oxygen isotope composition of Hercynian granites and pre-Hercynian gneisses from the Schwarzwald, SW.-Germany. *Contrib. Mineral. Petrol.*, **83**, 320–9.

Hogan, C. J. (1989) Signals from the Big Bang. *Nature*, **339**, 15.

Holland, H. D. (1978) *The Chemistry of the Atmosphere and Oceans*, John Wiley & Sons, New York, 352 pp.

Holland, H. D. (1984) *The Chemical Evolution of the Atmosphere and the Oceans*. Princeton Univ. Press, Princeton, NJ, 528 pp.

Hollander, D. J., McKenzie, J. A. and Hsü, K. J. (1993) Carbon isotope evidence for unusual plankton blooms and fluctuations of surface water CO2 in 'Strangelove Ocean' after terminal Cretaceous event. *Palaeogeogr., Palaeoclim., Palaeoecol.*, **104**, 229–37.

Holmes, A. (1946) An estimate of the age of the Earth. *Nature*, **157**, 680–4.

Holser, W. T., Schönlaub, H.-P., Attrep, M. Jr *et al.* (1989) A unique geochemical record at the Permian/Triassic boundary. *Nature*, **337**, 39–44.

Horibe, Y. and Oba, T. (1972) Temperature scales of aragonite-water and calcite-water systems. *Fossile*, **23/24**, 69–79.

Houtermans, F. G. (1946) Die Isotopenhäufigkeiten im natürlichen Blei und das Alter des Urans. *Naturwissenschaften*, **33**, 185–186.

Houtermans, F. G. (1953) Determination of the age of the Earth from the isotopic composition of meteoritic lead. *Nuovo Cimento*, **10**, 1–623.

Hoyle, F. (1994) *Home is Where the Wind Blows: Chapters from a Cosmological Lifer*, W. H. Freeman and Company.

Hsü, K. J. and McKenzie, J. A. (1991) Carbon-isotope anomalies at era boundaries; global catastrophes and their ultimate cause, in *Global Catastrophes in Earth History; An Interdisciplinary Cong. on Impact, Volcanism and Mass Mortality* (eds V. L. Sharpton and P. D. Ward), *Geol. Soc. Am., Colorado, Spec. Pap.*, **247**, pp. 61–70.

Hubberten, H.-W. (1990) Die Fraktionierung der Schwefelisotope bei der Entstehung und Veränderung der ozeanischen Kruste. *Chem. Erde*, **50**, 39–57.

Hubberten, H.-W., Katz-Lehnert, K. and Keller, J. (1988) Carbon and oxygen isotope investigations in carbonatites and related rocks from the Kaiserstuhl, Germany. *Chemical Geology*, **70**, 257–74.

Hubberten, H.-W., Nielsen, H. and Puchelt, H. (1975) The enrichment of ^{34}S in the solfataras of the Nea Kameni Volcano, Santorini Archipelago, Greece. *Chem. Geol.*, **16**, 197–205.

Hübner, H. (1986) Isotope effects of nitrogen in the soil and biosphere, in *Handbook of Environmental Isotope Geochemistry*, Volume Two (eds P. Fritz and J.-Ch. Fontes), Elsevier, pp. 361–425.

Hulston, J. R. (1978) Methods of calculating isotopic fractionation in minerals. *DSIR Bull.*, **220**, 211–9.

Huneke, J. C., Armstrong, J. T. and Wasserburg, G. J. (1983) FUN with PANURGE: high mass resolution ion microprobe measurements of Mg in Allende inclusions. *Geochim. Cosmochim. Acta*, **47**, 1635–50.

Hurford, A. J. and Green, P. F. (1982) A user's guide to fission-track dating calibration. *Earth Planet. Sci. Lett.*, **59**, 343–54.

Hurford, A. J. and Green, P. F. (1983) The zeta age calibration of fission-track dating. *Isotope Geoscience*, **1**, 258–317.

Hurford, A. J. and Hammerschmidt, K. (1985) ^{40}Ar/^{39}Ar and K/Ar dating of the Bishop and Fish Canyon Tuffs: calibration ages for fission-track dating standards. *Chem. Goel.*, **58**, 23–32.

Hurley, P. M. and Rand, J. R. (1969) Predrift continental nuclei. *Science*, **164**, 1229–42.

Ikeya, M. (1975) Dating a stalactite by electron paramagnetic resonance. *Nature*, **225**, 48–50.

Ikeya, M. (1977) Electron spin resonance dating and fission-track detection of Petrolna stalagmite. *J. Anthropol.*, **4**, 152–68.

Ikeya, M. (1978) Electron spin resonance as a method of dating. *Archaeometry*, **20**, 147–58.

Ikeya, M. and Miki, T. (1980) Electron spin dating of animal and human bones. *Science*, **207**, 977–9.

Inoue, T. and Tanaka, S. (1976) ^{10}B in marine sediments. *Earth Planet. Sci. Lett.*, **29**, 155–60.

Israelson, C., Buchardt, B., Funder, S. and Hubberten, H.-W. (1994) Oxygen and carbon isotope composition of Quaternary bivalve shells as a water mass indicator: last interglacial and Holocene, East Greenland. *Palaeogeogr., Palaeoclim., Palaeoecol.*, **111**, 119–34.

Izett, G. A. and Obradovich, J. D. (1992a) Laser-fusion ^{40}Ar/^{39}Ar ages of Australasian tektites. *Lunar Planet. Sci. XXIII* (abstract submitted to the 23rd Lunar and Planetary Science Conference), pp. 593–4.

Izett, G. A. and Obradovich, J. D. (1992b) Laser-fusion $^{40}Ar/^{39}Ar$ ages of Australasian tektite cores and flanges. *EOS, Trans. AGU*, **73(43)**, Fall Meeting Suppl., 328.

Izett, G. A. and Obradovich, J. D. (1994) $^{40}Ar/^{39}Ar$ age constraints for the Jaramillo Normal Subchron and the Matuyama-Brunhes geomagnetic boundary. *J. Geophys. Res.*, **99**, 2925–34.

Izett, G. A. and Wilcox, R. E. (1982) Map showing localities and inferred distributions of the Huckleberry Ridge, Mesa Falls and Lava Creek ash beds (Pearlette family ash beds) of Pliocene and Pleistocene age in the Western United States and southern Canada. US Geol. Surv., Misc. Inv. Map, I–1325.

Jacobsen, S. B. and Harper, C. L. Jr (1993) Constraints on the differentiation of the Earth from the coupled $^{146,147}SM - ^{142,143}Nd$ systematics. *Lunar Planet. Sci. XXIV* (abstract submitted to the 24th Lunar and Planetary Science Conference), 709–10.

Jacobsen, S. B. and Wasserburg, G. J. (1981) Sm/Nd isotopic systematics in Angra dos Reis. *Lunar Planet. Sci. XII* (abstract submitted to the 12th Lunar and Planetary Science Conference), pp. 500–2.

Jacoby, G. H. (1994) The universe in crisis. *Nature*, **373**, 741–2.

Jahn, B.-M., Gruau, G. and Glickson, A. Y. (1982) Komatiites of the Onverwacht Group, S. Africa: REE geochemistry, Sm/Nd age and mantle evolution. *Contrib. Mineral. Petrol.*, **80**, 25–40.

Jansen, J. B. H. and Schuiling, R. D. (1976) Metamorphism on Naxos: petrology and geothermal gradients. *Am. J. Sci.*, **276**, 1225–53.

Jasper, J. P. and Gagosian, R. B. (1989) Glacial-interglacial climatically forced $\delta^{13}C$ variations in sedimentary organic matter. *Nature*, **342**, 60–2.

Jasper, J. P. and Gagosian, R. B. (1990) The sources and deposition of organic matter in the late Quaternary Pigmy Basin, Gulf of Mexico. *Gegochim. Cosmochim. Acta*, **54**, 1117–32.

Jasper, J. P. and Gagosian, R. B. (1993) The relationship between sedimentary organic carbon isotopic composition and organic biomarker compound concentration. *Geochim. Cosmochim. Acta*, **57**, 167–86.

Jaupart, C. and Provost, A. (1985) Heat focusing, granite gneiss and inverted metamorphic gradients in continental collision zones. *Earth Planet. Sci. Lett.*, **73**, 385–97.

Javoy, M. (1977) Stable isotopes and geothermometry. *J. Geol. Soc.*, **133**, 609–36.

Joachimski, M. M., Buggisch, W. and Anders, T. (1994) Mikrofazies, Conodontenstratigraphie und Isotopengeochemie des Frasne/Famenne-Grenzprofils Wolayer Gletscher (Karnische Alpen). *Abh. Geol. B.-A.*, **50**, 183–95.

Jull, A. J. T. and Donahue, D. J. (1993) ^{14}C terrestrial ages of two lunar meteorites ALHA 81005 and EET 87521. *Lunar Planet. Sci. XXIII* (Abstract submitted to the 23rd Lunar and Planetary Science Conference), pp. 637–8.

Kahn, M. I., Oba, T. and Ku, T.-L. (1981) Paleotemperatures and the glacially induced changes in the oxygen-isotope composition of sea water during late Pleistocene and Holocene time in Tanner Basin, California. *Geology*, **9**, 485–90.

Kamen, M. D. (1963) Early history of carbon-14. *Science*, **140**, 584–90.

Kaplan, I. R. and Rittenberg, S. C. (1964) Microbiological fractionation of sulfur isotopes. *J. Gen. Microbiol.*, **34**, 195–212.

Karlen, I., Olsson, I. U., Kalkberg, P. and Kilicci, S. (1966) Absolute determination of the activity of two ^{14}C dating standards. *Arkiv Geofysik*, **6**, 465–71.

Karlstrom, K. E. and Williams, M. L. (1994) The case for simultaneous deformation, metamorphism and plutonism: an example from Proterozoic rocks in central Arizona. *J. Struct. Geol.*, **XX**, 000–000.

Kaufman, A. (1993) An evaluation of several methods for determining $^{230}Th/U$ ages in impure carbonates. *Geochim. Cosmochim. Acta*, **57**, 2303–17.

Kaufman, A. J., Hayes, J. M. and Klein, C. (1990) Primary and diagenetic controls of isotopic compositions of iron-formation carbonates. *Geochim. Cosmochim. Acta*, **54**, 3461–73.

Kaufman, A., Yechieli, Y. and Gardosh, M. (1992) Reevaluation of the lake-sediment chronology in the Dead Sea basin, Israel, based on new $^{230}Th/U$ dates. *Quaternary Res.*, **38**, 292–304.

Kaufman, A. J., Hayes, J. M., Knoll, A. H. and Germs, G. J. B. (1991) Isotopic compositions of carbonates and organic carbon from upper Proterozoic successions in Namibia: stratigraphic cariation and the effects of diagenesis and metamorphism. *Precamb. Res.*, **49**, 301–27.

Keeling, C. D. (1958) The concentration and isotopic abundances of atmospheric carbon dioxide in rural areas. *Geochim. Cosmochim. Acta*, **13**, 322–34.

Keeling, C. D., Mook, W. G. and Tans, P.P. (1979) Recent trends in the $^{13}C/^{12}C$ ratio of atmospheric carbon dioxide. *Nature*, **277**, 121–2.

Keeling, R. F. and Shertz, S. R. (1992) Seasonal and interannual variations in atmospheric oxygen and implications for the global carbon cycle. *Nature*, **358**, 723–7.

Keith, M. L. and Anderson, G. M. (1963) Radiocarbon dating: fictitious results with mollusk sheels. *Science*, **141**, 634–7.

Keith, M. L. and Weber, J. N. (1964) Carbon and oxygen isotopic composition of selected limestones and fossils. *Geochim. Cosmochim. Acta*, **28**, 1787–1816.

Keigwin, L. D. Jr (1980) Palaeoceanographic change in the Pacific at the Eocene-Oligocene boundary. *Nature*, **287**, 722–5.

Keller, R. A., Bomse, D. S. and Cremers, D. A. (1981) Pushing isotopic selectivity to one part in 10^{14}. *Laser Focus*, 75–80.

Kelley, S. and Bluck, B. J. (1989) Detrital mineral ages from the Southern Uplands using ^{40}Ar-^{39}Ar laser probe. *J. Geol. Soc. London*, **146**, 401–3.

Kennett, J. P. and Shackleton, N. J. (1976) Oxygen isotope evidence for the development of the psychrosphere 39 m.y. ago. *Nature*, **260**, 513–5.

Kerridge, J. F. (1985) Carbon, hydrogen and nitrogen in carbonaceous chondrites: abundances and isotopic compositions in bulk samples. *Geochim. Cosmochim. Acta*, **49**, 1707–14.

Khlopin, V. G., Gerling, E. K. and Baranovskaya, N. V. (1947) The presence of certain stable products of the spontaneous disintegration of uranium in nature. *Bull. Acad. Sci. Chim.*, 599–604. (in Russian).

Killingley, J. S. (1983) Effects of diagenetic recrystallization on $^{18}O/^{16}O$ values of deep-sea sediments. *Nature*, **301**, 594–7.

Killingley, J. S. and Newman, W. A. (1982) ^{18}O fractionation in barnacle calcite: a barnacle paleotemperature equation. *J. Mar. Res.*, **40**, 893–901.

Kirshner, R. P. (1994) The Earth's elements. *Sci. Am.*, **271(4)**, 37–43.

Kistler, R. W. and Peterman, Z. E. (1973) Variations in Sr, Rb, K, Na and initial $^{87}Sr/^{86}Sr$ in Mesozoic granitic rocks and intruded wall rocks in Central California. *Bull. Geol. Soc. Am.*, **84**, 3489–512.

Kistler, R. W. and Peterman, Z. E. (1978) Reconstruction of crustal blocks of California on the basis of initial strontium isotopic compositions of Mesozoic granitic rocks. *US Geol. Surv. Prof. Paper*, **1071**, 17 pp.

Kleeman, J. D. (1988) Constraints from granite intrusives and felsic extrusives on the tectonics of the southern New England orogen in the late Palaeozoic-early Triassic, in *The New England Orogen Tectonics and Metallogenesis* (ed. J. D. Kleeman), Dept. of Geology and Geophysics, University of New England, pp. 129–33.

Kobayashi, S. and Watabe, N. (1959) *The Study of Pearls.* Gihodo Press, Tokyo.

Koerberl, C., Bottomley, R. J. Glass, B. P. *et al.* (1989) Geochemistry and age of Ivory Coast tektites. *Meteoritics*, **24**, 287.

Kolodny, Y. and Raab, M. (1988) Oxygen isotopes in phosphatic fish remains from Israel: paleothermometry of tropical Cretaceous and Tertiary shelf waters. *Palaeogeogr., Palaeoclim., Palaeoecol.*, **64**, 59–67.

Kolodny, Y., Luz, B. and Navon, O. (1983) Oxygen isotope variations in phosphate of biogenic apatites, I Fish bone apatite-rechecking the rules of the game. *Earth Planet. Sci. Lett.*, **64**, 393–404.

Krantz, D. E. (1990) Mollusk-isotope records of Plio-Pleistocene marine paleoclimate, U. S. Middle Atlantic coastal plain. *Palaios*, **5**, 317–35.

Krantz, D. E., Williams, D. F. and Jones, D. S. (1987) Ecological and paleoenvironmental information using stable isotope profiles from living and fossil molluscs. *Palaeogeogr., Palaeoclim., Palaeoecol.*, **58**, 249–66.

Kralik, M. (1982) Rb–Sr age determination on Precambrian carbonate rocks of the Carpentarian McArthur Basin, Northern Territories, Australia. *Precambrian Res.*, **18**, 157–70.

Kreulen, R. (1980) CO_2-rich fluids during regional metamorphism on Naxos (Greece): carbon isotopes and fluid inclusions. *Am. J. Sci.*, **280**, 745–71.

Kreulen, R. and van Beek, P. C. J. M. (1983) The calcite–graphite isotope thermometer: data on graphite-bearing marbles from Naxos, Greece. *Geochim. Cosmochim. Acta*, **47**, 1527–30.

Krishnaswami, S. and Lal, D. (1982) Deep-sea sedimentation: Processes and chronology, in *Nuclear and Chemical Dating Techniques* (ed. L. A. Currie), Washington, pp. 363–88.

Krishnaswami, S., Turekian, K. K. and Bennett, J. T. (1984) The behavior of ^{232}Th and the ^{238}U decay chain nuclides during magma formation and volcanism. *Geochim. Cosmochim. Acta*, **48**, 505–11.

Krogh, T. E. (1982a) Improved accuracy of U–Pb zircon dating by selection of more concordant fractions using a high gradient magnetic separation technique. *Geochim. Cosmochim. Acta*, **46**, 631–5.

Krogh, T. E. (1982b) Improved accuracy of U–Pb zircon ages by the creation of more concordant system using an air abrasion technique. *Geochim. Cosmochim. Acta*, **46**, 637–49.

Krogh, T. E., Kamo, S. L., Sharpton, V. L. *et al.* (1993) U–Pb ages of single shocked zircons linking distal K/T ejecta to the Chicxulub crater. *Nature*, **366**, 731–4.

Kromer, B. and Becker, B. (1993) German oak and pine calibration, 7200–9400 B. C. *Radiocarbon*, **35(1)**, 125–35.

Kröner, A. and Todt, W. (1988) Single zircon dating constraining the maximum age of the Barberton Greenstone Belt, southern Africa. *J. Geophys. Res.*, **93**, 15329–37.

Kronfeld, J. and Rosenthal, E. (1981) Uranium isotopes as a natural tracer in the waters of the Bet Shean-Harod valleys, Israel. *J. Hydrol.*, **50**, 179–90.

Kroopnik, P. (1985) The distribution of ^{13}C of ΣCO_2 in the world oceans. *Deep Sea Res.*, **32**, 57–84.

Ku, T.-L. and Broecker, W. S. (1967) Rates of sedimentation in the Arctic Ocean, in *Progress in Oceanography*, Vol. 4 (ed. M. Sears), Pergamon Press, New York, pp. 95–104.

Kuhn, M. W., Davis, S. N., Bentley, H. W. and Zito, R. (1984) Measurements of thermal neutrons in the subsurface. *Geophys. Res. Lett.*, **11**, 607–10.

Kunk, M. J., Snee, L. W., French, B. M. *et al.* (1993) Preliminary ^{40}Ar/^{39}Ar age spectrum and laser probe dating of the MI core of the Manson Impact Structure, Iowa: a K–T boundary crater candidate. *Lunar Planet. Sci. XXIV* (abstract submitted to the 24th Lunar and Planetary Science Conference), pp. 835–6.

Kurie, F. N. D. (1934) A new mode of disintegration induced by neutrons. *Phys. Rev.* **45**, 904–5.

Kuroda, P. K., Edwards, R. R. and Ashizawa, F. Y. (1956) Radiochemical measurements of the natural fission rate of uranium and the natural occurrence of short-lived iodine isotopes. *J. Chem. Phys.*, **25**, 603–9.

Lakatos, S. and Miller, D. S. (1972) Evidence for the effect of water content on fission-track annealing in volcanic glass. *Earth Planet. Sci. Lett.*, **14**, 128–30.

Lal, D. and Peters, S. (1967) Cosmic ray produced radioactivity on the Earth, in *Handbuch der Physik 46/2* (ed. K. Sitte), Springer Verlag, Berlin, pp. 551–612.

Lal, D. and Somanyjulu, B. L. K. (1984) Some aspects of the geochemistry of silicon isotopes. *Tectonophysics*, **105**, 383.

Lal, D. and Suess, H. E. (1968) The radioactivity of the atmosphere and hydrosphere. *Ann. Rev. Nucl. Sci.*, **18**, 407–37.

Lal, D., and Muralli, A. V., Rajan, R. S. *et al.* (1968) Techniques for the proper revelation and viewing of etch-tracks in meteoritic and terrestrial minerals. *Earth Planet. Sci. Lett.*, **5**, 111–9.

Lal, D., Jull, A. J. T., Donahue, D. J. *et al.* (1990) Polar ice ablation rates measured using *in situ* cosmogenic ^{14}C. *Nature*, **346**, 350–2.

Lalou, C., Reyss, J. L. and Brichet, E. (1993) Actinide-series disequilibrium as a tool to establish the chronology of deep-sea hydrothermal activity. *Geochim. Cosmochim. Acta*, **57**, 1221–31.

Lanphere, M. A., Dalrymple, G. B., Fleck, R. J. and Pringle, M. S. (1990) Intercalibration of mineral standards for K–Ar and ^{40}Ar/^{39}Ar age measurements. *EOS, Trans. AGU*, **71**, 1658.

Lanzetta, K. M. (1994) Intergalactic matters. *Nature*, **370**, 15–6.

Lavrukhina, A. K. and Ustinova, G. K. (1993) On possible Mn-53 heterogeneity in the early solar system. *Lunar Planet. Sci. XXIV* (abstract submitted to the 24th Lunar and Planetary Science Conference), pp. 855–6.

Lawrence, J. R. and Taylor, H. P. (1971) Deuterium and oxygen-18 correlation: clay minerals and hydroxides in Quaternary soils compared to meteoric waters: *Geochim. Cosmochim. Acta*, **35**, 993–1003.

Le Bel, L., Y. D. and Sheng, J. F. (1984) Granitic evolution of the Xihuanshan–Dangping tungsten-bearing system. *Teschermaks Mineral. Petrogr. Mitt.*, **33**, 149–67.

Lee, T. and Papanastassiou, D. A. (1978) Calcium isotopic anomalies in the Allende meteorite. *Astrophys. J. Lett.*, **220**, L21–5.

Lee, T., Ku, T. -L., Lu, H. -L. and Chen, J. -C. (1993) First detection of fallout Cs–135 and potential applications of $^{137}Cs/^{135}Cs$ ratios. *Geochim. Cosmochim. Acta*, **57**, 3493–7.

Le Fort, P. (1981) Manaslu leucogranite: a collision signature of the Himalaya. A model for its genesis and emplacement. *J. Geophys. Res.*, **86**, 10545–68.

Leutwin, F. and Sonet, J. (1974) Geochronological studies in the southern Black Forest. *N. Jb. Mineral. Abh.*, **121**, 252–71.

Libby, W. F. (1955) *Radiocarbon Dating*, 2nd edn, University of Chicago Press, Chicago, 175 pp.

Lin, Y. T, Goresy, A. and Fang., H. (1991) Ca–Al-rich inclusions in Ningqian (CV3) chondrite: evidence for primordial high enrichment in Re in Pt-group element nuggets. *Meteoritics*, **26**, 364–5.

Lingenfelter, P. E. (1963) Production of carbon-14 by cosmic ray neutrons. *Rev. Geophys.*, **1**, 35–55.

Linick, T. W. (1980) Bomb produced carbon-14 in the surface water of the Pacific Ocean. *Radiocarbon*, **22**, 599–606.

Livio, M., Hollowell, D., Weiss, A. and Truran, J. W. (1989) The anthropic significance of the existence of an excited state of ^{12}C. *Nature*, **340**, 281–4.

Locker, G. L. (1933) Neutrons from cosmic ray stösse. *Rev. Phys.*, **44**, 779–81.

Long, A., Eastoe, C. J., Kaufman, R. S. *et al.* (1993) High-precision measurement of chlorine stable isotope ratios. *Geochim. Cosmochim. Acta*, **57**, 2907–912.

Long, D. G. F. (1993) Oxygen and carbon isotopes and event stratigraphy near the Ordovician–Silurian boundary, Anticosti Island Quebec. *Palaeogeogr., Palaeoclim., Palaeoecol.*, **104**, 49–59.

Longinelli, A. (1984) Oxygen isotopes in mammal bone phosphate: a new tool for paleohydrological and paleoclimatological research? *Geochim. Cosmochim. Acta*, **48**, 385–90.

Longinelli, A. and Nuti, S. (1973) Revised phosphate–water isotopic temperature scale. *Earth Planet. Sci. Lett.*, **19**, 373–6.

Loosli, H. H. and Oeschger, H. (1968) Detection of ^{39}Ar in atmospheric argon. *Earth Planet. Sci. Lett.*, **5**, 191–8.

Lorius, C., Jouzel, J., Ritz, C. *et al.* (1985) A 150,000–year climatic record from Antarctic ice. *Nature*, **316**, 591–6.

Lowenstam, H. A. (1961) Mineralogy, O–18/O–16 ratios and strontium and magnesium contents of recent and fossil brachiopods and their bearing on the history of the oceans. *J. Geol.*, **69**, 241–60.

Lowenstam, H. A. and Epstein, S. (1959) Cretaceous paleotemperatures as determined by the oxygen isotope method, their relations to and the nature of rudistid reefs. *Int. Geol. Congr., 20th, Mexico, 1956, Symp. Cretaceous*, pp. 65–76.

Luck, J. M. and Allègre, C. J. (1983) ^{187}Re–^{187}Os systematics in meteorites and cosmological consequences. *Nature*, **302**, 130–2.

Luck, J. M., Birck, L. and Allègre, C. J. (1980) ^{187}Re–^{187}Os systematics in meteorites: early chronology of the solar system and age of the galaxy. *Nature*, **283**, 256–9.

Lugmair, G. W. and Marti, K. (1978) Lunar initial $^{143}Nd/^{144}Nd$: differential evolution of the lunar crust and mantle. *Earth Planet. Sci. Lett.*, **39**, 349–57.

Luz, B., Kolodny, Y. and Horowitz, M. (1984) Fractionation of oxygen isotopes between mammalian bone phosphate and environmental drinking water. *Geochim. Cosmochim. Acta*, **48**, 1689–93.

MacFadden, B. J. (1992) *Fossil Horses: Systematics, Paleobiology and Evolution of the Family Equidae.* Cambridge University Press, New York, 369 pp.

MacFadden, B. J., Wang, Y., Cerling, T. E. and Anaya, F. (1994) South American fossil mammals and carbon isotopes: a 25 million-year sequence from the Bolivian Andes. *Palaeogeogr., Palaeoclim., Palaeoecol.*, **107**, 257–68.

Mackensen, A., Grobe, H., Hubberten, H. -W. *et al.* (1989) Stable isotope stratigraphy from the Antarctic continental margin during the last one million years. *Marine Geol.*, **87**, 315–21.

Mackensen, A., Hubberten, H. -W., Bickert, T. *et al.* (1993) The $\delta^{13}C$ in benthic foraminiferal tests of *Fontbotia wuellerstorfi* (Schwager) relative to the $\delta^{13}C$ of dissolved inorganic carbon in southern ocean deep water: implications for glacial ocean circulation models. *Paleoceanography*, **8**, 587–610.

MacNamara, J. and Thode, H. H. (1950) The isotopes of xenon and krypton in pitchblende and the spontaneous fission of uranium. *Phys. Rev.*, **80**, 471–2.

Macpherson, C. and Mattey, D. (1994) Carbon isotope variations of CO_2 in Central Lau Basin basalts and ferrobasalts. *Earth Planet. Sci. Lett.*, **121**, 263–76.

Maddox, J. (1992) The anthropic view of nucleosynthesis. *Nature*, **355**, 107.

Magenheim, A. J., Spivack, A. J., Volpe, C. and Ransom, B. (1994) Precise determination of stable chlorine isotopic ratios in low-concentration natural samples. *Geochim. Cosmochim. Acta*, **58**, 3117–21.

Maloszewski, P. and Zuber, A. (1983) Theoretical possibilities of the ^3H–^3He method in investigations of groundwater systems. *Catena*, **10**, 189–198.

Manhes, G., Minster, J. F. and Allègre, C. J. (1978) Comparative uranium–thorium–lead and rubidium–strontium study of St Severin amphoterite: consequence for early solar system chronology. *Earth Planet. Sci. Lett.*, **39**, 14–24.

Mankinen, E. A. and Dalrymple, G. B. (1979) Revised geomagnetic polarity time scale for the interval 0–5 m.y. B. P. *J. Geophys. Res.*, **84**, 615–26.

Mariotti, A. (1983) Atmospheric nitrogen is a reliable standard for natural abundance measurements. *Nature*, **303**, 685–7.

Mariotti, A., Lancelot, C. and Billen, G. (1984) Natural isotopic composition of nitrogen as a tracer of origin for suspended organic matter in the Scheldt Estuary. *Geochim. Cosmochim. Acta*, **48**, 549–55.

Marshall, B. D. and DePaolo, D. J. (1982) Precise age determinations and petrogenetic studies using the K/Ca method. *Geochim. Cosmochim. Acta*, **46**, 2537–45.

Marti, K. (1967) Mass-spectrometric detection of cosmic-ray-produced ^{81}Kr in meteorites and the possibility of Kr–Kr dating. *Phys. Rev. Lett.*, **18**, 264–6.

Maruèjol, P., Cuney, M. and Turpin, L., (1990) Magmatic and hydrothermal R. E. E. fractionation in the Xihuashan granites (SE China). *Contrib. Mineral. Petrol.*, **104**, 668–80.

Mathews, D. E. and Hayes, J. M. (1978) Isotope-ratio-monitoring gas chromatography mass spectrometry. *Anal. Chem.*, **50**, 1465–73.

Mathews, G. J. (1991) Finding the proton drip line. *Nature*, **351**, 348–9.

Mathews, G. J. (1992) Flashing giants forge flourine. *Nature*, **359**, 18–9.

Mathews, G. J. and Cowan, J. J. (1990) New insights into the astrophysical r-process. *Nature*, **345**, 491–4.

McConnaughey, T. (1989a) ^{13}C and ^{18}O isotopic disequilibrium in biological carbonates: I. Patterns. *Geochim. Cosmochim. Acta*, **53**, 151–62.

McConnaughey, T. (1989b) ^{13}C and ^{18}O isotopic disequilibrium in biological carbonates: II. In vitro simulation of kinetic isotopic effects. *Geochim. Cosmochim. Acta*, **53**, 163–71.

McCrea, J. M. (1950) The isotopic chemistry of carbonates and a paleotemperature scale. *J. Chem. Phys.*, **18**, 849–57.

McCulloch, M. T. (1994) Primitive $^{87}Sr/^{86}Sr$ from Archean barite and conjecture on the Earth's age and origin. *Earth Planet. Sci Lett.*, **126**, 1–13.

McCulloch, M. T., de Laeter, J. R. and Rosman, K. J. R. (1976) The isotopic composition and elemental abundance of lutetium in meteorites and terrestrial samples and the ^{176}Lu cosmochronometer. *Earth Planet. Sci. Lett.*, **28**, 308–22.

McDougall, I. and Harrison, T. M. (1988) *Geochronology and Thermochronology by the $^{40}Ar/^{39}Ar$ Method.* Oxford University Press, Oxford, New York, 212 pp.

McNutt, R. H., Crocket, J. H., Clark, A. *et al.* (1975) Initial $^{87}Sr/^{86}Sr$ ratio of plutonic and volcanic rocks of the central Andes between latitudes 26° and 29° South. *Earth Planet. Sci. Lett.*, **27**, 305–13.

Measures, C. I. and Edmond, J. M. (1982) Beryllium in the water column of the central North Pacific. *Nature*, **297**, 51–3.

Mehta, P. P. and Rama, (1970) Annealing effects in muscovite and their influence on dating by fission track method. *Earth Planet. Sci. Lett.*, **7**, 82–6.

Mensing, T. M., Faure, G., Jones, L. M. *et al.* (1984) Petrogenesis of the Kirkpatrick Basalt, Solo Nunatak, northern Victoria Land, Antarctica, based on isotopic compositions of strontium, oxygen and sulfur. *Contrib. Mineral. Petrol.*, **87**, 101–8.

Meshceryakov, R. P. and Stolbov, M. Y. (1967) Measurement of the isotopic composition of calcium in natural materials. *Geochemistry (USSR)*, **4**, 1001–3.

Michael, H. N. and Ralph, E. K. (1970) Correction factors applied to Egyptian radiocarbon dates from the era before Christ, in *Radiocarbon Variations and Absolute Chronology* (ed. I. Olsson), Almqvist and Wiksell, Stockholm, and John Wiley & Sons, New York, pp. 109–19.

Miller, D. M., Goldstein, S. L. and Langmuir, C. H. (1994) Cerium/lead and lead isotope ratios in arc magmas and their enrichment of lead in the continents. *Nature*, **368**, 514–7.

Misawa, K., Tatsumoto, M., Dalrymple, G. B. and Yanai, K. (1993) An extremely low U/Pb source in the Moon: U-Th-Pb, Sm-Nd, Rb-Sr and $^{40}Ar/^{39}Ar$ isotopic systematics and age of lunar meteorite Asuka 88157. *Geochim. Cosmochim. Acta*, **57**, 4687–702.

Miura, Y. and Nagao, K. (1991) Noble gases and ^{81}Kr–Kr exposure ages of non-Antarctic ordinary chondrites: an attempt to measure terrestrial ages of Antarctic meteorites. *Proc. NIPR Symp. Antarctic Meteorites*, **5**, 298–309.

Miura, Y., Nagao, and Fujitani, T. (1993) ^{81}Kr terrestrial ages and grouping of Yamato eucrites based on noble gases and chemical composition. *Geochim. Cosmochim. Acta*, **57**, 1857–66.

Monié, P. Montigny, R. and Maluski, H. (1992) Age burdigalien de la tectonique ductile extensive dans le massif de l'Edough (Kabylies, Algérie). Données radiometriques $^{39}Ar/^{40}Ar$. *Bull. Soc. Geol. Fr.*, **163**, 571–84.

Monster, J., Appel, P. W. U., Thode, H. G. *et al.* (1979) Sulfur isotope studies in early Archean sediments from Isua, West Greenland: implications to the antiquity of bacterial sulfate reduction. *Geochim. Cosmochim. Acta*, **43**, 405–13.

Montgomery, C. G. and Montgomery, D. D. (1939) The intensity of neutrons of thermal energy in the atmosphere at sea level. *Phys. Rev.*, **56**, 10–2.

Mook, W. G. (1968) Geochemistry of the stable carbon and oxygen isotopes of natural waters in the Netherlands. Dissertation, Groningen, 157 pp.

Mook, W. G. (1986) ^{13}C in atmospheric CO_2. *Netherlands J. Sea Res.*, **20**, 211–23.

Mook, W. G. (1980) Carbon-14 in hydrogeological studies, in *Handbook of Environmental Isotope Geochemistry*, Vol. 1A (ed P. Fritz and J. -C. Fontes), Elsevier, Amsterdam, pp. 49–74.

Mook, W. G. and van der Hoek, S. (1983) The N_2O correction in the carbon and oxygen isotopic analyses of atmospheric CO_2. *Isotope Geosci.*, **1**, 237–42.

Moorbath, S. and Welke, H. (1969) Lead isotope studies on igneous rocks from the Isle of Skye, northwestern Scotland. *Earth Planet. Sci. Lett.*, **5**, 217–30.

Moore, L., Heald, E. F. and Filliben, J. J. (1978) An isotopic fractionation model for the multiple filament thermal ion source. *Adv. Mass. Spectrom.*, **7A**, 448–74.

Morton, J. P. (1985) Rb–Sr dating of diagenesis and source age of clays in Upper Devonian black shale of Texas. *Bull. Geol. Soc. Am.*, **96**, 1043–9.

Mrini, Z., Rafi, A., Duthou, J. -L. and Vidal, P. (1992) Chronologie Rb–Sr des granitoïdes hercyniens du Macro: consequences. *Bull. Soc. Geol. Fr.*, **163(3)**, 281–91.

Munnich, K. O. (1957) Messung des ^{14}C Gehaltes von hartem Grundwasser. *Naturwissenschaften*, **44**, 32.

Naeser, C. W. (1967) The use of apatite and sphene for fission track age determination. *Bull. Geol. Soc. Am.*, **78**, 1523–6.

Naeser, C. W. and Faul, H. (1969) Fission track annealing in apatite and sphene. *J. Geophys. Res.*, **74**, 705–10.

Naidin, D. P., Teis, R. V. and Chupakhin, M. A. (1956) The Cretaceous Crimean fauna. *Geokhimiya*, **8**, 752.

Nakai, S., Halliday., A. N., Kesler, S. L. and Jones, H. D. (1990) Rb–Sr dating of sphalerites from Tennessee and the genesis of Mississippi Valley type ore deposits. *Nature*, **346**, 354–7.

Nakamura, N., Unrich, D. M. and Tatsumoto M. (1982) Origin and evolution of the Nakhla meteorite inferred from the Sm-Nd and U–pb systematics and REE, Ba, Sr, Rb and K abundances. *Geochim. Cosmochim. Acta*, **46**, 1555–73.

Naqvi, W. A., Charles, C. D. and Fairbanks, R. G. (1994) Carbon and oxygen isotopic records of benthic foraminifera from the Northeast Indian Ocean: implications on glacial-interglacial atmospheric CO_2 changes. *Earth Planet. Sci. Lett.*,**121**, 99–110.

Navon, O. and Wasserburg, G. J. (1985) Self-shielding in O_2– a possible explanation for oxygen isotopic anomalies in meteorites. *Earth Planet. Sci. Lett.*, **73**, 1–16.

Nelson, A. R., Atwater, B. F., Bobrowsky, P. T. *et al.* (1995) Radiocarbon evidence for extensive plate-boundary rupture about 300 years ago at the Cascadia subduction zone. *Nature*, **378**, 372–4.

Ness, G. S., Levi, S. and Couch, R. (1980) Marine magnetic anomaly timescales for the Cenozoic and Late Cretaceous: a précis, critique, and synthesis. *Rev. Geophys. Space Phys.*, **18**, 753–70.

Newman, S., Finkel, R. C. and MacDougall, J. D. (1983) ^{230}Th–^{238}U disequilibrium systematics in oceanic tholeiites from 21° on the East Pacific Rise. *Earth Planet. Sci. Lett.*, **65**, 17–33.

Niederer, F. R. and Papanastassiou, D. A. (1984) Ca isotopes in refractory inclusions *Geochim. Cosmochim. Acta*, **48**, 1279–93.

Niemeyer, S. (1979) I–Xe dating of silicate and troilite from 1AB iron meteorites. *Geochim. Cosmochim. Acta*, **47**, 843–60.

Nier, A. O. (1940) A mass spectrometer for routine isotope abundance measurements. *Rev. Sci. Instr.* ,**18**, 398–411.

Nishiizumi, K., Arnold, J. R., Elmore, D. *et al.* (1983) ^{36}Cl and ^{53}Mn in Antarctic meteorites and ^{10}Be–^{36}Cl dating on Antarctic ice. *Earth Planet. Sci. Lett.*, **62**, 407–17.

Nishiizumi, K., Winterer, E. L., Kohl, C. P. *et al.* (1989) Cosmic ray production rates of ^{10}Be and ^{26}Al in quartz from glacially polished rocks. *J. Geophys. Res.*, **94**, 17907–15.

Nohda, S. and Wasserburg, G. J. (1986) Trends of Sr and Nd isotopes through time near the Japan Sea in northeastern Japan. *Earth Planet. Sci. Lett.*, **78**, 157–67.

Notsu, K. (1983) Strontium isotope composition in volcanic rocks from the Northeast Japan arc. *J. Volcanol. Geothern. Res.*, **18**, 531–48.

Nozaki, Y. (1986) ^{226}Ra-^{222}Rn-^{210}Pb systematics in seawater near the bottom of the ocean. *Earth Planet. Sci. Lett.*, **80**, 36–40.

Nyquist, L. E. (1977) Lunar Rb–Sr chronology. *Phys. and Chem. of the Earth*, **10**, 103–42.

Nyquist, L. E. (1982) Do oblique impacts produce Martian meteorites? *J. Geophys. Res.*, **88**, suppl., A779–84.

Nyquist, L. E., Shih, C. -Y., Wooden, J. L. *et al.* (1979). The Sr and Nd isotopic record of Apollo 12 basalts: implications for lunar geochemical evolution. *Proc. 10th Lunar Planet Sci Conf., Geochem. Cosmochim. Acta*, 77–114.

Oeschger, H. (1982) The contribution of radioactive and chemical dating to the understanding of the environmental system, in *Nuclear and Chemical Dating Techniques* (ed. L. A. Currie), pp. 5–42.

Ohmoto, H. and Rye, R. O. (1979) Isotopes of sulfur and carbon, in *Geochemistry of Hydrothermal Ore Deposits*, 2nd edn (ed. H. L. Barnes), John Wiley & Sons, New York, 798 pp.

Olive, K. A. and Schramm, D. N. (1992) Astrophysical ^{7}Li as a product of big bang nucleosynthesis and galactic cosmic-ray spallation. *Nature*, **360**, 439–42.

Oliver, G. J. H. and Kelley, S. (1993) ^{40}Ar–^{39}Ar fusion ages from the Polish Sudetes: Variscan tectonothermal reworking of Caledonian protoliths. *N. Jb. Geol. Paläont. Mh.*, **6**, 321–34.

O'Neil, J. R., Shaw, S. E. and Flood, R. H. (1977) Oxygen and hydrogen isotope compositions as indicators of granite gneiss in the New England Batholith, Australia. *Contrib. Mineral. Petrol.*, **62**, 313–28.

Osmond, J. K. (1979) Accumulation models of ^{230}Th and ^{231}Pa in deep sea sediments. *Earth Sci. Rev.*, **15**, 95–151.

Oversby, V. M. and Gast, P. W. (1968) Lead isotope compositions and uranium decay series in Recent volcanic rocks. *Earth Planet. Sci. Lett.*, **5**, 199–206.

Owen, H. R. and Schaeffer, O. A. (1955) The isotope abundances of chlorine from various sources. *J. Am. Chem. Soc.*, **77**, 898–904.

Owen, L. B. (1974) Age determinations by the lutetium-176/hafnium-176 method. Unpublished PhD dissertation, Ohio State Univ., Columbus, Ohio, 300 pp.

Owen, L. B. and Faure, G. (1974) Simultaneous determination of hafnium and zirconium in silicate rocks by isotope dilution. *Anal. Chem.*, **46**, 1323–6.

Ozima, M., Zashu, S., Takigami, Y. and Turner, G. (1989) Origin of the anomalous ^{40}Ar–^{30}Ar age of Zaire cubic diamonds: excess ^{40}Ar in pristine mantle fluids. *Nature*, **337**, 226–9.

Pal, D. K., Tuniz, C., Moniot, R. K. *et al.* (1982) Beryllium-10 in Australian tektites: evidence for a sedimentary precursor. *Science*, **218**, 787–9.

Palmer, M. R. and Elderfield, H. (1985) Variations in the Nd isotopic composition of foraminifera from Atlantic Ocean sediments. *Earth Planet. Sci. Lett.*, **73**, 299–305.

Papanastassiou, D. A. and Wasserburg, G. J. (1969) Initial strontium isotopic abundances and the resolution of small time difference in the formation of planetary objects. *Earth Planet. Sci. Lett.*, **5**, 361–76.

Papanastassiou, D. A. and Wasserburg, G. J. (1973) Rb-Sr ages and initial Sr in basalts from Apollo 15. *Earth Planet. Sci. Lett.*, **17**, 324–37.

Papanastassiou, D. A. and Wasserburg, G. J. (1978) Strontium isotopic anomalies in the Allende meteorite. *Geophys. Res. Lett.*, **5**, 595–8.

Parker, R. L. (1967) *Data of Geochemistry: Composition of the Earth's Crust*, US Geol. Surv. Prof. Paper, 440-D.

Patchett, P. J. (1980) Sr isotopic fractionation in Allende chondrules: a reflection of solar nebula processes. *Earth Planet. Sci. Lett.*, **50**, 181–8.

Patchett, P. J. (1983) Importance of the Lu–Hf isotopic system in studies of planetary chronology and chemical evolution. *Geochim. Cosmochim. Acta*, **47**, 81–91.

Patchett, P. J. and Tatsumoto, M. (1980) Lu–Hf total-rock isochron for the eucrite meteorites. *Nature*, **288**, 571–4.

Patchett, P. J., Kouvo, O., Hedge, C. E. and Tatsumoto, M. (1981) Evolution of continental crust and mantle heterogeneity: evidence from Hf isotopes paleoecology. *Contrib. Mineral. Petrol.*, **78**, 279–97.

Patterson, W. P. and Walter, L. M. (1993) Syndepositional diagenesis of modern platform carbonates: evidence from isotopic and minor element data. *Geology*, **22**, 127–30.

Pearson, P., Shackleton, N. J. and Hall, M. A. (1993) Stable isotope paleoecology of middle Eocene planktonic foraminifera and multi–species isotope stratigraphy, DSDP Site 523, South Atlantic. *J. Foraminiferal Res.*, **23**, 123–40.

Peebles, P., James, E., Schramm, D. N. *et al.* (1994) The evolution of the universe. *Sci. Am.*, **271(4)**, 29–33.

Penrose, R. (1989) *The Emperor's New Mind*, Oxford University Press.

Pepin, R. O. and Phinney, D. (1976) The formation interval of the Earth. *Lunar Sci. VII (A)*, pp. 682–4.

Perry, E. C. Jr, Monster, J. and Reimer, T. (1971) Sulfur isotopes in Swaziland system barites and the evolution of the Earth's atmosphere. *Science*, **171**, 1015–6.

Peterman, Z. E. and Hedge, C. E. (1971) Related strontium isotopic and chemical variations in oceanic basalts. *Bull. Geol. Soc. Am.*, **82**, 493–500.

Pettingill, H. S., Patchett, P. J., Tatsumoto, M. and Moorbath, S. (1981) Lu–Hf total-rock ages for Amîtsoq gneisses, West Greenland. *Earth Planet. Sci. Lett.*, **55**, 150–6.

Pettini, M., Boksberg, A. and Hunstead, R. W. (1989) in *The Epoch of Galaxy Formation* (eds C. S. Frenk *et al.*), Kluwer, Dordrecht, pp. 107–12.

Pierce, M. J., Welch, D. L., McClure, R. D. *et al.* (1994) The Hubble constant and Virgo cluster distance from observations of Cepheid variables. *Nature*, **371**, 385–9.

Pirrie, D. and Marshall, J. D. (1990a) High-paleolatitude late Cretaceous paleotemperatures: new data from James Ross Island, Antarctica. *Geology*, **18**, 31–4.

Pirrie, D. and Marshall, J. D. (1990b) Diagenesis of *Inoceramus* and Late Cretaceous paleoenvironmental geochemistry: a case study from James Ross Island, Antarctica. *Palaios*, **5**, 336–45.

Podosek, F. A. (1994) A noble endeavour. *Nature*, **368**, 290–1.

Podosek, F. A., Pier, J., Nitho, O. *et al.* (1988) Normal potassium, inherited argon in Zaire cubic diamonds. *Nature*, **334**, 607.

Polevaya, N. L., Titov, N. E. Belyaev, V. S. and Sprintsson, V. D. (1958) Application of the Ca method in the absolute age determination of sylvites. *Geochemistry (USSR)*, **8**, 897–906.

Popp, B. N., Podosek, F. A., Brannon, J. C. *et al.* (1986) $^{87}Sr/^{86}Sr$ ratios in Permo-Carboniferous sea water from the analyses of well-preserved brachiopod shells. *Geochim. Cosmochim. Acta*, **50**, 1321–8.

Premo, W. R. (1993) Nd-Sr signature of Pierre Shale: a target material at the Manson impact site and source of the Haitian tektites (K-T boundary)? *Lunar Planet. Sci. XXIV* (abstract submitted to the 24th Lunar and Planetary Science Conference), pp. 1099–100.

Prinz, M., Weisberg, M. K., Clayton, R. N. and Mayeda, T. K. (1993) Oxygen isotopic relationships between the LEW85332 carbonaceous chondrite and CR chondrites. *Lunar Planet. Sci. XXIV* (abstract submitted to the 24th Lunar and Planetary Science conference), pp. 1185–6.

Prombo, C. A. and Clayton, R. N. (1993) Nitrogen isotopic compositions of iron meteorites. *Geochim. Cosmochim. Acta*, **57**, 3749–61.

Puchelt, H. and Hubberten, H. -W. (1980) Preliminary results of sulfur isotope investigations on deep sea drilling project cores from legs 52 and 53. Initial Rep. DSDP 51, 52, 53, Part 2, pp. 1145–8.

Puchelt, H., Sabels, B. R. and Hoering, T. C. (1971) Preparation of sulfur hexafluoride for isotope geochemical analysis. *Geochim. Cosmochim. Acta*, **35**, 625–8.

Pujol, C. and Grazzini, C. V. (1989) Palaeoceanography of the last deglaciation in the Alboran Sea (Western Mediterranean); stable isotopes and planktonic foraminiferal records. *Mar. Micropaleont.*, **15**, 153–79.

Purser, K. H., Russo, C. J., Liebert, R. B. *et al.* (1982) The application of electrostatic tandems to ultrasensitive mass spectrometry and nuclear dating, in *Nuclear and Chemical Dating Techniques: Interpreting the Environmental Record* (ed. L. A. Currie), Washington, pp. 45–73.

Quade, J., Cerling, T. E. and Bowman, J. R. (1989) Systematic variations in the carbon and oxygen isotopic composition of pedogenic carbonate along elevation transects in the southern Great Basin, United States. *Geol. Soc. Am. Bull.*, **101**, 464–75.

Quade, J., Solounias, N. and Cerling, T. E. (1994) Stable isotope evidence from paleosol carbonates and fossil teeth in Greece for forest or woodlands over the past 11 Ma. *Palaeogeogr., Palaeoclim., Palaeoecol.*, **107**, 41–53.

Quay, P. D., Tilbrook, B. and Wong, C. S. (1992) Oceanic uptake of fossil fuel CO_2: carbon-13 evidence. *Science*, **256**, 74–79.

Raisbeck, G. M., Yiou, F., Fruneau, M. *et al.* (1978) Measurement of ^{10}Be in 1000 and 5000 year-old Antarctic ice. *Nature*, **275**, 731–2.

Raisbeck, G. M., Yiou, F., Fruneau, M. *et al.* (1979) ^{10}Be concentration and residence time in the oceans surface layer. *Earth Planet. Sci. Lett.*, **43**, 237–40.

Raisbeck, G. M., Yiou, F., Fruneau, M. *et al.* (1980) ^{10}Be concentration and residence time in the deep ocean. *Earth Planet. Sci. Lett.*, **51**, 257–78.

Rajan, R. S. and Tammane, A. S. (1982) Evidence for ^{244}Pu fission tracks in hibonites from Murchison carbonaceous chondrite. *Earth Planet. Sci. Lett.*, **58**, 129–35.

Ralph, E. K. (1971) Carbon-14 dating, in *Dating Techniques for the Archaeologist* (ed. H. N. Michael and E. K. Ralph), M. I. T Press, Cambridge, Mass., pp. 1–48.

Rau, G. H. (1991) Another recipe for bomb ^{14}C dilution. *Nature*, **350**, 116.

Ravizza, G. (1990) Rhenium–osmium isotope geochemistry of modern and ancient organic-rich sediments. Unpublished PhD dissertation, Yale University.

Reeh, N., Oerter, H., Letréguilly, A. *et al.* (1991) A new, detailed ice-age oxygen-18 record from the ice-sheet margin in central West Greenland. *Palaeogeogr., Palaeoclim., Palaeoecol., (Global and Planetary Change)*, **90**, 373–83.

Rees, C. E. (1978) Sulphur isotope measurements using SO_2 and SF_6. *Geochim. Cosmochim. Acta*, **42**, 383–89.

Reimer, G. M., Storzer, D. and Wagner, G. A. (1970) Geometry factor in fission-track counting. *Earth Planet. Sci. Lett.*, **9**, 401–4.

Reynolds, J. H. (1953) Natural variations in the isotopic constitution of silicon. *Proc. Conf. Nucl. Processes in Geologic Settings*, pp. 64–71.
Reyss, J. L., Yokoyama, Y. and Guichard, G. (1981) Production cross sections of ^{26}Al, ^{22}Na, ^{7}Be from argon and ^{10}Be, ^{7}Be from nitrogen: implications for production rates of ^{26}Al and ^{10}Be in the atmosphere. *Earth Planet. Sci. Lett.*, **53**, 203–10.
Reyss, J. L., Yokoyama, Y. and Tanaka, S. (1976) Aluminium-26 in deep sea sediments. *Science*, **193**, 1119–21.
Richards, D. (1988) Order and chaos in strong fields. *Nature*, **336**, 518–9.
Richards, T. W. and Lembert, M. E. (1914) Atomic weight of lead of radioactive origin. *J. Am. Chem. Soc.*, **36**, 1329–44.
Rieley, G., Collier, R. J., Jones, D. M. *et al.* (1991) Sources of sedimentary lipids deduced from stable carbon-isotope analyses of individual compounds. *Nature*, **352**, 425–7.
Rink, W. J., Schwartz, H. P., Smith, F. H. and Radovčić, J. (1995) ESR ages for Krapina hominids. *Nature*, **378**, 24.
Ripley, E. M. (1981) Sulfur isotopic studies of the Dunka Road Cu–Ni deposit, Duluth Complex, Minnesota. *Econ. Geol.*, **76**, 610–20.
Robbins, G. A. (1992) Radiogenic argon diffusion in muscovite under hydrothermal conditions, MS Thesis, Providence, Rhode Island, Brown University, 42 pp.
Robbins, J. A. (1978) Geophysical and geochemical applications of radioactive lead, in *The Biogeochemistry of Lead in the Environment*, Vol. 1A (ed. J. O. Niagru), Elsevier, Amsterdam.
Robbins, J. A., Edgington, D. N. and Kemp, A. L. W. (1978) Comparative ^{210}Pb, ^{137}Cs and pollen geochronologies of sediments from Lakes Ontario and Erie. *Quaternary Res.*, **10**, 256.
Robinson, B. W. (1995) Sulphur isotope standards, in Reference and intercomparision materials for stable isotopes of light elements,. IAEA-TECDOC-825, pp. 39–45.
Romanek, C. S. and Grossman, E. L. (1989) Stable isotope profiles of Tridacna maxima as environmental indicators. *Pslaios*, **4**, 402–13.
Romanek, C. S., Jones, D. S., Williams, D. F. *et al.* (1987) Stable isotopic investigation of physiological and environmental changes recorded in shell carbonate from the giant clam *Tridanca maxima. Marine Biol.*, **94**, 385–93.
Rosholt, J. N., Emiliani, C., Geiss, *J. et al.* (1961) Absolute dating of deep-sea cores by the Pa^{231}/Th^{230} method. *J. Geol.*, **69**, 162–85.
Ross, P. J. and Martin, E. A. (1972) Rapid procedure for preparing gas samples for nitrogen-15 determination. *Analyst*, **95**, 817–22.
Rostek, F., Ruhland, G., Bassinot, F. C. *et al.* (1993) Reconstructing sea surface temperature and salinity using δ^{18}O and alkenone records. *Nature*, **364**, 319–21.
Rubin, K. H., Macdougall, J. D. and Perfit, M. R. (1994) ^{210}Po–^{210}Pb dating of recent volcanic eruptions on the sea floor. *Nature*, **368**, 841–4.
Rubinson, M. and Clayton, R. N. (1969) Carbon-13 fractionation between aragonite and calcite. *Geochim. Cosmochim. Acta*, **33**, 997–1002.
Rumbaugh, L. H. and Locker, G. L. (1936) Neutrons and other heavy particles in cosmic radiation of the stratosphere. *Phys. Rev.*, **49**, 855.
Rumble, D. III (1982) Stable isotope fractionation during metamorphic devolatilization reactions, in *Characterization of Metamorphism through Mineral Equilibria. Rev. Mineral.*, **10**, 327–53.
Russ, J., Hyman, M., Schafer, H. J. and Rowe, M. W. (1990) Radiocarbon dating of prehistoric rock paintings by selective oxidation of organic carbon. *Nature*, **348**, 710–1.
Russell, R. D. and Farquhar, R. M. (1960) *Lead Isotopes in Geology*. Interscience Publishers, New York.
Russell, W. A. and Papanastassiou, D. A. (1978) Calcium isotope fractionation in ion exchange chromatography. *Anal. Chem.*, **50**, 1151–4.
Russell, W. A., Papanastassiou, D. A. and Tombrello, T. A. (1978) Ca isotope fractionation on the Earth and other solar system materials. *Geochim. Cosmochim. Acta*, **42**, 1075–90.
Rye, R. O., Schuiling, R. D., Rye, D. M. and Jansen, J. B. H. (1976) Carbon, hydrogen and oxygen isotope studies of the regional metamorphic complex at Naxos, Greece. *Geochim. Cosmochim. Acta*, **40**, 1031–49.
Saito, T. and Hattori, A. (1980) ^{15}N natural abundance in oceanic suspended particulate matter. *Nature*, **283**, 752–4.

Sakai, H., Casadevall, T. J. and Moore, J. G. (1982) Chemistry and isotope ratios of sulfur in basalts and volcanic gases at Kilauea volcano, Hawaii. *Geochim. Cosmochim. Acta*, **46**, 720–38.

Samson, S. D. and Alexander, E. C. (1987) Calibration of the interlaboratory $^{40}Ar/^{39}Ar$ dating standard., MMhb-1. *Chem. Geol.*, **66**, 27–34.

Savin, S. M. and Epstein, S. (1970) The oxygen and hydrogen isotope geochemistry of clay minerals. *Geochim. Cosmochim. Acta*, **34**, 25–42.

Savin, S. M. and Yeh, H. W. (1981) Stable isotopes in ocean sediments, in *The Sea*, Vol. 7 (ed. C. Emiliani), John Wiley & Sons, New York, pp. 1521–54.

Savin, S. M., Douglas, R. G. and Stehli, F. G. (1975) Tertiary marine paleotemperatures. *Bull. Geol. Soc. Am.*, **86**, 1499–510.

Scaillet, S., Féraud, G., Lagabrielle, Y. *et al.* (1990) $^{40}Ar/^{39}Ar$ laser-probe dating by step heating and spot fusion of phengites from the Dora Marina nappe of the western Alps, Italy. *Geology*, **18**, 741–4.

Schärer, U. Rong-Hua, X. and Allègre, C. J. (1986) U–(Th)–Pb systematics and ages of Himalayan leucogranites, south Tibet. *Earth Planet. Sci. Lett.*, **77**, 35–48.

Schidlowski, M. (1982) Stable isotopes and the evolution of life: an overview, in *Stable Isotopes* (eds. H.-L. Schmidt, H. Förstel and K. Heinzinger), Elsevier Scientific Publ. Co., Amsterdam, pp. 95–100.

Schidlowski, M. (1988) A 3,800-million-year isotopic record of life from carbon in sedimentary rocks. *Nature*, **333**, 313–8.

Schimmelmann, A. and DeNiro, M. J. (1986) Stable isotopic studies on chitin. III. The D/H and $^{18}O/^{16}O$ ratios in arthropod chitin. *Geochim. Cosmochim. Acta*, **50**, 1485–96.

Schleser, G. H. (1990) Investigations of the $\delta^{13}C$ pattern in leaves of *Fagus sylvatica* L. *J. Exp. Botany*, **41**, 565–72.

Schleser, G. H. (1995) Parameters determining carbon isotope ratios in plants, in *Problems of Stable Isotopes in Tree-Rings, Lake Sediments and Peat Bogs as Climatic Evidence for the Holocene* (eds B. Frenzel, B. Stauffer and M. M. Weiss.), *Paläoklimaforsch./Paleoclim. Res.*, Vol. 15, European Science Foundation, pp. 71–96.

Schleser, G. H. and Pohling, R. (1980) $\delta^{13}C$-record in forest soil using a rapid method for preparing carbon dioxide samples. *Int. J. Appl. Radiat. Isot.*, **31**, 769–73.

Schleser, G. H., Bernhardt, K.-G. and Hurka, H. (1989) Climatic adaptability of populations of *Diplotaxis erucoides* D. C. (Brassicaceae) from Sicily, based on leaf morphology, leaf anatomy and $\delta^{13}C$ studies. *Int. J. Biometeorol.* **33**, 109–18.

Schwartzman, D. W. and Volk, T. (1989) Biotic enhancement of weathering and habitability of Earth. *Nature*, **340**, 457–460.

Shackleton, N. J. (1968) Depth of pelagic foraminifera and isotopic changes in Pleistocene oceans. *Nature*, **218**, 79–80.

Shackleton, N. J. and Kennett, J. P. (1975) Paleotemperature history of the Cenozoic and initiation of Antarctic glaciation: oxygen and carbon isotope analyses in DSDP sites 277, 279 and 281. Initial Rep. DSDP, **29**, pp. 743–55.

Shackleton, N. J. and Opdyke, N. D. (1973) Oxygen isotope and paleomagnetic stratigraphy of equatorial Pacific core V28–V39: Oxygen isotope temperatures and ice volumes on a 10^5 and 10^6 year scale. *Quaternary Res.*, **3**, 39–55.

Shackleton, N. J., Berger, A. and Peltier, W. R. (1990) An alternative astronomical calibration of the lower Pleistocene timescale based on ODP Site 677. *Trans. Roy. Soc. Edinburgh, Earth Sci.*, **81**, 251–61.

Shackleton, N. J., Corfield, R. M. and Hall, M. A. (1985) Stable isotope data and the ontogeny of Paleocene planktonic foraminifera. *J. Foraminiferal Res.*, **15**, 321–36.

Shackleton, N. J., Duplessy, J.-C., Arnold, M. *et al.* (1988) Radiocarbon age of last glacial Pacific deep water. *Nature*, **335**, 708–11.

Sharma, T. and Clayton, R. N. (1965) Measurement of O^{18}/O^{16} ratios ot total oxygen of carbonates. *Geochim. Cosmochim. Acta*, **29**, 1347–53.

Sharma, Y. P., Lal, N., Bal, K. D. *et al.* (1980) Closing temperatures of different fission-track clocks. *Contrib. Mineral. Petrol.*, **72**, 335–6.

Sharp, Z. D. (1990) A laser-based microanalytical method for the *in situ* determination of oxygen isotope ratios of silicates and oxides. *Geochim. Cosmochim. Acta*, **54**, 1353–7.

Shih, Y. N. and Taylor, H. P. (1969) Oxygen and carbon isotope studies of contact metamorphism of carbonate rocks. *J. Petrol.*, **10**, 307–31.

Shih, C-Y., Nyquist, L. E. and Wiesmann, H. (1993) K–Ca chronology of lunar granites. *Geochim. Cosmochim. Acta*, **57**, 4827–4841.

Shih, C-Y., Nyquist, L. E., Bogard, D. D. *et al.* (1982) Chronology and petrogenesis of young achondrites, Shergotty, Zagami and ALHA 77005: late magmatism on a geologically active planet. *Geochim. Cosmochim. Acta*, **46**, 2323–44.

Shih, C-Y., Wiesmann, H., Garrison, D. H. *et al.* (1993) Chronology of lunar granite 12033, 576: resetting the Rb-Sr and K-Ca isochrons. *Lunar Planet. Sci. XXIV* (abstract submitted to the 24th Lunar and Planetary Science Conference), pp. 1295–6.

Shih, C-Y., Nyquist, L. E., Bogard, D. D. and Wiesmann, H. (1994) K–Ca and Rb–Sr dating of two lunar granites: relative chronometer resetting. *Geochim. Cosmochim. Acta*, **58**, 3101–16.

Shimokawa, K., Imai, N. and Hirota, H. (1984) Dating of a volcanic rock by electron spin resonance. *Isotope Geoscience*, **2**, 365–73.

Shimoyama, S. and Hamano, T. (1988) The effect of oxygen-deficient water on the molluscan thanatocoenosis in Hakata Bay. Estratto da Atti del Quarto Simposio di Ecologia e Paleoecologia delle Comunita Bentoniche, Sorrento, 753–72.

Shukolyukov, Yu. A. (1970) *The Decay of Uranium Nuclei in Nature*, Atomizdat Publisky, Moscow (in Russian).

Silk, E. C. H. and Barnes, R. S. (1959) Examination of fission fragment tracks with an electron microscope. *Phil. Mag., Ser.* **6**, **4**, 970.

Sillen, A. and Lee-Thorp, J. A. (1994) Trace element and isotopic aspects of predator–prey relationships in terrestrial foodwebs. *Palaeogeogr., Palaeoclim., Palaeoecol.*,**107**, 243–55.

Silvermann, S. R. (1951) The isotope geology of oxygen. *Geochim. Cosmochim. Acta*, **2**, 26–42.

Simon, K. (1990) Hydrothermal alteration of Variscan granites, southern Schwartzwald, Federal Republic of Germany. *Contrib. Mineral. Petrol.*, **105**, 177–96.

Simoneit, B. R. T., Schoell, M., Dias, R. F. and Radler De Aquino Neto, F. (1993) Unusual carbon isotope compositions of biomarker hydrocarbons in a Permian tasmanite. *Geochim. Cosmochim. Acta*, **57**, 4205–11.

Singhvi, A. K., Sharma, Y. P and Agrawal, D. P. (1982) Thermoluminescence dating of sand dunes in Rajasthan, India. *Nature*, **195**, 313–5.

Smith, D. B. Wearn, P. L. Richards, H. J. and Rowe, P. C. (1970) Water movement in the unsaturated zone of high and low permeability strata by measuring natural tritium. *Isotope Hydrology*, IAEA, Vienna, pp. 73–87.

Smith, J. N. and Walton, A. (1980) Sediment accumulation rates and geochronologies measured in the Saguenay Fjord suing the ^{210}Pb dating method. *Geochim. Cosmochim. Acta*, **44**, 225.

Smith, T. M. and Dorobek, S. L. (1993) Alteration of early-formed dolomite during shallow to deep burial: Mississippian Mission Canyon Formation, central to southwestern Montana. *Geol. Soc. Am. Bull.*, **105**, 1389–99.

Sobotovich, E. V. (1961) Possibility of determining the absolute age of the granites of the Terksey Ala-Tau by the lead included in them. *Akad. Nauk. SSSR Kom. Opredeleniya Absolyut. Vozrasta Geol. Formatsii Trudysess*, **9**, 269–80.

Soddy, F. (1914) *The Chemistry of the Radio Elements*, Part II, Longmans, Green, London, 46 pp.

Southon, J. R., Nelson, D. E., Korteling, R. *et al.* (1982) Techniques for the direct measurement of natural beryllium-10 and carbon-14 with a tandem accelerator, in *Nuclear and Chemical Dating Techniques* (ed. L. A. Currie), pp. 75–87.

Sowers, T., Bender, M. and Raynaud, D. (1993) Elemental and isotopic composition of occluded O_2 and N_2 in polar ice. *J. Geophys. Res.*, **94**, 5137–50.

Spell, T. and McDougall, I. (1992) Revisions to the age of the Brunhes-Matuyama boundary and the Pleistocene geomagnetic polarity timescale. *Geophys. Res. Lett.*, **19**, 1181–4.

Spell, T., Harrison, M. T. and Wolff, J. A. (1990) ^{40}Ar/^{39}Ar dating of the Bandelier Tuff and San Diego Canyon ignimbrites, Jemez Mountains, New Mexico: temporal constraints on magmatic evolution *J. Volcanol. Geotherm. Res.*, **43**, 175–193.

Spero, H. J. and Lea, D. W. (1993) Intraspecific stable isotope variability in the planktic foraminifer *Globigerinoides sacculifer:* results from laboratory experiments. *Marine Micropal.*, **22**, 193–232.

Stacey, J. S. and Kramers, J. D. (1975) Approximation of terrestrial lead isotope evolution by a two-stage model. *Earth Planet. Sci. Lett.*, **26**, 207–21.

Stacey, J. S. and Stern, T. W. (1973) Revised tables for the calculation of lead isotope ages. US Dept. of Commerce, Natl Tech. Info. Service, Springfield, VA 22151, PB 20919, 35 pp.

Stahl, W. J. (1979) Carbon isotopes in petroleum geochemistry, in *Lectures in Isotope Geology* (eds E. Jäger and J. C. Hunziker), Springer-Verlag, Berlin, pp. 274–83.

Stakes, D. and Moore, W. S. (1991) Evolution of hydrothermal activity on the Juan de Fuca Ridge: observations, mineral ages and Ra isotope ratios. *J. Geophys. Res.*, **96**, 21739–52.

Staudacher, T. and Allègre, C. J. (1982) Terrestrial xenology. *Earth Planet. Sci. Lett.*, **60**, 389–406.

Steiger, R. H. and Jäger, E. (1977) Subcommission on Geochronology: convention on the use of decay constants in geo-and cosmochronology. *Earth Planet. Sci. Lett.*, **36**, 359–62.

Stein, M., Wasserburg, G. J., Aharon, P. *et al.* (1993) IMS U–series dating dating and stable isotopes of the last interglacial even in Papua New Guinea. *Geochim. Cosmochim. Acta*, **57**, 2541–54.

Stenhouse, M. J. and Baxter, M. S. (1977) Bomb ^{14}C and human radiation burden. *Nature*, **267**, 825–7.

Stern, T. W., Goldich, S. S. and Newell, M. F. (1966) Effects of weathering on the U-Pb ages of zircon from the Morton Gneiss, Minnesota. *Earth Planet. Sci. Lett.*, **1**, 369–71.

Steuber, A. M., Pushkar, P. and Hetherigton, E. A. (1984) A strontium isotope study of Smackover brines and associated solids, southern Arkansas. *Geochim. Cosmochim. Acta*, **48**, 1637–49.

Stevens, G. R. and Clayton, R. N. (1971) Oxygen isotope studies on Jurassic and Cretaceous belemnites from New Zealand and their biogeographic significance. *NZ J. Geol. Geophys.*, **14**, 829–97.

Storzer, D. (1970) Fission-track dating of volcanic glass and the thermal history of rocks. *Earth Planet. Sci. Lett.*, **8**, 55–60.

Storzer, D. and Poupeau, G. (1973) Age-plateau des verres et mineraux par la méthode des traces de fission, *Compt. Rend. Acad. Sci., Paris*, **276D**, 137–9.

Strain, J. A., Townsend, P. D., Jassemnjad, B. and McKeever, S. W. S. (1986) Emission spectra of meteorites during thermoluminescence. *Earth Planet. Sci. Lett.*, **77**, 14–9.

Strauss, H. and Schieber, J. (1990) A sulfur isotope study of pyrite genesis: The Mid-Proterozoic Newland Formation, Belt Supergroup, Montana. *Geochim. Cosmochim. Acta*, **54**, 197–204.

Stuiver, M., Long., A. and Kara, K. S. (eds) (1993) Calibration 1993. Radiocarbon, **35**(1), 244 pp.

Suess, H. E. (1995) Radiocarbon concentration in modern wood. *Science*, **122**, 415–7.

Suttle, C. A., Chan, A. M. and Cottrell, M. T. (1990) Infection of phytoplankton by viruses and reduction of primary productivity. *Nature*, **347**, 467–9.

Sweeney, R. E., Liu, K. K. and Kaplan, I. R. (1978) Oceanic nitrogen isotopes and their uses in determining the source of sedimentary nitrogen, in *Stable Isotope in Earth Science* (ed. B. W. Robinson), Dept. Sci. Ind. Res. Bull. **220**, New Zealand.

Talent, J. A., Mawson. R., Andrew, A. S. *et al.* (1993) Middle Paleozoic extinction events: faunal and isotopic data. *Palaeogeogr., Palaeoclim., Palaeoecol.*, **104**, 139–52.

Tans, P. P, Fung, I. Y. and Takahashi, T. (1990) Observational constraints on the global atmospheric CO_2 budget. *Science*, **247**, 1431–8.

Tanaka, T. and Masuda, A. (1982) The La–Ce geochronometer. A new dating method. *Nature*, **300**, 515–7.

Tarutani, T., Clayton, R. N. and Mayeda, T. K. (1969) The effect of polymorphism and magnesium substitution on oxygen isotope fractionation between calcium carbonate and water. *Geochim. Cosmochim. Acta*, **33**, 987–96.

Tatsumoto, M., Unruh, D. M. and Patchett, P. J. (1981) U–Pb and Lu–Hf systematics of Antarctic meteorites. *Proc. 6th Symp. Antarctic Meteorites*, National Institute for Polar Research, Tokyo, pp. 237–49.

Taylor, A. P. and Epstein, S. (1962) Relationship between $^{18}O/^{16}O$ ratios in coexisting minerals of igneous and metamorphic rocks. *Bull. Geol. Soc. Am.*, **73**, 461–80.

Taylor, H. P. (1968) The oxygen isotope geochemistry of igneous rocks. *Contrib. Mineral. Petrol.*, **19**, 1–71.

Taylor, H. P. (1977) Water/rock interactions and the origin of H_2O in granitic batholiths. *J. Geol. Soc. London.* **133**, 509–58.

Taylor, H. P. (1978) Oxygen and hydrogen isotope studies of plutonic granitic rocks. *Earth Planet. Sci. Lett.*, **38**, 177–210.

Taylor, H. P. and Epstein, S. (1966) Oxygen isotope studies of Ivory Coast tektites and impactite glass from the Bosumtwi crater, Ghana. Science, **153**, 173–75.

Taylor, H. P. and Epstein, S. (1969) Correlations between O^{18}/O^{16} ratios and chemical composition of tektites. *J. Geophys. Res.*, **74**, 6834–44.

Taylor, K. C., Lamorey, G. W., Doyle, G. A. *et al.* (1993) The 'flickering switch' of late Pleistocene climatic change. *Nature*, **361**, 432–6.

Taylor, P. (1988) Why continents look too young. *Nature*, **336**, 713.

Teis, R. V., Chupakhin, M. S. and Naidin, D. P. (1957) Determination of paleotemperatures from the isotopic composition of oxygen in calcite of certain Cretaceous fossil shells from Crimea. *Geochemistry*, **4**, 323–9.

Thierstein, H. R. and Berger, W. H. (1978) Injection events in ocean history. *Nature*, **276**, 461–6.

Thode, H. G. (1970) Sulfur isotope geochemistry and fractionation between coexisting minerals. *Mineral. Soc. Am. Spec. Pap.*, **3**, 133–44.

Thode, H. G., Cragg, C. B., Hulston, J. R. and Rees, C. E. (1971) Sulphur isotope exchange between sulphur dioxide and hydrogen sulphide. *Geochim. Cosmochim. Acta*, **35**, 35–45.

Thomson, J. J. (1912) Further experiments on positive rays. *Phil. Mag.*, **24**, 209–53.

Thomson, J. J. (1914) Rays of positive electricity. *Proc. Roy. Soc. London, Ser S*, **89**, 1–20.

Thurber, D. L. (1962) Anomalous $^{234}U/^{238}U$ in nature. *J. Geophys. Res.*, **67**, 4518–20.

Tilton, G. R. (1960) Volume diffusion as a mechanism for discordant lead ages. *J. Geophys. Res.*, **65**, 2933–45.

Toots, H. and Voorhies, M. R. (1965) Strontium in fossil bones and the reconstruction of food chains. *Science*, **149**, 854–5.

Tudge, A. P. (1960) A method of analysis of oxygen isotopes in orthophosphate, its use in the measurement of paleotemperatures. *Geochim. Cosmochim. Acta*, **18**, 31–93.

Tudge, A. P. and Thode, H. G. (1950) Thermodynamic properties of isotopic compounds of sulphur. *Can. J. Res.*, **B28**, 567–78.

Tuli, J. K. (1985) *Nuclear Card Wallets.* National Nuclear Data Center, Brookhaven National Laboratory, Upton, NY 11973.

Turekian, K. K. (1976) *Oceans*, Prentice-Hall Inc., Englewood Cliffs, NJ, 149 pp.

Turekian, K. K. and Wedepohl, K. H. (1961) Distribution of the elements in some major units of the Earth's crust. *Geol. Soc. Am. Bull.*, **72**, 175–96.

Turner, S., Foden, J., Sandiford, M. and Bruce, D. (1993) Sm–Nd isotopic evidence for the provenance of sediments from the Adelaide Fold Belt and southeastern Australia with implications for episodic crustal addition. *Geochim. Cosmochim. Acta*, **57**, 1837–56.

Tuttas, D. (1992) The measurement of osmium isotope ratios using negative thermal ionization. Finnigan MAT, *Application News*, No. **1**, 1–7.

Urey, H. C. (1947) The thermodynamic properties of isotopic substances. *J. Chem. Soc.*, 1947, 562–81.

Urey, H. C. (1948) Oxygen isotopes in nature and in the laboratory. *Science*, **108**, 489–96.

Urey, H. C., Lowenstam, H. A., Epstein, S. and McKinney, C. R. (1951) Measurements of paleotemperatures and temperatures of the Upper Cretaceous of England, Denmark, and the southeastern United States. *Bull. Geol. Soc. Am.*, **62**, 399–416.

Usdowski, E. and Hoefs, J. (1993) Oxygen isotope exchange between carbonic acid, bicarbonate, carbonate, and water: a re-examination of the data of McCrea (1950) and an expression for the overall partitioning of oxygen isotopes between the carbonate species and water. *Geochim. Cosmochim. Acta*, **57**, 3815–18.

Usdowski, E., Michaelis, J., Böttcher, M. B. and Hoefs, J. (1991) Factors for the oxygen isotope equilibrium fractionation between aqueous CO_2, carbonic acid, bicarbonate, carbonate, and water (19°C) *Z. Phys. Chem.*, **170**, 237–49.

Valladas, H., Geneste, J. M., Joren, S. L. and Chadelle, J. P. (1986) Thermoluminescence dating at Le Moustier (Dordogne, France). *Nature*, **322**, 452–5.

Valladas, H., Cachier, H., Maurice, P. *et al.* (1992) Direct radiocarbon dates for prehistoric paintings at the Altamira, El Castillo and Niaux caves. *Nature*, **357**, 68–70.

Valley, J. W. and O'Neil, J. R. (1981) $^{13}C/^{12}C$ exchange between graphite and calcite: a possible thermometer in Grenville marbles. *Geochim. Cosmochim. Acta*, **45**, 411–9.

Veizer, J. and Hoefs, J. (1976) The nature of O^{18}/O^{16} and C^{13}/C^{12} secular trends in sedimentary carbonate rocks. *Geochim. Cosmochim. Acta*, **40**, 1387–95.

Veizer, J., Fritz, P. and Jones, B. (1986) Geochemistry of brachiopods: oxygen and carbon isotopic records of Paleozoic oceans. *Geochim. Cosmochim. Acta*, **50**, 1679–96.

Veizer, J., Compston, W., Hoefs, J. and Nielsen, H. (1982) Mantle buffering of the early oceans. *Naturwissenschaften*, **69**, 173–80.

Von Gunten, H. R. (1969) Distribution of mass in spontaneous and neutron-induced fission. *Actinide Rev.*, **1**, 275–98.

Wachter, E. A. and Hayes, J. M. (1985) Exchange of oxygen isotopes in carbon dioxide-phosphoric acid systems. *Chemical Geology (Isotope Geoscience Section)*, **52**, 365–74.

Wada, H. (1977) Isotopic studies of graphite in metamorphosed carbonate rocks of central Japan. *Geochem. J.*, **11**, 183–97.

Wanger, G. A. and Reimer, G. M. (1972) Fission track tectonics: the tectonic interpretation of fission track apatite ages. *Earth Planet. Sci. Lett*, **14**, 263–8.

Walczyk, T., Hebeda, E. H. and Heumann, K. G. (1991) Osmium isotope ratio measurements by negative thermal ionization mass spectrometry (NTI-MS): improvement in precision and enhancement in emissions by introducing oxygen or freons into the ion source. *Fresenius' Z. Anal. Chem.*, **341**, 537–41.

Wallace, R. K. and Woosley, S. E. (1984) in *High Energy Transients in Astrophysics* (ed. S. E. Woosley), American Institute of Physics, New York, pp. 319–21.

Wang, Y. and Cerling, T. E. (1994) A Model of fossil tooth and bone diagenesis: implications for paleodiet reconstruction from stable isotopes. *Palaeogeogr., Palaeoclim., Palaeoecol.*, **107**, 281–9.

Wang, Y., Cerling, T. E. and MacFadden, B. J. (1994) Fossil horses and carbon isotopes: new evidence for Cenozoic dietary, habitat, and ecosystem changes in North America. *Palaeogeogr., Palaeoclim., Palaeoecol.*, **107**, 269–79.

Wasserburg, G. J. (1954) K40–Ar40 dating, in *Nuclear Geology* (ed. H. Faul), John Wiley & Sons, New York, p. 341.

Wasserburg, G. J., Jacobsen, S. B., DePaolo, D. J. *et al.* (1981) Precise determination of Sm/Nd ratios, Sm and Nd isotopic abundances in standard solutions. *Geochim. Cosmochim. Acta*, **45**, 2311–23.

Weber, J. N. (1968) Fractionation of the stable isotopes of carbon and oxygen in calcareous marine invertebrates–the Asteroidea, Ophiuroidea and Crinoidea. *Geochim. Cosmochim. Acta*, **32**, 33–70.

Weber, J. N. and Raup, D. M. (1966) Fractionation of the stable isotopes of carbon and oxygen in marine calcareous organisms–the Echi oidea. I. Variation of ^{13}C and ^{18}O content within individuals. *Geochim. Cosmochim. Acta*, **30**, 681–703.

Wefer, G. (1985) Die Verteilung stabiler Isotope in Kalkschalen mariner Organismen. *Geol. Jb.*, **A82**, 111 pp.

Weidman, C. R. and Jones, G. A. (1993) A shell-derived time history of bomb ^{14}C on Georges Bank and its Labrador Sea implications. *J. Geophys. Res.*, **98**, 14577–88.

Weinbruch, S., Zinner, E. K., El Goresy, A. *et al.* (1993) Oxygen isotopic composition of individual olivine grains from the Allende meteorite. *Geochim. Cosmochim. Acta*, **57**, 2649–61.

West, D. P. and Lux, D. R. (1993) Dating mylonitic deformation by the ^{40}Ar-^{39}Ar method: an example from the Norumbega Fault Zone, Maine. *Earth Planet. Sci. Lett.*, **120**, 221–37.

Whelan, J. F. and Rye, R. O. (1978) Stable isotope studies in the Balmat–Edwards Zn–Pb district, New York, *Geol. Soc. Am. Abstr. with Progr.*, **10**, 515.

White, W. M. and Patchett, P. J. (1984) Hf–Nd–Sr isotopes and incompatible element abundances in island arcs: implications for magma origins and crust-mantle evolution. *Earth Planet. Sci. Lett.*, **67**, 167–85.

Whitford, D. J. (1975) Strontium isotopic studies of the volcanic rocks of the Sunda arc, Indonesia, and their petrogenetic importance. *Geochim. Cosmochim. Acta*, **39**, 1287–302.

Wiesberg, L. H. G. and Tavares, T. M. (1987) The $^{13}C/^{12}C$ record in wood of palmtrees. *Geochim. Cosmochim. Acta*, **51**, 1783–6.

Wintle, A. G. (1973) Anomalous fading of thermoluminescence in mineral samples. *Nature*, **245**, 143–4.

Wintle, A. G. (1985) Stability of TL signal from fine grains in loess. *Nucl. Tracks*, **10**, 725–30.

Wintle, A. G. and Huntley, D. J. (1980) Thermoluminesence dating of ocean sediments. *Can. J. Earth. Sci.*, **17**, 348–60.

Wintle, A. G. and Westgate, J. A. (1986) Thermoluminescence age of Old Crow tephra in Alaska. *Geology*, **14**, 594–7.

Woodruff, F., Savin, S. M. and Douglas, R. G. (1980) Miocene stable isotope record: a detailed deep Pacific Ocean study and its paleoclimatic implications. *Science*, **212**, 665–8.

Woosley, S. E. and Weaver, T. A. (1984) in *High Energy Transients in Astrophysics* (ed. S. E. Woosley), American Institute of Physics, New York, pp. 273–97.

Wright, I. (1993) Ingrained evidence of origin. *Nature*, **365**, 786–7.

Wright, I. P., Boyd, S. R., Franchi, I. A. and Pillinger, C. T. (1988) High-precision determination of nitrogen stable isotope ratios at the sub-nanomol level. *J. Phys, Res.*, **21**, 865–75.

Xiao, Y. -K. and Zhang, C. -G. (1992) High precision isotopic measurement of chlorine by thermal ionization mass spectrometry of the Cs_2Cl^+ ion. *Int. J. Mass. Spectrom. Ion Proc.*, **116**, 183–92.

Yapp, C. J. and Poths, H. (1986) Carbon in natural goethites. *Geochim. Cosmochim. Acta*, **50**, 1213–20.

Yapp, C. J. and Poths, H. (1992) Ancient atmospheric CO_2 pressures inferred from natural goethites. *Nature*, **355**, 342–4.

Yin, Q. Z., Jagoutz, E., Verkhovskiy, A. B. and Wanke, H. (1993) 187Os–186Os and 187Os–188Os method of dating: and introduction. *Geochim. Cosmochim. Acta*, **57**, 4119–28.

Yokoyama, Y. F., Guichard, G., Reyss, J. -L. and Van, N. H. (1978) Oceanic residence times of dissolved beryllium and aluminium deduced from cosmogenic tracers ^{10}Be and ^{26}Al. *Science*, **201**, 1016–7.

York, D. (1984) Cooling histories from ^{40}Ar/^{39}Ar age spectra: implications for Precambrian plate tectonics. *Ann. Rev. Earth Planet. Sci.*, **12**, 383–409.

Young, E. D. and Rumble, D. III (1993) The origin of correlated variations in *in-situ* ^{18}O/^{16}O and elemental concentrations in metamorphic garnet from southeastern Vermont, USA. *Geochim. Cosmochim. Acta*, **57**, 2585–97.

Yui, T. -F., Yeh, H. -W. and Lee, C. W. (1990) A stable isotope study of serpentinization in the Fengtien ophiolite, Taiwan. *Geochim. Cosmochim. Acta*, **54**, 1417–26.

Zachos, J. C., Arthur, M. A. and Dean, W. E. (1989) Geochemical evidence for suppression of pelagic marine productivity at the Cretaceous/Tertiary boundary. Nature, **337**, 61–4.

Zeller, E. J., Wray, J. L. and Daniels, F. (1955) Thermoluminescence induced by pressure and crystallization. *J. Chem. Phys*, **23**, 2187.

Zeller, E. J., Wray, J. L. and Daniels, F. (1957) Factors in age determination of carbonate sediments by luminescence. *Bull. Am. Assoc. Petrol. Geol.*, **40**, 140–52.

Zheng, Y., Hong-Fei, H. and Lian-Fang, Y. (1993) Carbon and oxygan isotope event markers near the Frasnian-Famennian boundary, Luoxiu section, South China. *Palaeogeogr., Palaeoclim., Palaeoecol.*, **104**, 97–104.

Zhong, S. and Mucci, A. (1993) Calcite precipitation in seawater using a constant addition technique: a new overall reaction kinetic expression. *Geochim. Cosmochim. Acta*, **57**, 1409–17.

Zimmerman, U., Münnich, K. O. and Roether, W. (1967) Downward movement of soil moisture traced by means of hydrogen isotopes, in *Isotope Techniques in the Hydrologic Cycle*, Am. Geophys. Union, Washington, DC, pp. 28–36.

Index

Author Index